Sustainable Construction Materials

Related titles

Sustainable Construction Materials: Copper Slag
(ISBN: 978-0-08-100986-4)

Sustainable Construction Materials: Sewage Sludge Ash
(ISBN: 978-0-08-100987-1)

Sustainable Construction Materials: Municipal Incinerated Bottom Ash
(ISBN: 978-0-08-100997-0)

Woodhead Publishing Series in Civil and Structural Engineering

Sustainable Construction Materials: Glass Cullet

Ravindra K. Dhir OBE

Jorge de Brito

Gurmel S. Ghataora

Chao Qun Lye

ELSEVIER

WP
WOODHEAD PUBLISHING
An imprint of Elsevier

Woodhead Publishing is an imprint of Elsevier
The Officers' Mess Business Centre, Royston Road, Duxford, CB22 4QH, United Kingdom
50 Hampshire Street, 5th Floor, Cambridge, MA 02139, United States
The Boulevard, Langford Lane, Kidlington, OX5 1GB, United Kingdom

Library of Congress Cataloging-in-Publication Data
A catalog record for this book is available from the Library of Congress

British Library Cataloguing-in-Publication Data
A catalogue record for this book is available from the British Library

ISBN: 978-0-08-100984-0 (print)
ISBN: 978-0-08-100990-1 (online)

Publisher: Matthew Deans
Acquisition Editor: Gwen Jones
Editorial Project Manager: Charlotte Cockle
Production Project Manager: Poulouse Joseph
Cover Designer: Alan Studholme

Typeset by TNQ Books and Journals

This book is dedicated to
Singapore
where it all began
&
Our families
for their unwavering support

Contents

Author Profiles			**xi**
Preface			**xiii**
1	**Introduction**		**1**
		Synopsis	**1**
	1.1	Background	**2**
	1.2	Sustainable Construction Materials	**3**
	1.3	Glass Cullet	**6**
	1.4	Layout and Contents	**9**
		References	**10**
2	**Methodology**		**15**
		Synopsis	**15**
	2.1	Introduction	**16**
	2.2	Literature Search and Appraisal	**17**
	2.3	Building the Data Matrix	**27**
	2.4	Analysis, Evaluation and Modelling of Data	**28**
	2.5	Dissemination	**29**
	2.6	Conclusions	**32**
		Reference	**33**
3	**Production and Properties of Glass Cullet**		**35**
		Synopsis	**35**
	3.1	Introduction	**36**
	3.2	Market Opportunities for Glass Cullet Use	**40**
	3.3	Types of Waste Glass and Collection Schemes	**41**
	3.4	Classification and Handling of Unprocessed Waste Glass	**44**
	3.5	Processing Waste Glass into Glass Cullet	**47**
	3.6	Chemical Properties of Glass Cullet	**53**
	3.7	Physical Properties of Glass Cullet	**62**
	3.8	Conclusions	**76**
		References	**78**
4	**Use of Ground Glass Cullet as Cement Component**		**97**
		Synopsis	**97**
	4.1	Introduction	**98**
	4.2	General Information	**98**

4.3	Cementitious Characteristics of Ground Glass Cullet	**99**
4.4	Fresh Concrete Properties	**103**
4.5	Mechanical Properties	**111**
4.6	Permeation	**128**
4.7	Durability	**134**
4.8	Environmental Impact	**148**
4.9	Case Studies	**149**
4.10	Conclusions	**152**
	References	**154**
5	**Use of Glass Cullet as a Sand Component**	**167**
	Synopsis	**167**
5.1	Introduction	**168**
5.2	General Information	**168**
5.3	Fresh Properties	**170**
5.4	Strength Properties	**176**
5.5	Deformation	**185**
5.6	Permeation	**190**
5.7	Durability - Chemical Attack	**196**
5.8	Durability - Physical Attack	**205**
5.9	Environmental Impact	**211**
5.10	Case Studies	**213**
5.11	Conclusions	**214**
	References	**215**
6	**Use of Glass Cullet as Filler Aggregate**	**231**
	Synopsis	**231**
6.1	Introduction	**232**
6.2	Fresh Properties	**232**
6.3	Strength	**236**
6.4	Deformation Properties	**241**
6.5	Permeation	**243**
6.6	Durability	**246**
6.7	Environmental Impact and Case Studies	**251**
6.8	Conclusions	**252**
	References	**253**
7	**Use of Glass Cullet in Geotechnical Applications**	**257**
	Synopsis	**257**
7.1	Introduction	**258**
7.2	General Information	**258**
7.3	Geotechnical Characteristics of Glass Cullet	**259**
7.4	Mechanical Properties	**269**
7.5	Durability	**281**
7.6	Environmental Impact	**283**

7.7 Case Studies 284
7.8 Guidance on Using Glass Cullet 286
7.9 Conclusions 288
References 290

8 Use of Glass Cullet in Road Pavement Applications 297
Synopsis 297
8.1 Introduction 298
8.2 General Information 298
8.3 Unbound Applications 300
8.4 Hydraulically Bound Applications 302
8.5 Bituminous Bound Applications 304
8.6 Environmental Impact 317
8.7 Case Studies 318
8.8 Conclusions 320
References 321

9 Use of Glass Cullet in Ceramics and Other Applications 327
Synopsis 327
9.1 Introduction 328
9.2 Glass Cullet Use in Ceramics 328
9.3 Glass Cullet as Aesthetic Finishing Material 371
9.4 Glass Cullet as Filtration Medium 371
9.5 Glass Cullet as Epoxy Composite 373
9.6 Glass Cullet as Glass Fibres 373
9.7 Glass Cullet as Blast Abrasive 374
9.8 Glass Cullet as Paint Filler 375
9.9 Glass Cullet in Elastomeric Roof Coating 375
9.10 Other Applications 376
9.11 Conclusions 376
References 378

10 Environmental Impact, Case Studies and Standards and Specifications 389
Synopsis 389
10.1 Introduction 390
10.2 Environmental Impact 390
10.3 Case Studies 400
10.4 Standards and Specifications 409
10.5 Conclusions 424
References 425

11 Epilogue 435
References 437

Index 439

Author Profiles

Ravindra Kumar Dhir OBE is an honorary professor of concrete engineering at the University of Birmingham, United Kingdom; adjunct professor at Trinity College Dublin, Ireland, and professor emeritus of concrete technology at the University of Dundee, United Kingdom, where he held the position of founding director of the Concrete Technology Unit (1998–2008) and developed it into an internationally acknowledged Centre of Excellence. His approach to research is visionary and creative, and by working closely with industry, he ensures a meaningful dissemination of his research into practice. He has won many recognitions, awards and honours, including the Order of the British Empire, Officer Rank, for services to concrete technology, from the Queen (1989); Secretary of State for Trade and Industry for innovative partnership with industry (1989 and 1990) and honorary fellowships from the Institute of Concrete Technology, United Kingdom, and Indian Concrete Institute. He has served on numerous technical committees, including as President of the Concrete Society (2009–10) and on the editorial board of the *Magazine of Concrete Research*.

Jorge de Brito is a full professor of civil engineering in the Department of Civil Engineering, Architecture and Georesources of the Instituto Superior Técnico (University of Lisbon), where he graduated and obtained his MSc and PhD degrees. His main research topic is sustainable construction, with emphasis on the use of recycled aggregates in concrete and mortars. He has participated in 22 competitively financed research projects and supervised 22 PhD theses and 160 MSc theses. He is the author of five previous books and around 300 papers in referenced international journals. He is a member of the editorial boards of 25 international journals and of the following scientific/professional organizations: CIB, FIB, RILEM, IABMAS, IABSE.

Gurmel S. Ghataora is a senior lecturer at the University of Birmingham, United Kingdom, with over 40 years' experience in the geotechnical engineering industry and academia. He has been at the university for over 25 years, where, in addition to his teaching, he is the head of postgraduate admissions for all taught programmes in civil engineering. He has supervised over 60 research students and authored over 190 articles. His interests include the geotechnics of earth roads and railways and sustainable construction materials. Amongst his many inventions are zero-compaction trench backfill made from as-dug materials, fibre reinforcement of clayey soils, improving the stiffness of railway tracks without removing the track, a novel railway drainage system and remediating acidity in colliery spoil using waste materials. Current studies include the durability of concrete made with sustainable construction materials as a component of both cement and aggregate. He is a member of the IOM^3 and sits on two TRB committees.

Chao Qun Lye, PhD, is a graduate of the National University of Singapore and obtained his PhD from the University of Birmingham, United Kingdom, for his research in the subject area of the *Use of Recycled and Secondary Aggregates in Concrete: Deformation Properties*. He has recently been appointed as a senior technical manager for ready-mix concrete with G&W Group in Singapore. He holds a strong interest in sustainability and innovation applied to the use of cement additions such as fly ash and ground granulated blast furnace slag and the use of recycled and secondary materials in concrete, geotechnics and road pavements.

Preface

Sustainability is now commonly referred to in the construction sector, zero waste scenarios are frequently floated, a great deal of research has been undertaken in the use of recycled and secondary materials (RSM) and standards and codes of practice are becoming more sympathetic to their adoption; however, a clear view of the potential for RSM use and how this may affect performance of structures still remains to be established. This is important and needed to absorb RSM within the present hierarchy of construction materials.

The use of RSM requires a clear understanding of their characteristics and the potential for required applications. This can be problematic as the variability of the material can be high, though this is not unusual, as well-established materials such as Portland cement and naturally occurring sand and gravel and crushed-rock aggregates are also known for their high variations at individual plants and even more so between plants. Material processing and design procedures can help to minimize variability. Why then is the construction industry slow in adopting the use of the new breed of materials arising from wastes, such as copper slag from material extraction processes, sewage sludge ash and municipal incinerated bottom ash from the incineration of sewage sludge and municipal solid wastes, glass cullet from domestic and industrial glass waste and recycled aggregate arising from demolition and excavation wastes? It can be argued that the inertia in accepting the use of RSM is due to two main reasons:

1. Research has not come together to exploit the present knowledge of RSM and their potential use.
2. A robust case for the value-added use of RSM has not yet been made.

This book, as part of a series of five, brings together the global research information published in English, dealing with glass cullet production and characteristics and its potential for use as a cement component and aggregate in concrete, geotechnical and road pavement applications, including the related case studies, standards and environmental impacts. The data analysed and evaluated for the book were sourced globally from 751 publications contributed by 1402 authors, from 513 institutions in 51 countries, over a time period from 1970 to 2017.

The main purpose of the book, which is aimed at academics, researchers, design engineers, specifiers and contractors and is structured in an incisive and easy-to-follow manner, is to bring out what is known and can be considered for use, and at the same time to avoid unnecessary repetitive research and waste of resources.

In completing this work, the authors gratefully acknowledge the help of many individuals at different stages of the work, but would like particularly to thank Edwin Trout of the Concrete Society, UK, for his help with sourcing of the literature and C.J. Lynn and Abdurrahman A. Elgalhud of the University of Birmingham, UK, for their help with some aspects of the work. Rui V. Silva of the Instituto Superior Técnico, Universidade de Lisboa, Portugal, is specially thanked for his help with preparing Chapters 3 and 9.

Ravindra K. Dhir OBE
Jorge de Brito
Gurmel S. Ghataora
Chao Qun Lye

Introduction

1

Main Headings

- Background
- Sustainable construction materials
- Glass cullet
- Layout and contents

Synopsis

The experience, the collaborative industrial research projects and their dissemination to the point of use that have established the grounds for this work of producing a series of five books are described in this chapter. The role of secondary and recycled construction materials in achieving sustainable development leading to sustainability is highlighted. This book, the fourth in the series, deals with glass cullet. An introduction to the material is provided, along with a brief description of the novel procedure of systematic analysis and evaluation used in developing the work. The structure of the book, in terms of the layout and contents, is also described.

Keywords: Sustainable development, Sustainable construction materials, Glass cullet, Book layout and contents.

Sustainable Construction Materials: Glass Cullet. https://doi.org/10.1016/B978-0-08-100984-0.00001-1

1.1 Background

The basis of this book stems from years of active research undertaken in close collaboration with the construction industry since 1988, involving small- to medium-sized enterprises, national/multinational companies, charities and government departments and a commitment to the dissemination of knowledge, as well as an active and decisive involvement in promoting sustainability and the use of sustainable materials in the construction sector.

The work has involved the undertaking of carefully planned and focused research to address some of the most challenging issues over the years, including sustainability in construction in general (Whyte et al., 2005); the sustainable use of natural resources to reduce CO_2 emissions, for example, by reducing the cement content of concrete mixes across all strength grades (Dhir and Hewlett, 2008; Dhir et al., 2000, 2004a, 2006), and the recycling of waste materials to conserve natural resources (Limbachiya et al., 2000; Dyer and Dhir, 2001; Paine et al., 2002; Dhir, 2006; Dyer et al., 2006; Paine and Dhir, 2010a).

Of note, an outreach programme was launched to share and transfer knowledge, in the form of organised seminars, workshops and conferences, during the period of 1988–2008 (Dhir and Green, 1990; Dhir et al., 2008, 2015) and, in doing so, a centre for the advancement of small- to medium-sized enterprises in the construction sector was established. This also included the initiation of the globalisation of concrete research and the formation of the UK–India (Newlands and Dhir, 2011) and Ireland–India research collaboration groups in 2008 and 2012, respectively, and the establishment of the UK–India Education Research Initiative (UKIERI) Concrete Congress in 2013 (Dhir et al., 2013, 2015).

Working at the forefront of cutting-edge research, in close partnership with a wide industrial base, also brought to light the fragmented and therefore often ineffective nature of the research that has generally been undertaken. Indeed, in the area of sustainable construction materials, this has stifled the rate of progress in realising the potential for developing greater adoption of these materials. As a response to this, an approach to research, *analytical systemisation*, has been developed to bring together and analyse and evaluate the published data in the global literature, to better understand and utilise the information.

Using this analytical systemisation method, the following selected successful comprehensive studies have been published:

- A study undertaken by Silva et al. (2014a) has provided a method for classifying recycled aggregates derived from construction demolition waste for use in concrete, which could help with their certification and boost stakeholders' confidence in their use. The same authors have produced a series of further studies assessing the effect of using recycled aggregates in concrete and geotechnical applications (Silva et al., 2014b, 2015a,b,c, 2016a,b).
- A series of studies aimed at assisting the design engineer in adopting the use of sustainable construction materials within the framework of existing design codes, such as Eurocode 2 (2004), has been published by Lye et al. (2015a, 2016a,b,c, and 2017). This work assessed the effect of coarse recycled concrete aggregate, glass cullet (GC) as a fine aggregate and copper slag as a fine aggregate on the modulus of elasticity, creep and shrinkage of concrete.

- On the carbonation, chloride ingress and associated corrosion of steel reinforcement in concrete made with cement incorporating fly ash, ground granulated blast furnace slag and limestone, complying with European Standard EN 197-1 (2011), the analysis and evaluation of global data have revealed some challenging facts about the performance of concrete and the accompanying impacts on sustainability that had hitherto not generally been appreciated (Lye et al., 2015b, 2016d; Elgalhud et al., 2017a,b).
- Similarly, in the area of incinerated ashes, a series of studies has used the analytical systemisation method in characterising sewage sludge ash and municipal incinerated bottom ash and assessing their environmental impacts and potential for use as components of cement or aggregate in mortar and concrete, geotechnics, road pavements and ceramic applications (Lynn et al., 2015, 2016a,b,c, 2017a,b).

The analytical systemisation method is proving to be an increasingly powerful tool in analysing and evaluating globally published experimental data on recycled and secondary materials, in terms of characterising the materials, establishing their potential applications and engineering performance across different disciplines, as well as addressing the important environmental impacts and sustainability issues. This approach has been adopted in developing a series of five books on sustainable construction materials, and the first, second and third dealing with copper slag (Dhir et al., 2016a), sewage sludge ash (Dhir et al., 2016b) and municipal incinerated bottom ash (Dhir et al., 2017), respectively, have been published.

This work, the fourth in the series, dealing with GC, which is obtained from processed waste glass, should serve as a useful resource for academics, researchers and practitioners, providing an up-to-date, comprehensive view of the research undertaken on GC and its use in construction, concrete, geotechnics, road pavements and ceramics and other applications, as well as the associated environmental impacts, case studies and issues related to standards and specifications. Of equal importance, this work should help to reduce wasteful repetitive studies and potentially spark new ideas and useful projects in areas of need.

1.2 Sustainable Construction Materials

Whilst it could be argued that the term 'sustainability' is now generally recognised, the wider implications of this are still difficult to comprehend. Alternatively, 'sustainable development' appears to be a much more straightforward and graspable expression, which is easier to appreciate. It is defined in the prominent United Nations' Brundtland report (1987) as 'development which meets the need of the present without compromising the ability of the future generations to meet their own needs.'

In this context, the ever-growing demand for building of infrastructure is fast assuming a central stage in national development, as a major consumer of natural sources of non-renewable materials and energy. This development is expected to affect increasingly environmental impacts in terms of CO_2 emissions, which can lead to subsequent climate change and temperature rises on the earth's surface, as well as having a major influence on social and economic conditions. The possible consequences in this respect are frightening, potentially leading ultimately to famine, floods, mass movement of people and the destruction of species (Stern, 2006). As such, it is not surprising that governments across the world look to the construction industry to play a major role in addressing the issues relating to sustainable development and therefore sustainability.

Along with the more efficient design, construction and operation of buildings, the growing use of recycled and secondary materials, which, for obvious reasons, are increasingly being addressed as sustainable construction materials, can also help to lower the environmental impact of construction work. For example, minimising the use of Portland cement, of which the current annual global production is around 4.2 billion tonnes (see Figure 1.1), can lead to significant reductions in CO_2 emissions. The use of GC in ground form as a component of cement is discussed in Chapter 4. Whilst this has potential to make some contribution in reducing CO_2 emissions, similar use of other waste materials can collectively make a significant contribution. Indeed, in this respect, EN 197-1 (2011) on common cements recognises several by-product materials as constituent materials of cement. Furthermore, it is interesting to note the total cement production in China, shown in Figure 1.1, which brings home the threat to sustainability that emerging countries will carry in future as the development of infrastructure in these countries, which accounts for nearly two-thirds of that in the world, begins to move full speed ahead.

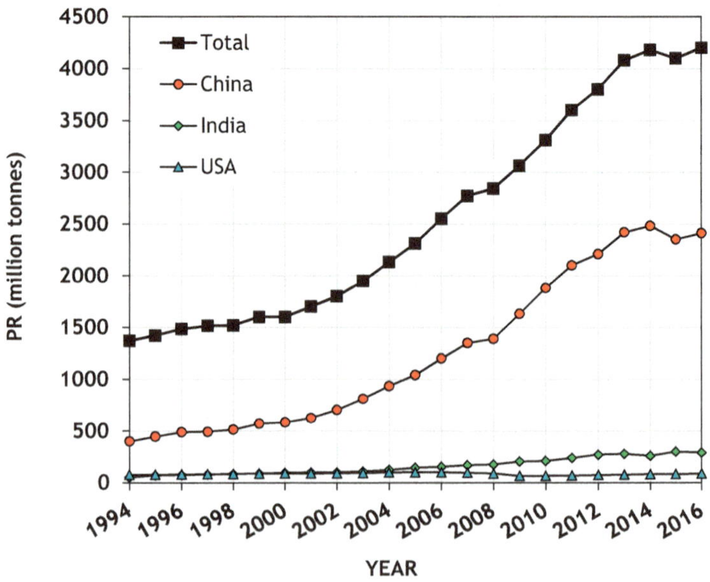

Figure 1.1 World cement production from 1994 to 2016.
Data taken from USGS (2017).

As another example, minimising the consumption of natural aggregates, of which the annual global production is currently around 50 billion tonnes and forecasted to increase further at the rate of 5% per annum, can be realised by developing the use of recycled and secondary aggregates (RSAs) in construction. Whilst this is perhaps generally appreciated, the pertinent question is how to change the mindset and accelerate the process of routinely specifying RSAs in the construction industry. Figure 1.2 clearly

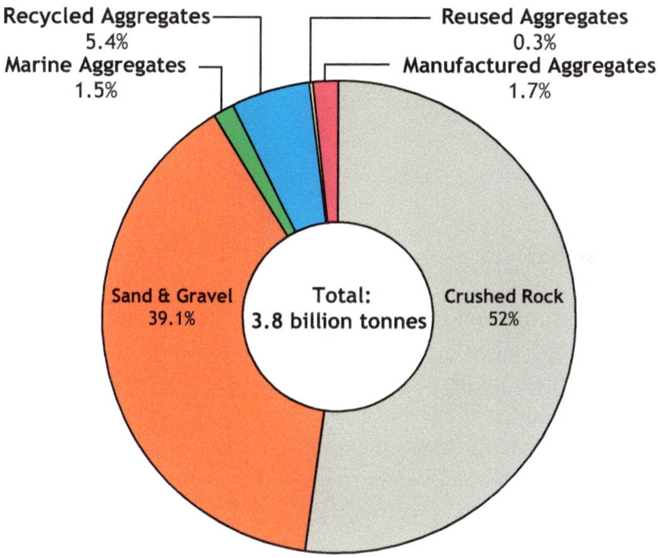

Figure 1.2 Aggregate production in 38 European countries and Israel in 2015. Data taken from UEPG (2017).

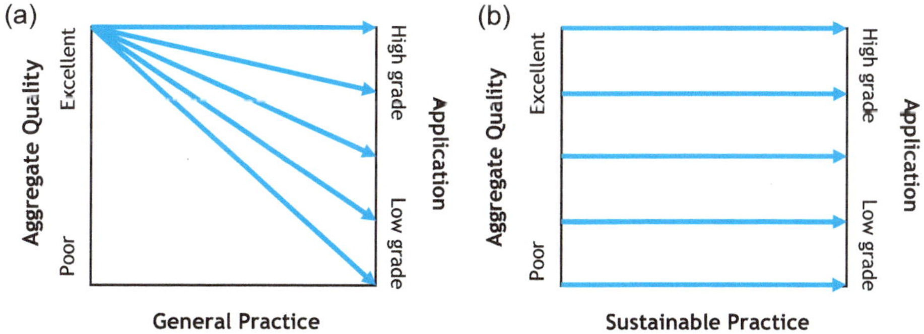

Figure 1.3 General and sustainable practices in dealing with aggregates. Adapted from Dhir et al. (2004b).

emphasises the need to develop the use of RSA materials. In this context, the quantity of manufactured aggregates used in 38 European nations amounts to only 1.5% of the total estimated production of RSAs. The numbers become even more daunting when one considers that the corresponding share of recycled aggregates arising from construction demolition and excavation waste used in this region stands at only 5.5%.

It is recognised that the national standards world over are moving towards facilitating the use of RSAs in construction, and the performance-based approach is being advanced (Paine and Dhir, 2010b; Collery et al., 2015). Figure 1.3 emphasises the pertinent point of sustainability as a simple workable philosophy that is easy to understand and points the way forward in adopting the sustainable use of construction materials by matching the material quality with the application demands.

1.3 Glass Cullet

Glass is an important material in human daily life. Its use can be traced as far back as about 2000 BC (De Jong et al., 2011a). As the properties of glass can be modified by adjusting its chemical composition, there is a record of at least 400,000 different types of glass formulations that have been developed (Sciglass, 2014). This highly versatile material, along with its uniqueness in transparency, chemical durability and mechanical properties, offers a diverse range of industrial applications in packaging, construction, transportation and electronics.

On a global basis, in 2015, about 101 million tonnes of glass was produced, and the glass production in the European Union (EU-28) accounted for about one-third of the total global production, making it the world's largest glass producer region (Glass Alliance Europe, 2015). The production of container glass, such as bottles, jars and flagons for use in food, drink, cosmetic and pharmaceutical packaging, is the largest sector in the EU-28 glass industry, occupying 63% of the total output (Figure 1.4). This is followed by flat glass, amounting to 29% of the total production, which is mainly used in the construction and automotive industries. Other sectors, including domestic glass to produce tableware and cookware, glass fibre for composite materials and special glass for cathode ray tubes, are relatively small and collectively account for less than 10% of the total glass industry.

Silica sand is essentially the primary virgin material used in glassmaking, blended together with fluxes, firing agents, stabilisers, colourants and other components for efficient melting and forming processes, as well as the desired product characteristics. Melting these raw materials to a homogeneous melt that forms a glass on cooling is energy intensive, demanding over 75% of the total energy consumption in glass production (Scalet et al., 2013). Whilst the energy demand

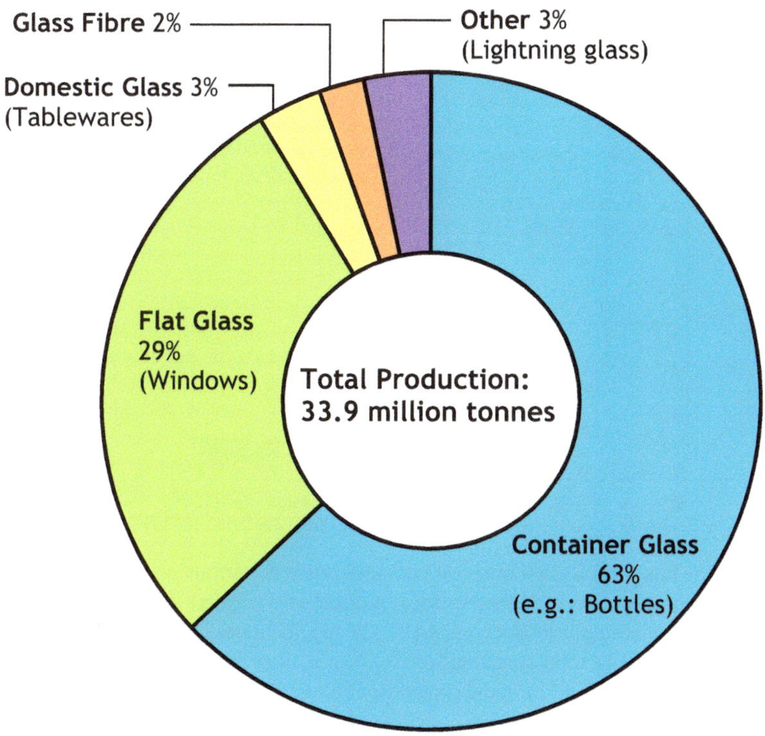

Figure 1.4 Glass production by sector in the European Union-28 in 2015.
Data taken from Glass Alliance Europe (2015).

can be reduced through the design of the furnace, the addition of processed waste glass, also known as GC, as part of the raw materials in glassmaking is one of the most viable options for saving energy. This process is known as closed-loop recycling, that is, GC being recycled back for re-melting to produce a new glass. Depending on the glass type, the typical amounts of GC used can vary greatly from 25% up to as high as 95%, but normally no more than 50% (De Jong et al., 2011b). On average, 1 tonne of GC can save 1.2 tonnes of virgin materials and 0.67 tonne of CO_2 emissions, for every tonne of glass produced (FEVE, 2016).

Figure 1.5 shows that during 2011–15, Europe, the United States, Australia, New Zealand, Brazil, Singapore and South Africa consumed 29.4 million tonnes of container glass, with an average recovery rate of 56%, which would be greatly less on a global basis. Due to the 60% glass recovery target set by the EU Packaging and Packaging Waste Directive (84/62/EC), Europe on a whole has the highest glass recovery rate (Figure 1.5); however, the rate differs markedly across the region, with Denmark the highest at 98% and Malta the lowest at 21%. The container GC recovery

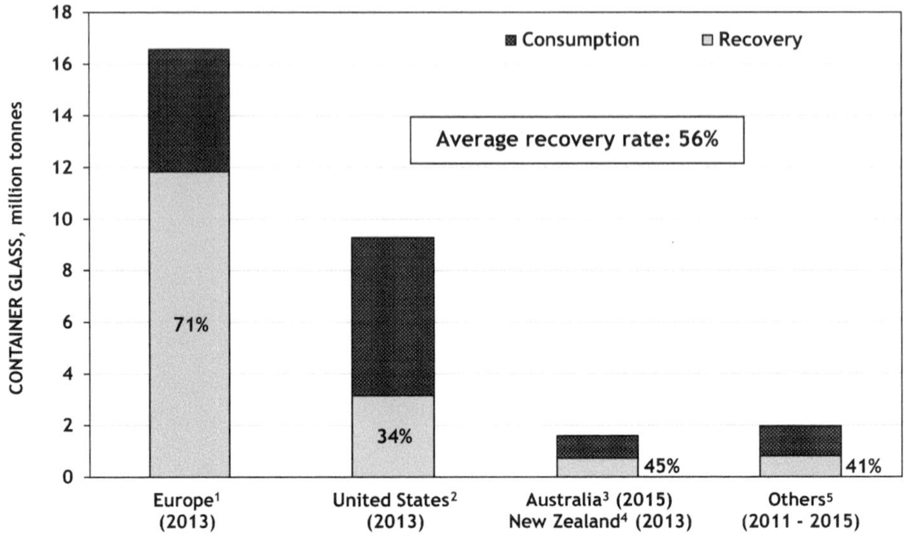

Figure 1.5 Consumption and recovery of container glass in various countries. [1]Based on FEVE (2015), consumption is defined as local production + imports – exports. [2]US EPA (2015), content in municipal solid waste. [3]APC (2015), local production + imports. [4]Glass Packaging Forum (2014), total consumption. [5]Others include Brazil, Singapore and South Africa: Brazil, CEMPRE (2013), local production; Singapore, NEA (2016), content in solid waste; South Africa, GreenCape (2015), local production + imports – exports.

rate of other regions (combined) is no more than 50%. Flat glass waste, on the other hand, generated from renovation and demolition of buildings in the EU-28 in 2013, amounted to about 1.5 million tonnes, but it is almost never recycled into new glass products (Hestin et al., 2016).

Although re-melting of GC clearly offers more environmentally friendly benefits than its use elsewhere, this method, in reality, is often restricted, especially for post-consumer GC, owing to the presence of impurities or contaminants, such as (Vieitez et al., 2011):

- Non-glass material components, such as metals, organics, ceramics and glass ceramics, which can cause defects in the final products.
- Commingling of different glass types, as the production of glass normally accepts only GC of similar types.
- Colour contamination, especially for the manufacture of flint glass, which has relatively low tolerance of colour contamination.

Proper sorting, processing and treating of GC can be a difficult and expensive exercise to undertake in practice. Additionally, other factors such as supply and demand of GC can also act as a barrier, for example, green GC is surplus in the United Kingdom, as a result of imports of large quantities of green container glass, as reported by Vieitez et al. (2011).

Thus, it is important to create alternative markets for GC in dealing with the remaining GC after re-melting and increasing its economic value. It has been suggested (Dhir and Dyer, 2003) that alternative markets for GC can be categorised into three broad groups, based on its:

- Physical properties, as a granular material in concrete, geotechnics, road pavements, filtration and abrasive industries.
- Chemical properties, as a cement component and raw material in the synthesis of industrial minerals.
- Properties at elevated temperatures, in ceramic applications and as a foamed material.

1.4 Layout and Contents

The book consists of 11 chapters.

This chapter introduces the nature and purpose of the work undertaken for this book, as well as providing the basic statistics on the production of glass and GC.

Details of the methodology, the analytical systemisation method, adopted in accomplishing the work, which involved bringing together the global knowledge on the characteristics of GC and its potential use in construction, are described in Chapter 2. This chapter explains how the exhaustive search of globally published literature in the English medium, consisting mainly, but not exclusively, of journal papers, conference papers and reports produced by public and private bodies, has been carried out. The manner in which the systematic analysis, evaluation and structuring of the published information dealing with the use of GC in various construction applications was carried out is also described.

Chapter 3 describes the production of GC as well as its chemical and physical characteristics, and thus provides a fundamental knowledge of the material for any user.

Finely ground GC possesses pozzolanic behaviour and its use as a cement component in concrete over a wide range of concrete strength is examined in Chapter 4. This chapter provides a wealth of information on the effects of ground GC on the performance of concrete in both the fresh state and the hardened state, including the durability of the concrete.

The effects of the use of GC as a component of sand (fine aggregate), up to 100%, on the fresh and hardened properties of concrete, including permeation and durability properties, as well as the potentially contentious subject of alkali–silica reaction expansion induced by GC, are discussed in Chapter 5.

Chapter 6 describes the use GC as a filler aggregate, of size less than 65 μm, and how this may influence the properties of both fresh and hardened concrete. The application of GC filler in self-compacting concrete is also discussed, as well as the risk of alkali–silica reaction developing in concrete due to the use of GC filler.

Chapter 7 deals with the use of GC as an aggregate (mostly as fine gravel and sand) in geotechnical applications, such as fill materials, covering the material characteristics, geotechnical properties, durability, environmental impact, case studies and guidance relating to the use of the material.

Chapter 8 presents the use of GC as an aggregate in road pavements such as unbound, hydraulically bound and bituminous bound applications, as well as the associated environmental impact and case studies.

The suitability of the use of GC in ceramic applications and a range of other potentially attractive markets, such as filtration, epoxy composites, paints, abrasives and roof coatings, is discussed in Chapter 9.

The associated environmental impacts due to the use of GC in all its applications as discussed in Chapters 4–8 are dealt with in Chapter 10, in which relevant case studies as well as standards and specifications are also discussed.

The epilogue, in essence, is presented as Chapter 11, providing the salient closing points emerging from this work.

References

APC, 2015. Australian Packaging Covenant Annual Report 2014–15. Australian Packaging Covenant Council, Sydney, Australia.

Brundtland G H, 1987. Report of the World Commissions on Environment and Development: Our Common Future. United Nations World Commission on Environment and Development. Available from: http://www.un-documents.net/our-common-future.pdf.

CEMPRE, 2013. Glass. Compromisso Empresarial para Reciclagem. Available from: http://cempre.org.br/artigo-publicacao/ficha-tecnica/id/6/vidrofrom.

Collery D J, Paine K A and Dhir R K, 2015. Establishing rational use of recycled aggregates in concrete: a performance related approach. Magazine of Concrete Research 67 (11), 559–574.

De Jong B, Beerkens R, Van Nijatten P and Le Bourhis E, 2011a. Glass, 1. Fundamentals. Ullmann's Encyclopaedia of Industrial Chemistry. Wiley-VCH, Weinheim, Germany.

De Jong B, Beerkens R, Van Nijatten P and Le Bourhis E, 2011b. Glass, 2. Production. Ullmann's Encyclopaedia of Industrial Chemistry. Wiley-VCH, Weinheim, Germany.

Dhir R K, 2006. Towards total use of fly ash in concrete construction. In: Proceedings of Coal Ash Technology Conference, United Kingdom Quality Ash Association (CD ROM), Birmingham, 15–17 May 2006.

Dhir R K and Dyer T D, 2003. Maximising opportunities for recycling glass. In: Limbachiya M and Roberts J (Eds.), Proceedings of the International Conference on Sustainable Waste Management and Recycling: Glass Waste. Thomas Telford, London, UK, pp. 1–13.

Dhir R K and Green J W, 1990. Protection of Concrete. E. & F. N. Spon, London. 1090 pp.

Dhir R K and Hewlett P C, 2008. Cement: a question of responsible use. Concrete 42 (7), 40–42 plus correspondence, Concrete 42 (10), 11–12.

Dhir R K, Tittle P A J and McCarthy M J, 2000. Role of cement content in specifications for durability of concrete. Concrete 34 (10), 68–76 Discussion: 35 (3), 2001, 11–13.

Dhir R K, McCarthy M J, Zhou S and Tittle P A J, 2004a. Role of cement content in specifications for concrete durability: cement type influences. Proceedings of the Institution of Civil Engineers – Structures and Buildings 157 (2), 113–127.

Dhir R K, Paine K A, Dyer T D and Tang M C, 2004b. Value-added recycling of domestic, industrial and construction arising as concrete aggregate. Concrete Engineering International 8 (1), 43–48.

Dhir R K, McCarthy M J, Tittle P A J and Zhou S, 2006. Role of cement content in specifications for concrete durability: aggregate type influences. Proceedings of the Institution of Civil Engineers – Structures and Buildings 159 (4), 229–242.

Dhir R K, Hewlett P, Csetenyi L and Newlands M D, 2008. Role for Concrete in Global Development. IHS BRE Press, Garston, Watford, UK. 910 pp.

Dhir R K, Singh S P and Goel S, 2013. Innovations in Concrete Construction. Excel India Publishers, New Delhi. 312 pp.

Dhir R K, Singh S P, Bedi R and Goel S, 2015. Concrete Research Driving Profit and Sustainability. Excel India Publishers, New Delhi, India.

Dhir R K, de Brito J, Mangabhai R and Lye C Q, 2016a. Sustainable Construction Materials: Copper Slag. Woodhead Publishing, Elsevier, Cambridge. 322 pp.

Dhir R K, Ghataora G S and Lynn C J, 2016b. Sustainable Construction Materials: Sewage Sludge Ash. Woodhead Publishing, Elsevier, Cambridge. 274 pp.

Dhir R K, de Brito J, Lynn C J and Silva R V, 2017. Sustainable Construction Materials: Municipal Incinerated Bottom Ash. Woodhead Publishing, Elsevier, Cambridge. 458 pp.

Dyer T D and Dhir R K, 2001. Chemical reactions of glass cullet used as a cement component. Journal of Materials in Civil Engineering 13 (6), 412–417.

Dyer T D, Dhir R K and Halliday J E, 2006. Influence of solid solutions on chloride leaching from waste forms. Proceedings of the Institution of Civil Engineers, Waste and Resource Management 159 (3), 131–139.

Elgalhud A A, Dhir R K and Ghataora G S, 2017a. Carbonation resistance of concrete: limestone addition effect. Magazine of Concrete Research 69 (2), 84–106.

Elgalhud A A, Dhir R K and Ghataora G S, 2017b. Chloride ingress of concrete: limestone addition effect. Magazine of Concrete Research. https://www.icevirtuallibrary.com/doi/abs/10.1680/jmacr.17.00177.

EN 197, 2011. Cement Composition, Specifications and Conformity Criteria for Common Cements. Comité Européen de Normalisation (CEN), Brussels, Belgium.

FEVE, 2015. Container Glass – Recycling Year 2013. The European Container Glass Federation, Brussels, Belgium. Available from: http://feve.org/glass-recycling-hits-73-eu/.

FEVE, 2016. Recycling: Why Glass Always Has a Happy CO_2 Ending. The European Container Glass Federation, Brussels, Belgium.

Glass Alliance Europe, 2015. EU Glass Production 2015. Glass Alliance Europe, Brussels, Belgium. Available from: http://www.glassallianceeurope.eu/en/statistical-data.

Glass Packaging Forum, 2014. Glass Recovery Holds Steady at 69% in 2013. Glass Packaging Forum, Auckland, New Zealand. Available from: http://www.glassforum.org.nz/news-recycling-rate/.

GreenCape, 2015. Waste Economy. GreenCape, Cape Town, South Africa, (Market Intelligence Report).

Hestin M, de Veron S and Burgos S, 2016. Economic study on recycling of building glass in Europe. Deloitte Sustainability, London, UK, p. 141.

Limbachiya M C, Leelawat T and Dhir R K, 2000. Use of recycled concrete aggregate in high-strength concrete. Materials and Structures 33 (9), 574–580.

Lye C Q, Koh S K, Mangabhai R and Dhir R K, 2015a. Use of copper slag and washed copper slag as sand in concrete: a state-of-the-art review. Magazine of Concrete Research 67 (12), 665–679.

Lye C Q, Dhir R K and Ghataora G S, 2015b. Carbonation resistance of fly ash concrete. Magazine of Concrete Research 67 (21), 1150–1178.

Lye C Q, Dhir R K and Ghataora G S, 2016a. Elastic modulus of concrete made with recycled aggregates. ICE Proceedings: Structures and Buildings 169 (5), 314–339.

Lye C Q, Dhir R K and Ghataora G S, 2016b. Creep strain of recycled aggregate concrete. Construction and Building Materials 102 (1), 244–259.

Lye C Q, Dhir R K and Ghataora G S, 2016c. Shrinkage of recycled aggregate concrete. Proceedings of the Institution of Civil Engineers – Structures and Buildings 169 (12), 867–891.

Lye C Q, Dhir R K and Ghataora G S, 2016d. Carbonation resistance of GGBS concrete. Magazine of Concrete Research 68 (18), 936–969.

Lye C Q, Dhir R K and Ghataora G S, 2017. Deformation properties of concrete using glass cullet fine aggregate. Proceedings - Institution of Civil Engineers: Structures and Buildings 170 (5), 321–335.

Lynn C J, Dhir R K, Ghataora G S and West R P, 2015. Sewage sludge ash characteristics and potential for use in concrete. Construction and Building Materials 98, 767–779.

Lynn C, Dhir R K and Ghataora G S, 2016a. Sewage sludge ash characteristics and potential for Use in bricks, tiles and ceramics. Water Science and Technology 74 (1), 17–29.

Lynn C J, Dhir R K and Ghataora G S, 2016b. Municipal incinerated bottom ash characteristics and potential for use as aggregate in concrete. Elsevier – Construction and Building Materials 127, 504–517.

Lynn C, Ghataora G S and Dhir R K, 2016c. Environmental impacts of MIBA in geotechnics and road applications. Institution of Civil Engineers Journal: Environmental Geotechnics. https://doi.org/10.1680/jenge.15.00029.

Lynn C J, Ghataora G S and Dhir R K, 2017a. Municipal incinerated bottom ash (MIBA) characteristics and potential for use in road pavements and geotechnical applications. International Journal of Pavement Research and Technology 10 (2), 185–201.

Lynn C J, Dhir R K and Ghataora G S, 2017b. Municipal incinerated bottom ash use as a cement component in concrete. Magazine of Concrete Research 69 (10), 512–525.

NEA, 2016. Waste Statistics and Overall Recycling. National Environment Agency, Singapore. Available from: http://www.nea.gov.sg/energy-waste/waste-management/waste-statistics-and-overall-recycling.

Newlands M D and Dhir R K, 2011. Concrete for high performance sustainable infrastructure. In: Proceedings of International UKIERI Concrete Congress, Concrete for 21st Century Construction. Shroff Publishers and Distributors, Pvt. Ltd., Mumbai, India. 294 pp.

Paine K A and Dhir R K, 2010a. Research on new applications for granulated rubber in concrete, Proceedings of the Institution of Civil Engineers. Construction Materials 163 (1), 7–17.

Paine K A and Dhir R K, 2010b. Recycled aggregates in concrete: a performance related approach. Magazine of Concrete Research 62 (7), 519–530.

Paine K A, Dhir R K and Doran V P, 2002. Incinerator bottom ash: engineering and environmental properties as a cement bound paving material. International Journal of Pavement Engineering 3 (1), 43–52.

Scalet B M, Garcia Munoz M, Sissa A Q, Roudier S and Delgado Sancho L, 2013. Best Available Techniques (BAT) Reference Document for Manufacture of Glass. European Commission Joint Research Centre, Seville, Spain. JRC Reference Report, Industrial Emissions Directive 2010/75/EU.

Sciglass, 2014. Glass Property Information System. Available from: http://www.sciglass.info/.

Silva R V, de Brito J and Dhir R K, 2014a. Properties and composition of recycled aggregates from construction and demolition waste suitable for concrete production. Construction and Building Materials 65, 201–217.

Silva R V, de Brito J and Dhir R K, 2014b. The influence of the use of recycled aggregates on the compressive strength of concrete: a review. European Journal of Environmental and Civil Engineering 19 (7), 825–849.

Silva R V, de Brito J and Dhir R K, 2015a. Tensile strength behaviour of recycled aggregate concrete. Construction and Building Materials 83, 108–118.

Silva R V, Neves R, de Brito J and Dhir R K, 2015b. Carbonation behaviour of recycled aggregate concrete. Cement and Concrete Composites 62, 22–32.

Silva R V, de Brito J, Neves R and Dhir R K, 2015c. Prediction of chloride ion penetration of recycled aggregate concrete. Materials Research 18 (2), 427–440.

Silva R V, de Brito J and Dhir R K, 2016a. Establishing a relationship between the modulus of elasticity and compressive strength of recycled aggregate concrete. Journal of Cleaner Production 112 (4), 2171–2186.

Silva R V, de Brito J, Evangelista L and Dhir R K, 2016b. Design of reinforced recycled aggregate concrete elements in conformity with Eurocode 2. Construction and Building Materials 105, 144–156.

Stern N, 2006. The Economics of Climate Change: The Stern Review. HM Treasury, London, UK. 662 pp.

UEPG, 2017. Estimates of Aggregates Production Data 2014. European Aggregates Association, Brussels, Belgium. http://www.uepg.eu/statistics/estimates-of-production-data/data-2015.

US EPA, 2015. Advancing Sustainable Materials Management: Fact and Figures 2013. United State Environmental Protection Agency, Washington DC, USA. Report EPA530-R-15–1002.

USGS, 2017. Mineral commodity Summaries. United States Geology Survey, Reston, VA, USA. Available from: http://minerals.usgs.gov/minerals/pubs/commodity/cement/.

Vieitez E R, Eder P, Villanueva A and Saveyn H, 2011. End-of-waste Criteria for Glass Cullet: Technical Properties. European Commission, Seville, Spain, (Final Report).

Whyte A, Dyer T D and Dhir R K, 2005. Best practicable environmental option (BPEO) for recycling demolition waste. In: Dhir R K, Dyer T D and Newlands M D (Eds.), Achieving Sustainability in Construction, pp. 245–252.

Methodology

2

Main Headings

- Literature search and appraisal
- Building the data matrix
- Analysis, evaluation and modelling of data
- Dissemination

Synopsis

For the reader to fully benefit from this work, the methodology adopted in preparing the base material for writing this book is described. This consists of three distinct tasks, undertaken in sequence. First, the globally published literature on the subject of glass cullet and its use in construction is thoroughly sourced and appraised. The second stage involves sorting of the literature, mining the data from the sourced publications and parking this information to build the data matrix. The third part of the work involves the systematic analysis and evaluation of the data.

Keywords: Glass cullet, Literature sourcing and appraisal, Data matrix, Data analysis and evaluation.

Sustainable Construction Materials: Glass Cullet. https://doi.org/10.1016/B978-0-08-100984-0.00002-3

2.1 Introduction

The work described in this book has been developed using an approach that is very different to the norm and is best suited to establishing what is already known, and how well it is known, in a field of study. It can further the value-added sustainable use of processed waste glass, commonly known as glass cullet (GC), in construction and at the same time help to minimise repetitive research and better channel the resources to advance the material's use. To realise this, a robust and clearly structured methodology, *analytical systemisation*, has been designed. To understand and achieve the full impact and benefit of the work presented in this book, a detailed description of the methodology is provided.

Figure 2.1 outlines the four main stages of the work, beginning with the sourcing and assembling of the base information from the published literature. As an indication of the sheer scale of the work, it would be useful to consider the efforts required to produce this publication, which involved four experts working over a prolonged period.

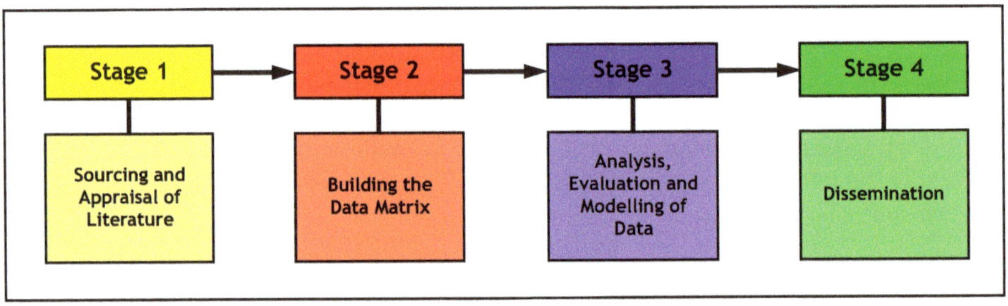

Figure 2.1 Outline of the main stages of the methodology.

This book is based on 751 publications on the production, characteristics and use of GC in construction. There was a large amount of information to be managed, with, where necessary, reference to an additional 261 works of authoritative persons and standards/codes of practice in construction, bringing the number of publications used to 1012.

All the 751 publications were vetted and sorted and the data therein extracted, to construct the complete data matrix. Thereafter, with the combined pool of extracted experimental results in hand, a fresh analysis, evaluation and modelling of the data were undertaken. To finish, the findings were carefully structured in this book, as the form to facilitate effective dissemination.

The book contains 11 chapters, covering first the processing and management of the residues, followed by the characteristics and use of GC in various applications, its environmental impact, case studies, relevant standards and codes of practice. Each chapter was assigned its own Excel file, containing as many as 25 separate sheets for the different subheadings. Individual sheets were subsequently populated with the extracted data, each containing hundreds to thousands of distinct data points. These sheets then formed the basis of the analysis, evaluation and modelling of the sourced data in developing this book.

2.2 Literature Search and Appraisal

Whilst it is recognised that the literature on the subject of GC and its use in construction has been published in many languages, for practicality, the global search of literature was limited to the material published in English. The main contribution has come from peer-reviewed journal papers, which provided a reputable source of information, covering most of the different relevant subject areas. Although more difficult to obtain, conference papers were also sourced. Reports produced from government bodies and private organisations have been included, where available. Additionally, there were several other minor sources of information that were used in completing this search, as detailed in Section 2.2.4.

2.2.1 Identifying and Sourcing Literature

The process of sourcing the literature was wide-reaching and thorough. A list of the relevant keywords covering the scope of the work, and the search engines and websites used for sourcing the literature, is provided in Table 2.1.

The literature search was undertaken until no further publication could be sourced and it could be judged assertively to be exhausted. This search policy proved to be rewarding, though a challenging and time-consuming exercise. To systematically catalogue the sourced literature and, thereafter, the information extracted from the publications, a data matrix was created in Excel, containing all the various subject areas. Once the search was concluded, the initial background information was logged to determine the nature of the sourced literature, including the year of publication and details of the authors in the form of their affiliated institution and country, as well as the publication type. A few points of interest emerging from this exercise are discussed below.

2.2.2 Publication Timeline

In total, 751 GC publications were sourced and used for data mining and developing a data matrix to prepare for writing this book; these were published over a period of 48 years, from 1970 to 2017. Amongst the early work published in this subject area was the paper produced by Malisch et al. (1970) from the University of Missouri–Rolla, about bituminous bound application, published in *Highway Research Record*, Issue 307. In this study the GC used was obtained from container glass, which was crushed to sizes less than 12.7 mm and was reported to be flat and elongated. As Figure 2.2 shows, there was little literature published during the first period of 25 years from 1970 to 1994, though of the 18 publications produced during this period, 14 were from the United States, two from the United Kingdom and one from Australia and one from Malta.

Table 2.1 Keywords and search engines and websites used

(a) Keywords Used	
• Glass cullet	• Unbound
• Waste glass	• Hydraulically bound
• Ground glass cullet	• Bituminous bound
• Glass filler	• Ceramics
• Glass powder	• Glass-ceramics
• Properties	• Bricks
• Characteristics	• Tiles
• Production	• Glaze
• Processing	• Porcelain
• Waste management	• Abrasive
• Aggregate	• Epoxy composite
• Concrete	• Filtration
• Cement	• Glass fibres
• Pozzolanic	• Leaching
• Geotechnical applications	• Environment
• Fill/backfill	• Case studies
• Road pavements	• Field studies
(b) Search Engines and Websites Used	
• Academic Search Complete	• Construction Information Service
• American Concrete Institute	• ProQuest
• American Society of Civil Engineers	• Researchgate
• ASTM	• RILEM
• BASE	• Sagepub
• British Standards Online	• ScienceDirect
• EBSCOhost	• Science.gov
• Engineering Village	• Scientific.net
• Google	• Scopus
• Google Scholar	• SpringerLink
• JSTOR	• Taylor & Francis Online
• Inderscience Online	• Web of Knowledge
• Ingenta Connect	• Web of Science
• Institute of Civil Engineering	• Wiley Online Library

In 1996, Clean Washington Center published a number of publications as best practice guides on the use of GC in concrete and geotechnical applications (as aggregate) and other applications such as paving tiles, epoxy and blasting abrasive.

A sudden increase from 8 to 27 papers published in 2001 was due to the 16 relevant papers published in the proceedings of the symposium on recycling and reuse of GC organised by the Concrete Technology Unit at the University of Dundee (Figure 2.2). Likewise, the sharp increase from 17 to 49 papers published in 2004 was due to the 20 relevant papers published from a single conference held in London. However, the high number of publications in 2013, 2014 and 2015, with the highest number of 78 papers, are all mainly journal papers.

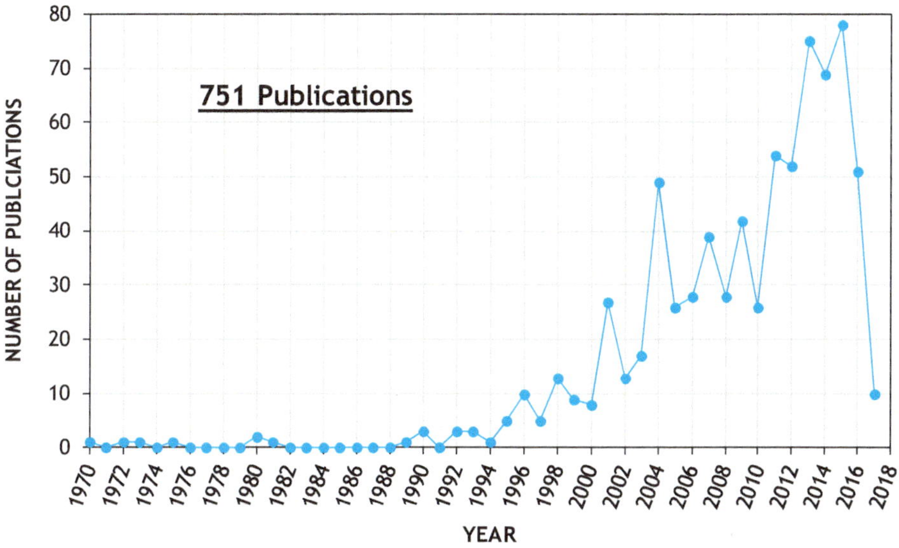

Figure 2.2 Distribution of publications by year.

2.2.3 Global Publication Status

The country-wise distribution of the published literature, based on all the authors of each publication, not only the first author, has been logged and this information is presented in Figure 2.3. This shows that the distribution of publications amongst the 51 countries, with 40% each from Europe and Asia, that have published their research has tended to concentrate in a few countries. The United States and the United Kingdom each have over 100 publications, accounting for one-third of the total publications, and together with Australia and Italy, each having more than 50 publications, these four countries account for over 50% of the total publications.

The publication timeline for the top 10 countries, with a minimum of 20 publications each, is different, as shown in Figure 2.4. The United States started publishing in 1970, but showed a steady increase only from 1995, with a publishing rate of 6.7 papers per annum. The United Kingdom started to publish in the year 1980, and showed a steep increase in publishing rate of 11 papers per annum during 2003–09, but the rate dropped thereafter to 3.7 per annum. Australia also started publishing in 1980, but became active only in 2006, having a rate of publishing of 4.3 per annum. Italy started to publish in 2001, with a steady rate of 2.9 per annum. Canada started to publish in 1995, but with a steady rate of 3.1 papers per annum from 2007. China and India started publishing in 2005 and 2009, with a steady rate of 3.5 and 4.7 papers per annum from 2009 to 2011, respectively. Hong Kong started to publish in 2002 and had a peak rate of publishing 7.3 papers per annum for a short period of 2011–13. Taiwan and Turkey stated to publish in 2002 and 2001, respectively, and had a steady rate of publishing of 2.3 and 2.1 papers per annum from 2006 to 2008.

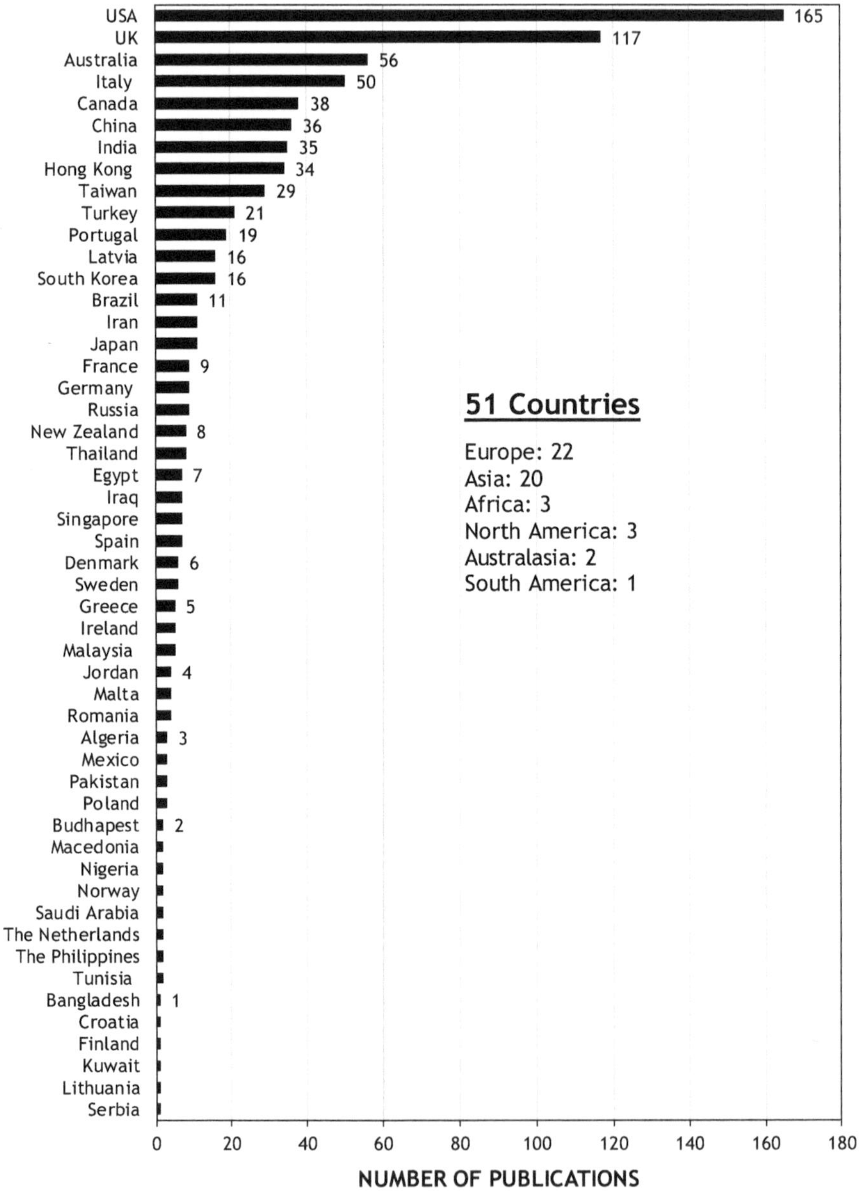

Figure 2.3 Distribution of publications by country.

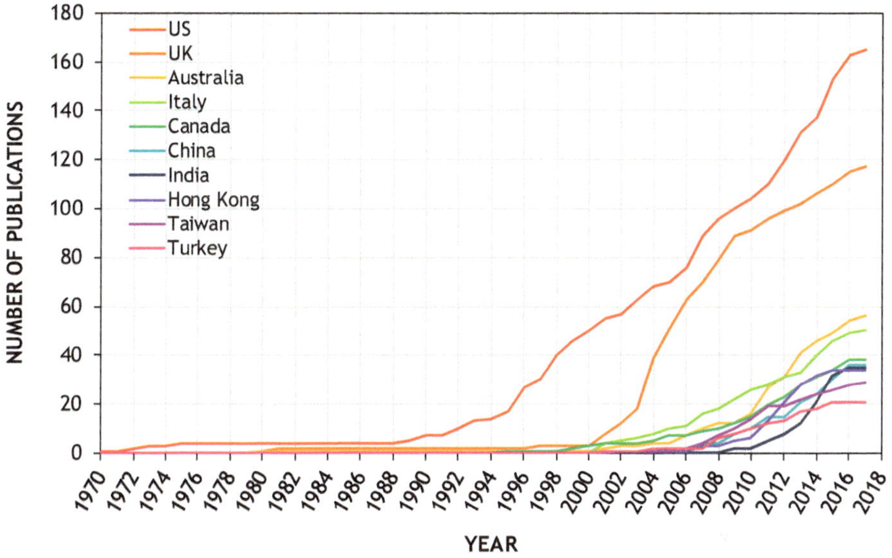

Figure 2.4 Cumulative rate of publications on glass cullet produced per country.

2.2.4 Publication Types

Knowing where the sourced literature has been published is another important part of the process of evaluating the overall credentials of the research. As can be seen from Figure 2.5, nearly 60% of the publications on GC are journal papers, with conference papers and reports both contributing about the same amount, 16% and 13%, respectively, of the total published material. There are also a good number of specifications-related publications (6%). Together, these four types of publications accounted for 93% of the total literature sourced. It would be expected that the research published in these source types would generally be of a reasonably high standard. There were also smaller amounts of additional research information found (7%) in the form of online theses, bulletins, magazines, books, fact sheets, online articles, digests and presentation slides.

Investigating further into the nature of the biggest publication type, a staggering number of 173 different journals were found to contain information on the subject of GC. This stretched across the fields of engineering, material sciences and environmental sciences. Journals with a minimum of 5 papers are listed in Table 2.2, with 7 of 12 journals published by Elsevier. *Construction and Building Materials* have published the highest number of papers (56), followed by *Cement and Concrete Composites* (30). Of the remaining 161 journals, 7 have published four papers each, 11 have published three papers, 24 have published two papers and 119 have published only one paper.

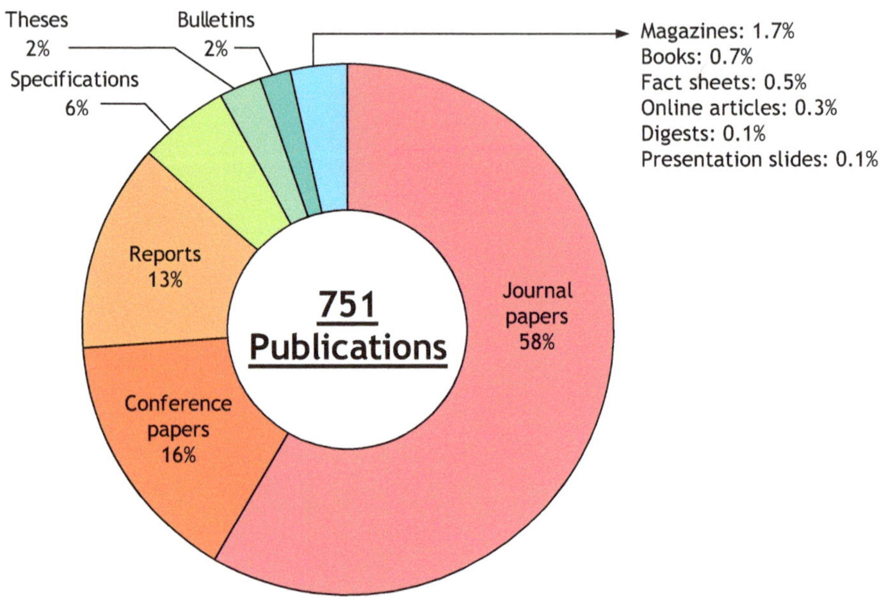

Figure 2.5 Distribution of publications by nature.

Table 2.2 Main journals publishing on glass cullet

Journal	Number of Publications	Time Period
Construction and Building Materials	56	2007–16
Cement and Concrete Composites	30	2003–13
Ceramics International	20	2002–16
Waste Management	19	2002–14
Cement and Concrete Research	16	1995–14
Journal of Cleaner Production	15	2007–16
Materials and Structures	14	2004–16
Journal of Materials in Civil Engineering	10	1998–15
ACI Materials Journal	8	2000–15
Journal of Hazardous Materials	8	2007–13
Glass and Ceramics	5	2013–15
Journal of Geotechnical and Geoenvironmental Engineering	5	2006–08

2.2.5 Researchers Involved

The background information gathered from the literature on the subject of GC and its use in construction showed that 1402 authors have published in this area, though Table 2.3 has been limited to only 20 authors who have contributed a minimum of seven publications. In addition to these, there are 10 authors who have contributed six publications, 9 authors with five publications, 18 with four, 97 with three, 261 with two and a staggering number of 987 authors who have published one.

Table 2.3 Key researchers on glass cullet

Author	Country	Years	Number of Publications
Poon, C.S.	Hong Kong	2007–15	26
Ling, T.C.	Hong Kong	2010–15	21
Dhir, R.K.	United Kingdom	2001–17	18
Tagnit-Hamou, A.	Canada	2004–16	14
Meyer, C.	United States	1996–03	14
Dyer, T.D.	United Kingdom	2001–14	13
Wang, H.	Taiwan	2008–16	13
Bernardo, E.	Italy	2003–15	11
Arulrajah, A.	Australia	2011–13	9
Grubb, D.	United States	2004–08	9
Tang, A.	United Kingdom	2003–09	9
Byars, E.	United Kingdom	2003–07	8
Disfani, M.M.	Australia	2011–14	8
Kou, C.S.	Hong Kong	2009–13	8
Wartman, J.	United States	2004–08	8
Zhu, H.Y.	United Kingdom	2003–07	8
Ali, M.M.	Australia	2011–14	7
Barbieri, L.	Italy	2001–15	7
Kara, P.	Latvia	2012–16	7
Lin, K. -L.	Taiwan	2007–17	7

There are several interesting points that emerge from Table 2.3.

- C. Meyer from Columbia University, New York, USA, appears to be the first of the active researchers, working in the area of using GC as a fine aggregate in concrete, with the first conference paper published in 1996.
- C.S. Poon (2007–15) and T.C. Ling (2010–15), both from the University of Hong Kong Polytechnic, Hong Kong, have collaborated in many works relating to GC and together they are in the top two positions in terms of the number of publications they have produced. This is followed by R.K. Dhir (2001–07) from the University of Dundee, UK.
- Most of the researchers listed in Table 2.3 appear to have published consistently since 2000, but only R.K. Dhir (2001–17), A. Tagnit-Hamou (2004–16), T.D. Dyer (2001–14), E. Bernardo (2003–15) and L. Barbieri (2001–15) have remained active in the field for more than 10 years.
- E. Arulrajah (2011–13), M.M. Ali (2011–14) and M.M. Disfani (2011–14), all from Swinburne University of Technology, Australia, started later than the others and also appear to have remained active for a short duration.

2.2.6 Institutions and Organisations Involved

A staggeringly high number of institutions and organisations, 513 worldwide, have been involved in research in the area of GC and its use in construction. Table 2.4 provides a list of the 25 main institutions and organisations that have contributed a minimum of five publications on the subject. Of the institutions listed in the table, eight have a double-digit number of publications. The other institutions not listed have four or fewer, with as many as 346 having published just one. The domination of European and North American (mainly the United States) institutions, and a few key organisations from Asia and Australia, in this specific subject area is clearly visible.

2.2.7 Subject Area Distribution

In the main, the sourced literature has been categorised under eight main subject areas, as shown in Figure 2.6. On the production and properties of GC (Chapter 3), data were sourced from 274 publications, and the analysis, evaluation and synthesis work made reference to 16 standards and specifications and 14 other supplementary publications.

Owing to its chemical composition, when finely ground, GC has the potential for use as a component of cement, with 154 publications, 19 standards and 26 other references used in Chapter 4.

As GC can be produced as a fine aggregate, there has been a natural tendency to develop its use as a granular material in the areas of concrete, geotechnics and road pavements, with primary preference as a component of fine aggregate in concrete. This is reflected in the number of research publications produced in the area of concrete for its use as fine aggregate, as discussed in Chapter 5, in which 203 publications, 28 standards and 9 supplementary references are referred. There is also

Table 2.4 Institutions and organisations on glass cullet

Institution/Organisation	Country	Number of Publications
Hong Kong Polytechnic University	Hong Kong	28
University of Dundee	United Kingdom	20
Columbia University	United States	14
Riga Technical University	Latvia	14
University of Sherbrooke	Canada	14
National Kaohsiung University of Applied Sciences	Taiwan	13
Swinburne University of Technology	Australia	12
University of Padua	Italy	12
Clean Washington Center	United States	8
Drexel University	United States	8
Pennsylvania State University	United States	8
PennDOT	United States	7
WRAP	United Kingdom	7
Clarkson University	United States	6
Michigan State University	United States	6
National University of Singapore	Singapore	6
Sardar Vallabhbhai National Institute of Technology	India	6
Schnabel Engineering North	United States	6
Technical University of Lisbon	Portugal	6
University of Toulouse	France	6
National Ilan University	Taiwan	5
University of Modena and Reggio Emilia	Italy	5
University of Erlangen-Nuremberg	Germany	5
University of Malta	Malta	5
University of Sheffield	United Kingdom	5

some interest for the use of GC as a filler aggregate, as discussed in Chapter 6, which has been based on 21 publications, 6 standards and 6 supplementary references. The numbers of publications, standards and supplementary references for the use of GC in geotechnical applications, presented in Chapter 7, are 71, 18 and 15, respectively. The corresponding numbers for the use of GC in road pavement applications, presented in Chapter 8, are 52, 10 and 11.

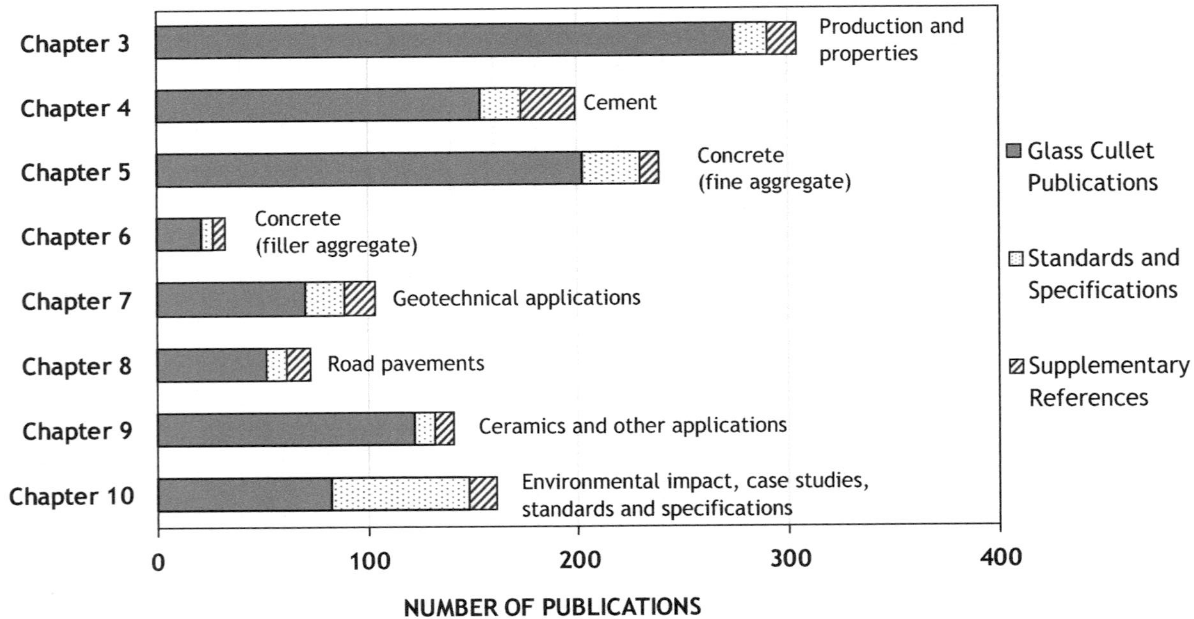

Figure 2.6 Glass cullet publications, standards and specifications and supplementary references used in Chapters 3–10.

In addition, both the chemical and the physical characteristics of GC appear to have made the material a sufficiently attractive option to explore its potential use in a number of alternative applications, including as raw feed for the manufacture of various ceramic products, as a filtration medium and as a blasting abrasive, among others, with 122 publications, 10 standards and 9 supplementary references used in Chapter 9.

Environmental issues and case studies relating to the material account for 83 publications, 65 standards and specifications and 13 other supplementary references, which are also discussed in Chapter 10.

2.3 Building the Data Matrix

This work consists of two main tasks required to facilitate the subsequent process: the systematic analysis and evaluation of the experimental data, followed by the structuring and modelling of the analysed work. Similar to laying the foundation of a building, it is extremely important to set a solid base for this work. This is done through the initial sorting of the literature and the meticulous data mining and parking of the experimental results. Although it may be seen as repetitive, and at times laborious and tiresome due to the sheer size of the task involved, the work demands a keen attention to detail, as the thoroughness of the process can greatly affect the quality and reliability of the findings.

2.3.1 Initial Sorting of Literature

This stage of the work is very much like the post office sorting the mail to deliver letters. It serves as the foundation and needs to be carried out correctly. Each publication must be thoroughly vetted and allocated to specific relevant subject areas, such as concrete, geotechnics and ceramics. The publications are then sorted into further subdivisions in each subject area; an example of this is shown in Figure 2.7 on the use of GC in geotechnical applications.

2.3.2 Data Mining and Parking

The next stage consists of identifying and extracting both qualitative descriptive information in the text and quantitative results in tables and figures, making use of the software package Plot Digitizer when required, from the sourced publications, for each subject area. The data matrix was formed through this process of data mining and parking. A partial screen capture of a sample of the data matrix is shown in Figure 2.8, for data on the particle size distribution of GC fine aggregate.

			Characteristics														
Authors	Year	Country	Chemical Composition	Particle Size Distribution	Post compaction	Cu	Cc	Shape	Flakiness Index	Specific Gravity	Water Absorption	Los Angeles Abrasion	Aggregate Crushing Value	Soundness	pH	Organic Content	Atterberg Limit
Ali and Arulrajah	2012	Australia		X	X	X	X			X		X			X	X	
Ali et al	2011a	Australia		X	X	X	X			X		X			X	X	
Ali et al	2011b	Australia		X	X	X	X			X		X			X	X	
Amlashi et al	2015	Canada															
Arabani et al	2012	Iran		X													
Archibald et al	1995	Canada															
Archibald et al	1999	Canada															
Arnold et al	2008	New Zealand		X													
ARRB Group	2010	Australia															
Arulrajah et al	2013	Australia		X	X	X	X			X	X	X			X	X	
Arulrajah et al	2014	Australia		X	X	X	X			X	X	X			X	X	
Banba and Nakamura	1984	Japan	X														
Basari	2012	Turkey		X						X							X
Blewett & Woodward	2000	UK											X				
Bo and Arulrajah	2011	Canada															
Chiou & Chen	2013	Taiwan															
Clean Washington Centre	1996 a	USA															
Clean Washington Centre	1996 b	USA															
Clean Washington Centre	1996 c	USA															
Cosentino et al	1995	USA		X		X	X			X							
Dames & Moore (CWC)	1993	USA		X				X		X		X					
Davidovic et al	2012	Serbia															
Dept. of Environment & Climate Change NSW	2007	Australia															

Figure 2.7 A partial screen capture of the initial sorting of literature showing the research subjects covered in Chapter 7 on geotechnical applications.

2.4 Analysis, Evaluation and Modelling of Data

This step involves the critical assessment of the globally published experimental results on GC and its use in construction. Using Excel, the data were assembled in a manner allowing a great deal of flexibility in the analysis and evaluation and where possible, the development of models, whilst at the same time retaining a very close connection with the results. The analysis and evaluation process proved to be very demanding, with no magic recipe or straightforward set strategy. The exercise was very much dependent on the nature of the available results and the knowledge and experience of the assessor, requiring sensitivity and attention to detail in the handling of the data, whilst retaining a pragmatic and imaginative touch.

The immediate problem one faces with the analysis and evaluation of the global data is the large amount of variation in the test results obtained by different researchers, and this must be assessed carefully. This variability can be controlled, to some extent, by working with relative values, with respect to the reference test material, usually comparing GC with accepted construction materials. The data were analysed systematically using varying approaches depending on the volume of data, nature of the subject (e.g., chemical

Mechanical Properties																	Durability			Others						Applications					
Compaction	California Bearing Ratio	Permeability	Direct Shear	Triaxial shear	Unconfined Comp. Str	Repeated load triaxial	Resistance R-Value	Stress-Strain	Elastic Modulus	Secant Modulus	Constraint modulus	Resilient Modulus	Permanent Strain	Consolidation	Earthquake-induced uplift	Thermal Conductivity	Freeze Thaw	Frost Susceptibility	Degradation	Lightweight Fill	Raw Mat for Lightweight	Mine backfill	Environmental Impacts	Case Studies	Guide	Soil improvement	structural fill	non-structural fill	trench backfill	pipe surround	landfill cover
X	X	X		X				X																							
X	X	X	X																												
X	X	X	X									X	X																		
X	X		X		X																										
					X			X	X													X				x					
					X																	X				x					
X						X																									
																														x	
X	X	X		X																						x					
X	X	X		X								X	X																		
																						X	X								
X			X																							x					
		X	X	X					X																						
																					X										
																									X						
																									X		x				
																									X						
	X	X	X	X				X		X													X								
X	X	X	X	X		X	X					X				X											x		x	x	x
X	X															X															
																							X	X							

Figure 2.7 Cont'd

composition of GC, strength activity index, deformation of concrete, compactability of soil, Marshall stability bituminous mix and leaching tests), application of the material and test parameters involved. In addition, reference was made to the current standards and specifications when necessary, to assess the products for compliance.

One example of this work is shown in Figure 2.9, in which a total of 139 shrinkage results for concrete made with GC-sand as a replacement for natural sand were considered. Box-and-whisker plots were adopted to identify the outliers and show their distribution. A trend line was obtained for GC-sand replacement levels from 0% to 100% using polynomial regression analysis. The particle size distribution of GC-sand and that of the reference natural sand were also compared to investigate how this can affect the performance of GC-sand on the shrinkage of concrete.

2.5 Dissemination

The findings emerging from the analysis, evaluation and modelling of the combined experimental results are structured in an incisive and easy-to-digest manner that can be usable for researchers and practitioners. The work is disseminated in written form as part of a series of books on sustainable construction materials,

Authors	Year	Glass Type	
			Sieve Size, (mm)
Abdallah & Fan	2014	Soda lime	% Passing
			Sieve Size, (mm)
Abendeh et al.	2015(a)	Soda lime	
			Sieve Size, (mm)
Abendeh et al.	2015(b)	Soda lime	
			Sieve Size, (mm)
Aghabaglou et al.	2015	Not specified	
			Sieve Size, (mm)
Berry et al.	2011	Not specified	
			Sieve Size, (mm)
Borhan & Bailey	2014	Soda lime	
			Sieve Size, (mm)
Dhir et al.	2005a	Soda lime	
			Sieve Size, (mm)
Disfani et al.	2011	Soda lime (fine)	
			Sieve Size, (mm)
Disfani et al.	2011	Soda lime (medium)	
			Sieve Size, (mm)
Guo et al.	2015	Soda lime	
			Sieve Size, (mm)
Isler	1984	Soda lime	
			Sieve Size, (mm)
Lee	2011	Not specified	A
			Sieve Size, (mm)
Lee	2011	Not specified	B
			Sieve Size, (mm)
Lee	2011	Not specified	C
			Sieve Size, (mm)
Lee	2011	Not specified	D
			Sieve Size, (mm)
Lee et al.	2013	Soda lime	**A**
			Sieve Size, (mm)
Lee et al.	2013	Soda lime	**B**
			Sieve Size, (mm)
Lee et al.	2013	Soda lime	**C**
			Sieve Size, (mm)

Figure 2.8 A partial screen capture of data mining and parking showing the glass cullet particle size distribution results taken from various studies.

Particle size distribution (% passing)								
4.75	**2.36**	**1.18**	**0.6**	**0.3**	**0.15**	**0.075**	**pan**	
100	99.83	64.04	40.97	20.95	9.63	0	0	
19.5	**12.5**	**9.5**	**4.75**	**2.36**	**0.85**	**0.425**	**0.075**	
100	95	88	51	31	11	6	2	
19.5	**12.5**	**9.5**	**4.75**	**2.36**	**0.85**	**0.425**	**0.075**	
100	95	88	51	31	11	6	2	
4	**2**	**1**	**0.5**	**0.3**	**0.125**			
100	85	60	35	25	15			
9.5	**4.75**	**2.36**	**0.85**	**0.425**	**0.075**	**0**		
100	99	88	48	21	8	0		
4.75	**2.36**	**1.18**	**0.6**	**0.3**	**0.15**	**0**		
100	91.1	70	5	0.2	0.2	0		
7	**4**	**2**	**1**	**0.5**	**0.3**	**0.15**		
53	29	12	7	2	1	0.1		
4.75	**2.36**	**1.18**	**0.6**	**0.5**	**0.3**	**0.15**	**0.075**	**0.05**
100	91	62	39	28	20	10	5	3
10	**7**	**4.75**	**2.36**	**1.18**	**0.6**	**0.5**	**0.3**	**0.15**
100	91	74	46	29	19	15	12	8
4.75	**2.36**	**1.18**	**0.6**	**0.3**	**0.15**			
100	90	55	30	12	7			
9.5	**4.75**	**1.18**	**0.6**	**0.3**	**0.15**	**0.075**	**0**	
100	100	76	54	28	8	4	0	
10	**4.75**	**2.36**	**1.18**	**0.6**	**0.3**	**0.15**		
100	100	88	51	25	10	0		
10	**4.75**	**2.36**	**1.18**	**0.6**	**0.3**	**0.15**		
100	100	100	58	29	11	0		
10	**4.75**	**2.36**	**1.18**	**0.6**	**0.3**	**0.15**		
100	100	100	100	49	18	0		
10	**4.75**	**2.36**	**1.18**	**0.6**	**0.3**	**0.15**		
100	100	100	100	100	86	0		
10	**4.75**	**2.36**	**1.18**	**0.6**	**0.3**	**0.15**		
100	100	88	51	26	9	0		
10	**4.75**	**2.36**	**1.18**	**0.6**	**0.3**	**0.15**		
100	100	100	58	28	10	0		
10	**4.75**	**2.36**	**1.18**	**0.6**	**0.3**	**0.15**		
100	100	100	100	49	18	0		
10	**4.75**	**2.36**	**1.18**	**0.6**	**0.3**	**0.15**		

Figure 2.8 Cont'd

30% GC	
30	269
30	150
30	140
30	0
30	-1
30	-4
30	-6
30	-16
30	-16
30	-37

40% GC	
40	-19
40	-23
40	-37

45% GC	
45	-9

50% GC	
50	50
50	18
50	8
50	6
50	0
50	0
50	-2
50	-5
50	-7
50	-7
50	-11
50	-13
50	-14
50	-14
50	-14
50	-15
50	-17
50	-26
50	-31
50	-22
50	-33

60% GC	
60	-12
60	3

70% GC	
70	10

75% GC	
75	73
75	22
75	-11
75	-12
75	-12
75	-14
75	-14
75	-15
75	-17
75	-17
75	-22
75	-24
75	-30
75	-34
75	-68

80% GC	
80	-23.7

100% GC	
100	41
100	36
100	32
100	27
100	17
100	53
100	38
100	32
100	25
100	20
100	13
100	0
100	0
100	-4
100	-8
100	-10
100	-10
100	-11
100	-14
100	-14
100	-14
100	-18
100	-18
100	-20
100	-22
100	-22
100	-24
100	-25
100	-25
100	-27
100	-27
100	-29
100	-30
100	-38

Figure 2.9 A partial screen capture of analysis and evaluation showing the effect of glass cullet (GC) as a natural sand replacement on the shrinkage of concrete.

published by Elsevier. With the novel approach undertaken, it is hoped that this book can contribute towards establishing a more widespread practical use of GC as a sustainable construction material, stimulating forward-thinking research and reducing repetitive work.

2.6 Conclusions

So that the reader can understand and benefit from this work, the clearly structured methodology that has been designed and adopted to develop the base material to enable writing this book has been described. This methodology is in three distinct parts, which follow in sequence.

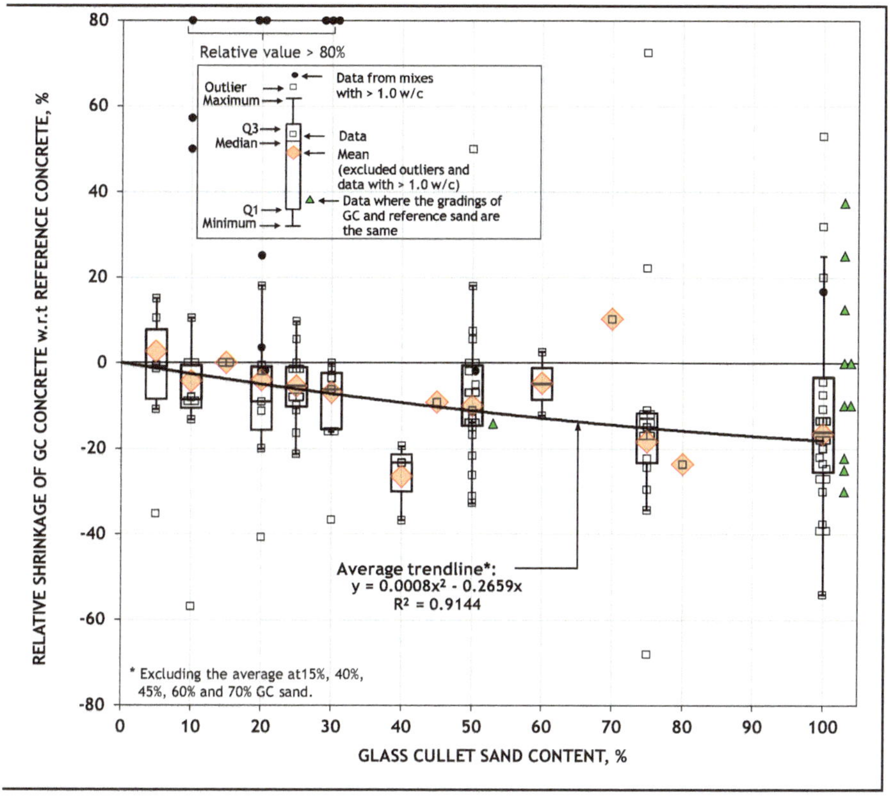

Figure 2.9 Cont'd

The first part deals with the procedures used in sourcing and appraising the literature on the subject of GC and its use in construction. The next step involves sorting of the literature and the subsequent mining and parking of the data in a well-defined and orderly manner. Finally, the data are analysed and evaluated as part of the critical assessment of the combined experimental results to determine the emerging findings, which are then presented in a manner that can be clearly understood and disseminated.

Reference

Malisch W R, Day D E and Wixson B G, 1970. Use of domestic waste glass as aggregate in bituminous concrete. Highway Research Record (307), 1–10.

Production and Properties of Glass Cullet

3

Main Headings

- Introduction
- Market opportunities for glass cullet use
- Types of waste glass and collection schemes
- Classification and handling of unprocessed waste glass
- Processing waste glass into glass cullet
- Chemical properties of glass cullet
- Physical properties of glass cullet
- Concluding remarks

Synopsis

This chapter, which in many ways can be viewed as the foundation chapter for what follows, begins with describing briefly the barriers to the use of waste glass, as well as its market opportunities, instead of sending it to the landfill, together with the related financial and environmental benefits. The types of waste glass and the collection schemes in operation and how these may influence the quality of the final product and the classification and handling of unprocessed waste glass are dealt with. This is followed by the processing of waste glass into recyclable glass cullet for the primary glass market and secondary markets, with specific focus on the use of glass cullet in the construction market and ceramics. The chemical and physical properties of glass cullet, mainly relating to the use of glass cullet as a fine aggregate, as a filler or cement in mortar and concrete, and in geotechnical and road pavement applications, are discussed, based on widely published data over a period of time.

Keywords: Waste glass, Glass cullet, Sustainability, Financial and environmental impacts, Barriers, Markets, Collection scheme, Processing, Chemical properties, Physical properties.

Sustainable Construction Materials: Glass Cullet. https://doi.org/10.1016/B978-0-08-100984-0.00003-5

3.1 Introduction

Owing to its high potential for recycling, the glass industry was amongst the first to start collecting and recycling its packaging and achieved one of the highest recycling rates. Glass from containers, such as bottles and jars, can be recycled almost completely and indefinitely, without any loss in its quality or purity (Glass Packaging Institute, 2015).

The Glass Packaging Institute (2015), which represents the North American glass container industry, has estimated that over a tonne of natural resources are saved for every tonne of used glass recycled back to glass melt for making glass products. In addition, incorporation of 10% glass cullet (GC) in the glass manufacturing process can result in a reduction of energy costs of about 2%–3%, depending on the GC distribution and mixing systems operating at the glass manufacturing plants. This in turn can greatly affect the homogenisation of the materials into a uniform mixture, such that the glass melting regime is not disrupted, nor is the thermal uniformity (Efremenkov, 2015; Efremenkov and Subbotin, 2015). Furthermore, in terms of potential environmental impact, the recycling of six tonnes of GC of container glass results in a reduction of one tonne of carbon dioxide released into the atmosphere. As the glass melting time in modern glassmaking furnaces depends on the GC content in the batch (Beletskii et al., 2013), it is possible to reduce, even further, the cost of the end products and CO_2 emissions by incorporating greater amounts of GC in the manufacture of soda lime–silicate commercial containers. This also includes CG from crushed liquid crystal display (LCD) glass, even though it is an aluminoborosilicate glass, with significant strontium oxide content (Kim et al., 2014b).

According to Eurostat (2016), the total amount of waste glass produced in the European Union (28 countries), in 2012, was about 18.3 million tonnes. Figure 3.1, which presents the total amount of waste glass produced and treated in the European Union, shows that, despite the notably high quantities of waste glass generated in some cases, it is almost always completely recovered. Indeed, in comparison to other recyclable materials, waste glass has one of the highest recovery rates, at an average rate of 79% in the 28 countries of the European Union for which the data are presented in Figure 3.1. Although some countries have set out to recover and add value to waste glass by using it as backfill material (mainly the Czech Republic, Germany and Estonia), most have decided to recycle and use it in the production of new glass-based products.

Figure 3.1 also shows that, in some cases, the amount of recovered glass in a country can be even higher than the amount produced during a given period, for example, in Finland, Czech Republic, The Netherlands, Spain and Germany. Although not explained, if all values are correct, the differences, most likely, represent additional efforts to find and add value to waste glass produced in previous years.

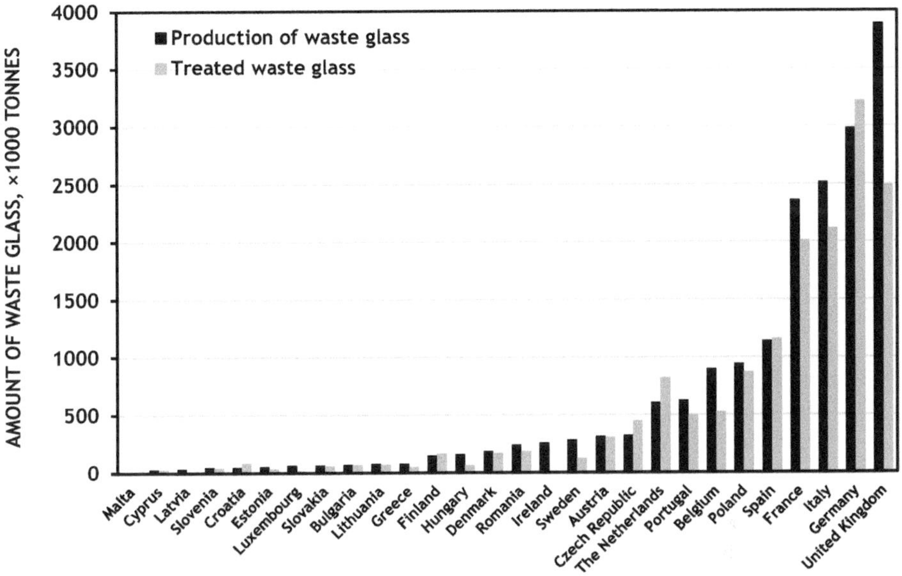

Figure 3.1 National figures for the waste glass generated and treated in 2012. Data taken from Eurostat (2016).

The various types of glass are also discussed in this chapter, as well as existing waste collection schemes, which may influence the characteristics and contamination levels of the resulting GC and its end-use application. Furthermore, with the aim of increasing the added value of GC in a way that it becomes a recyclable material in several markets, the quality control stage at the glass processors' gate and beneficiation procedures of waste glass into cullet are also discussed.

In view of the large amounts of waste glass produced yearly and its high potential for recycling back into glass manufacture and use in a wide range of secondary markets, such as concrete (fine aggregate, filler aggregate and cement), geotechnical and road pavement applications, as well as ceramics applications and others, the chemical and physical properties of GC, which in many cases can be the deciding factor in developing the use of GC in the secondary markets, are dealt with in some detail.

3.1.1 Financial and Environmental Benefits

It is widely known that recycling waste glass in the production of new glass-based products will lead to a lower volume of the material to be disposed in landfills and, at the same time, will reduce the procuring of natural raw materials to produce glass and the environmental impact of the glass manufacturing industry (Glass Packaging Institute, 2015). Furthermore, in the production of container and flat glass, replacing raw materials with GC allows a more efficient production process and, consequently, a reduction in carbon emissions. This is also true of other types of glass products; for example, recycling end-of-life cathode ray tubes (CRTs) in a closed-loop approach may allow savings of up to 1 kg CO_2 emission per kilogram of CRTs produced. Furthermore, environmental benefits would also be observed if the CRTs were to be redirected to the production of materials meant for other applications, such as the production of flat screens (Rocchetti and Beolchini, 2014).

Notwithstanding the benefits that can be accrued, it must also be recognised that the environmental benefits of recycling waste glass will be influenced by the existing local conditions and the technological know-how that can effectively aid the development of GC use. For example, the environmental impact of the transportation distance can outweigh the benefits of glass incorporation in closed-loop recycling or 'downcycling' (Meylan et al., 2014). Even when GC batches show a higher level of contamination, making them unusable for closed-loop recycling, some of the environmental benefits can still be observed with the use of GC as a natural aggregate replacement.

Furthermore, as specifications for its use in several construction applications are fairly flexible, adopting the use of GC as an alternative to natural aggregates in the construction industry becomes an ideal outlet for waste glass that is unsuitable for use in the production of new glass-based products (Holcroft and Pudner, 2007). Indeed, even though recycling waste glass may have significant costs in terms of segregation of materials and beneficiation procedures, it is widely agreed that significant financial benefits can be reaped from recycling it (WRAP, 2009b). These benefits can begin as soon as the GC is diverted from the landfill, for which no landfill charges are paid. Furthermore, with the increasing awareness of the potential for recycling waste glass and how this route can be a cheaper alternative to natural raw materials, landfill taxation for GC can be expected to continue to increase, thereby strengthening the economic feasibility of recycling waste glass.

3.1.2 Existing Barriers to Reuse and Recycling of Glass Cullet

A series of constraints, which hinder the recycling of waste glass, has been addressed by several researchers, though at the initial stages this happened mainly in the United Kingdom, the United States, Australia and New Zealand. This covered various aspects, namely, availability of the material, presence of contamination in the waste material, inefficiency of facilities for collection and processing of GC, financial issues involving the whole GC recycling process and public awareness, motivation and perception of GC waste material and its potential for recycling and the benefit of this to sustainability (Barlow, 2001; Coakley et al., 2002; Snow and Dickinson, 2005; Holcroft and Pudner, 2007; Hurley, 2003; Ralph, 2006; Ralph and Prince, 2006; USEPA, 1992; WRAP, 2009a).

For example, with well-thought-through kerbside schemes, carried out alongside bottle banks, it is quite possible that much of the final waste glass will be unsorted in terms of its colour (Holcroft and Pudner, 2007). This reduces the availability of GC suitable for recycling in glass production. Although the nonavailability of good-quality GC has been identified as a potential barrier to its recycling, overcoming this difficulty in practice, to a large extent, would be influenced by the end-use application of the material (USEPA, 1992).

Naturally, specifications for the characteristics of GC are more restrictive when collected for recycling in the manufacturing of new glass-based products, which require the material to be of a small size range, to exhibit a specific colour and to contain very little contamination of non-glass components (organic, ferrous and nonferrous metals). However, these aspects do not have such relevance when considering the use of GC as a construction aggregate, and accordingly the specifications are normally nowhere near as limiting as those for the glass-based applications.

Significant progress has been made in some countries and GC is collected in suitable form for recycling back to the manufacture of glass-based products and, failing this, for use in other less demanding applications, including as an aggregate in the construction industry. However, it must be noted that, at present, on a global scale, the provision of suitable facilities for the collection, sorting and processing of GC is at a level that renders a greater proportion of GC unsuitable for use in glass manufacturing and in the construction industry. Naturally, this problem is more evident in developing countries than in developed ones, where the issues involved are more complex and challenging.

Other main barriers to the wider recycling of waste glass are brought about by widely varying financial considerations. For example, some of the wastes are not properly sorted, and the processing cost of GC, in such circumstances, can be particularly steep due to the high contamination removal and colour-based separation costs. This can give rise to an increase in the price of GC to the extent that it could even exceed the price of the raw materials normally used in the manufacture of glass-based products, thereby making it less competitive (WRAP, 2009a). Without cost benefit, commercial entities would be reluctant to recycle GC as an alternative component and, instead, would use raw materials that they already know and which are readily available.

In addition, the low landfill charges practiced in some countries do not encourage waste glass recycling (Barlow, 2001; Snow and Dickinson, 2005) to a point that it is cheaper to landfill GC than to recycle it (Ralph, 2006; Ralph and Prince, 2006). Furthermore, the distance from some waste glass collection sites to processing plants may be such that transportation costs and environmental impacts of the final product may become unfavourable to recycling glass (Snow and Dickinson, 2005; Dacombe et al., 2004).

Furthermore, because of the general public's misconception of what types of glass and non-glass materials should be deposited at collection sites, the resulting waste is often a mixture of unsorted glass containing non-glass constituents capable of affecting the quality of the end-use application (Coakley et al., 2002). To obtain a suitably sorted GC, based on the colour and composition of the material, a high degree of quality control is required to be in place right from the beginning of the process. Though this could be achieved easily by the adoption of comprehensive recycling schemes, in most cases, there has been little governmental support and funding available to offset any of the associated costs (Snow and Dickinson, 2005).

The direct result of this has been the production of mixed waste with various types of glass, which needs further processing for it to be used as part of the raw feed for producing glass-based products, or in a mixed-glass waste stream with other recyclable materials, or, in the worst-case scenario, it remains unsuitable for use and can only be sent for landfill.

Some suggestions have been made, although localised and specific, for flat glass collection in the United Kingdom (Hurley, 2003) and for container glass in the Otago region in New Zealand (Snow and Dickinson, 2005), specifically on how some of the aforementioned barriers can be overcome, by substantially improving reclamation, reuse and recycling of waste glass, which include:

- The development of simple and practical rules at the planning stage to maximise reuse/recycling of quality glass-based materials and products at the end of their life.
- Allowing sufficient time when dismantling glass-based units, to ensure that those components are removed in a good condition and remain capable of being reused later on.
- Promoting recognised trade associations for reclamation, which would increase awareness of the construction industry's activities and encourage the reclamation of glass-based units.
- The reclamation industry developing long-term strategies to expand their successful activities to the commercial and construction sectors.
- Increased government involvement and support from commercial bodies to increase the reclamation industry's capacity to recover, refurbish and reuse quality glass units.
- Pressure from the local population and proactive engagement by ruling regional entities, which has been found to play a vital coordinating role in waste reduction and higher segregation of materials.
- 'Green' marketing, communicating previous experiences and demonstrating the feasibility of using GC at an industrial scale, which may positively influence contractors' and clients' perception regarding the use of recycled materials.

3.2 Market Opportunities for Glass Cullet Use

Two approaches are practiced when recycling GC, i.e., closed-loop and open-loop recycling (Enviros, 2004; Holcroft and Pudner, 2007). The former consists of processing waste glass in a manner such that the resulting GC may return to glass manufacturers to be re-melted into products similar to the original (Figure 3.2(a)). This type of recycling, which is associated with primary markets, is preferred, as waste glass can be repeatedly recycled and replace other raw materials, resulting in new glass with potentially lower energy consumption costs and lower carbon footprint, reported as 320 CO_2 kg savings compared with a nil savings in aggregate use.

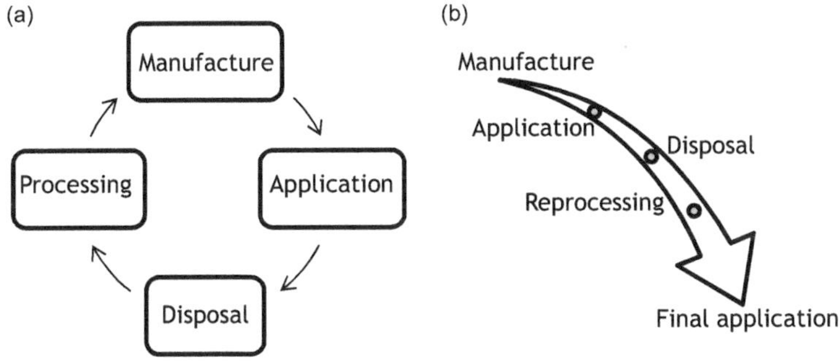

Figure 3.2 (a) Closed- and (b) open-loop recycling approaches.

An open-loop recycling approach for GC that cannot be returned to glass manufacturers for economic or quality reasons is usually associated with a single use in long-life secondary or alternative applications (Figure 3.2(b)). The reason it is recycled once is that the glass-based constituents are mixed together with a significant amount of other materials that are considered as contaminants in the glass recycling process, thereby making it extremely complex and financially unsustainable to segregate them. The main alternative market for the use of GC that cannot be easily recycled into new glass is as a construction aggregate (Holcroft and Pudner, 2007).

Table 3.1 contains the main applications for the use of GC in primary markets and alternative markets, with special emphasis on those within the construction industry. The potential end markets for processed waste glass are governed by the amount of contaminants and the quality requirements set out by processing plant operatives and clients. To minimise environmental impacts and optimise the retail prices of collected and processed container and flat glass, contamination must be kept to a minimum throughout the recycling process.

Other alternative markets in which GC may also be used include the replacement of zeolites in cationic exchange, as the rooting medium in hydroponics, as the filtration media for drinking water, as blast abrasive media and as filler for paint and plastic (Enviros, 2004; Snow and Dickinson, 2005; Wright, 2003).

3.3 Types of Waste Glass and Collection Schemes

3.3.1 Types of Waste Glass

Glass is a versatile material and can be used in diversified applications, and waste glass is expected to take up many different forms. Table 3.2 lists the main types of waste glass, categorised based on the original glass application, with information on their basic composition, their original use and the likely product types that need to be handled at the end of their life cycle for the recycling process to begin.

Table 3.1 Glass cullet primary and secondary markets

Use	Application	Replaced Products	Potential Benefits
Primary Markets[a]			
Feed stock	• Container • Flat glass • Fibreglass	• Raw materials	• Low-cost raw material • Energy savings • Reduced CO_2 emissions • Increased furnace life
Secondary Markets[b]			
Construction aggregate	• Concrete • Fill • Asphalt • Pipe bedding • Decorative	• Crushed rock • Gravel • Sand	• High resistance to impact/compression • Good drainage properties • Colourful, decorative
Flooring	• Terrazzo • Synthetic marble • Resin composite	• Crushed rock	• Colourful • Glass–resin composite • Highly resistant
Cement	• Concrete	• Portland cements • Pozzolanas	• Pozzolanic reaction
Flux/binder	• Bricks • Pottery • Ceramics	• Clay • Mineral fluxes	• Reduced firing temperature and time • Reduced hydrogen fluoride emissions
Foamed glass insulation	• Construction	• Virgin glass materials	• Low cost

[a]Data taken from CapitaSymonds (2007), Coakley et al. (2002), Enviros (2004), Holcroft and Pudner (2007), Hurley (2003), Ralph (2006), Ralph and Prince (2006), USEPA (1992) and WRAP (2007).
[b]Data taken from WRAP (2007), Pascoe et al. (2001), Wright (2003), Enviros (2004), Snow and Dickinson (2005), Barlow (2001), CapitaSymonds (2007), Cresswell (2007), Coakley et al. (2002), Holcroft and Pudner (2007), Searles (2004), Ralph (2006), Ralph and Prince (2006), Stainer (2006) and BG (2014).

The vast majority of generated waste glass comes from container and flat glass. However, even though both container glass and flat glass are essentially of similar composition (i.e., soda lime glass), handling these two products at the end of their life cycle and preparing them for recycling will need two different collection and processing approaches. The container glass, because of the diversity of packaging products, with different colours, sizes, shapes and contamination, is much more difficult to recycle compared with flat glass, in both glass-based products and other alternative applications. Notwithstanding this, the quantities involved for recycling both glass types are high and, as waste glass production is expected to increase and the concept of sustainability gains greater importance, the economics-related challenges of their collection and processing would need also to be addressed as a matter of priority.

Table 3.2 Types of sourcing and collection of waste glass

Type of Waste Glass	Description
Container	Usually made of 'soda lime glass,' which is produced by melting silica sand (SiO_2), soda ash (Na_2CO_3), limestone ($CaCO_3$) and other minor additions; the amounts of each constituent vary slightly depending on the application Waste glass from containers, bottles, jars, domestic tableware and lighting products
Flat glass	Also made of soda lime glass, flat glass is categorised into two types: preconsumer and postconsumer Preconsumer flat glass consists of off-cuts from glazing units' manufacture; this uncontaminated material is generally 100% recyclable and may be put back into flat glass melting furnaces Postconsumer flat glass consists of old windows and doors that have been replaced or have come from demolished buildings, as well as from end-of-life vehicles
Cathode ray tubes (CRTs)	CRT glass is manufactured from a wide range of glass consisting of SiO_2, Na_2O, CaO and other components for colouring, oxidising and protecting from X-rays (K_2O, MgO, ZnO, BaO, PbO) CRT glass is used in TV screens and computer monitors
Lead glassware	Contains relatively high amounts of lead oxide (PbO) The end-use applications include decorative glassware, ceramic glazes and vitreous enamels, high refractive index optical glass, radiation shielding, high electrical resistance and glass solders and sealants
Pyrex	Heat-resistant and low thermal-expansion borosilicate glass, containing up to 12% boric oxide (B_2O_3) Used for laboratory and household glassware
Fibreglass	Commonly composed of aluminoborosilicate glass Used in fibre-optic cables, insulation and reinforced components

Data taken from Andela and Zavitz (2004), Snow and Dickinson (2005), ESYS Consulting Ltd (2006), Holcroft and Pudner (2007), SEPA (2003), Ralph (2006), Ralph and Prince (2006), USEPA (1992), Hurley (2003), Leong and Hurley (2004), WRAP (2007), Enviros (2004), Wright (2003) and Kang and Schoenung (2005).

CRTs and LCDs are used in the production of electronic components, such as televisions, computers and mobile phones. These types of glass normally contain significant amounts of hazardous substances such as lead and cadmium and thus their safe disposal and processing are necessary to avoid contamination of air, soil and groundwater (Kang and Schoenung, 2005). For example, a recently developed option for CRT glass recovery involves a process of crushing, acid washing and water rinsing (Ling and Poon, 2011b). The recovered CRT glass has potential for broad construction applications (Andreola et al., 2007). Recovered LCD glass may be used in the production of new monitors as well as in alternative applications, such as additives for metallurgical processes (Lee, 2013) and construction aggregates (Wang et al., 2013).

The production of lead glassware and Pyrex is very much smaller compared with that of container and flat glass. These glass types contain some specific chemical components, often toxic types, to impart certain characteristics to the final products. For example, lead is added to increase the reflective index of decorative glass items. Therefore, waste glass that arises from these types of glass also requires special treatments for final disposal or reusing in new products.

In addition, fibreglass, commonly made from aluminoborosilicate glass, together with epoxy resin and metals (Long et al., 2010), is also used in electronic components such as printed circuit boards (PCBs). After the recovery of reusable metals (Handout-10, 2012), fibreglass can be recovered, at a high-purity state, by decomposing thermoset resins of PCBs at a temperature up to 600°C (Long et al., 2010; Zheng et al., 2009) (Table 3.2).

3.3.2 Waste Glass Collection Schemes

Several methods that can be applied for the collection of waste glass for recycling have been proposed, and an example of these is given in Table 3.3. As a rule, the selection of the collection scheme will depend on the area where the collection is taking place, as well as the logistical and financial characteristics of the organisations involved (ESYS Consulting Ltd, 2006).

An analysis of the literature on the matter showed that facilities for the collection of waste container glass are generally provided by organisations such as local authorities/government bodies, private contractors and community organisations. Owing to the relatively higher volume and potential recyclability of container and flat glass, greater efforts have been directed to these types of waste glass.

A kerbside collection recycling scheme is normally regarded as a better option compared with bring-site collection, as the recycling/recovery rates from kerbside schemes are expected to be considerably higher than those in bring sites (Dacombe et al., 2004). However, it is a fact that, whilst some of these centres apply separation of glass bottles/containers on the basis of their colour, for financial, logistical and time constraint reasons, this is normally not the case for kerbside collection (Holcroft and Pudner, 2007). This approach would be economically viable only when the revenue differentials in the market for colour-sorted glass were to increase substantially (ESYS Consulting Ltd, 2006).

3.4 Classification and Handling of Unprocessed Waste Glass

The Publicly Available Standard (PAS) 101, entitled 'PAS 101 Recovered Container Glass: Specification for Quality and Guidance for Good Practice in Collection' (BSI, 2003) in the United Kingdom, sets out maximum permissible contamination limits for organic material and ferrous and non-ferrous metals of waste glass gathered by the glass collector prior to processing.

Table 3.3 Collection schemes for waste glass

Type of Collection	Description
Kerbside collection	Containers are given to householders to segregate, as much as possible, household wastes from recyclable materials. Since waste glass can be separated from other materials at the source, thereby minimising its contamination, this collection system is normally the preferred approach by reprocessors.
Bring-site collection	Glass or bottle banks, placed in strategic locations based on the local population's size, are used to successfully sort glass based on its colour upon depositing.
Commercial collection	Retail establishments, like restaurants and bars, are key producers of empty bottles and containers. Though these locations would allow ideal colour-based collection schemes, in practice, due to lack of space and time constraints, the generated glass is often of mixed colour.
Flat glass collection	Normally, flat glass cullet is collected at the source. Some manufacturers of flat glass collect large amounts of clear flat glass directly from their own customers using reverse haul logistics, and smaller companies collect and segregate window waste for supply to a cullet processor or send it to a flat glass manufacturer.
Electrical glass collection	CRTs are made of two parts, the funnel and the panel. The funnel, which is made of leaded glass, has limited reuse and is usually sent back to companies that manufacture CRTs. The panel glass is lead free and, after being separated from the funnel, can be reused in alternative uses, except for container or flat glass recycling.

CRT, cathode ray tube.
Based on Holcroft and Pudner (2007), WRAP (2007) and Wright (2003).

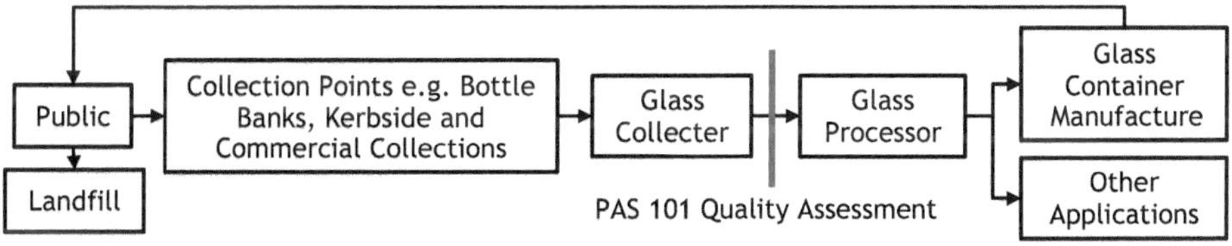

Figure 3.3 Application of PAS 101 in the container glass recycling process.
Adapted from BSI (2003).

Figure 3.3 shows the application phase of PAS 101 in the quality grading assessment of the container glass recycling process, which has been adopted in the United Kingdom (BSI, 2003). The guidance seeks to provide good practice rules for the collection and delivery of waste glass to glass processors. It comprises a grading system for unprocessed waste glass, such as premium, intermediate, or low grade, specifying product quality requirements in each case (Hartley, 2004).

To ensure that collectors apply proper segregation to the waste glass with subsequent reduction of the complexity of the beneficiation procedures carried out by the processor, various criteria based on colour, amount of contamination and particle size need to be enforced at the processor's gate (Table 3.4). Of the four classes of collected glass, classes A and B relate to colour-separated glass, with class B having less stringent specifications for colour and contamination levels. Classes C and D relate to colour-mixed glass, with class D having less stringent specifications for contamination levels. The grade of each unprocessed waste glass load is recommended to be assigned upon delivery to the processor, as quality control, by means of a visual inspection of a tipped load (Hartley, 2004).

Table 3.4 Criteria for classification of collected container glass

	Grade			
Requirement	**A**	**B**	**C**	**D**
Condition				
State	Whole/broken	Whole/broken	Not compacted	Compacted
Colour	Separated	Separated	Mixed	Not available
Colour				
Clear	≤4% other colours	≤6% other colours	>6% other colours	As per grade C
Brown	≤5% other colours	≤15% other colours	>15% other colours	As per grade C
Green	≤5% other colours	≤30% other colours	>30% other colours	As per grade C
Contamination, %[a]				
Organic	≤0.5	As per grade A	≤1.0	≤3.0
Inorganic[b]	≤0.025	As per grade A	As per grade A	≤0.005
Ferrous	≤0.1	As per grade A	≤0.2	≤3.0
Nonferrous	≤0.2	As per grade A	≤0.4	≤3.0
Particle size	Not intentionally crushed or broken	As per grade A	As per grade A	Subject to agreement between processor and collector

[a]Inorganic and other glass types subject to agreement between processor and collector. No medical, toxic or hazardous materials allowed. All percentages are by mass of total material.
[b]Proposed limit for inorganic content (waste glass fraction<10mm).
Based on BSI (2003) and Hartley (2004).

However, should an initially identified clear cullet load be contaminated, it may be downgraded to mixed cullet, resulting in a considerably lower price being paid for the load, to cover additional processing costs and also because the potential market value of the resulting material will decrease (WRAP, 2008).

3.5 Processing Waste Glass into Glass Cullet

One of the most important steps at the end of the life cycle of glass is the beneficiation process, in which attempts are made to convert waste glass into a valuable glass resource (GC) that is suitable for use in several applications and markets. The main aim of this is to increase the awareness and demand of GC in the market, which calls for high quality control during the production stages of the material (Tsai and Krogmann, 2013).

The collected waste glass has to go through a series of processes, mainly of contamination removal, before the resulting GC can be circulated back into the market. The processes that waste glass is subjected to normally consist of a visual inspection, crushing and screening stages and ferrous and nonferrous metal removal. Figure 3.4 presents the recommended minimum processing steps required for different types of waste flat glass to produce GC for each specific end-use market (WRAP, 2007).

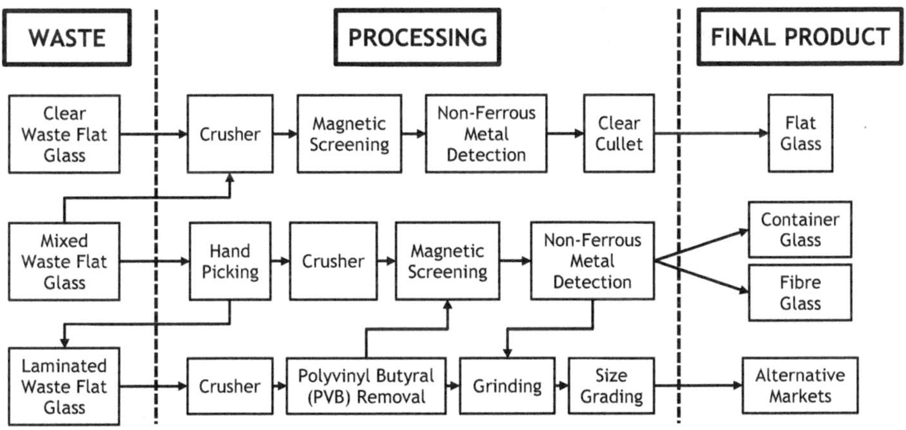

Figure 3.4 Recommended minimum processing steps to produce glass cullet for the end-use market.
Adapted from WRAP (2007).

3.5.1 Visual Inspection and Manual Separation

Upon arrival at the recycling plant, and before any contamination removal processing, selection/segregation of the waste glass is carried out to ensure that proper input material is introduced in the correct sequence. Assuming that the input material is primarily composed of waste glass only, the selection is based on the physical appearance of the waste glass, such as colour (for waste container glass) and lamination condition (for waste flat glass), to produce high-quality end products (CWC, 1996a,c). Additional contamination removal stages must be applied in the presence of contaminants such as (Fisk, 2014):

- Clay/ceramics (coffee mugs, pottery)
- Pyroceram (high-temperature glass-ceramics)
- Carborundum (silicon carbide abrasives)
- Corundum (alumina abrasives, refractory raw materials, sand blast materials)
- Metals (aluminium alloys and nonferromagnetic stainless steel alloys)
- Chromite stones in the glass
- Fuse cast alumina–zirconia–silica refractory stone
- Bonded alumina–zirconia–silica refractory stones
- Alumina refractory stones
- Zircon refractory stones
- Tin oxide refractory stones

During the visual inspection/handpicking stage, waste glass passes through a conveyor belt and any visible foreign materials are manually removed, which allows an initial assessment of the quality and adequacy of the material (CWC, 1996d). Despite the potentially high separation ability of some processing devices, constituents such as ceramics, porcelain and Pyrex are still not easily identified and cannot be removed because of their similar density and clarity compared to glass (Ralph, 2006). For this reason, to achieve a high quality control during separation, a manual removal stage is strongly recommended.

3.5.2 Crushers

Waste glass may be crushed into smaller sizes using the equipment briefly described in Table 3.5. As reported in the literature, it is possible to use a cage mill-style crusher, which effectively impacts and breaks glass whilst leaving the contaminants undamaged, thus enabling it to be easily screened (Egosi, 1992). The use of a stirred mill detritor is also possible for grinding GC into rounder particles with low consumption of energy (Pascoe et al., 2001).

Table 3.5 Types of glass-crushing equipment

Type	Description
Hammer mill	This normally consists of a series of hammers attached to pivots and fixed to a rotating shaft. The input waste glass is broken by the swinging hammers. The pivots allow the transfer of the impact energy to the glass. Oversized glass particles are recirculated until they are reduced to the targeted size.
Rotating disk and breaker bar	This is a series of irregularly shaped disks spaced at regular intervals along a rotating shaft. The disks crush the waste glass against stationary breaker bars. The spacing between the disks and the breaker bars determines the size of the resulting glass cullet.
Rotating drum and breaker plate	Waste glass is crushed by several bars installed on a spinning drum. The bars propel the glass at a stationary breaker plate and the spacing between them defines the size of the cullet produced.
Rotating breaker bar	Similar to the hammer mill crusher, this crusher consists of numerous longitudinal bars fixed to a rotating shaft, but the breaker bars are not allowed to pivot at the shaft.
Vertical shaft impactor	A rotating device is used to propel waste glass at impact plates around the inside of a vertical casing. The size of the cullet produced is governed by the velocity of the material and the geometry of the thrower and impact surfaces.
Helically fluted rollers	Two interlocking helical rollers spin on adjacent counterrotating shafts. Waste glass is crushed between the interlocking rollers and the space between them governs the cullet's size.

Adapted from CWC (1996b).

3.5.3 Screening and Contamination Removal

After the waste glass is crushed into smaller sizes using the equipment presented in Table 3.5, it is sent to vibrating screens, where further foreign materials are removed. During the screening process, the glass passes through vacuum ducts, where lightweight materials such as papers and plastics can be lifted. Similarly, polyvinyl butyral in laminated flat glass may be removed by means of an air-sifting procedure once it has been separated from the glass by crushing.

Ferrous metal contaminants are removed by self-cleaning electromagnets, whilst nonferrous metals such as aluminium, brass, copper and lead are detected and subsequently removed with the use of an eddy current system.

Other separating technologies have been introduced in an attempt to separate the contaminants and also the waste glass by colour with the use of a high-energy laser light system and charge-coupled device (CCD) camera system (Mayer, 2004). When the waste glass goes across a laser curtain in the laser light system, if there is a break or a decrease in the laser light (i.e., due to a difference in transparency from opaque contaminants), it will signal the information with the exact position and an air-jet impulse removes the contaminant into a discharge chute. Using a high-resolution CCD camera system, the colour of glass can be inspected within milliseconds and, depending on the required parameters, the specific colours are classified to either accept the material or reject it. The elimination of the contaminant takes place using air-jet impulses similar to those of the laser light system.

In view of the potential existence of impurities or hazardous contaminants, it is important that waste glass is handled and treated properly, depending on its intended application. For example, Tsai et al. (2009) recommended that the leachate of waste glass stockpiles should not be released into surface water. Furthermore, the amount and type of impurities in the waste glass must be evaluated according to their potential risk of impairing the quality of the resulting material. Owing to the occasional significant presence of heavy metals in waste glass and to avoid subsequent downcycling of the material to progressively lower-grade applications, Chen et al. (2006b) developed a method for obtaining silicate glass with a high SiO_2 purity via the extraction of heavy metal ions in coloured glass using glass phase separation. Similarly, experimental research was conducted to extract silica contents of waste glass through alkaline extractive treatment. The potential use of extracted products and residue has been suggested by Mavilia et al. (2001a,b).

In another study, Glüsing and Conradt (2001) evaluated and determined the amount of contamination by means of dissolution kinetics, which can be used to formulate acceptance thresholds for particle sizes of ceramic and metallic constituents, in such a way that glass producers can exert enhanced quality control rather than relying solely on the detection of glass defects.

3.5.4 Guidance for Glass Cullet Handling and Classification

Guidance and specifications for the quality control of GC relating to its applications were proposed by the Clean Washington Center (CWC) in the United States in 1996 and the Waste & Resources Action Programme in the United Kingdom in 2004. These publications are suitable for processors when refining waste glass into cullet and for users when specifying the desired product, in terms of its particle size, colour, contamination and moisture, for different applications.

The CWC (1996c) developed a report containing methods for sampling and testing recycled glass, as well as several specifications for classifying GC into different grades suitable for various industries. Table 3.6 provides a description of the various types of GC in different grades. After processing, depending on the grade of the material, GC can be sent to the glass container, fibreglass or construction aggregate industries.

Table 3.6 Specifications for grades of glass cullet

GC Grade and Type	Description
A–Processed B–Unprocessed	Glass container industry, food and beverage containers (soda lime–silica) only[a]
C–Fine D–Coarse	Fibreglass industry, food and beverage containers (soda lime–silica) only[a]
E–Construction	Aggregate industry, food and beverage containers (soda lime–silica) and other types of glass
F–Open specification	Shape and type of glass shall be determined by negotiating parties and may include postconsumer or postindustrial glass in any form (i.e., containers, Pyrex, crystal, ovenware, plate glass, light bulbs, ceramics, art glass, mirror glass)
H–Bottles	Various industries, postconsumer food and beverage containers (soda lime–silica) only, in the shape of whole and/or broken bottles[a]

GC, glass cullet.
[a]Other types of glass are considered contaminants and may provide justification for load rejection based on agreement between negotiating parties and discretion of the buyer. Contaminants include Pyrex, crystal, ovenware, plate glass, windows, light bulbs, ceramics, art glass, mirror glass, among others.
Based on Clean Washington Center (1996c).

Table 3.7 presents the requirements for GC for each of the grades and applications as shown in Table 3.6. The selection and processing of GC for the construction aggregate industry are obviously much less demanding than for any other industry. Other applications require a strict colour-based segregation or have restricting limitations for the amounts of contaminants with physical properties and appearance similar to those of container glass (e.g., Pyrex, crystal, ovenware, plate glass, windows). However, GC for construction purposes (grade E), apart from the size limitation (the maximum particle size must be 19 mm and no more than 5% should pass the 75-μm sieve), mainly requires a maximum debris content of 5%, which is defined as any non-glass material that does not show sufficient compatibility with the remainder material and the end-use application (i.e., labels, caps, cork, food residue, aluminium or tin foil, plastic, metal). In addition, for environmental reasons, the average content of lead (Pb) for five different samples must not exceed a total of 80 parts per million. There is also a requirement that the material should be in a free-flowing state and a restriction on the presence of medical or hazardous materials, in any amount, and, at the discretion of the buyer, the supply load may be rejected.

In addition, for GC to be used in concrete and asphalt (BS EN 13437, 2003), a BSI publication PD CEN/TR 13688 (2008) also specifies a series of requirements, which are essentially similar to those of CWC (1996c), though a little more detailed and with different limits.

Table 3.7 Selected glass cullet requirements

	Grade						
Requirement	**A**	**B**	**C**	**D**	**E**	**F**	**H**
Size	10–19 mm	≤25% passing 19 mm	<6.3 mm; ≤15% passing 0.074 mm	<6.3 mm; ≤15% passing 0.074 mm	<19 mm; ≤5% passing 0.074 mm	Negotiation[a]	N.A.
Colour							
Clear, %	95–100	95–100	Predominant colour	Predominant colour	N.A.	Negotiation[a]	95–100
Brown, %	90–100	90–100	Predominant colour	Predominant colour	N.A.		90–100
Green, %	70–100	70–100	Predominant colour	Predominant colour	N.A.		70–100
Mixed, %	–	–	≤25 Brown	≤25 Brown	N.A.		Note[a]
Moisture	Note[b]	Note[b]	≤0.5%	≤0.5%	Note[b]	Negotiation[a]	Note[b]
Contamination							
Organic, %	≤0.2/0.4[c]	≤0.5	≤0.1	≤0.1	Note[d]	Negotiation[a]	≤0.5
Ferrous metals, %	≤0.005	Note[e]	≤0.005	≤0.005			Note[e]
Nonferrous metals, %	Note[f]		≤0.01	≤0.01			

[a]Requirement based on agreement between negotiating trading parties.
[b]No visible drainage of liquid when tipped. No visible snow or ice. Material shall be noncaking and free flowing.
[c]Included glass packaging but overall limit depends on GC colour, 0.2 for clear and 0.4 for brown and green.
[d]Average maximum 5% debris content, which is defined as any non-glass material (i.e., labels, caps, cork, food residue, aluminium or tin foil, plastic, metal). The average lead content of five samples shall not exceed a total of 80 ppm.
[e]Glass packaging materials (e.g., closures, foil) present in quantities equivalent to normal amounts inherent to glass packaging.
[f]Not to exceed five particles per truckload and two particles per truckload upon initial visual inspection.
Based on Clean Washington Center (1996c).

Table 3.8 Specified glass cullet requirements for various applications

	Application				
Requirement	**Ceramic Sanitary Ware Production**	**Fluxing Agent in Brick Manufacture**	**Sports Turf and Related Applications**	**Water Filtration Media**	**Abrasive**
Colour	✗	✗	✗	✓	✗
Contamination	✓	✓	✗	✓	✓
Particle size	✓	✓	✓	✓	✓

✓, requirement specified; ✗, requirement not specified.
Based on WRAP (2004).

Table 3.8 summarises the requirements for quality control of GC in terms of its particle size, colour and degree of contamination in other applications, namely as:

1. Fluxing agent in ceramics, sanitary ware and brick manufacture
2. Granular material for the turf of sport-related activities, abrasive media and water filtration systems

3.6 Chemical Properties of Glass Cullet

As previously mentioned, GC can be categorised based on its chemical composition, e.g., soda lime, aluminosilicate, borosilicate and lead glass. These are the common glass types that have been studied for their potential use as sustainable construction materials, for example, in concrete (as cement, filler or fine aggregate), geotechnical and road pavement applications.

3.6.1 Chemical Composition

Depending on the applications, the properties of glass, such as chemical durability, optical transparency and thermal expansion, can be formulated by modifying its chemical composition. Table 3.9 lists the oxide composition of ground GC, for which information was taken from 165 glass samples tested in the studies undertaken in 33 countries the world over and reported since 2000, except for one research study in 1996. Silica, reported as silicon dioxide (SiO_2), is the main constituent of glass, with an average of about 60%–75% of the total composition (Table 3.9). Other components are added to impart the desired properties to the glass.

Soda lime glass is the most widely used commercial glass for the production of containers, windows and lighting devices. The basic raw materials for making soda lime glass consist of sand (principal source of SiO_2), soda ash (source of Na_2O) and limestone (source of CaO). Soda ash is used as a major fluxing agent to lower the

Table 3.9 Chemical composition of various types of glass

	Soda Lime Glass			Aluminosilicate Glass			Borosilicate Glass			Lead Glass		
Oxide	**Mean, %**	**SD, %**	**Range, %**	**Mean, %**	**SD, %**	**Range, %**	**Mean, %**	**SD, %**	**Range, %**	**Mean, %**	**SD, %**	**Range, %**
SiO_2	71.0	2.9	56.5–78.8	62.5	0.9	60.4–64.7	75.1	3.7	69.1–84.0	60.7	8.2	50.5–69.1
Al_2O_3	2.0	1.9	0.1–11.8	16.8	0.3	16.2–17.2	1.8	0.3	1.0–2.0	1.5	1.2	0–3.0
Fe_2O_3	0.4	0.7	0–7.4	9.4	0.1	9.1–9.4	0.15	0.04	0.02–0.17	0.2	0.1	0–0.2
CaO	9.8	2.4	1.3–18.0	5.0	4.9	2.6–7.5	1.8	0.6	0.4–2.1	1.8	1.4	0.12–3.7
MgO	2.3	1.5	0–10.3	0.47	0.5	0.2–1.4	0.08	0.22	0–0.60	0.8	0.8	0–1.9
MnO	0.04	0.04	0–0.10	–	–	–	0.01	–	–	–	–	–
Na_2O	13.0	2.6	6.1–20.0	0.61	0.1	0.60–0.64	4.3	1.5	3.3–8.0	4.2	3.0	0–5.5
K_2O	1.0	1.7	0–10.0	1.0	0.6	0.1–1.4	1.6	1.6	0.9–5.4	5.8	4.3	0.8–10.1
PbO	0.33	0.41	0.01–0.78	–	–	–	0.13	0.14	0–0.25	22.1	3.8	18.3–27.4
B_2O_3	3.2	4.3	0.2–11.6	10.5	0	–	16.6	0	–	–	–	–
TiO_2	0.13	0.38	0.01–2.89	0.01	0	–	0.10	0.14	0.01–0.20	0.05	0.05	0.01–0.08
P_2O_5	0.04	0.02	0–0.10	0.01	0	–	0.02	–	–	–	–	–
BaO	0.81	1.22	0.05–2.95	–	–	–	–	–	–	1.7	1.5	0–2.9

	Soda Lime Glass			Aluminosilicate Glass			Borosilicate Glass			Lead Glass		
Oxide	Mean, %	SD, %	Range, %	Mean, %	SD, %	Range, %	Mean, %	SD, %	Range, %	Mean, %	SD, %	Range, %
SrO	0.04	0.02	0.02–0.05	0.8	0	–	–	–	–	2.1	–	–
SO_3	0.27	0.43	0–3.10	–	–	–	0.03	0.04	0–0.05	–	–	–

SD, standard deviation.
Data taken from Abendeh et al. (2015a,b), Afshinnia and Rangaraju (2015), Al-Akhras (2012), Aly et al. (2012), Anagnostopoulos et al. (2009), Bajad and Modhera (2009), Bajad et al. (2014, 2011a,b), Bajad et al. (2012a,b), Bignozzi et al. (2009, 2015), Byars et al. (2004), Calmon et al. (2014), Carsana et al. (2014), Chaïd et al. (2015), Chen and Wong (2015), Chen et al. (2006a,b), Chidiac and Mihaljevic (2011), Cholleti (2005), Corbu et al. (2013), Dhir et al. (2009), Du and Tan (2014a,b, 2015), Dyer and Dhir (2001, 2010), Federico and Chidiac (2009), Georgiadis et al. (2007), Girbes et al. (2004), Guo et al. (2015), Huang et al. (2015), Idir et al. (2009, 2010a, 2011), Jain and Neithalath (2010), Jangid and Saoji (2014), Kamali and Ghahremaninezhad (2015, 2016), Kara (2013a,b, 2015), Kara et al. (2012), Karamberi and Moutsatsou (2005), Karamberi et al. (2004), Khmiri et al. (2012, 2013), Kim et al. (2014a,b, 2015), Korjakins et al. (2009, 2010, 2011, 2012), Kou and Xing (2012), Kou and Poon (2013), Lee et al. (2013), Limbachiya (2009), Lin et al. (2009), Ling and Poon (2014b), Liu et al. (2013), Madandoust and Ghavidel (2013), Matos and Sousa-Coutinho (2012), Matos et al. (2015), Matos and Sousa-Coutinho (2016a), Miranda Jr. et al. (2014), Mirzahosseini and Riding (2014), Nassar and Soroushian (2011, 2012a,b), Neithalath (2008), Neithalath and Schwarz (2009), Nunes et al. (2013), Omran and Tagnit-Hamou (2016), Oyekan and Oyelade (2011), Özkan and Yüksel (2008), Parghi and Alam (2016), Park et al. (2004), Patagundi and Prakash (2015), Pavoine et al. (2014), Pedersen (2004), Polley (1996), Poutos et al. (2008), Rajabipour et al. (2012), Sagoe-Crentsil et al. (2001), Saccani and Bignozzi (2010), Schwarz and Neithalath (2008), Schwarz et al. (2007, 2008), Seju et al. (2015), Serpa et al. (2013), Shafaatian et al. (2013), Sahmenko et al. (2010), Shayan and Xu (2004, 2006), Shi and Wu (2005), Shi et al. (2005), Siad et al. (2016), Singh et al. (2014), Sobolev et al. (2007), Soroushian (2012), Spathi et al. (2015), Su and Chen (2002), Tagnit-Hamou and Bengougam (2012), Taha and Nounu (2008, 2009), Tang et al. (2005), Topcu et al. (2008), Turgut (2008a, 2013), Turgut and Yahlizade (2009), Vaitkevicius et al. (2014), Wang (2009a, 2011), Wang and Chen (2010), Wang and Hou (2011), Wang and Huang (2010), Wang et al. (2009, 2014, 2016), Wattanapornprom and Stimannaithum (2015), Yilmaz and Degirmenci (2010) and Yuksel et al. (2013).

melting point of sand for economic reasons. Limestone acts as a stabiliser to improve the chemical durability of the glass. A portion of the Na_2O and CaO is sometimes replaced by potassium oxide (K_2O) and magnesium oxide (MgO) in some soda lime glass products (Scalet et al., 2013). The addition of aluminium oxide (Al_2O_3) in silica melt increases its viscosity, which is an important feature for optimum melting, forming and annealing in glassmaking (De Jong et al., 2011).

Other additives such as transition metal oxides, with a concentration of 0.5%–5%, are used to impart colour to the end products (De Jong et al., 2011). Some of the common colouring elements used include cadmium(II) for yellow, orange and red colour; cobalt(III) for green colour; iron(II) for bluish-green colour and copper for light blue colour (Scalet et al., 2013).

In the production of borosilicate glass, the inclusion of boric oxide (B_2O_3) as a replacement for alkali (Na_2O and K_2O) improves the thermal resistance of the glass (De Jong et al., 2011). This glass is known for its use in laboratory equipment and ovenware.

Aluminosilicate glass has high content of Al_2O_3 at 16%–17% (Table 3.9). As this glass can withstand high temperatures, the materials are normally used in halogen lamp glass and cooking utensils. Aluminoborosilicate, one of the variations of aluminosilicate glass containing B_2O_3, has a low coefficient of thermal expansion, which is normally used in flat panel displays for electronics products (Schott, 2014).

Lead glass typically contains over 20% lead oxide (PbO), which is an important additive to increase the refractive index of decorative glass items. Lead glass is also used in CRTs for television sets owing to the radiation-shielding ability of PbO.

3.6.2 Cementitious Properties

There are two important chemical reactions in cementation processes, namely the hydration and pozzolanic reactions. Although both reactions form cementitious compounds, the former takes place in the presence of water, whilst the latter requires water and calcium hydroxide ($Ca(OH)_2$), which is released from the hydration of Portland cement (PC). Materials possessing pozzolanic behaviour are normally of a siliceous or aluminosilicate nature. A notable example of pozzolanic material is fly ash, a by-product of coal combustion, which has been accepted for use as one of the common cements since 2000 (EN 197-1, 2011) and is now commonly used in the construction industry in combination with Portland clinker as a component of cement to enhance the properties of concrete.

The abundance of SiO_2 present in glass, along with small quantities of the major constituents of PC such as CaO, Al_2O_3 and Fe_2O_3 (Table 3.9), suggests that the material can potentially be used as a cement addition. The chemical compositions of various glass types given in Table 3.9 are further assessed for these four main oxides, arranged in ascending order by year of publication from 1984 to 2016, as presented in Figure 3.5. The minimum sum of SiO_2, Al_2O_3 and Fe_2O_3 contents of 70% for fly ash specified in BS EN 450-1 (2012) and class F fly ash in ASTM C618 (2015) is also shown for comparison.

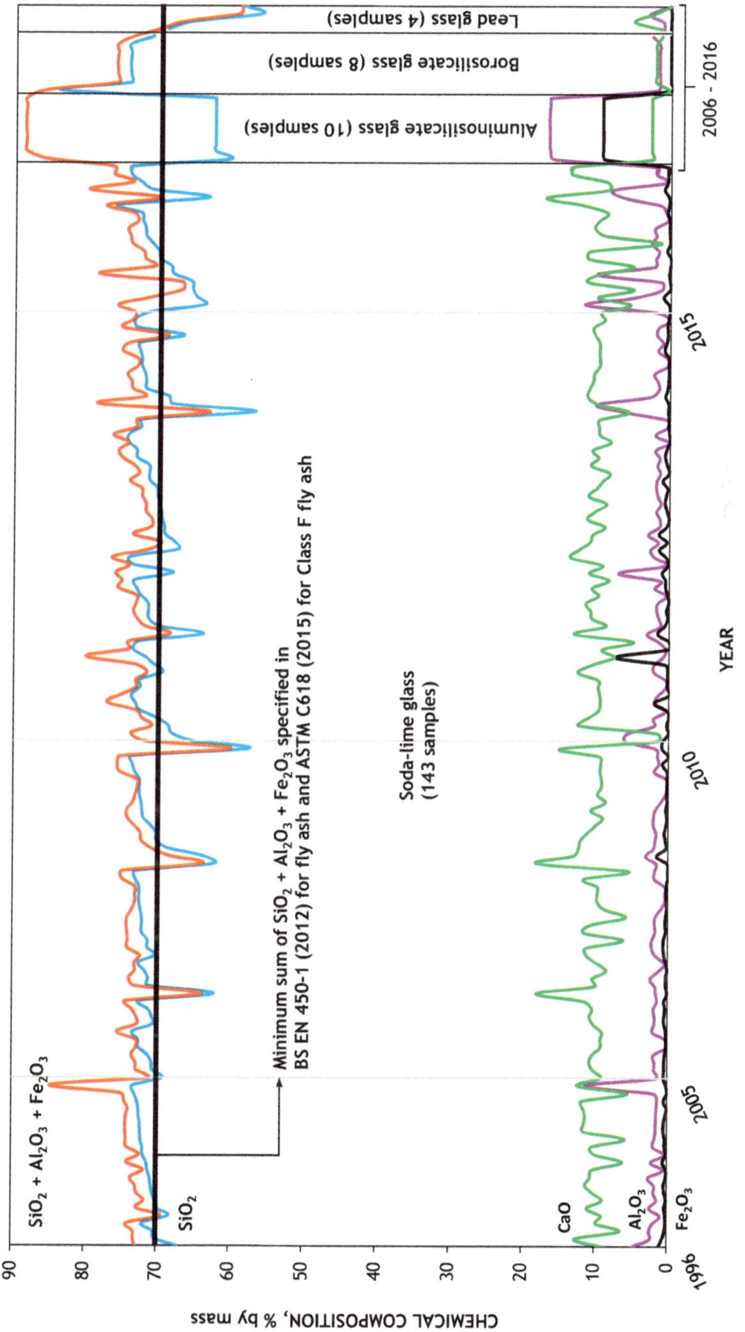

Figure 3.5 Changes in major oxides in various glass samples from 1996–2016. Data used is the same as in Table 3.9.

It can be seen from Figure 3.5 that the SiO_2 content of all the glasses is high, varying from 56% to 84%. On average, the SiO_2 content of soda lime and borosilicate glass is slightly more than 70%, and only 2 of 143 samples of the former have a value less than 60%. Aluminosilicate glass has a relatively low but consistent SiO_2 content, showing an average value of 62% (Figure 3.5). Although small in sample size, the SiO_2 content of lead glass fluctuates significantly from 56% to 69%.

The presence of CaO in glass may provide the materials with some degree of hydraulic behaviour. Soda lime glass normally contains an appreciable amount of CaO content with an average value of nearly 10% (Figure 3.5), which is greater than that of typical class F fly ash. The highest and lowest CaO contents of soda lime glass are 18.0% (seen in two samples) and about 1.5% (seen in two samples), respectively. The CaO contents of aluminosilicate, borosilicate and lead glass are about the same, generally no more than 5% (Figure 3.5).

As to be expected, the Al_2O_3 content of aluminosilicate glass is the highest (average of 16.7%) among all types of glass (Figure 3.5). Soda lime glass generally has Al_2O_3 contents up to 3%, but about 8% of the samples, especially those in the recent studies, showed a value greater than 5% (Figure 3.5). Both borosilicate and lead glass have Al_2O_3 content less than 3% of their total composition.

Figure 3.5 also shows that, apart from aluminosilicate glass, the Fe_2O_3 contents of all the other types of glass (with an exception of one sample of soda lime glass) are only about 0.5%. The average Fe_2O_3 content of aluminosilicate glass is 9.4%.

Using the chemical composition requirements for fly ash as a guide, it can be seen from Figure 3.5 that the sum of the SiO_2, Al_2O_3 and Fe_2O_3 contents of aluminosilicate glass and borosilicate glass are consistently greater than the minimum content of 70%. Although aluminosilicate glass has a comparatively low SiO_2 content, its sum of oxides is the highest because of its high content of Al_2O_3. Most samples of soda lime glass show that the sum of oxides is higher than 70% and that only 6% of the samples do not comply with the requirement (Figure 3.5). The sum of oxides of lead glass is less than 70%, suggesting that the material may not be suitable for use as a replacement for PC.

Given that the sum of SiO_2, Al_2O_3 and Fe_2O_3 contents of most of the glass samples used in the studies reported since 1996 is greater than 70%, these oxides are presented in a ternary diagram (Figure 3.6), which also allows for comparison with PC and other established cement additions such as fly ash, silica fume, metakaolin and ground granulated blast furnace slag (GGBS). Glass is positioned near the apex of the triangle and close to silica fume, as both materials have abundant SiO_2 content. A low level of CaO content in glass hides any visible sign of latent hydraulic properties like those of GGBS. However, it is evident from Figure 3.6 that the composition of the three main oxides in glass is comparable to that of pozzolanic cements. Further evaluation of the cementitious properties of ground GC, such as water required for standard consistence, initial and final setting times, soundness and strength activity index, are discussed in Section 4.3 of Chapter 4.

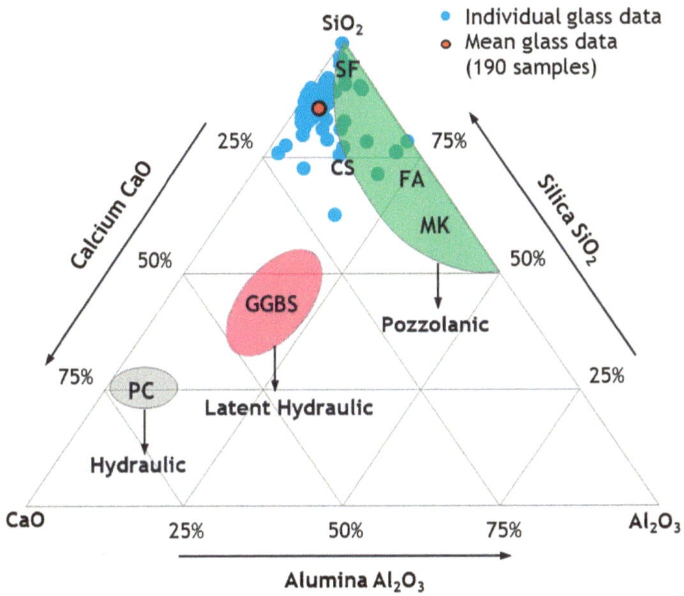

Figure 3.6 Ternary plot of the SiO_2, Al_2O_3 and CaO contents of glass in comparison to other cements. *CS*, copper slag; *GGBS*, ground granulated blast furnace slag; *MK*, metakaolin; *PC*, Portland cement; *FA*, fly ash; *SF*, silica fume.

Data taken from Abendeh et al. (2015a,b), Afshinnia and Rangaraju (2015), Al-Akhras (2012), Aly et al. (2012), Anagnostopoulos et al. (2009), Bajad and Modhera (2009), Bajad et al. (2014, 2011a,b), Bajad et al. (2012a, 2012b), Bignozzi et al. (2009, 2015), Borhan and Bailey (2014), Byars et al. (2004), Calmon et al. (2014), Carsana et al. (2014), Chaïd et al. (2015), Chen and Wong (2015), Chidiac and Mihaljevic (2011), Cholleti (2005), Corbu et al. (2013), Corinaldesi et al. (2005), Dhir et al. (2009, 2016), Du and Tan (2014a,b), Du and Tan (2015), Dyer and Dhir (2001, 2010), Federico and Chidiac (2009), Georgiadis et al. (2007), Girbes et al. (2004), Guo et al. (2015), Huang et al. (2015), Idir et al. (2009, 2010a, 2011), Ismail and Al-Hashmi (2009), Jain and Neithalath (2010), Jangid and Saoji (2014), Kamali and Ghahremaninezhad (2015, 2016), Kara (2013a, 2015), Kara et al. (2012), Karamberi and Moutsatsou (2005), Karamberi et al. (2004), Khmiri et al. (2012, 2013), Kim et al. (2014a,b), Kim et al. (2015), Korjakins et al. (2009, 2010, 2012), Kou and Xing (2012), Kou and Poon (2013), Lee and Lee (2016), Lee et al. (2013), Limbachiya (2009), Limbachiya et al. (2012), Lin et al. (2009), Ling and Poon (2014a,b), Liu et al. (2013), Madandoust and Ghavidel (2013), Maschio et al. (2013), Matos and Sousa-Coutinho (2012), Matos et al. (2015), Matos and Sousa-Coutinho (2016a), Metwally (2007), Mirzahosseini and Riding (2014, 2015), Nassar and Soroushian (2011, 2012a,b), Neithalath (2008), Neithalath and Schwarz (2009), Nunes et al. (2013), Omran and Tagnit-Hamou (2016), Oyekan and Oyelade (2011), Özkan and Yüksel (2008), Parghi and Alam (2016), Park et al. (2004), Patagundi and Prakash (2015), Pavoine et al. (2014), Pedersen (2004), Polley (1996), Poutos et al. (2008), Rajabipour et al. (2012), Saccani and Bignozzi (2010), Sahmenko et al. (2014), Schwarz and Neithalath (2008), Schwarz et al. (2007, 2008), Seju et al. (2015), Serpa et al. (2013), Shafaatian et al. (2013), Sahmenko et al. (2010), Shao and Lehoux (2001), Shao et al. (2000), Sharifi et al. (2015), Shayan and Xu (2004, 2006), Shi (2009), Shi and Wu (2005), Shi et al. (2005), Siad et al. (2016), Singh et al. (2014), Sobolev et al. (2007), Soroushian (2012), Spathi et al. (2015), Su and Chen (2002), Tagnit-Hamou and Bengougam (2012), Taha and Nounu (2008, 2009), Tang et al. (2005), Terro (2006), Tognonvi et al. (2015), Topcu et al. (2008), Turgut (2008a, 2013), Turgut and Yahlizade (2009), Vaitkevicius et al. (2014), Wang (2009a, 2011), Wang and Chen (2010), Wang and Hou (2011), Wang and Huang (2010), Wang et al. (2009, 2014, 2016), Wattanapornprom and Stimannaithum (2015), Yilmaz and Degirmenci (2010) and Yuksel et al. (2013).

3.6.3 Sulphate Content

Excessive sulphate in concrete, due to the exaggerated addition of gypsum (used for controlling the setting time of cement in production), sulphate-containing aggregates or contamination, can cause internal sulphate attack, leading to expansion and eventually the deterioration of the concrete.

The sulphate content (expressed as SO_3) of glass is low and normally no greater than 0.5%. Based on the data reported in the studies examined, soda lime glass may have SO_3 content varying from less than 0.01% to 3.1%, with an average value of 0.27% (Table 3.9). For borosilicate glass, its SO_3 content is almost negligible (Table 3.9). In general, unless contaminated, the SO_3 content from using GC in construction applications is unlikely to exceed the maximum existing specified limits, given that the maximum limits for SO_3 content for fly ash use are 3.0% and 5.0% in BS EN 450-1 (2012) and ASTM C618 (2015), respectively.

3.6.4 Alkali Content

Two metal alkali oxides, namely sodium oxide (Na_2O) and potassium oxide (K_2O), are of particular importance in terms of the cement and concrete chemistry, as they can become involved in alkali–silica reaction (ASR), which can be harmful to concrete (see Section 3.6.6). The soluble alkalis in cement are usually expressed as sodium oxide equivalents (Na_2O_{eq}), which are the sum of Na_2O and 0.658 times K_2O, tested in accordance with ASTM C114 (2015) and BS EN 196-2 (2013).

As soda ash (source of Na_2O) and potash (source of K_2O) are added as fluxes to lower the melting point of glass formers, it is expected that the Na_2O and K_2O contents can be quite high. Indeed, based on Table 3.9, the average total content of Na_2O and K_2O in glasses, in descending order, is soda lime glass at 14.0%, lead glass at 10.0%, borosilicate glass at 5.9% and aluminosilicate glass at 1.6%.

Although soda lime glass is high in alkalis, Schwarz and Neithalath (2008) showed that only a small fraction of the alkalis is released, measured in a powder form using flame emission spectroscopy and electrical conductivity studies. This should also apply to other types of glass, although the results are based on soda lime glass only. Similarly, an assessment of the alkali released potential of powdered soda lime glass conducted by Dumitru et al. (2010) also confirmed that most of the alkalis are bound within the glass structure, and the available alkali and acid-soluble alkali contents are very low and not greater than 0.6% and 0.3% Na_2O_{eq}, respectively.

Therefore, in general, it can be assumed that the alkalis released from GC used either as cement addition or as fine aggregate should not give rise to any significant alkali contribution to a cement-based system such as concrete. Whilst this may help with developing the application of GC in construction, it alone should not mean that cement-based materials, such as mortars and concrete, would be free of ASR when GC is used, as discussed in Section 3.6.6.

3.6.5 Loss on Ignition

Loss on ignition (LOI) determines the reduction in weight of cement when heated at 950°C. This is primarily attributed to the presence of water and carbon dioxide in the cement, which reacts with free lime and magnesia (Neville, 1995). High LOI is undesirable as it can imply poor performance of cement. The maximum LOI value for PC is limited to 3.0% (type I) and 5.0% (CEM I) in ASTM C150 (2016) and BS EN 197-1 (2011), respectively. For fly ash, the LOI value is an indication of unburnt carbon content and is fixed at 5% as the most stringent maximum limit in BS EN 450-1 (2012) for both fly ash N and S types and at 6% in ASTM C618 (2015) for class F fly ash.

The LOI of ground glass cullet (GGC), mainly in samples made of soda lime and few made of aluminosilicate glass, and that of the corresponding reference PC are shown in Figure 3.7, together with the limits for PC and fly ash given in the standards. It is evident from Figure 3.7 that the LOI of GC is very low, with an average value of 0.75%, which is far less than the limits set for PC and fly ash. However, the presence

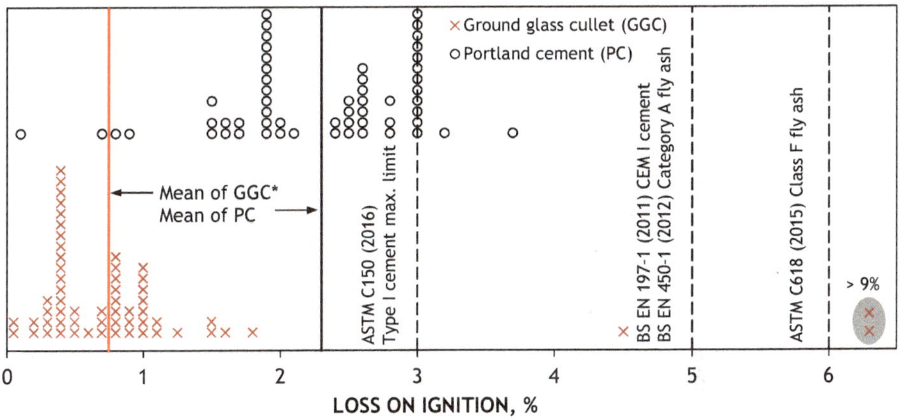

Figure 3.7 Loss on ignition of ground glass cullet and reference Portland cement. *Excluding values >9%.

Data taken from Afshinnia and Rangaraju (2015), Al-Akhras (2012), Anagnostopoulos et al. (2009), Bajad et al. (2014, 2011a,b), Bajad et al. (2012a, 2012b), Chaïd et al. (2015), Chidiac and Mihaljevic (2011), Girbes et al. (2004), Guo et al. (2015), Huang et al. (2015), Idir et al. (2011, 2010a), Jain and Neithalath (2010), Jang et al. (2015), Jangid and Saoji (2014), Kamali and Ghahremaninezhad (2015, 2016), Kim et al. (2015, 2014a), Kou and Poon (2009), Lam et al. (2007), Lee et al. (2013), Limbachiya (2009), Ling and Poon (2012a,b, 2014a,b), Ling et al. (2012), Maier and Durham (2012), Maraghechi et al. (2012), Maschio et al. (2013), Matos and Sousa-Coutinho (2016a,b, 2012), Matos et al. (2015), Metwally (2007), Nassar and Soroushian (2011, 2013, 2012a,b), Neithalath (2008), Neithalath and Schwarz (2009), Nunes et al. (2013), Omran and Tagnit-Hamou (2016), Özkan and Yüksel (2008), Parghi and Alam (2016), Park and Lee (2004), Pavoine et al. (2014), Pedersen (2004), Rajabipour et al. (2012), Schwarz et al. (2008), Seju et al. (2015), Serpa et al. (2013), Shayan and Xu (2004), Siad et al. (2016), Soroushian (2012), Tognonvi et al. (2015), Turgut (2008b, 2013), Turgut and Yahlizade (2009), Vaitkevicius et al. (2014), Wattanapornprom and Stimannaithum (2015), Yilmaz and Degirmenci (2010), Zhao and Wei (2011) and Zhao et al. (2013a,b).

of organic impurities (Calmon et al., 2014) and contaminant residue after washing (Yilmaz and Degirmenci, 2010) in GC could have resulted in a value higher than 9% in some cases. This emphasises the importance of keeping the GC free of contamination. Although the LOI values of PC generally comply with the requirements of ASTM C150 (2016) and BS EN 197-1 (2011), its average value (2.3%) is higher than that of GGC (Figure 3.7).

3.6.6 Alkali–Silica Reaction

Certain types of natural rocks such as chert, limestone and greywacke are known to be reactive, as they contain reactive forms of silica, which can cause damage to concrete due to ASR. An alkali–silica gel is formed when alkalis from an internal source (e.g., cement, aggregate and admixture) or an external source (e.g., deicing salts) react with the reactive silica. When sufficient moisture is present, the alkali–silica gel absorbs water and swells. The increase in volume of the gel can cause internal pressure and expansion in concrete, leading to cracking.

Numerous research studies have been undertaken on the ASR in mortar and concrete containing GC, in ground and granular forms, and these will be discussed in Section 4.7.5 of Chapter 4 and Section 5.7.5 of Chapter 5, respectively. Briefly, the use of GC-sand in PC-based materials may lead to ASR-induced expansion because of the amorphous nature of GC, and the intensity of this reaction can be affected by the type, colour and particle size of the GC used. The ASR expansion has been shown to be minimised with the use of pozzolanic materials such as fly ash, GGBS, metakaolin and, indeed, GC itself in a finely ground form.

3.7 Physical Properties of Glass Cullet

As most consumer glass has a flat form, such as container glass and flat glass, in consequence, upon crushing, the resulting particles, unless reduced to finer-sized fractions, would take the form of extremely elongated and flaky particles, and such crushed GC will be unsuitable for most construction applications. Therefore, the majority of GC recycled within the construction industry is as sand or ground to finer sizes. In this section, with some exceptions, most of the physical properties of GC are assessed in its fine size fraction, with a maximum particle size of 4 or 5 mm, which is considered to be best suited to its intended use as a fine aggregate in concrete, geotechnical and road pavement applications.

3.7.1 Physical Appearance

The shape and surface texture of GC have a direct influence on the properties of the resulting products. For example, particle shape affects the particle packing of the material, formation of air voids, flow and shear strength, whilst surface texture affects the bond strength between the GC particles and the binder.

The type of particle shape and surface texture of GC have been reported in some of the studies undertaken using the material in a granular form as a fine aggregate and ground form as a cement constituent. The data gathered from those studies were collated as shown in Table 3.10. As to be expected, granular GC is considered to have an angular shape and smooth surface texture. Compared with natural sand, which is normally somewhat rounded at the edges, the angularity of GC particles can provide a better particle interlocking effect, but at the same time it can adversely affect their packing and consequently the porosity-related properties of concrete. The smooth surface texture of GC, though it reduces interparticle friction and may weaken the bond at the interfacial transition zones between aggregate and cement paste or bitumen, in cementitious and bituminous mixtures, respectively, this has not been studied thoroughly based on the results analysed when developing Chapter 4 (concrete) and Chapter 7 (road pavement).

In contrast to shape and texture, a considerably greater number of studies appear to have been reported on the different colour types of GC used to produce sand, with clear (sometimes known as flint), green and amber (or yellowish brown) being the most commonly used (Table 3.10). However, other than the effect of GC-sand colour on the

Table 3.10 Glass shape, surface texture and colour

Property	Description	Granular Form (as Sand)	Ground Form (as Cement)
Shape	**Angular**	Lee et al. (2013), Turgut (2013) and Wang and Huang (2010)	Mirzahosseini and Riding (2014, 2015), Nassar and Soroushian (2011, 2012a,b), Parghi and Alam (2016) and Schwarz et al. (2007)
	Angular and sharp	Tan and Du (2013)	n.a.
	Angular and flat	Lee et al. (2013)	Liu (2011)
	Irregular and sharp	Limbachiya (2009)	n.a.
	Irregular and round	Ling and Poon (2011a)	n.a.
Surface texture	**Smooth**	De Castro and de Brito (2013), Lee et al. (2013), Limbachiya (2009), Ling and Poon (2011a,b, 2014b), Tan and Du (2013) and Wang and Huang (2010)	n.a.

Continued

Table 3.10 Continued

Property	Description	Granular Form (as Sand)	Ground Form (as Cement)
Colour	**Flint**	Abdallah and Fan (2014), Breakspere et al. (1980), De Castro and de Brito (2013), Du and Tan (2013), Ganiron (2012), Karamberi and Moutsatsou (2005), Kou and Poon (2009), Ling and Poon (2012a, 2013), Liu (2011), Malik et al. (2013, 2014), Mirzahosseini and Riding (2014), Park et al. (2004), Tan and Du (2013), Topcu et al. (2008), Wang and Chen (2010), Wright (2012), Wright et al. (2013), Yuksel et al. (2013) and Ismail and Al-Hashmi (2009)	Borosnyoi et al. (2013), Chaïd et al. (2015), Kara (2013a) and Liu (2011)
	Green	Du (2011), Du and Tan (2013), Karamberi and Moutsatsou (2005), Kou and Poon (2009), Lee (2011), Lee et al. (2011, 2013), Limbachiya et al. (2012), Ling and Poon (2014a), Ling et al. (2011), Liu (2011), Maier and Durham (2012), Park et al. (2004), Tan and Du (2013), Tang et al. (2005), Topcu et al. (2008) and Yuksel et al. (2013)	Borosnyoi et al. (2013), Kara (2013a), Liu (2011), Mirzahosseini and Riding (2014, 2015), Nwaubani and Poutos (2013) and Parghi and Alam (2016)
Colour	**Amber**	Du (2011), Du and Tan (2013), Du and Tan (2014b), Karamberi and Moutsatsou (2005), Kou and Poon (2009), Koh (2014), Maier and Durham (2012), Maraghechi et al. (2012), Oliveira et al. (2008), Park et al. (2004), Phillips et al. (1972), Tan and Du (2013), Topcu et al. (2008) and Yuksel et al. (2013)	Borosnyoi et al. (2013) and Kara (2013a)
	Blue	Ling and Poon (2011a)	n.a.
	Mixed	Breakspere et al. (1980), Du (2011), Dumitru et al. (2010), Guo et al. (2015), Idir et al. (2010a), Lam et al. (2007), Limbachiya (2009), Ling and Poon (2012a), Ling and Poon (2013), Maraghechi et al. (2012), Rajabipour et al. (2012), Taha and Nounu (2008), Taha and Nounu (2009), Wright (2012) and Wright et al. (2013)	Dumitru et al. (2010), Nassar and Soroushian (2012a,b), Taha and Nounu (2008) and Taha and Nounu (2009)

development of ASR in concrete (Chapter 5), little information is available regarding the colour effect on the engineering performance of concrete. The green GGC, on the other hand, has been shown to be highly pozzolanic compared with other colours and its use in a mixture with PC has also been shown to lead to higher total heat of hydration (Chapter 4).

The colour effect of GC, and possibly when used in a coarse fraction, can be potentially exploited in the architectural applications of concrete, in the form finishes and decorative exposed glass aggregate concrete (Byars et al., 2004). This has potential for developing interesting and attractive concrete products.

3.7.2 Particle Size Distribution of Glass Cullet Fine Aggregate

Particle size distribution (PSD), or grading, of a granular material provides information on the different-sized particles present in the material. It is expressed in the form of percentage by mass of particles passing a range of sieves. The PSD of aggregate has a profound influence on particle packing and consequently the air voids formed upon compaction of the material. A continuously graded aggregate is always considered to be desirable in most construction applications, as it potentially offers a densely packed structure with minimum air voids. However, occasionally, a gap-graded aggregate, in which some of the size fractions are missing, may be used to achieve specific finishes and/or performance.

For GC to be used as a fine aggregate, the material needs to be crushed and sieved to specified grading requirements. In this section, the PSD of GC fine aggregates of 95 samples used in 53 studies, undertaken in 11 countries, related to concrete research reported since 2000 are examined with reference to the CEN and ASTM standards for the use of fine aggregate in concrete, geotechnical and road pavement applications.

(a) Concrete Applications

In the latest standard for aggregates for use in concrete, BS EN 12620:2002+A1 (2008), the description of the fineness of fine aggregate is based on the percentage passing at the 0.5-mm sieve, which differs from the previous standard, BS 882 (1992), in which the description is based on grading limits. Dhir et al. (2005c) developed three grading limits for fine aggregate by combining the definitions given in the two standards, which were considered to be more useful in describing the fineness of fine aggregate. Figure 3.8 shows the PSD of GC fine aggregate together with the proposed combined grading limits based on Dhir et al. (2005c).

Figure 3.8 shows that the large majority of the PSD of GC fine aggregate used in the studies (shaded in grey) generally falls within the coarse limits and the lower side of the medium limits. In only a few cases the PSD of GC lies within the fine limits. Taking this as a broad measure, overall, it would be fair to argue that the PSD of GC fine aggregate used in concrete research has mostly tended to be coarse (Figure 3.8). In some cases, albeit fewer in number, the PSD of GC has been widely coarse. Such materials should not be used without the utmost care in

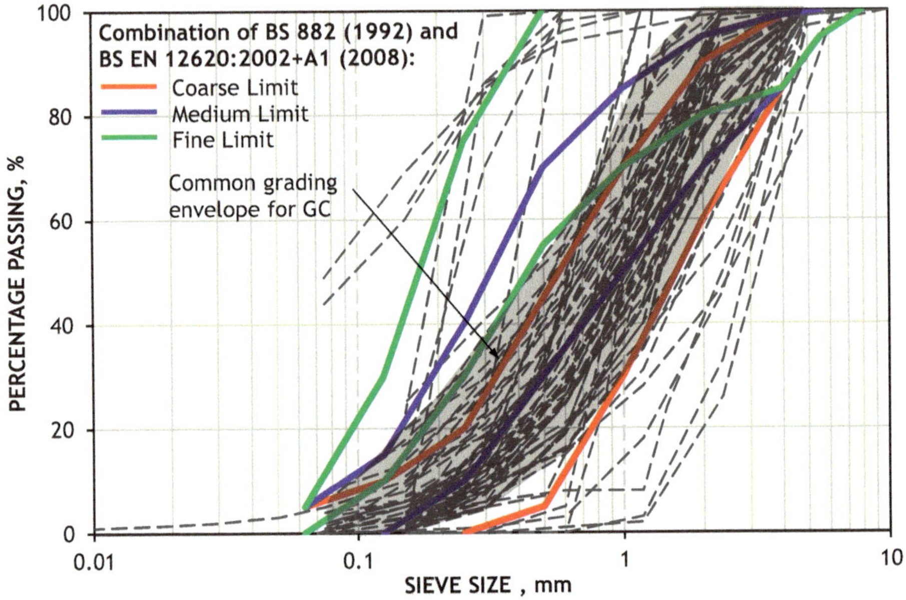

Figure 3.8 Particle size distribution of glass cullet (*GC*) as a fine aggregate with the combined limits of BS EN 12620:2002+A1 (2008) and BS 882 (1992).
Data taken from Abdallah and Fan (2014), Abendeh et al. (2015a,b), Aghabaglou et al. (2015), Berry et al. (2011), Borhan and Bailey (2014), Dhir et al. (2005a, 2009), Disfani et al. (2012), Du and Tan (2014b), Guo et al. (2015), Isler (2012), Ismail and Al-Hashmi (2009), Kou and Poon (2009, 2013), Lam et al. (2007), Lee (2011), Lee et al. (2013, 2008, 2011), Lim (2014), Limbachiya et al. (2012), Ling and Poon (2014a,2011a,b), Ling and Poon (2012a,b,c), Ling and Poon (2013, 2014c,d), Ling et al. (2011, 2012), Miranda Jr. et al. (2014), Oliveira et al. (2013, 2008), Penacho et al. (2014), Poon and Chan (2007), Poutos et al. (2008), Rajabipour (2012), Romero et al. (2013), Saccani and Bignozzi (2010), Soyer et al. (2010), Su and Chen (2002), Taha and Nounu (2008, 2009), Turgut (2008b, 2013), Turgut and Yahlizade (2009), Wang (2009a,b), Wang and Huang (2010), Wang et al. (2014), Yuksel et al. (2013), Zhao and Wei (2011) and Zhao et al. (2013a,b).

designing concrete mixes, as it can easily lead to some undesirable effects, such as bleeding in fresh concrete and the presence of excessive voids in hardened concrete, consequently affecting the long-term performance of the concrete, because of the lack of sufficient fines.

The results shown in Figure 3.8 suggest that, although GC can be crushed into any size fraction, the material studied since the year 2000 has tended to be crushed and processed as fine aggregate with coarse grading and, at times, in much coarser form. As the data also showed little change with time in the crushing and processing of GC as sand, this would suggest that the material used in the studies has mostly been in the coarse sand zone with no clear evidence of the grading of the GC fine aggregate shifting towards the medium zone, where maximum particle packing is more likely to be achieved.

(b) Geotechnical Applications

The size definition of sand (fine aggregate) in geotechnics is different to that for concrete. In the unified soil classification system of ASTM D2487 (2011), sand is considered as material passing a 4.75-mm sieve and retained on a 75-µm sieve; for BS 5930 (2015), the corresponding sieve sizes are 2 mm and 63 µm, respectively.

Figure 3.9 shows the PSD of GC-sand samples used previously in Figure 3.8, together with the soil classifications defined in ASTM D2487 (2011) and in BS 5930 (2015) for comparison purposes. Based on ASTM D2487 (2011), a large majority of GC samples can be considered as 'clean sand' as they contain less than 5% fines (particles passing 75 µm). However, as the material generally has a high proportion of medium and coarse sand and given that the coefficient of uniformity (C_u), the ratio of particle size passing at 60% to that at 10%, of GC samples is mostly less than 6, the material can be classified as 'poorly graded sand' in accordance with ASTM D2487 (2011).

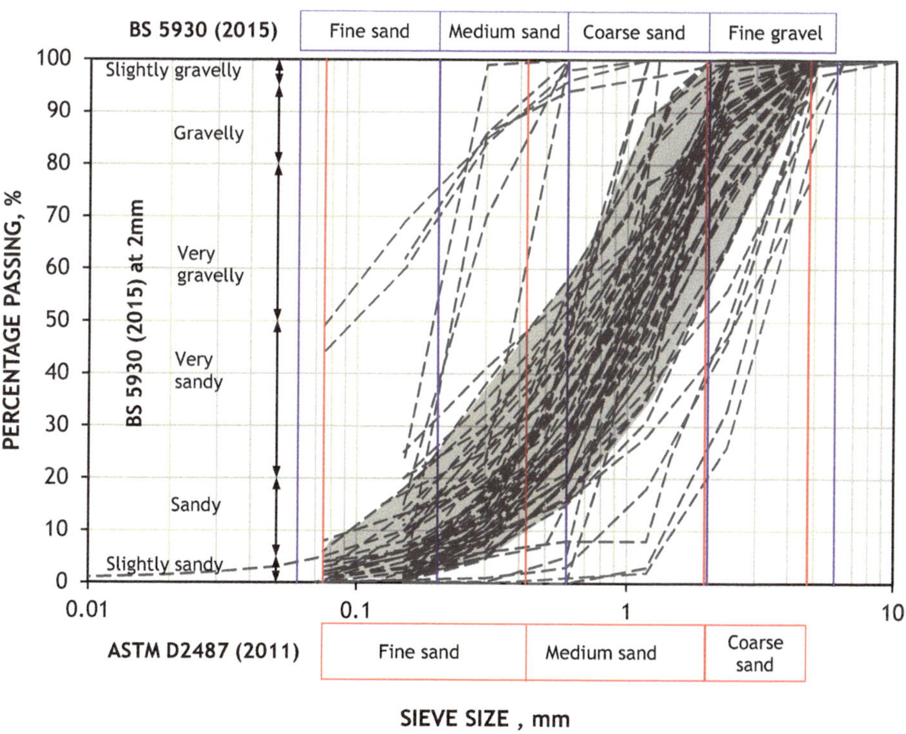

Figure 3.9 Particle size distribution of glass cullet as a fine aggregate with the soil classifications given in ASTM D2487 (2011) and BS 5930 (2015).

On the other hand, based on BS 5930 (2015), the GC samples in Figure 3.9 can be seen to contain two types of materials, namely, gravel and sand. Figure 3.9 shows that the majority of the materials are in the form of 'gravelly medium and coarse sand' and 'very gravelly medium and coarse sand.' The remaining small number of samples can be classified as 'very sandy fine gravel' and on the whole, the GC samples normally contain no more than 10% fine sand particles.

(c) Road Pavement Applications

In road pavement applications, the PSD requirement for aggregate varies depending upon the maximum nominal aggregate size, the desired aggregate grading characteristics (well graded or gap graded) and the intended use (in base, binder course or surface course). Two fine aggregate grading limits for asphalt mixtures, one with 4-mm upper sieve size for asphalt concrete specified in BS EN 13108-1 (2016) and another as grading No. 4 in ASTM D1073 (2016), were selected to examine how well the grading of the GC samples previously plotted in Figure 3.8 may comply with the grading requirements for road pavement use. This is shown in Figure 3.10.

It can be seen that the processing of GC as a fine aggregate in terms of road construction applications since the year 2000 would be considered as tending to produce GC with PSD generally within the grading No. 4 limits as per ASTM D1073 (2016). On the other hand, as seen from Figure 3.10, to meet the requirements of BS EN 13108-1 (2016), further crushing would be required to increase the presence of particles of less than 1 mm in GC for use as a fine aggregate in road pavement applications.

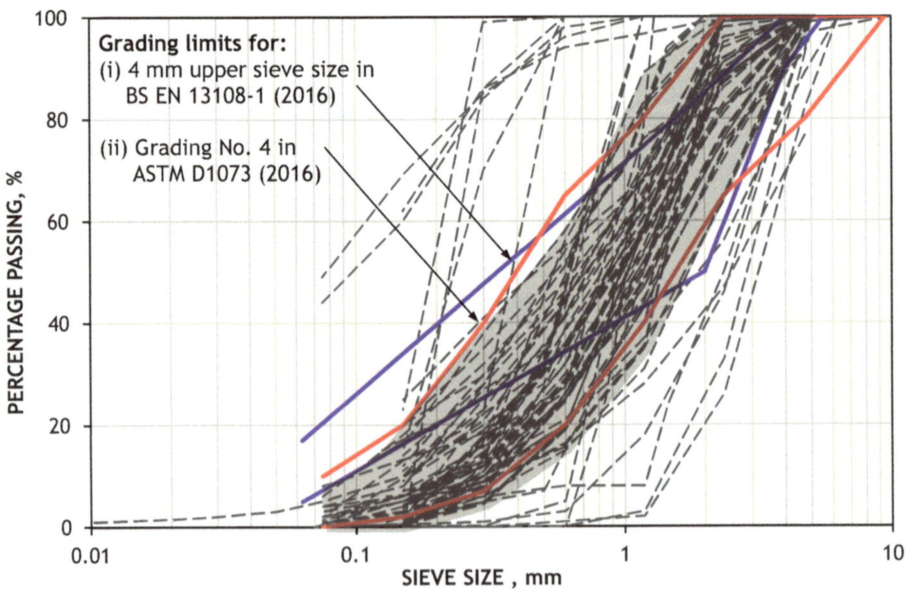

Figure 3.10 Particle size distribution of glass cullet as a fine aggregate with the grading limits for asphalt mixes in ASTM D1073 (2016) and BS EN 13108-1 (2016).

3.7.3 Fineness Modulus of Glass Cullet Fine Aggregate

The fineness modulus (FM) is another way of describing the PSD of fine aggregate. It is concise and easy to define and specifies fine aggregates for use in practice. It is calculated as the sum of the cumulative percentages retained on a series of specific sieves, divided by 100. Thus, the higher the FM value, the coarser the aggregate.

The FM values of GC fine aggregate and the corresponding natural fine aggregate, as reported, are shown plotted in Figure 3.11, together with the FM classifications for aggregates for use in concrete given in BS EN 12620:2002+A1 (2008) for reference. The FM of natural fine aggregate is mostly between 2.0 and 3.0, with an average value of 2.6, which is within the MF–CF (medium to coarse) range. For GC fine aggregate, its FM values show a wide spread, mostly ranging from 2.3 (medium fine) to 3.8 (very coarse), although a majority falls into the CF category. The average FM value of GC fine aggregate is 3.0, which is significantly coarser than natural fine aggregate.

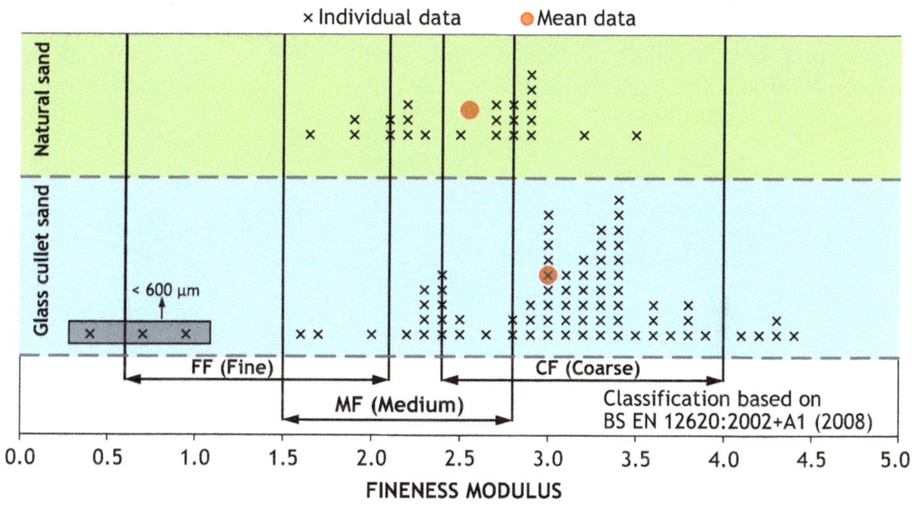

Figure 3.11 Fineness modulus for glass cullet sand and reference natural sand.
Data taken from Abdallah and Fan (2014), Aghabaglou et al. (2015), Berry et al. (2011), Chen et al. (2006a, 2011), Dhir et al. (2009), Guo et al. (2015), Huang et al. (2015), Isler (2012), Ismail and Al-Hashmi (2009), Kim and Soh (2002), Kou and Poon (2009, 2013), Lam et al. (2007), Lee et al. (2013), Limbachiya et al. (2012), Ling and Poon (2013, 2014c, 2011b), Ling and Poon (2012a,b, 2014a,b,c,d), Ling et al. (2011), Liu (2011), Maier and Derham (2012), Maschio et al. (2013), Miranda Jr. et al. (2014), Nassar and Soroushian (2012a,b, 2013), Oliveira et al. (2008), Park et al. (2004), Penacho et al. (2014), Poon and Chan (2007), Rajabipour et al. (2012), Romero et al. (2013), Saccani and Bignozzi (2010), Sharif et al. (2014), Shayan and Xu (2006, 2004), Soyer et al. (2010), Terro (2006), Topcu et al. (2008), Wang (2009a,b), Wang and Chen (2010), Wang and Huang (2010), Wang et al. (2015, 2014), Wright (2012), Wright et al. (2013, 2014), Yuksel et al. (2013) and Zhao et al. (2013a,b).

It should be noted that the GC fine aggregate with 0.40–0.95 FM value, highlighted in Figure 3.11, with a particle size less than 600 μm, was used in making mortar (Chen et al., 2006a; Maschio et al., 2013) and dry-mixed concrete blocks (Lee et al., 2013). Interestingly, the resulting mixes showed a significant increase in compressive strength, suggesting that GC crushed to this level of fineness can possibly exhibit appreciable pozzolanic reactivity.

3.7.4 Particle Size Distribution of Glass Cullet as Filler

Filler is a very fine aggregate, with most of its particles passing a 63-μm sieve (in Europe) and 75 μm (in the United States). Although the use of filler in the construction industry is beginning to have a role, at present it is normally used in self-compacting concrete and hot-mixed asphalt. The incorporation of filler can fill up the very small spaces between fine aggregates and even cement particles, thus reducing the voids of the mixes, and, in the case of asphalt mixes, it can also modify the properties of the asphalt binder.

The PSD of GC fillers that have been used to study its effect on the properties of concrete and the grading requirements for filler aggregate as specified in BS EN 12620:2002+A1 (2008) are provided in Figure 3.12. As the standard was first implemented only in 2002, all the GC filler samples used in the studies reported between 2009 and 2016 had at least 70% by mass passing a 63-μm sieve and conformed to the standard requirements (Figure 3.12).

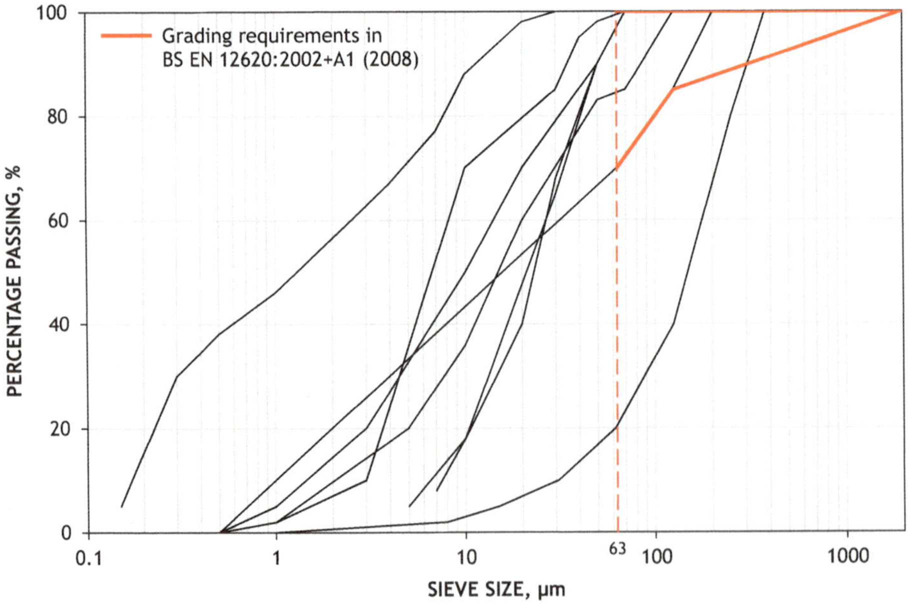

Figure 3.12 Particle size distribution of glass cullet as filler.
Data taken from Bignozzi et al. (2009), Dhir et al. (2009), Dyer and Dhir (2010), Korjakins et al. (2009, 2010), Klevbo (1998), Matos and Sousa-Coutinho (2016b) and Vaitkevicius et al. (2014).

However, the PSD of an early GC sample (Klevbo, 1998), undertaken before the introduction of the standard, with 20% by mass passing the 63-μm sieve (Figure 3.12), will now be considered to be too coarse to use as filler. Despite its benefits in designing concrete mixes, it is obvious from Figure 3.12 that the use of fillers in concrete, which came into recognition in 2002, in Europe has been slow to develop.

3.7.5 Fineness of Ground Glass Cullet as Cement

When used as a cement addition, the fineness of GGC can affect its pozzolanic reactivity, with the material becoming more reactive with increasing fineness. Like any other cementitious material, the fineness of GGC can also influence the performance of concrete in the fresh state.

Figure 3.13 shows the fineness of GGC (in black colour), together with the corresponding PC (in blue colour) in terms of their PSD, used in several studies. The red lines in the figure show the fineness requirements for fly ash, expressed as the percentage retained on a 45-μm sieve, specified in ASTM C618 (2015) and BS EN 450-1 (2012). In the former

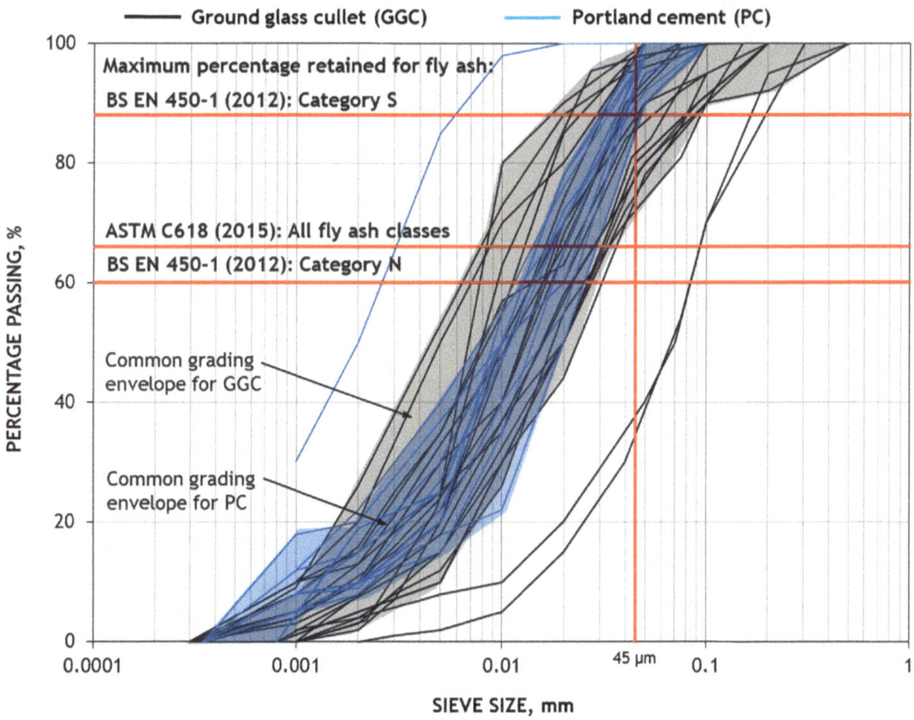

Figure 3.13 Particle size distribution of ground glass cullet.
Data taken from Afshinnia and Rangaraju (2015), Bajad et al. (2011a,b, 2012a,b), Calmon et al. (2014), Carsana et al. (2014), Chaïd et al. (2015), Dhir et al. (2005a), Du and Tan (2014a), Kim et al. (2014a), Matos and Sousa-Coutinho (2012, 2016b), Matos et al. (2015), Nassar and Soroushian (2012a,b), Neithalath and Schwarz (2009), Nunes et al. (2013), Omran and Tagnit-Hamou (2016), Pavoine et al. (2014), Sharifi et al. (2015), Schwarz et al. (2008) and Siad et al. (2016).

standard, the maximum sieve retention is limited to 34% by mass for all classes of fly ash. In the latter standard, two categories are defined; for category N (coarser particle) and for category S (finer particle), the maximum retention is set at 40% and 12% by mass, respectively. Apart from three studies in which either the GGC used was too coarse or the PC too fine, most of the PC samples had a PSD within the wider PSD envelope of GGC.

Except for two with about 40% by mass passing 45 μm (i.e., about 60% mass retention), the fineness of all GGC samples reported in the literature complies with the requirements for fly ash specified in ASTM C618 (2015) and BS EN 450-1 (2012). The two coarse GGC samples were from the studies investigating the effect of fineness on the hydration-related properties of concrete, showing improvement in both the engineering and the durability performance (Afshinnia and Rangaraju, 2015; Neithalath and Schwarz, 2009).

In addition, there has been some tendency to use finer GGC in recent years, with fineness complying with the 12% 45-μm sieve retention for type S fly ash in BS EN 450-1 (2012) (Matos et al., 2015; Matos and Sousa-Coutinho, 2016b; Omran and Tagnit-Hamou, 2016; Siad et al., 2016).

There has been a growing trend towards using the Blaine test for measuring the fineness of GGC in direct comparison to PC and to improve the credibility of GGC as a cementitious material. Figure 3.14 shows the frequency of Blaine fineness for GGC and that of the corresponding PC samples used in several studies undertaken since 2004–16.

For PC samples, the Blaine fineness varies from 200 to 500 m^2/kg, but typically is within 301–400 m^2/kg. For GGC, the material was ground to a wider range, with fineness varying within the range of <100–800 m^2/kg, but typically within 401–500 m^2/kg, which is finer than PC. These results also tend to suggest that GGC is at present very much at an experimental and development stage and is customised, whilst PC is a long-established industrial product, being produced in great masses in a ready-to-use form confirming standard specifications.

3.7.6 Specific Gravity

The specific gravity of materials is an important property when designing a mix for various reasons, for example, to ensure the stability of fresh concrete, to determine the compaction properties of soil and to calculate the voids of a bituminous mix. It also affects the weight of the final products. Thus, it is widely reported in the literature for characterising the materials.

The specific gravity of GC varies depending on its chemical composition. The reported data are shown plotted in Figure 3.15. As shown, the specific gravities of soda lime glass and aluminosilicate glass are close, with average values of 2.46 (range 2.15–2.65) and 2.54 (range 2.42–2.80), respectively. This makes the GC lighter than natural sand, with an average value of 2.60 and range of 2.45–2.68, and considerably lighter than PC, having an average value of 3.14 (range 3.0–3.2). The density of lead glass has been reported as an average of 3.04 with a range of 2.99–3.15, which is close to PC clinker. Thus, to maintain the volume-related properties such as yield and design of mixtures, the density of GC needs to be taken into account in a proposed application.

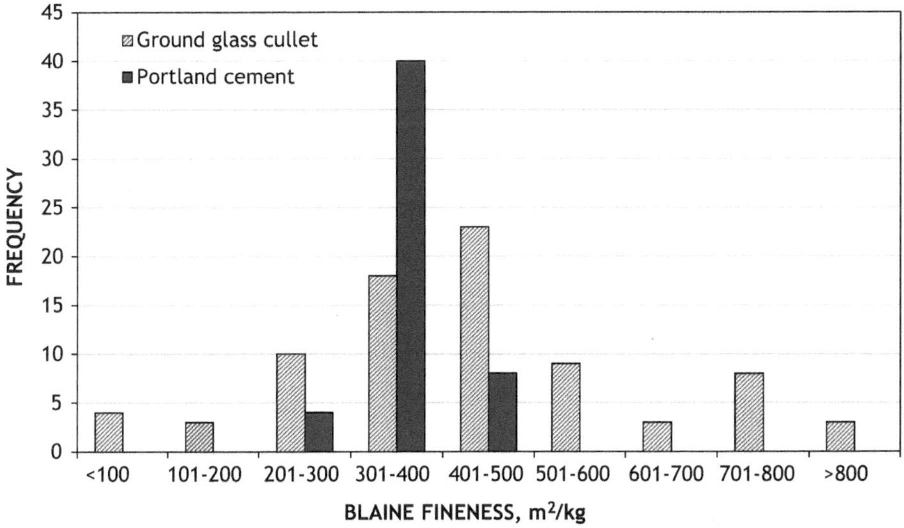

Figure 3.14 Fineness of ground glass cullet and reference Portland cement.
Data taken from Afshinnia and Rangaraju (2015), Al-Akhras (2012), Aly et al. (2011), Borosnyoi et al. (2013), Byars et al. (2004), Calmon et al. (2014), Carsana et al. (2014), Chaïd et al. (2015), Chen et al. (2006a), Corbu et al. (2013), Dahmardeh et al. (2015), Dumitru et al. (2013, 2010), Guo et al. (2015), Idir et al. (2010a,b, 2011, 2009), Jang et al. (2015), Kara (2013a,b), Khmiri et al. (2012, 2013), Kim et al. (2014a, 2015), Kou and Poon (2009), Kou and Xing (2012), Laldji et al. (2007), Lee and Lee (2016), Limbachiya (2009), Lin et al. (2009), Ling and Poon (2012b), Madandoust and Ghavidel (2013), Matos et al. (2015), Metwally (2007), Mirzahosseini and Riding (2015, 2014), Nasaar and Soroushian (2012a,b), Nassar and Soroushian (2011, 2012b, 2013), Niang et al. (2015), Omran and Tagnit-Hamou (2016), Özkan and Yüksel (2008), Parghi and Alam (2016), Park and Lee (2004), Pavoine et al. (2014), Priscilla and Naik (2014), Proshin et al. (2005), Schwarz and Neithalath (2008), Schwarz et al. (2007), Sekhar (2011), Sharifi et al. (2015), Shayan and Xu (2006), Shi and Wu (2005), Shi et al. (2005), Siad et al. (2016), Tagnit-Hamou and Bengougam (2012), Tognonvi et al. (2015), Tuncan et al. (2001), Turgut (2008b, 2013), Turgut and Yahlizade (2009), Vaitkevicius et al. (2014), Wang (2011), Wang and Hou (2011), Wang et al. (2014, 2016), Yilmaz and Degirmenci (2010), Zhao and Wei (2011) and Zhao et al. (2013b).

Lead glass has a relatively high specific gravity, with an average value of 3.04 (with a range of 2.99–3.15), which is close to that of Portland clinker. For completeness, the specific gravity data of the samples for which the glass type has not been reported is also provided, giving an average value of 2.63 and a higher range of 2.40–3.20, as expected.

3.7.7 Water Absorption

The water absorption of an aggregate is the function of its porosity and is commonly used to obtain an approximate measure of this. In the construction industry, highly absorptive aggregates are undesirable materials in any mix, as they will increase the water demand in concrete for a given workability (consistence), water in soil compaction and bitumen in asphalt mixes.

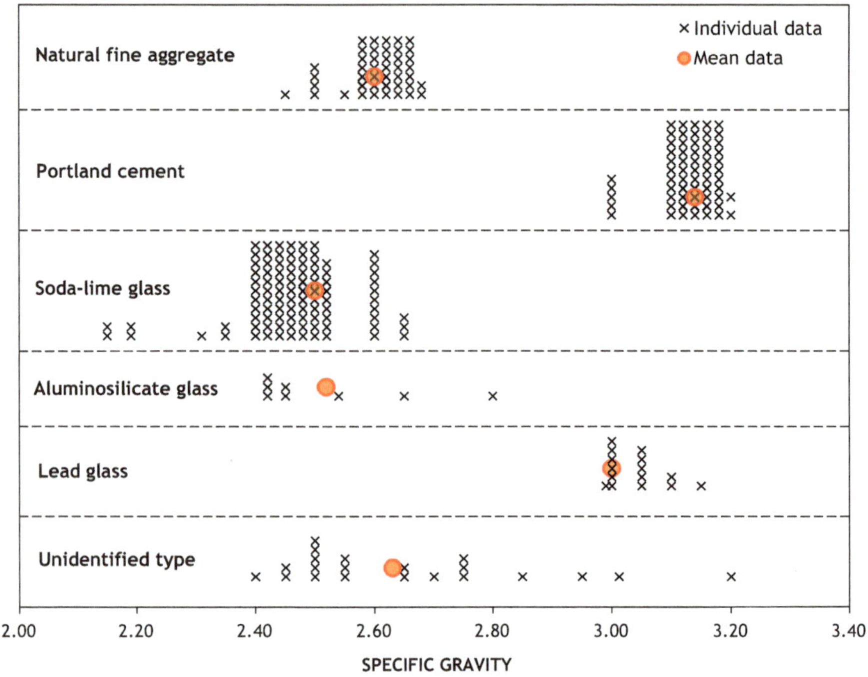

Figure 3.15 Specific gravity of various glasses in comparison with natural fine aggregate and Portland cement.
Data taken from Abendeh et al. (2015a,b), Al-Akhras (2012), Almesfer and Ingham (2014), AL-Saffar (2013), Aly et al. (2012), Anagnostopoulos et al. (2009), Aghabaglou et al. (2015), Altaf et al. (2013), Bhat and Rao (2014), Bajad et al. (2012a,b), Calmon et al. (2014), Cassar and Camilleri (2012), Chaïd et al. (2015), Chen et al. (2006a), Chen and Wong (2015), Corinaldesi et al. (2005), De Castro and de Brito (2013), Dhir et al. (2005a), Disfani et al. (2012), Du and Tan (2014b), Dumitru et al. (2010), Georgiadis et al. (2007), Idir et al. (2009, 2010a, 2011), Jain and Neithalath (2010), Jangid and Saoji (2014), Kamali and Ghahremaninezhad (2015, 2016), Kim et al. (2015), Kou and Poon (2009), Kou and Xing (2012), Huang et al. (2015), Ismail and Al-Hashmi (2009), Kim et al. (2014a), Lee and Lee (2016), Jang et al. (2015), Lin et al. (2009), Ling and Poon (2011b, 2012a,b,c, 2014a,b,c), Ling et al. (2011, 2012), Dumitru et al. (2013), Kim and Soh (2002), Klevbo (1998), Laldji et al. (2007), Lee (2011), Liu (2011), Lam et al. (2007), Lee et al. (2013), Limbachiya (2009), Malik et al. (2013, 2014), Matos and Sousa-Coutinho (2012), Matos et al. (2015), Matos and Sousa-Coutinho (2016b), Metwally (2007), Mirzahosseini and Riding (2014, 2015), Maier and Durham (2012), Maschio et al. (2013), Mitra et al. (2016), Narayana and Mailar (2015), Nassar and Soroushian (2011, 2012a,b, 2013), Neithalath (2008), Neithalath and Schwarz (2009), Niang et al. (2015), Nunes et al. (2013), Omran and Tagnit-Hamou (2016), Özkan and Yüksel (2008), Oliveira et al. (2008), Priscilla and Naik (2014), Parghi and Alam (2016), Park et al. (2004), Park and Lee (2004), Peyvandi et al. (2013), Polley (1996), Polley et al. (1998), Poon and Chan (2007), Proshin et al. (2005), Romero et al. (2013), Rajabipour et al. (2012), Salehuddin (2012), Schwarz and Neithalath (2008), Schwarz et al. (2007, 2008), Seju et al. (2015), Serpa et al. (2015), Shafaatian et al. (2013), Shao and Lehoux (2001), Shao et al. (2000), Sharif et al. (2014), Sharifi et al. (2015), Soyer et al. (2010), Shayan and Xu (2006), Shi and Wu (2005), Shi et al. (2005), Siad et al. (2016), Singh et al. (2014), Su and Chen (2002), Tagnit-Hamou et al. (2015), Tagnit-Hamou and Bengougam (2012), Taha and Nounu (2008, 2009), Tang et al. (2005), Tejaswi et al. (2015), Tuncan et al. (2001), Tognonvi et al. (2015), Turgut (2008b, 2013), Turgut and Yahlizade (2009), Türkmen and Fındık (2010), Wattanapornprom and Stimannaithum (2015), Wright et al. (2013, 2014), Wang (2009a,b, 2011), Wang and Chen (2010), Wang and Huang (2010), Wang and Hou (2011), Wang et al. (2009, 2014, 2016), Yilmaz and Degirmenci (2010), Zhao and Wei (2011) and Zhao et al. (2013a,b).

Table 3.11 Water absorption of glass measured in granular form

	Water Absorption, %	
Glass Type	**Mean**	**Range**
Soda lime	0.18	0–0.48
Aluminosilicate	0.40	0.36–0.45
Lead	0.10	0–0.40
Unidentified	0.23	0–0.48

Data taken from Abendeh et al. (2015a,b), Aghabaglou et al. (2015), Chen and Wong (2015), Chen et al. (2006a), Corinaldesi et al. (2005), De Castro and de Brito (2013), Dhir et al. (2005b,a), Du and Tan (2014b), Dumitru et al. (2010), Huang et al. (2015), Kim and Soh (2002), Kim et al. (2014a), Kou and Poon (2009), Lam et al. (2007), Lee (2011), Lee et al. (2013), Ling and Poon (2011b, 2012a,b), Ling and Poon (2014a,b,c,d), Ling et al. (2012), Liu (2011, 2009), Maier and Durham (2012), Maschio et al. (2013), Nassar and Soroushian (2012a,b), Omran and Tagnit-Hamou (2016), Park et al. (2004), Park and Lee (2004), Polley et al. (1998), Poon and Chan (2007), Rajabipour et al. (2012), Serpa et al. (2015), Su and Chen (2002), Taha and Nounu (2008, 2009), Tuncan et al. (2001), Turgut and Yahlizade (2009), Wang (2009a,b), Wang and Chen (2010), Wang and Huang (2010), Wang et al. (2015, 2014), Wright et al. (2013), Zhao and Wei (2011) and Zhao et al. (2013b,a).

As glass is essentially an impermeable material, its maximum water absorption value has been reported in numerous studies to be less than 0.5%, with many results being 0% or close to it (Table 3.11). In comparison, the water absorption of natural fine aggregate is high, varying from less than 1.0% to –3.0%. Thus, the use of GC fine aggregate as a replacement for natural fine aggregate, particularly in concrete applications, can be expected to reduce its permeation properties, preventing the ease of transportation of moisture and harmful ions, and thus increase its durability.

3.7.8 Modulus of Elasticity/Hardness

The basic requirement of an aggregate for use in construction is that the material must be strong; it does not deform easily under load and has good wear-resistance properties. These attributes can be expressed in different forms and require specific tests on aggregates such as aggregate crushing value (now withdrawn), Los Angeles abrasion (BS EN 1097-2, 2010) and micro-Deval tests (BS EN 1097-1, 2011). Although such tests have not been commonly performed on GC, the modulus of elasticity and hardness of glass can be good general indications of the mechanical properties of GC. The literature using GC in various construction applications has not provided any test data on modulus of elasticity and hardness. Table 3.12 has been compiled for completeness concerning the modulus of elasticity and hardness of various types of glass and the natural sand.

Table 3.12 suggests that, in general, GC has slightly better deformation properties than natural sand, whilst its hardness is essentially of a similar order and, therefore, in this respect, its use in construction applications should not present any problem.

Table 3.12 Modulus of elasticity and hardness of different types of glass and natural fine aggregate

Material	Modulus of Elasticity, GPa	Hardness, Mohs
(a) Glass		
Soda-lime	70–74	5.0–7.0
Aluminosilicate	83–91	
Lead	58–65	
Borosilicate	64–89	
(b) Natural Fine Aggregate		
Quartz/Quartzite	55–85	5.5–7.0

Data taken from Le Bourhis (2008), De Jong et al. (2011), Neville et al. (1983) and Kogel et al. (2006).

3.8 Conclusions

The bulk of the waste glass generated is of the soda lime type from container and flat glass and, as expected, much developmental work for recycling glass has been based on such materials. However, in recent years, waste electronics glass, in the form of CRTs, LCDs and PCBs, has been produced in increasingly greater amounts and consequently has drawn considerable attention in developing applications for such materials.

It is widely acknowledged, and has been materialised in practice for a long time, that with well-developed recycling schemes, waste glass can be channelled successfully into the production of glass and glass-based products. Furthermore, it is also now understood that, in principle, processed waste glass can also potentially be developed for use in the construction, ceramics and other markets. The environmental and sustainability benefits of diverting waste glass from the landfill and, instead, using the material as a replacement for natural resources are also well appreciated, though the economic case has not been always well placed. Indeed, governments and local authorities the world over have been addressing the various issues involved in overcoming the barriers to the wider development of the GC market. In this regard the role of proper segregation during collection of waste glass, resulting in minimising contamination of the material, would finally increase the client and specifier confidence in the feasibility of using GC.

The recommended minimum processing steps involved in the production of GC for recycling include visual inspection and hand-picking of contaminated materials, screening and removal of ferrous and nonferrous metallic materials. However, owing to the general lack of information concerning the types of glass that may be present in a waste stream, glass-based contaminants, such as Pyrex, ceramics and lead glass, are often encountered mixed with the rest of the waste. This can pose problems in assembling suitable GC for use in the production of new glass products such as containers and flat glass. Contamination removal techniques, which can identify and mechanically remove the contamination, should be developed for further beneficiation of waste glass. As the glass collectors can often present poorly colour-segregated waste glass or may even present batches contaminated with relatively high quantities of non-glass constituents, additional quality control steps should be applied on delivery to the glass processor (PAS 101 interface), so that the GC is properly graded and priced as per its true cost, taking account of additional processing costs and potential market value.

Some waste glass batches cannot be thoroughly processed into high-quality GC for technical, economic and environmental reasons, and the resulting GC may present varying characteristics that may affect the properties of the end-use application. For this reason, the CWC proposed clear and concise criteria for several characteristics of GC (based on colour, moisture and contamination of non-glass materials), which can be easily determined and applied with subsequent certification for use as a feasible replacement for raw material in various applications.

For the most part, the chemical composition of GC is dominated by its SiO_2 content; other oxides are present to impart certain specific properties in the final products. Except for lead glass, the sum of SiO_2, Al_2O_3 and Fe_2O_3 of soda lime, aluminosilicate and borosilicate GC, as used in the published research since 1996, has been found to be greater than the minimum of 70% specified for fly ash to be considered as potentially reactive pozzolanic material.

The SO_3 content and LOI of GC are reported to be less than 3% and 1%, respectively. Whilst not exceeding the SO_3 maximum limit for fly ash, for soda lime glass the range for SO_3 content may considered to be high at 0%–3.10%, with a mean value of 0.27%, standard deviation 0.43% and coefficient of variation 159%. The concentration of alkalis released from GGC is reported to be low. However, ASR of GC fine aggregate in PC concrete, which depends on GC particle size, type and colour, can be a durability issue. It has been shown, though, that the risk of induced expansion in the form of ASR can be nullified with the use of pozzolanic cements, interestingly, including GGC.

GC, available mainly in clear (flint), green and amber colours, is an angular material having a smooth surface texture and excellent permeation properties. The material can be crushed to size fractions of sand, filler and cement with PSD complying with specifications for use in concrete, geotechnical and road pavement applications. The grading of manufactured GC fine aggregate in the reported literature has tended to be on the coarse side compared with natural sand, with no explanation provided for it and, more importantly, with no consideration of its impact on particle packing and the performance of the end products.

The average FM of the manufactured GC-sand in the research literature has tended to be within the 2.3–3.8 range, with an average value of 3.0, compared with a 2.0–3.0 range with an average of 2.6 for the corresponding natural sands. GC use as a filler in both concrete and asphalt mixes is also becoming gradually recognised. The information available suggests that a considerably high proportion of GC fillers used comply with the requirement of passing a 63-μm sieve. Used in the ground form at a cement fraction level, GC has tended to comply with the requirements for fly ash in ASTM C618 (2015) and BS EN 450-1 (2012). In addition, in direct comparison to PC, the data on Blaine fineness show that GGC has been tested to a much wider range of 100–800 m^2/kg and typically 401–500 m^2/kg, compared with the corresponding fineness ranges of 200–500 and 301–400 m^2/kg for PC.

The exceptionally low porosity and consequently the excellent permeation properties, namely absorption, permeability and diffusion, of GC crushed for use as sand, filler or cement, where applicable, can be of considerable benefit in terms of potentially lowering the water demand of mortar and concrete mixtures, because of its exceptionally low water absorption, in the fresh state and in the long term, improving their durability. Though GC has lower density compared with sand and most cementitious materials, its modulus of elasticity and hardness are essentially similar to those of sand.

References

Abdallah S and Fan M, 2014. Characteristics of concrete with waste glass as fine aggregate replacement. International Journal of Engineering and Technical Research (IJETR) 2 (6), 11–17.

Abendeh R, Baker M B, Salem Z A and Ahmad H, 2015a. The feasibility of using milled glass wastes in concrete to resist freezing-thawing action. International Journal of Civil, Environmental, Structural, Construction and Architectural Engineering 9 (8), 981–984.

Abendeh R, Ramadan K, Haddad R and Baker M, 2015b. Freezing thawing resistance of concrete incorporating glass waste. Journal of Materials Science Research 4 (4), 19–25.

Afshinnia K and Rangaraju P R, 2015. Influence of fineness of ground recycled glass on mitigation of alkali– silica reaction in mortars. Construction and Building Materials 81, 257–267.

Aghabaglou M A, Tuyan M and Ramyar K, 2015. Mechanical and durability performance of concrete incorporating fine recycled concrete and glass aggregates. Materials and Structures 48 (8), 2629–2640.

Al-Akhras N, 2012. Performance of glass concrete subjected to freeze-thaw cycling. The Open Construction & Building Technology Journal 6, 392–397.

Almesfer N and Ingham J, 2014. Effect of waste glass on the properties of concrete. Journal of Materials in Civil Engineering 26 (11), 1–6.

AL-Saffar N A, 2013. The effect of filler type and content on hot asphalt concrete mixtures properties. Al-Rafadain Engineering 21 (6), 88–100.

Altaf S N, Asmat K M, Seemab G M, Aalam K M and Azim K F, 2013. Suitability of Waste Glass Powder as a Partial Replacement of Cement in Fibre Reinforced Concrete. Diploma Report. Civil Engineering Department, A.I.A.R. Kalsekar Polytechnic, Mumbai, India. 142 pp.

Aly M, Hashmi M S J, Olabi A G, Messeiry M and Hussain A I, 2011. Effect of nano clay particles on mechanical, thermal and physical behaviours of waste- glass cement mortars. Materials Science and Engineering a 528 (27), 7991–7998.

Aly M, Hashmi M S J, Olabi A G, Messeiry M, Abadir E F and Hussain A I, 2012. Effect of colloidal nano-silica on the mechanical and physical behaviour of waste- glass cement mortar. Materials and Design 33, 127–135.

Anagnostopoulos N, Sideris K and Georgiadis A, 2009. Mechanical characteristics of self-compacting concretes with different filler materials, exposed to elevated temperatures. Materials and Structures 42 (10), 1393–1405.

Andela C and Zavitz A, 2004. Environmental mandates vs. Economic realities: can both be satisfied? In: Proceedings of the International Conference on Sustainable Waste Management and Recycling: Glass Waste. Kingston University, London, UK, pp. 48–54.

Andreola F, Barbieri L, Corradi A and Lancellotti I, 2007. CRT glass state of the art: a case study: recycling in ceramic glazes. Journal of the European Ceramic Society 27 (2), 1623–1629.

ASTM C114, 2015. Standard Test Methods for Chemical Analysis of Hydraulic Cement. ASTM International, West Conshohocken, PA, USA.

ASTM C150, 2016. Standard Specification for Portland Cement. ASTM International, West Conshohocken, PA, USA.

ASTM C618, 2015. Standard Specification for Coal Fly Ash and Raw or Calcined Natural Pozzolan for Use in Concrete. ASTM International, West Conshohocken, PA, USA.

ASTM D1073, 2016. Standard Specification for Fine Aggregate for Asphalt Paving Mixtures. ASTM International, West Conshohocken, PA, USA.

ASTM D2487, 2011. Standard Practice for Classification of Soils for Engineering Purposes (Unified Soil Classification System). ASTM International, West Conshohocken, PA, USA.

Bajad M N and Modhera C D, 2009. Experimental investigations in developing concrete containing waste glass powder as pozzolana. Journal of Information, Knowledge and Research in Civil Engineering 1 (1), 32–67.

Bajad M N, Modhera C and Desai A, 2011a. Effect of glass on strength of concrete subjected to sulphate attack. International Journal of Civil Engineering Research and Development (IJCERD) 1 (2), 1–12.

Bajad M N, Modhera C and Desai A, 2011b. Influence of a fine glass powder on strength of concrete subjected to chloride attack. International Journal of Civil Engineering & Technology 2 (2), 1–12.

Bajad M N, Modhera C and Desai A, 2012a. Effects of sodium chloride and magnesium sulphate on glass concrete. IUP Journal of Structural Engineering 5 (3), 28–47.

Bajad M N, Modhera C and Desai A, 2014. Factors affecting the properties of conglasscrete. IOSR Journal of Mechanical and Civil Engineering 11 (2), 43–48.

Bajad M N, Modhera C D and Desai A K, 2012b. Resistance of concrete containing waste glass powder against $MgSO_4$ attack. NBMCW 1–16.

Barlow D, 2001. Glass recycling - the UK scene. In: Proceeding of the International Conference on Recycling and Reuse of Glass Cullet. University of Dundee, Scotland, UK, pp. 11–14.

Beletskii B I, Trofimov A N, Sventskaya N V, Zhuravleva N V and Pilepenko O S, 2013. Analysis of sheet-glass production in large-tonnage glassmaking furnaces. Glass and Ceramics 69 (9–10), 295–298.

Berry M, Stephens J and Cross D, 2011. Performance of 100% fly ash concrete with recycled glass aggregate. ACI Materials Journal 108 (4), 378–384.

BG, 2014. UK glass Manufacturing Sector Decarbonization Roadmap to 20 British Glass.

Bhat V V and Rao N B, 2014. Influence of glass powder on the properties of concrete. International Journal of Engineering Trends and Technology 16 (5), 196–199.

Bignozzi M C, Saccani A and Sandrolini F, 2009. Matt waste from glass separated collection: an eco-sustainable addition for new building materials. Waste Management 29 (1), 329–334.

Bignozzi M C, Saccani A, Barbieri L and Lancellotti I, 2015. Glass waste as supplementary cementing materials: the effects of glass chemical composition. Cement and Concrete Composites 55, 45–52.

Borhan T and Bailey C, 2014. Structural behaviour of basalt fibre reinforced glass concrete slabs. Materials and Structures 47 (1), 77–87.

Borosnyoi A, Kara P, Mlinarik L and Kase K, 2013. Performance of waste glass powder (WGP) supplementary cementitious material (SCM) – workability and compressive strength. Epitoanyag - Journal of Silicate Based and Composite Materials 65 (3), 90–94.

Breakspere R, Heath P, Jones J and Morgan R, 1980. New developments in waste glass utilization. Resources, Conservation and Recycling 3 (3), 233–248.

BS 882, 1992. Specification for Aggregates from Natural Sources for Concrete. BSI, London, UK.

BS 5930, 2015. Code of Practice for Ground Investigations. BSI, London, UK.

BS EN 1097-1, 2011. Tests for Mechanical and Physical Properties of Aggregates. Determination of the Resistance to Wear (Micro-Deval). BSI, London, UK.

BS EN 1097-2, 2010. Tests for Mechanical and Physical Properties of Aggregates. Methods for the Determination of Resistance of Fragmentation. BSI, London, UK.

BS EN 12620:2002+A1, 2008. Aggregates for Concrete. BSI, London, UK.

BS EN 13108-1, 2016. Bituminous Mixtures. Material Specifications. Asphalt Concrete. BSI, London, UK.

BS EN 196-2, 2013. Method of Testing Cement. Chemical Analysis of Cement. BSI, London, UK.

BS EN 197-1, 2011. Cement. Composition, Specifications and Conformity Criteria for Common Cements. BSI, London, UK.

BS EN 450-1, 2012. Fly Ash for Concrete. Definition, Specifications and Conformity Criteria. BSI, London, UK.

BS EN 13437, 2003. Packaging and Material Recycling. Criteria for Recycling Methods. Description of Recycling Processes and Flow Chart. BSI, London, UK.

BSI, 2003. Summary of PAS 101-Recovered Container Glass - Specification for Quality and Guidance for Good Practice in Collection. British Standards Institution.

Byars E A, Morales-Hernandez B and HuiYing Z, 2004. Waste glass as concrete aggregate and pozzolan Laboratory and industrial projects. Concrete 38 (1), 41–44.

Calmon J L, Sauer A S, Vieira G L and Teixeira J E S L, 2014. Effects of windshield waste glass on the properties of structural repair mortars. Cement and Concrete Composites 53, 88–96.

CapitaSymonds, 2007. Survey of Arisings and Use of Alternatives to Primary Aggregates in England, 2005-Other Materials. Capita Symonds, London, UK.

Carsana M, Frassoni M and Bertolini L, 2014. Comparison of ground waste glass with other supplementary cementitious materials. Cement and Concrete Composites 45, 39–45.

Cassar J and Camilleri J, 2012. Utilisation of imploded glass in structural concrete. Construction and Building Materials 29, 299–307.

Chaïd R, Kenai S, Zeroub H and Jauberthie R, 2015. Microstructure and permeability of concrete with glass powder addition conserved in the sulphatic environment. European Journal of Environmental and Civil Engineering 19 (2), 219–237.

Chen C H, Huang R, Wu J K and Yang C C, 2006a. Waste E - glass particles used in cementitious mixtures. Cement and Concrete Research 36 (3), 449–456.

Chen D, Masui H, Miyoshi H, Akai T and Yazawa T, 2006b. Extraction of heavy metal ions from waste coloured glass through phase separation. Waste Management 26 (9), 1017–1023.

Chen M J and Wong Y D, 2015. Porous asphalt mixture with a combination of solid waste aggregates. Journal of Materials in Civil Engineering 27 (6), 1–9.

Chen S, Chang C, Wang H and Huang W, 2011. Mixture design of high performance recycled liquid crystal glasses concrete (HPGC). Construction and Building Materials 25 (10), 3886–3892.

Chidiac S E and Mihaljevic S N, 2011. Performance of dry cast concrete blocks containing waste glass powder or polyethylene aggregates. Cement and Concrete Composites 33, 855–863.

Cholleti R N R, 2015. Use of Waste Glass as a Supplementary Cementitious Material (MSc thesis). Department of Civil and Environmental Engineering, The University of Utah, USA. 64 pp.

Coakley T, Hogan J and Cunningham D, 2002. A Strategy for Developing Recycling Markets in Ireland. Environmental Protection Agency, Wexford, Ireland.

Corbu O, Chira N, Constantinescu H and Szilagyi H, 2013. Ecological concrete by use of waste glass. In: 13th SGEM Geo Conference on Nano, Bio and Green – Technologies for a Sustainable Future, pp. 411–418.

Corinaldesi V, Gnappi G, Moriconi G and Montenero A, 2005. Reuse of ground waste glass as aggregate for mortars. Waste Management 25 (2), 197–201.

Cresswell D, 2007. Characterisation of Mineral Wastes, Resources and Processing Technologies – Integrated Waste Management for the Production of Construction Material. Department for Environment, Food & Rural Affairs.

CWC, 1996a. Best Practices in Glass Recycling - Sampling Procedures for Recycled Glass Construction Aggregate. Clean Washington Center.

CWC, 1996b. Best Practices in Glass Recycling - Types of Glass Crushing Equipment. Clean Washington Center.

CWC, 1996c. Methods for Sampling and Testing Recycled Glass. Clean Washington Center.

CWC, 1996d. Visual Inspection for Glass Construction Aggregate. Clean Washington Center.

Dacombe P J, Banks C J, Krivtsov V and Heaven S, 2004. Use of energy footprint analysis to determine the best options for management of glass from household waste. In: Proceedings of the International Conference on Sustainable Waste Management and Recycling: Glass Waste. Kingston University, London, UK, pp. 265–272.

Dahmardeh S R, Moghaddam M H M, Sargazi M S and Moghaddam S R M, 2015. Assessment the potential use of milled glass as Portland cement in self-compacting concrete. Applied Mathematics in Engineering, Management and Technology 3 (2), 345–353.

De Castro S and de Brito J, 2013. Evaluation of the durability of concrete made with crushed glass aggregates. Journal of Cleaner Production 41, 7–15.

De Jong B, Beerkens R, Van Nijatten P and Le Bourhis E, 2011. Glass, 1. Fundamentals. In: Ullmann's Encyclopaedia of Industrial Chemistry. Wiley-VCH, Weinheim, Germany.

Dhir R, Dyer T, Tang M, Cui Y and Wang L, 2005a. Towards maximising the use of glass cullet in concrete. Concrete 32–34.

Dhir R K, Paine K and Caliskan S, 2005b. Research information Digest 2: recycled glass. Concrete 44–46.

Dhir R, Dyer T and Tang M, 2009. Alkali-silica reaction in concrete containing glass. Materials and Structures 42 (10), 1451–1462.

Dhir R K, Dyer T, Tang M, Cui Y and Wang L, 2005c. Realising a High-value Sustainable Recycling Solution to the Glass Cullet Surplus Closing the Loop. Concrete Technology Unit, University of Dundee, UK. 155 pp.

Dhir R K, de Brito J, Mangabhai R and Lye C Q, 2016. Sustainable Construction Materials: Copper Slag. Woodhead Publishing, Elsevier, Cambridge. 322 pp.

Disfani M M, Arulrajah A, Bo M W and Sivakugan N, 2012. Environmental risks of using recycled crushed glass in road applications. Journal of Cleaner Production 20 (1), 170–180.

Du H, 2011. Glass Sand Concrete and "Sandless Concrete" (Ph.D. thesis). Department of Civil and Environmental Engineering, National University of Singapore, Singapore. 234 pp.

Du H and Tan K H, 2013. Use of waste glass as sand in mortar: Part II – alkali– silica reaction and mitigation methods. Cement and Concrete Composites 35 (1), 118–126.

Du H and Tan K H, 2014a. Concrete with recycled glass as fine aggregates. ACI Materials Journal 111 (1), 47–57.

Du H and Tan K H, 2014b. Waste glass powder as cement replacement in concrete. Journal of Advanced Concrete Technology 12 (11), 468–477.

Du H and Tan K, 2015. Transport properties of concrete with glass powder as supplementary cementitious material. ACI Materials Journal 112 (3), 429.

Dumitru I, Song T, Bornstein B, Brooks P and Moss J, 2013. Field trials using recycled glass as natural sand replacement and powdered glass as cementitious materials replacement in concrete pavement. In: Third International Conference on Sustainable Construction Materials and Technologies, pp. 1–9.

Dumitru I, Song T, Caprar V, Brooks P and Moss J, 2010. Incorporation of recycled glass for durable concrete. In: Second International Conference on Sustainable Construction Materials and Technologies, Ancona, Italy, pp. 1–9.

Dyer T and Dhir R, 2001. Chemical reactions of glass cullet used as cement component. Journal of Materials in Civil Engineering 13 (6), 412–417.

Dyer T D and Dhir R K, 2010. Evaluation of powdered glass cullet as a means of controlling harmful alkali- silica reaction. Magazine of Concrete Research 62 (10), 749–759.

Efremenkov V V, 2015. Improvement of glass batch conveyer lines. Glass and Ceramics 72 (7–8), 280–284.

Efremenkov V V and Subbotin K Y, 2015. Particularities of computing cullet consumption in the production of glass containers. Glass and Ceramics 72 (5–6), 184–186.

Egosi N G, 1992. Mixed broken glass processing solutions. In: Proceedings of the National Convention on the Utilization of Waste Materials in Civil Engineering Construction. American Society of Civil Engineers, New York, USA, pp. 71–80.

EN 197-1, 2011. Cement Composition, Specifications and Conformity Criteria for Common Cements. Comité Européen de Normalisation (CEN), Brussels, Belgium.

Enviros, 2004. Recycled Glass Market Study & Standards Review – 2004 Update. Waste & Resources Action Programme, Banbury, Oxon, UK.

ESYS Consulting Ltd, 2006. The Costs & Operational Implications of Kerbside Glass Collections. The Waste & Resources Action Programme, Banbury, Oxon, UK.

Eurostat, 2016. Waste Statistics in Europe. Available from: epp.eurostat.ec.europa.eu.

Federico L M and Chidiac S E, 2009. Waste glass as a supplementary cementitious material in concrete – critical review of treatment methods. Cement and Concrete Composites 31 (8), 606–610.

Fisk J T, 2014. Glass production losses originating from contaminants in cullet and raw materials. In: Proceedings of the 74th Conference on Glass Problems, Columbus, Ohio, USA, pp. 15–24.

Ganiron Jr. T, 2012. Recycled window glass for non-load bearing walls. International Journal of Innovation, Management and Technology 3 (6), 725–730.

Georgiadis A S, Anagnostopoulos N S and Sideris K K, 2007. Mechanical characteristics of self-compacting concretes produced with different filler materials. In: 5th International RILEM Symposium on Self-compacting Concrete, Ghent, Belgium, pp. 611–618.

Girbes I, Suesta C, Tendero M J L and Belena I, 2004. Recycled glass cullet as partial Portland cement replacement. In: Proceedings of the International Conference on Sustainable Waste Management and Recycling: Glass Waste, London, UK. Thomas Telford, London, UK, pp. 142–148.

Glass Packaging Institute GPI, 2015. Glass recycling facts. GPI Glass Packaging Institute. Available from: http://www.gpi.org.

Glüsing A K and Conradt R, 2001. Dissolution kinetics of impurities in recycled cullet. In: Proceedings of the International Symposium on Recycling and Reuse of Glass Cullet. University of Dundee, Scotland, UK, pp. 29–41.

Guo M, Chen Z, Ling T C and Poon C S, 2015. Effects of recycled glass on properties of architectural mortar before and after exposure to elevated temperatures. Journal of Cleaner Production 101, 158–165.

Handout-10, 2012. Printed Circuit Board Recycling Methods. Workshop Materials on WEEE Management in Taiwan).

Hartley A, 2004. Recovered Container Glass: Development of Test Methods and Inorganic Contamination Limits. Waste & Resources Action Programme, Banbury, Oxon, UK.

Holcroft C and Pudner M, 2007. Glass market development assessments for the English regions and the nations. Waste & Resources Action Programme, Banbury, Oxon, UK.

Huang L, Wang H and Wang S, 2015. A study of the durability of recycled green building materials in lightweight aggregate concrete. Construction and Building Materials 96, 353–359.

Hurley J, 2003. Research into Waste Glass, Window and Door Frames from the Demolition and Replacement Window Industries. Waste & Resources Action Programme, Banbury, Oxon, UK.

Idir R, Cyr M and Tagnit-Hamou A, 2009. Use of waste glass as powder and aggregate in cement-based materials. In: SBEIDCO-1st International Conference on Sustainable Built Environment Infrastructures in Developing Countries, ENSET, Algeria, pp. 109–116.

Idir R, Cyr M and Tagnit-Hamou A, 2010a. Use of fine glass as ASR inhibitor in glass aggregate mortars. Construction and Building Materials 24 (7), 1309–1312.

Idir R, Cyr M and Tagnit-Hamou A, 2010b. Use of waste glass in cement-based materials. In: Proceedings of the International Conference Waste Materials in Construction (Wascon 2009), Lyon, France, pp. 1–7.

Idir R, Cyr M and Tagnit-Hamou A, 2011. Pozzolanic properties of fine and coarse color- mixed glass cullet. Cement and Concrete Composites 33, 19–29.

Isler J W, 2012. Assessment of Concrete Masonry Units Containing Aggregate Replacements of Waste Glass and Rubber Tire Particles (MSc thesis). University of Colorado Denver, USA. 108 pp.

Ismail Z Z and AL-Hashmi E A, 2009. Recycling of waste glass as a partial replacement for fine aggregate in concrete. Waste Management 29 (2), 655–660.

Jain J A and Neithalath N, 2010. Chloride transport in fly ash and glass powder modified concretes – influence of test methods on microstructure. Cement and Concrete Composites 32 (2), 148–156.

Jang H, Jeon S, So H and So S, 2015. Properties of different particle size of recycled TFT- LCD waste glass powder as a cement concrete binder. International Journal of Precision Engineering and Manufacturing 16 (12), 2591–2597.

Jangid J B and Saoji A, 2014. Experimental investigation of waste glass powder as the partial replacement of cement in concrete production. IOSR Journal of Mechanical and Civil Engineering 55–60.

Kamali M and Ghahremaninezhad A, 2015. Effect of glass powders on the mechanical and durability properties of cementitious materials. Construction and Building Materials 98, 407–416.

Kamali M and Ghahremaninezhad A, 2016. An investigation into the hydration and microstructure of cement pastes modified with glass powders. Construction and Building Materials 112, 915–924.

Kang H Y and Schoenung J M, 2005. Electronic waste recycling: a review of U.S. infrastructure and technology options. Resources, Conservation and Recycling 45 (4), 368–400.

Kara P, 2013a. Application of recycled concrete aggregates containing waste glass powder/ suspension and bottom ash as a cement component in concrete. In: Engineering a Concrete Future: Technology, Modeling and Construction. Israel, pp. 109–112.

Kara P, 2013b. The influence of waste glass slurry on the properties of concrete. International Journal of Application or Innovation in Engineering and Management (IJAIEM) 2 (8), 325–330.

Kara P, 2015. Performance of lamp glass waste powder (LGWP) as supplementary cementitious material (SCM) – viscosity and electrical conductivity. Journal of Silicate Based and Composite Materials 67 (1), 12–18.

Kara P, Korjakins A and Kovalenko K, 2012. The usage of fluorescent waste glass powder in concrete. The Construction Specifier 13, 26.

Karamberi A and Moutsatsou A, 2005. Participation of coloured glass cullet in cementitious materials. Cement and Concrete Composites 27 (2), 319–327.

Karamberi A, Kerinis M and Moutsatsou A, 2004. Innovative reuse of glass cullet in cement products. Glass Waste 1, 133–140.

Khmiri A, Chaabouni M and Samet B, 2013. Chemical behaviour of ground waste glass when used as partial cement replacement in mortars. Construction and Building Materials 44, 74–80.

Khmiri A, Samet B and Chaabouni M, 2012. Assessement of the waste glass powder pozzolanic activity by different methods. International Journal of Research and Reviews in Applied Sciences 10 (2), 322–328.

Kim J, Moon J, Shim J W, Sim J, Lee H and Zi G, 2014a. Durability properties of a concrete with waste glass sludge exposed to freeze-and- thaw condition and de- icing salt. Construction and Building Materials 66, 398–402.

Kim J, Yi C and Zi G, 2015. Waste glass sludge as a partial cement replacement in mortar. Construction and Building Materials 75, 242–246.

Kim K, Kim K and Hwang J, 2014b. Thin film transistor-liquid crystal display cullet: a raw material for production of commercial soda lime silicate glasses. Journal of Cleaner Production 79, 276–282.

Kim W and Soh Y, 2002. Properties of unsaturated polyester mortars using crushed waste glass. Journal of Asian Architecture and Building Engineering 1 (1), 7–12.

Klevbo G, 1998. Recycled glass as filler in ready-mixed concrete. In: Proceedings of the 12th European Ready Mixed Concrete Congress, Lisbon, pp. 673–680.

Kogel J E, Trivedi N C, Baker J M and Krukowski S T, 2006. Industrial Minerals and Rocks. Society for Mining, Metallurgy and Exploration, Littleton, CO, USA.

Koh C J, 2014. Characterisation of shape of fine recycled crushed coloured glass and the effect on the properties of structural concrete when used as a fine aggregate replacement (Ph.D. thesis). University of Wolverhampton, UK. 301 pp.

Korjakins A, Šahmenko G and Būmanis G, 2009. Bore-silicate glass waste of lamp as a micro-filler for concrete. The Construction Specifier 2 (10), 132–139.

Korjakins A, Šahmenko G and Būmanis G, 2010. Recycling of bore-silicate glass waste by utilization as filler in concrete. In: 10th International Scientific Conference VSU, pp. 13–18.

Korjakins A, Šahmenko G, Bajare D and Būmanis Ģ, 2011. Effect of ground glass fineness on physical and mechanical properties of concrete. In: Proceedings of the 10th International Congress for Applied Mineralogy (ICAM), Springer, Berlin, pp. 395–402.

Korjakins A, Genadik S and Bumanis G, 2012. Utilisation of borosilicate glass waste as a micro-filler for concrete. Journal of Civil Engineering and Architecture 6 (6), 876–883.

Kou S C and Poon C S, 2009. Properties of self-compacting concrete prepared with recycled glass aggregate. Cement and Concrete Composites 31 (2), 107–113.

Kou S C and Poon C S, 2013. A novel polymer concrete made with recycled glass aggregates, fly ash and metakaolin. Construction and Building Materials 41, 146–151.

Kou S C and Xing F, 2012. The effect of recycled glass powder and reject fly ash on the mechanical properties of fibre-reinforced ultrahigh performance concrete. Advances in Materials Science and Engineering 2012, 1–8.

Laldji S, Fares G and Tagnit-Hamou A, 2007. Glass frit for concrete structures: a new, alternative cementitious material. Canadian Journal of Civil Engineering 34 (7), 793–802.

Lam C S, Poon C S and Chan D, 2007. Enhancing the performance of pre- cast concrete blocks by incorporating waste glass – ASR consideration. Cement and Concrete Composites 29 (8), 616–625.

Le Bourhis E, 2008. Glass: Mechanics and Technology. Wiley-VCH, Weinheim, Germany.

Lee C, Chen C, Huang R and Wu J, 2008. Properties of concrete made with recycled epoxy/glass particles. Journal of the Chinese Institute of Engineers 31 (6), 1061–1067.

Lee C T, 2013. Production of alumino-borosilicate foamed glass body from waste LCD glass. Journal of Industrial and Engineering Chemistry 19 (6), 1916–1925.

Lee G, 2011. Effects of Using Waste Glass with Different Particle Sizes to Replace Sand in Pre-cast Concrete Blocks (Ph.D. thesis). Department of Civil and Structural Engineering, The Hong Kong Polytechnic University, Hong Kong. 113 pp.

Lee G, Ling T, Wong Y and Poon C, 2011. Effects of crushed glass cullet sizes, casting methods and pozzolanic materials on ASR of concrete blocks. Construction and Building Materials 25 (5), 2611–2618.

Lee G, Poon C S, Wong Y L and Ling T C, 2013. Effects of recycled fine glass aggregates on the properties of dry- mixed concrete blocks. Construction and Building Materials 38, 638–643.

Lee S and Lee J J, 2016. Performance of cementitious composites incorporating ground TFT-LCD waste glass. Journal of Testing and Evaluation 44 (1), 213–221.

Leong M S and Hurley J, 2004. Flat glass waste in the domestic window replacement industry. In: Proceedings of the International Conference on Sustainable Waste Management and Recycling: Glass Waste. Kingston University, London, UK, pp. 249–255.

Lim A, 2014. The Use of Waste Glass as a Partial Substitution for Fine Aggregate in Concrete Design. School of Engineering and IT, Charles Darwin University. 44 pp.

Limbachiya M C, 2009. Bulk engineering and durability properties of washed glass sand concrete. Construction and Building Materials 23 (2), 1078–1083.

Limbachiya M, Meddah M S and Fotiadou S, 2012. Performance of granulated foam glass concrete. Construction and Building Materials 28, 758–768.

Lin K L, Huang W, Shie J L, Lee T C, Wang K S and Lee C H, 2009. The utilization of thin film transistor liquid crystal display waste glass as a pozzolanic material. Journal of Hazardous Materials 163 (2), 916–921.

Ling T and Poon C, 2011a. Properties of architectural mortar prepared with recycled glass with different particle sizes. Materials and Design 32 (5), 2675–2684.

Ling T and Poon C, 2011b. Utilization of recycled glass derived from cathode ray tube glass as fine aggregate in cement mortar. Journal of Hazardous Materials 192 (2), 451–456.

Ling T and Poon C, 2012a. A comparative study on the feasible use of recycled beverage and CRT funnel glass as fine aggregate in cement mortar. Journal of Cleaner Production 29–30, 46–52.

Ling T and Poon C, 2012b. Feasible use of recycled CRT funnel glass as heavyweight fine aggregate in barite concrete. Journal of Cleaner Production 33, 42–50.

Ling T and Poon C, 2012c. Stress–strain behaviour of fire exposed self- compacting glass concrete. Fire and Materials 37 (4), 297–310.

Ling T and Poon C, 2013. Effects of particle size of treated CRT funnel glass on properties of cement mortar. Materials and Structures 46 (1), 25–34.

Ling T and Poon C, 2014a. Use of recycled CRT funnel glass as fine aggregate in dry-mixed concrete paving blocks. Journal of Cleaner Production 68, 209–216.

Ling T and Poon C, 2014b. Feasible use of large volumes of GGBS in 100% recycled glass architectural mortar. Cement and Concrete Composites 53, 350–356.

Ling T and Poon C, 2014c. Use of CRT funnel glass in concrete blocks prepared with different aggregate-to- cement ratios. Green Materials 2 (1), 43–51.

Ling T and Poon C, 2014d. High temperatures properties of barite concrete with cathode ray tube funnel glass. Fire and Materials 38 (2), 279–289.

Ling T, Poon C and Kou S, 2011. Feasibility of using recycled glass in architectural cement mortars. Cement and Concrete Composites 33, 848–854.

Ling T, Poon C and Kou S, 2012. Influence of recycled glass content and curing conditions on the properties of self-compacting concrete after exposure to elevated temperatures. Cement and Concrete Composites 34, 265–272.

Liu M, 2009. Wider Application of Additions in Self-compacting Concrete (Ph.D. thesis). Department of Civil, Environmental and Geomatic Engineering, University College London, London, UK. 392 pp.

Liu M, 2011. Incorporating ground glass in self-compacting concrete. Construction and Building Materials 25 (2), 919–925.

Liu S, Xie G and Rao M, 2013. Effect of waste glass powder on properties and microstructure of ultrahigh performance cement based materials. Materials Research Innovations 17, 210–214.

Long L, Sun S, Zhong S, Dai W, Liu J and Song W, 2010. Using vacuum pyrolysis and mechanical processing for recycling waste printed circuit boards. Journal of Hazardous Materials 177 (1), 626–632.

Madandoust R and Ghavidel R, 2013. Mechanical properties of concrete containing waste glass powder and rice husk ash. Biosystems Engineering 116, 113–119.

Maier P L and Durham S A, 2012. Beneficial use of recycled materials in concrete mixtures. Construction and Building Materials 29, 428–437.

Malik M I, Bashir M, Ahmad S, Tariq T and Chowdhary U, 2013. Study of concrete involving use of waste glass as partial replacement of fine aggregates. IOSR Journal of Engineering 3 (7), 8–13.

Malik M I, Manzoor A, Ahmad B, Asima S, Ali R and Bashir M, 2014. Positive potential of partial replacement of fine aggregates by waste glass (<600 micron) in concrete. International Journal of Civil Engineering & Technology 5 (11), 146–153.

Maraghechi H, Shafaatian S, Fischer G and Rajabipour F, 2012. The role of residual cracks on alkali silica reactivity of recycled glass aggregates. Cement and Concrete Composites 34 (1), 41–47.

Maschio S, Tonello G and Furlani E, 2013. Recycling glass cullet from waste CRTs for the production of high strength mortars. The Journal of Wealth Management 2013, 1–8.

Matos A M and Sousa-Coutinho J, 2012. Durability of mortar using waste glass powder as cement replacement. Construction and Building Materials 36, 205–215.

Matos A M and Sousa-Coutinho J, 2016a. ASR and sulphate performance of mortar containing industrial waste. Structural Concrete 17 (1), 84–95.

Matos A M and Sousa-Coutinho J, 2016b. Waste glass powder in cement: macro and micro scale study. Advances in Cement Research 28 (7), 423–432.

Matos A, Ramos T, Nunes S and Sousa-Coutinho J, 2015. Durability enhancement of SCC with waste glass powder. Materials Research 1–8.

Mavilia L, Corigliano F, Dhir R K, Limbachiya M C and Dyer T D, 2001a. High added value products from off-quality waste glass. In: Proceedings of the International Symposium on Recycling and Reuse of Glass Cullet. University of Dundee, Scotland, UK, pp. 63–73.

Mavilia L, Di Bella G, Corigliano F, Dhir R K, Limbachiya M C and Dyer T D, 2001b. Characterization and valorization of the solid residue from glass extraction with an alkali solution. In: Proceedings of the International Symposium on Recycling and Reuse of Glass Cullet. University of Dundee, Scotland, UK, pp. 55–61.

Mayer P, 2004. State of art separation - technology for the processing of recycled glass. In: Proceedings of the International Conference on Sustainable Waste Management and Recycling: Glass Waste. Kingston University, London, UK, pp. 55–58.

Metwally I M, 2007. Investigations on the performance of concrete made with blended finely milled waste glass. Advances in Structural Engineering 10 (1), 47–53.

Meylan G, Ami H and Spoerri A, 2014. Transitions of municipal solid waste management. Part II: Hybrid life cycle assessment of Swiss glass-packaging disposal. Resources, Conservation and Recycling 86, 16–27.

Miranda Jr. E J P, Bezerra H J C L, Piliti F S and Paiva A E M, 2014. Increasing the compressive strength of Portland cement concrete using flat glass powder. Materials Research 17, 45–50.

Mirzahosseini M and Riding K A, 2014. Effect of curing temperature and glass type on the pozzolanic reactivity of glass powder. Cement and Concrete Research 58, 103–111.

Mirzahosseini M and Riding K A, 2015. Influence of different particle sizes on reactivity of finely ground glass as supplementary cementitious material (SCM). Cement and Concrete Composites 56, 95–105.

Mitra S, Srimany A and Dey B G, 2016. A practical study on effective utilisation of waste glass in concrete. International Journal of Science, Engineering and Technology Research (IJSETR) 5 (3), 694–696.

Narayana D and Mailar G, 2015. Experimental investigation on the properties and durability of partially replaced recycled aggregates, glass powder concrete for cost effective construction. International Journal of Research in Engineering & Advanced Technology 2 (6), 135–141.

Nassar R and Soroushian P, 2011. Field investigation of concrete incorporating milled waste glass. Journal of Solid Waste Technology and Management 37 (4), 307–319.

Nassar R and Soroushian P, 2012a. Strength and durability of recycled aggregate concrete containing milled glass as partial replacement for cement. Construction and Building Materials 29, 368–377.

Nassar R and Soroushian P, 2012b. Green and durable mortar produced with milled waste glass. Magazine of Concrete Research 64 (7), 605–615.

Nassar R and Soroushian P, 2013. Use of milled waste glass in recycled aggregate concrete. Proceedings - Institution of Civil Engineers 166 (5), 304–315.

Neithalath N and Schwarz N, 2009. Properties of cast-in-place concrete and precast concrete blocks incorporating waste glass powder. The Open Construction & Building Technology Journal 3, 42–51.

Neithalath N, 2008. Quantifying the effects of hydration enhancement and dilution in cement pastes containing coarse glass powder. Journal of Advanced Concrete Technology 6 (3), 397–408.

Neville A M, 1995. Properties of Concrete, fourth ed. Longman, Essex, UK. 844 pp.

Neville A M, Dilger W H and Brooks J J, 1983. Creep of Plain and Structural Concrete. Construction Press, London, UK.

Niang A, Roy N and Tagnit-Hamou A, 2015. Structural behavior of concrete incorporating glass powder used in reinforced concrete columns. Journal of Structural Engineering 141 (3), 1–10.

Nunes S, Matos A M, Duarte T, Figueiras H and Sousa-Coutinho J, 2013. Mixture design of self- compacting glass mortar. Cement and Concrete Composites 43, 1–11.

Nwaubani S O and Poutos K I, 2013. The influence of waste glass powder fineness on the properties of cement mortars. International Journal of Application or Innovation in Engineering and Management (IJAIEM) 2 (2), 110–116.

Oliveira L A P, Gomes J P C and Santos P, 2008. Mechanical and durability properties of concrete with ground waste glass sand. In: 11 DBMC International Conference on Durability of Building Materials and Components, Istanbul, Turkey, pp. 1–8.

Oliveira R, De Brito J and Veiga R, 2013. Incorporation of fine glass aggregates in renderings. Construction and Building Materials 44, 329–341.

Omran A and Tagnit-Hamou A, 2016. Performance of glass-powder concrete in field applications. Construction and Building Materials 109, 84–95.

Oyekan G L and Oyelade O A, 2011. Crushed waste glass as a partial replacement of cement in normal concrete production with sugar added as an admixture. Journal of Engineering and Applied Sciences 6 (6), 369–372.

Özkan O and Yüksel İ, 2008. Studies on mortars containing waste bottle glass and industrial by-products. Construction and Building Materials 22 (6), 1288–1298.

Parghi A and Alam M S, 2016. Physical and mechanical properties of cementitious composites containing recycled glass powder (RGP) and styrene butadiene rubber (SBR). Construction and Building Materials 104, 34–43.

Park S and Lee B, 2004. Studies on expansion properties in mortar containing waste glass and fibers. Cement and Concrete Research 34 (7), 1145–1152.

Park S B, Lee B C and Kim J H, 2004. Studies on mechanical properties of concrete containing waste glass aggregate. Cement and Concrete Research 34 (12), 2181–2189.

Pascoe R D, Barley R W and Child P R, 2001. Autogenous grinding of glass cullet in a stirred mill. In: Proceedings of the International Symposium on Recycling and Reuse of Glass Cullet. University of Dundee, Scotland, UK, pp. 15–27.

Patagundi B R and Prakash K B, 2015. Use of glass powder as mineral admixture in steel fibre reinforced concrete. International Journal of Research in Engineering. Science and Technologies (IJRESTs) 1 (1), 25–30.

Pavoine A, Harbec D, Chaussadent T, Tagnit-Hamou A and Divet L, 2014. Impact of alternative cementitious material on mechanical and transfer properties of concrete. ACI Materials Journal 111 (3), 251–261.

PD CEN/TR 13688, 2008. Packaging. Material Recycling. Report on Requirements for Substances and Materials to Prevent a Sustained Impediment to Recycling. BSI, London, UK.

Pedersen B, 2004. Alkali-reactive and Inert Fillers in Concrete. Rheology of Fresh Mixtures and Expansive Reactions (Ph.D. thesis). Department of Structural Engineering, Norwegian University of Science and Technology, Norway. 292 pp.

Penacho P, de Brito J and Veiga R, 2014. Physico-mechanical and performance characterization of mortars incorporating fine glass waste aggregate. Cement and Concrete Composites 50, 47–59.

Peyvandi A, Soroushian P and Nassar R, 2013. Recycled glass concrete. Concrete International 35 (1), 29–32.

Phillips J C, Cahn D and Keller G, 1972. Refuse glass aggregate in Portland cement concrete. In: Proceeding of the Third Mineral Waste Utilization Symposium, Chicago, USA, pp. 385–390.

Polley C, 1996. The Effects of Waste Glass Aggregate on the Strength and Durability of Portland Cement Concrete (MSc thesis). Department of Civil and Environmental Engineering, University of Wisconsin, Madison, USA. 174 pp.

Polley C, Cramer S M and Cruz D L, 1998. Potential for using waste glass in Portland cement concrete. Journal of Materials in Civil Engineering 10 (4), 210–219.

Poon C and Chan D, 2007. Effects of contaminants on the properties of concrete paving blocks prepared with recycled concrete aggregates. Construction and Building Materials 21 (1), 164–175.

Poutos K H, Alani A M, Walden P J and Sangha C M, 2008. Relative temperature changes within concrete made with recycled glass aggregate. Construction and Building Materials 22 (4), 557–566.

Priscilla M and Naik P A, 2014. Strength and durability study on recycled aggregate concrete using glass powder. International Journal of Engineering Trends and Technology 11 (5), 259–264.

Proshin A, Demjanova V, Kalashnicov D and Grintsova O, 2005. Super heavy high-strength concrete for protection against radiation. Asian Journal of Civil Engineering (Building and Housing) 6 (1–2), 67–73.

Rajabipour F, Shafaatian S and Wright J, 2012. Utilising Coal Fly Ash and Recycled Glass in Developing Green Concrete Materials. The Pennsylvania State University, PA, USA. PUS-2010-07, 51 pp.

Rajabipour F, Shafaatian S and Wright J, 2012. Utilizing Coal Fly Ash and Recycled Glass in Developing Green Concrete Materials. The Pennsyvania State University, Washington DC, USA. Report No.: DTRT07-G-0003, 51 pp.

Ralph R, 2006. Queensland Used Glass Sourcing Study. Queensland Environmental Protection Agency, Australia.

Ralph R and Prince A, 2006. Market Development Study Used Glass. Western Australia Waste Management Board, Australia.

Rocchetti L and Beolchini F, 2014. Environmental burdens in the management of end-of-life cathode ray tubes. Waste Management 34 (2), 468–474.

Romero D, James J, Mora R and Hays C D, 2013. Study on the mechanical and environmental properties of concrete containing cathode ray tube glass aggregate. Waste Management 33 (7), 1659–1666.

Saccani A and Bignozzi M C, 2010. ASR expansion behavior of recycled glass fine aggregates in concrete. Cement and Concrete Research 40 (4), 531–536.

Sagoe-Crentsil K, Brown T and Taylor A, 2001. Guide Specification: Recycled Glass as Sand Replacement in Premix Concrete. CSIRO Building Construction and Engineering, Victoria, Australia. 44 pp.

Sahmenko G, Toropovs N, Sutinis M and Justs J, 2014. Properties of high performance concrete containing waste glass micro-filler. Key Engineering Materials 604, 161–164.

Sahmenko G, Korjakins A, Bumanis G, 2010. Concrete with microfiller obtained from recycled lamp glass. In: the 10th International Conference on Modern Building Materials, Structures and Techniques, Vilnius, Lithuania, 5 pp.

Salehuddin N S D, 2012. Compressive Strength of Concrete Using Glass Powder as a Filler Replacement (BE thesis). Faculty of Civil Engineering & Earth Resources, Universiti Malaysia Pahang, Malaysia. 16 pp.

Scalet B M, Garcia Munoz M, Sissa A Q, Roudier S and Delgado Sancho L, 2013. Best Available Techniques (BAT) Reference Document for Manufacture of Glass. European Commission Joint Research Centre, Seville, Spain. JRC Reference Report, Industrial Emissions Directive 2010/75/EU.

Schott, 2014. Technical Glasses: Physical and Technical Properties. SCHOTT North America, Duryea, PA, USA. 74 pp.

Schwarz N and Neithalath N, 2008. Influence of a fine glass powder on cement hydration: comparison to fly ash and modelling the degree of hydration. Cement and Concrete Research 38 (4), 429–436.

Schwarz N, Cam H and Neithalath N, 2008. Influence of a fine glass powder on the durability characteristics of concrete and its comparison to fly ash. Cement and Concrete Composites 30 (6), 486–496.

Schwarz N, Dubois M and Neithalath N, 2007. Electrical conductivity based characterization of plain and coarse glass powder modified cement pastes. Cement and Concrete Composites 29 (9), 656–666.

Searles C, 2004. London remade - innovative recyclables market development programme. In: Proceedings of the International Conference on Sustainable Waste Management and Recycling: Glass Waste. Kingston University, London, UK, pp. 241–247.

Seju D D, Makwana A J, Vankar D K, Patel B D and Patel V M, 2015. Effect of glass powder on compressive strength of concrete (BE thesis). Department of Applied Mechanics, Government Engineering College. 31 pp.

Sekhar S, 2011. Studies on relationship of water- powder ratio and compressive strength of glass fibre self-compacted concrete. i-Manager's Journal on Civil Engineering 1 (4), 44–48.

SEPA, 2003. The National Waste Plan 2003. Scottish Environmental Protection Agency, Scotland, UK.

Serpa D, De Brito J and Pontes J, 2015. Concrete made with recycled glass aggregates: mechanical performance. ACI Materials Journal 112 (1), 29–38.

Serpa D, Santos Silva A, De Brito J, Pontes J and Soares D, 2013. ASR of mortars containing glass. Construction and Building Materials 47, 489–495.

Shafaatian S M, Akhavan A, Maraghechi H and Rajabipour F, 2013. How does fly ash mitigate alkali-silica reaction (ASR) in accelerated mortar bar test (ASTM C1567)? Cement and Concrete Composites 37, 143–153.

Shao Y and Lehoux P, 2001. Feasibility of using ground waste glass as a cementitious material. In: Proceedings of the International Syposium: Recycling and Reuse of Glass Cullet, Dundee, Scotland, UK, pp. 209–219.

Shao Y, Lefort T, Moras S and Rodriguez D, 2000. Studies on concrete containing ground waste glass. Cement and Concrete Research 30, 91–100.

Sharif M, Mehmood A, Yousaf M, Ghaffar A and Goraya R, 2014. Performance of concrete containing waste powdered glass as partial replacement of fine aggregate. NED University Journal of Research 11 (4), 43–49.

Sharifi Y, Afshoon I and Firoozjaie Z, 2015. Fresh properties of self- compacting concrete containing ground waste glass microparticles as cementing material. Journal of Advanced Concrete Technology 13 (2), 50–66.

Shayan A and Xu A, 2004. Value-added utilisation of waste glass in concrete. Cement and Concrete Research 34 (1), 81–89.

Shayan A and Xu A, 2006. Performance of glass powder as a pozzolanic material in concrete: a field trial on concrete slabs. Cement and Concrete Research 36 (3), 457–468.

Shi C and Wu Y, 2005. Mixture proportioning and properties of self- consolidating lightweight concrete containing glass powder. ACI Materials Journal 102 (5), 355–363.

Shi C, 2009. Corrosion of glasses and expansion mechanism of concrete containing waste glasses as aggregates. Journal of Materials in Civil Engineering 21 (10), 529–534.

Shi C, Wu Y, Riefler C and Wang H, 2005. Characteristics and pozzolanic reactivity of glass powders. Cement and Concrete Research 35 (5), 987–993.

Siad H, Lachemi M, Sahmaran M and Hossain K M A, 2016. Effect of glass powder on sulfuric acid resistance of cementitious materials. Construction and Building Materials 113, 163–173.

Singh G, Singh A K, Bhaskar A and Attree A S, 2014. A critical study of effectiveness of waste glass powder in concrete. International Archive of Applied Sciences and Technology 5 (3), 30–35.

Snow W and Dickinson J, 2005. Glass Mountains: Options for Glass Recycling in Otago. Envision, New Zealand.

Sobolev K, Turker P, Sololeva S and Iscioglu G, 2007. Utilization of waste glass in ECO-cement: strength properties and microstructural observations. Waste Management 27, 971–976.

Soroushian P, 2012. Towards Broad Use of Recycled Glass Concrete on Msu Campus. Michigan State University, Michigan. 48 pp.

Soyer E, Akgiray Ö, Eldem N Ö and Saatçi A M, 2010. Crushed recycled glass as a filter medium and comparison with silica sand. Clean - Soil, Air, Water 38 (10), 927–935.

Spathi C, Vandeperre L and Cheeseman C, 2015. Production of lightweight fillers from waste glass and paper sludge ash. Waste and Biomass Valorization 6 (5), 875–881.

Stainer L, 2006. Developing Markets for Container & Plate (Flat) Glass in Rural Areas. Waste & Resources Action Programme, Banbury, Oxon, UK.

Su N and Chen J S, 2002. Engineering properties of asphalt concrete made with recycled glass. Resources, Conservation and Recycling 35 (4), 259–275.

Tagnit-Hamou A and Bengougam A, 2012. The use of glass powder as supplementary cementitious material. Concrete International 34 (3), 56–61.

Tagnit-Hamou A, Soliman N, Omran A, Mousa M, Gauvreau N and Provencher M, 2015. Novel ultra- high- performance glass concrete. Concrete International 41–47.

Taha B and Nounu G, 2008. Properties of concrete contains mixed colour waste recycled glass as sand and cement replacement. Construction and Building Materials 22 (5), 713–720.

Taha B and Nounu G, 2009. Utilizing waste recycled glass as sand/cement replacement in concrete. Journal of Materials in Civil Engineering 21 (12), 709–721.

Tan K H and Du H, 2013. Use of waste glass as sand in mortar: Part I – fresh, mechanical and durability properties. Cement and Concrete Composites 35 (1), 109–117.

Tang M C, Wang L and Dhir R K, 2005. Engineering and durability of concrete containing waste glass. In: Dhir R K, Dyer T D and Newlands M D (Eds.), Achieving Sustainability in Construction. Thomas Telford, pp. 219–228.

Tejaswi S S, Rao R C, Vidya B and Renuka J, 2015. Experimental investigation of waste glass powder as partial replacement of cement and sand in concrete. IUP Journal of Structural Engineering 8 (4), 14–22.

Terro M J, 2006. Properties of concrete made with recycled crushed glass at elevated temperatures. Building and Environment 41 (5), 633–640.

Tognonvi M T, Zidol A, Aïtcin P and Tagnit-Hamou A, 2015. Aging of glass powder surface. Journal of Non-crystalline Solids 427, 175–183.

Topcu I B, Boga A R and Bilir T, 2008. Alkali–silica reactions of mortars produced by using waste glass as fine aggregate and admixtures such as fly ash and Li_2CO_3. Waste Management 28 (5), 878–884.

Tsai C L and Krogmann U, 2013. Material flows and energy analysis of glass containers discarded in New Jersey, USA. Journal of Industrial Ecology 17 (1), 129–142.

Tsai C L, Krogmann U and Strom P F, 2009. Handling leachate from glass cullet stockpiles. Waste Management 29 (4), 1296–1305.

Tuncan M, Karasu B and Yalcin M, 2001. The suitability for using glass and fly ash in Portland cement concrete. In: Proceedings of the Eleventh International Offshore and Polar Engineering Conference, Stavanger, Norway, pp. 146–152.

Turgut P and Yahlizade E, 2009. Research into concrete blocks with waste glass. International Journal of Civil & Environmental Engineering 1 (4), 203–209.

Turgut P, 2008a. Limestone dust and glass powder wastes as new brick material. Materials and Structures 41 (5), 805–813.

Turgut P, 2008b. Properties of masonry blocks produced with waste limestone sawdust and glass powder. Construction and Building Materials 22 (7), 1422–1427.

Turgut P, 2013. Fly ash block containing limestone and glass powder wastes. KSCE Journal of Civil Engineering 17 (6), 1425–1431.

Türkmen İ and Fındık S B, 2013. Several properties of mineral admixtured lightweight mortars at elevated temperatures. Fire and Materials 37 (5), 337–349.

USEPA, 1992. Markets for Recovered Glass. United States Environmental Protection Agency, Washington DC, USA.

Vaitkevicius V, Serelis E and Hilbig H, 2014. The effect of glass powder on the microstructure of ultra high performance concrete. Construction and Building Materials 68, 102–109.

Wang F, Gong H, Jia S, Zhang B and Zhang C, 2015. Research on waste glass concrete basic mechanical property. Applied Mechanics and Materials 711, 406–409.

Wang H and Chen J S, 2010. Mix proportions and properties of CLSC made from thin film transition liquid crystal display optical waste glass. Journal of Environmental Management 91 (3), 638–646.

Wang H and Hou T, 2011. A study of elevated temperatures on the strength properties of LCD glass powder cement mortars. Integrated Waste Management 1, 391–404.

Wang H and Huang W, 2010. Durability of self- consolidating concrete using waste LCD glass. Construction and Building Materials 24 (6), 1008–1013.

Wang H, 2009a. A study of the engineering properties of waste LCD glass applied to controlled low strength materials concrete. Construction and Building Materials 23 (6), 2127–2132.

Wang H, 2009b. A study of the effects of LCD glass sand on the properties of concrete. Waste Management 29 (1), 335–341.

Wang H, 2011. The effect of the proportion of thin film transistor– liquid crystal display (TFT–LCD) optical waste glass as a partial substitute for cement in cement mortar. Construction and Building Materials 25 (2), 791–797.

Wang H, Wang C, Chang S and Lin J, 2016. A study of engineering and electricity properties of cement mortar added with recycled materials and piezoelectric powders. Construction and Building Materials 113, 297–305.

Wang H, Zeng H and Wu J, 2014. A study on the macro and micro properties of concrete with LCD glass. Construction and Building Materials 50, 664–670.

Wang X, Lu X and Zhang S, 2013. Study on the waste liquid crystal display treatment: focus on the resource recovery. Journal of Hazardous Materials 244–245, 342–347.

Wang Z, Shi C and Song J, 2009. Effect of glass powder on chloride ion transport and alkali-aggregate reaction expansion of lightweight aggregate concrete. Journal of Wuhan University of Technology-materials Science Edition 24 (2), 312–317.

Wattanapornprom R and Stitmannaithum B, 2015. Comparison of properties of fresh and hardened concrete containing finely ground glass powder, fly ash, or silica fume. Engineering Journal 19 (3), 35–48.

WRAP, 2004. Introduction to PAS 102-Specification for Processed Glass for Selected Secondary End Markets. Waste & Resources Action Programme, Banbury, Oxon, UK.

WRAP, 2007. Waste Protocols Project - Flat Glass: A Technical Report on the Production of Processed Cullet from Waste Flat Glass. Waste & Resources Action Programme, Banbury, Oxon, UK.

WRAP, 2008. Collection of Flat Glass for Use in Flat Glass Manufacture - a Good Practice Guide. Waste & Resources Action Programme, Banbury, Oxon, UK.

WRAP, 2009a. Can You Recycle Your Flat Glass Waste. Waste & Resources Action Programme, Banbury, Oxon, UK.

WRAP, 2009b. Recycling in Construction - How to Segregate and Recycle Glass. Waste & Resources Action Programme, Banbury, Oxon, UK.

Wright J R, 2012. Examining Concrete Properties Containing Recycled Glass Cullet as a 100% Fine Aggregate Replacement (MSc thesis). Department of Civil and Environmental Engineering, The Pennsylvania State University, USA. 119 pp.

Wright J R, Cartwright C, Fura D and Rajabipour F, 2013. Fresh and hardened properties of concrete incorporating recycled glass as 100% sand replacement. Journal of Materials in Civil Engineering 26 (10), 1–11.

Wright L, 2003. Development of a Regional Waste Recovery/Processing Sector Wellington City Council. Ministry for the Environment & Ministry of Economic Development, Wellington, New Zealand.

Wright J R, Cartwright C, Fura D and Rajabipour F, 2014. Fresh and hardened properties of concrete incorporating recycled glass as 100% sand replacement. Journal of Materials in Civil Engineering 26 (10), 11 pp.

Yilmaz A and Degirmenci N, 2010. The combined usage of ground waste glass with an industrial by-product in manufacturing Portland cement mortar. Advances in Cement Research 22 (2), 73–80.

Yuksel C, Ahari R S, Ahari B A and Ramyar K, 2013. Evaluation of three test methods for determining the alkali-silica reactivity of glass aggregate. Cement and Concrete Composites 38, 57–64.

Zhao H and Wei S, 2011. Study of properties of mortar containing cathode ray tubes (CRT) glass as replacement for river sand fine aggregate. Construction and Building Materials 25 (10), 4059–4064.

Zhao H, Poon C S and Ling T C, 2013a. Utilizing recycled cathode ray tube funnel glass sand as river sand replacement in the high-density concrete. Journal of Cleaner Production 51, 184–191.

Zhao H, Poon C S and Ling T C, 2013b. Properties of mortar prepared with recycled cathode ray tube funnel glass sand at different mineral admixture. Construction and Building Materials 40, 951–960.

Zheng Y, Shen Z, Ma S, Cai C, Zhao X and Xing Y, 2009. A novel approach to recycling of glass fibers from nonmetal materials of waste printed circuit boards. Journal of Hazardous Materials 170 (2), 978–982.

Use of Ground Glass Cullet as Cement Component

4

Main Headings

- Cementitious characteristics
- Fresh properties
- Mechanical properties
- Permeation
- Durability
- Environmental impact
- Case studies

Synopsis

Finely ground glass cullet (GGC) exhibits pozzolanic properties. Its inclusion with Portland cement does not significantly change the characteristics of cement apart from prolonging the setting times. In the fresh state, the consistence and stability of concrete are almost unaffected, but the density is slightly reduced when GGC is used. It also reduces the temperature rise and heat of hydration of the concrete. Depending on the content and fineness of GGC, its use can maintain or improve the compressive strength of concrete. The relationship between compressive strength and tensile strength also remains unchanged. The use of GGC reduces the modulus elasticity of concrete and increases shrinkage. Other than carbonation, GGC is shown to generally improve the permeation and durability properties of concrete. The use of this material is safe in terms of environmental impact and it has been used in the concrete construction industry since the 2000s.

Keywords: Ground glass cullet, Cement characteristics, Concrete, Fresh properties, Mechanical properties, Permeation, Durability, Environmental impact, Case studies.

Sustainable Construction Materials: Glass Cullet. https://doi.org/10.1016/B978-0-08-100984-0.00004-7

4.1 Introduction

Cement is an important ingredient of concrete, which is second only to water in volume consumption terms globally. The cement manufacturing sector accounts for about 5% of carbon dioxide (CO_2) emissions (WBCSD, 2012), mainly during the production of Portland clinker, an essential ingredient used in the manufacturing of most modern cements, such as those mentioned in EN 197-1 (2011). Portland cement (PC) clinker is manufactured by grinding calcareous rocks (primary component) such as limestone or chalk and argillaceous rocks such as shale or clay and heating the mixture in a large rotary inclined kiln with temperature gradually rising along the length up to 1450°C. Carbon dioxide is emitted from two sources during the clinker manufacturing process (WBCSD, 2016):

(i) Calcination, whereby calcium carbonate coming from calcareous rocks is transformed into lime, with CO_2 released as a by-product, accounts for about 60% of the emitted CO_2.

(ii) Burning of fuels, the energy of which is needed to convert the raw materials to clinker, accounts for 40% of the emitted CO_2.

Depending on the energy efficiency of the production facilities and the choice of fuel, on average, 0.633 tonne of CO_2 was generated for every tonne of cement produced worldwide in 2014 (GNR, 2014). Substitution of PC with pozzolanic materials such as fly ash (FA), a residue from coal combustion, or latent hydraulic materials such as ground granulated blast slag (GGBS), a by-product from the iron industry, can reduce the volumes of clinker used, thus contributing to the reduction of CO_2 emissions from cement production.

Glass is an amorphous and mainly siliceous material, with 70% SiO_2 content. In its finely powdered form, ground glass cullet (GGC) exhibits pozzolanic properties and is suitable for use as a cement component. This attribute of GGC has been recognised probably since the early 1970s by researchers from the Colorado School of Mines research institute, who studied the pozzolanic properties of soda lime glass (container glass) ground to a size less than 44 μm (Pattengill and Shutt, 1973). However, it was not until the beginning of the 2000s when the interest in the subject became pronounced enough to develop the use of GGC in concrete, with the findings suggesting that its use as a replacement, in part, for PC is perfectly feasible.

4.2 General Information

Attributed to its amorphous and silica-rich nature, GGC undergoes a pozzolanic reaction with calcium hydroxide that is released during PC hydration, in the presence of moisture, producing calcium silicate hydrates, which enhance the performance of concrete (Dhir and Dyer, 2001, 2004). Its pozzolanic reactivity increases with increasing particle fineness (Shi and Zheng, 2007; Federico and Chidiac, 2009). In general, concrete made with GGC in combination with PC replacement up to 40%

has been reported to show an improvement in its strength properties, permeation and chloride resistance (Shekhawat and Aggarwal, 2014; Zheng, 2013). Additionally, and most importantly, Dhir and Dyer (2001) have confirmed that concrete containing GGC does not display the risk of detrimental alkali–silica reaction (ASR) expansion because of the presence, in abundance, of silica in GGC. Indeed, it has been shown that mortar made with GGC exhibits less ASR expansion than reference mortar made without GGC (Dhir and Dyer, 2001; Shi and Zheng, 2007).

4.3 Cementitious Characteristics of Ground Glass Cullet

This section covers the cementitious characteristics of GGC, with a focus on some selected important physical and chemical aspects, such as the water required for standard consistence, setting times, soundness and strength activity index. Readers are advised to refer to Chapter 3 for information on the chemical composition, particle size distribution and loss of ignition of GGC.

It should be noted that the fineness of GGC and PC reported in the literature has been expressed in various forms, Blaine fineness, particle mean size and percentage of particles retained on 45-μm sieve size being the most common. Given that the reactivity rate is affected by the surface area, for clarity, the fineness of GGC and the corresponding reference PC is compared based on their numerical difference, as given in Table 4.1. When the PC fineness is not available, the following typical values of PC are assumed: 350 m^2/kg Blaine fineness, 15-μm particle mean size and nil (0%) retention on 45-μm-size sieve.

4.3.1 Water Required for Standard Consistence

Consistence refers to the flow properties of a cementitious mixture in the fresh state. Cement pastes prepared for setting time and soundness tests are required to have a standard consistence, which is determined using the Vicat test. The standard consistence is achieved when the plunger of the apparatus penetrates into a cement paste to a depth of 10 ± 1 mm (in the case of ASTM C187, 2011) or 6 ± 2 mm (in the case of BS EN 196–3:2005+A1, 2008) below the original surface. The amount of water required for the standard consistence is expressed as a percentage by mass of cement.

Table 4.1 Comparison criteria for the fineness of ground glass cullet and Portland cement

Numerical Difference	Interpretation
More than +10%	GGC is finer than PC
Within ±10%	GGC has the same fineness as PC
More than −10%	GGC is coarser than PC

GGC, ground glass cullet; *PC*, Portland cement.

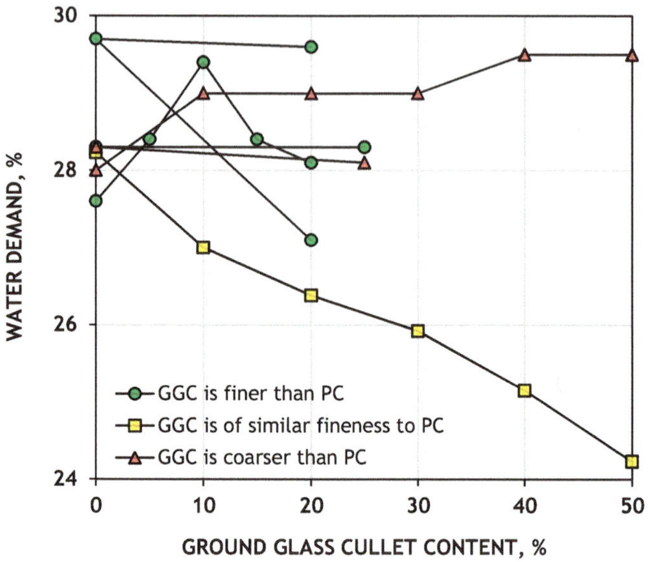

Figure 4.1 Influence of ground glass cullet (GGC) on the water required for standard consistency. *PC*, Portland cement.
Data taken from Byars et al. (2004b), Cui (2005), El-Alami (2004), Remarque et al. (2001, 2003), Sugita et al. (1997), Tognonvi et al. (2015), Wang (2011) and Yilmaz and Degirmenci (2010).

Figure 4.1 shows the water required for the standard consistence of cement paste made with up to 50% GGC as a component of PC. It can be seen from Figure 4.1 that different trends can be observed from the results that have been reported, but these changes are not likely to be related to the fineness of GGC used. Cui (2005) explained that the increase in the water demand of samples containing GGC can be due to its angular particle shape, because in a parallel test, FA, which largely consists of spherical particles, was observed to reduce the amount of water required to achieve standard consistence. However, Wang (2011) attributed the reduction in water demand to the low water absorption of GGC.

On the other hand, in the studies undertaken by Byars et al. (2004b) and El-Alami (2004), the results suggest that the difference in water requirement for the mortars made with 30% GGC, of fineness ranging from 240 to 862 m^2/kg, to that of reference mortar varied between −4.5% and +1.3%, but most of them were almost 0% (indicating no change).

Overall, although both the angularity of particle shape and the low water absorption of GGC can affect the water demand of cement paste for a given consistence, these two effects may cancel each other out and result in a small net change in the water demand, as shown in many studies (Byars et al., 2004b; El-Alami, 2004; Remarque et al., 2001, 2003; Tognonvi et al., 2015).

4.3.2 Initial and Final Setting Times

A cement paste is considered to be set when it loses its original plasticity and is no longer mobile. Cementitious materials such as FA and GGBS are commonly known to prolong the initial and final setting times of a mixture. The delay in the initial setting time of the cement due to the inclusion of FA is regulated by BS EN 450-1 (2012), in that the inclusion of 25% FA should not increase the initial setting time of the cement to twice that of the reference cement.

Figure 4.2 compares the initial setting times of cement containing soda lime and aluminosilicate GGC with the corresponding reference cement made without GGC, measured using BS EN 196–3:2005+A1 (2008) and ASTM C191 (2013). For soda lime GGC, the initial setting time of cement remains essentially unchanged for up to 20% content, beyond which, a small increase in initial setting time (25 min on average) is observed. It would also appear that the fineness of soda lime GGC does not have a significant influence on the initial setting time of cement (Figure 4.2).

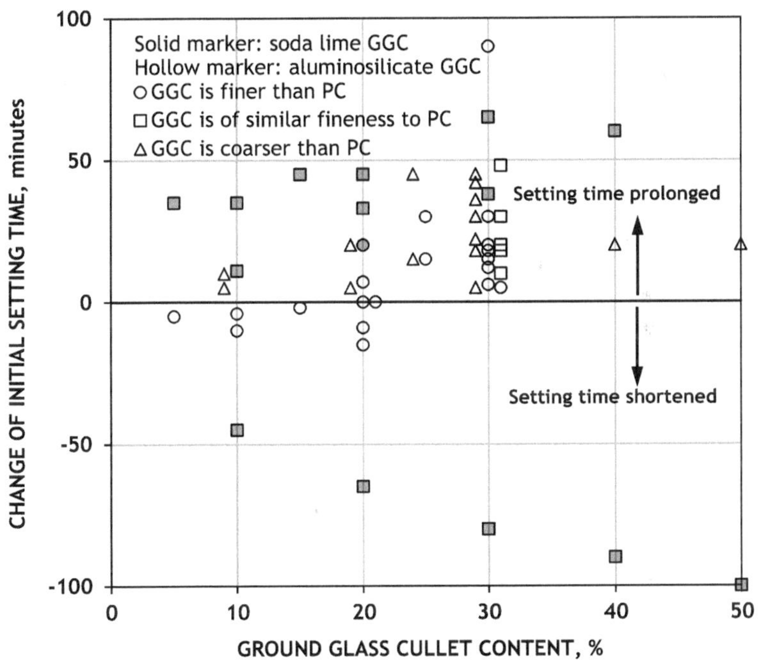

Figure 4.2 Change in initial setting time at various ground glass cullet (GGC) contents. *PC*, Portland cement.
Data taken from Byars et al. (2004b), Cui (2005), El-Alami (2004), Kamali and Ghahremaninezhad (2016), Lee and Lee (2016), Lin et al. (2009), Matos and Sousa-Coutinho (2012, 2016b), Matos et al. (2015), Remarque et al. (2001, 2003), Sugita et al. (1997), Tognonvi et al. (2015) and Wang (2011).

For aluminosilicate GGC, however, two completely opposite trends have been reported (Figure 4.2). It is, however, unlikely that the use of aluminosilicate GGC will shorten the initial setting time, because of the dilution of the tricalcium silicate (C_3S) and tricalcium aluminate (C_3A), which are responsible for the setting of cement. Thus, such results may be considered as outliers. Nevertheless, Figure 4.2 shows that aluminosilicate GGC tends to delay the initial setting time of cement more than soda lime GGC for a given replacement level.

When GGC is used in conjunction with FA as a PC replacement, with a combined content up to 25%, the initial setting time is found to increase by about 100 min longer than the corresponding reference cement (Yilmaz and Degirmenci, 2010). In general, GGC complies with the initial setting time requirement specified for FA in BS EN 450-1 (2012), as its inclusion at 25% or even at a higher content level does not exhibit a delay in setting time of more than 100 min.

It has also been observed that the effect of GGC on the initial setting time of cement reflects similarly on its final setting time (Cui, 2005; Kamali and Ghahremaninezhad, 2016; Lee and Lee, 2016; Lin et al., 2009; Matos and Sousa-Coutinho, 2012; Matos et al., 2015; Sugita et al., 1997; Tognonvi et al., 2015; Wang, 2011). However, as no limit on the FA effect on the final setting time of the resulting composite cement has been specified in BS EN 450-1 (2012), the prolonged final setting time due to the use of GGC should generally be of no real concern for compliance to the standard.

4.3.3 Soundness

The presence of free calcium oxide (CaO) and magnesium oxide (MgO) contents in cement can result in the expansion of a cement paste, potentially leading to cracking. This is due to the slow hydration process of these two compounds to form calcium hydroxide, $Ca(OH)_2$, and magnesium hydroxide, $Mg(OH)_2$, which have a larger volume than their original forms. Cements exhibiting excessive expansion are deemed to be unsound.

The soundness of cement containing up to 30% GGC (El-Alami, 2004; Matos and Sousa-Coutinho, 2012, 2016a,b; Matos et al., 2015) or up to 25% combination of GGC and FA (Yilmaz and Degirmenci, 2010) has been determined using the Le Chatelier test, which accounts only for the expansion caused by free CaO. All the specimens containing GGC showed a soundness value of not more than 1.5 mm, well below the maximum limit of 10 mm expansion for cement containing 30% FA as specified in BS EN 450-1 (2012).

4.3.4 Strength Activity Index

In the presence of moisture, GGC can undergo pozzolanic reaction with $Ca(OH)_2$ released during the hydration of PC to produce more calcium silicate hydrate, forming a denser cement paste structure and increasing the compressive strength of the mixture. As with FA and other pozzolanas, the pozzolanic activity of GGC can be quantified in terms of strength activity index: the ratio (in percentage) of the compressive strength of pozzolanic cement mortar to that of pure PC reference mortar.

The assessments of the strength activity index of pozzolanic material outlined in ASTM C311 (2013) in the United States and in BS EN 196-1 (2016) (or equivalent) in the United Kingdom (and Europe) are different. In the former test, the water/cement ratio of the reference mortar is fixed at 0.485 and the water content of the test mortar containing 20% pozzolanic material is adjusted for equivalent flow to that of the reference mortar. In the latter test, however, both the reference mortar and the test mortar containing 25% pozzolanic material are proportioned at a fixed 0.50 water/cement ratio. The strength activity index of GGC, predominantly soda lime glass, determined using these two methods is shown in Figure 4.3, together with the requirements specified for FA in ASTM C618 (2012) and BS EN 450-1 (2012) at different ages for comparison. It shall be mentioned that all the GGC mortars have a similar water/cement ratio compared to the reference mortar.

As to be expected, the fineness of GGC affects its rate of pozzolanic reaction and thus the strength activity index (Figure 4.3). When GGC is ground finer than PC, it has a greater specific surface for pozzolanic reaction to take place, resulting in a strength activity index higher than the minimum targets of 75% at 7 and 28 days specified in ASTM C618 (2012) and 75% and 85% at 28 and 90 days, respectively, specified in BS EN 450-1 (2012). The study of Byars et al. (2004b), investigating GGC of fineness ranging from 250 to 1000 m^2/kg, suggests that the pozzolanic effect of GGC increases, at a decreasing rate, as its fineness increases up to about 800 m^2/kg.

When GGC is ground to a fineness similar to that of PC, typically at about 350 m^2/kg, its strength activity index meets the requirements for 7 and 28 days set in ASTM C618 (2012) and 28 days in BS EN 450-1 (2012), but is likely to be less than the 85% target at 90 days of BS EN 450-1 (2012) (Figure 4.3(b)).

Coarse GGC exhibits relatively low pozzolanic effect. About half of the data in Figure 4.3(a) and (b) show that the strength activity index of GGC that is coarser than PC is below the minimum requirements. One set of the results as circled in Figure 4.3(a) is considered as an outlier, as the strength activity index seems too high for a coarse GGC, whilst all the coarse GGC results are less than 100%.

Overall, to comply with the strength activity index requirements set in the standards, GGC should be ground to a fineness similar to or higher (preferably) than that of the corresponding PC.

4.4 Fresh Concrete Properties

The duration of concrete in the fresh state is short, and lasts for 8–12 h depending on the mix design and weather conditions. However, the properties of fresh concrete can have a decisive influence on its hardened properties, including the long-term performance of concrete. The characteristics of cementitious materials, such as particle

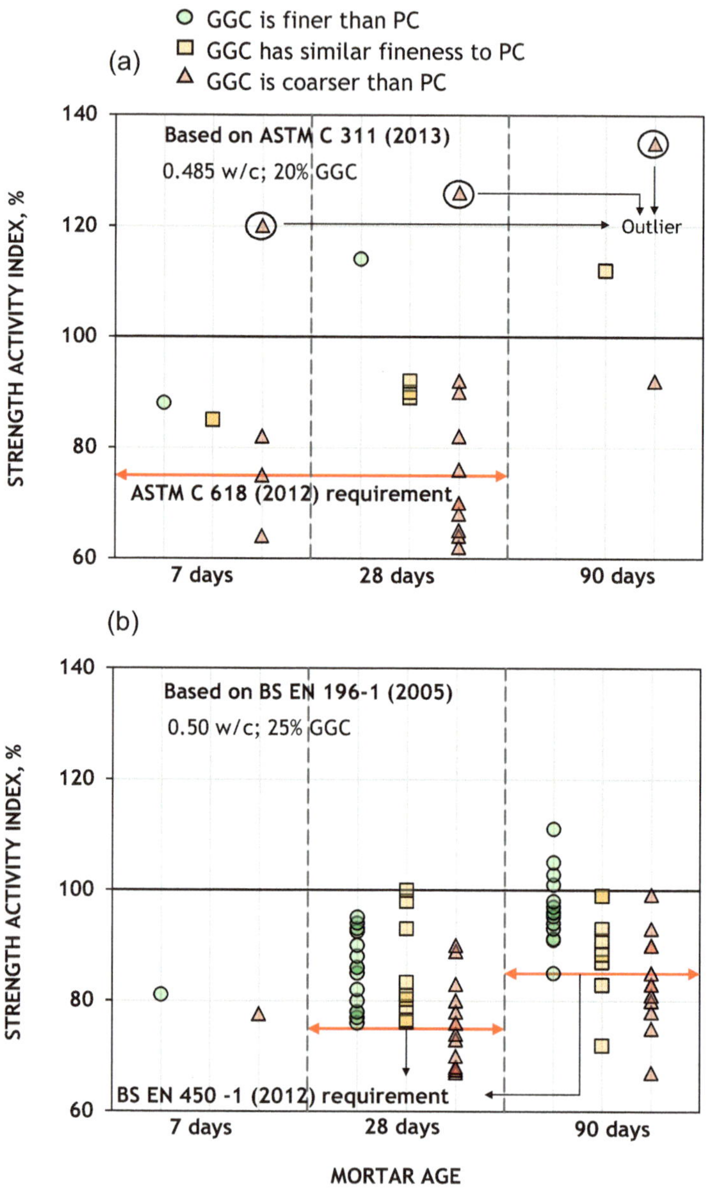

Figure 4.3 Strength activity index of ground glass cullet (GGC). (a) Based on ASTM C311 (2013). (b) Based on BS EN 196-1 (2005). *PC*, Portland cement.
Data taken from Afshinnia and Rangaraju (2015), Bignozzi et al. (2015), Byars et al. (2004b, 2004c), Dhir et al. (2005b), El-Alami (2004), Khmiri et al. (2012), Lin et al. (2008), Neithalath (2008), Parghi and Alam (2016), Pereira-de-Oliveira et al. (2007), Remarque et al. (2001), Shi et al. (2005) and Siad et al. (2016).

shape, fineness and chemical reactivity, can play a vital role in determining the fresh properties of concrete. For example, FA increases the consistence of a concrete mix owing to its spherical shape, which potentially leads to a reduction in water content, which can be used to recoup some reduction in strength by reducing the water/cement ratio whilst keeping the consistence of the mix constant. Thus, it is important to understand how the use of GGC as a cement component changes the fresh properties of concrete, to recognise areas of necessary development, which can be used to benefit the performance of concrete in both the fresh and the hardened states.

4.4.1 Consistence

Consistence, or the old terminology 'workability,' is the ability of a freshly mixed concrete to be handled, placed, compacted and finished without the loss of its homogeneity. Concrete with a good consistence allows for maximum density and minimum entrapped voids to be achieved from compaction. Several methods, with different degrees of sensitivity, have been developed to determine the consistence of concrete, such as the slump test, Vebe test, compacting factor test and flow table test. Of these, the slump test is the easiest to perform and also commonly adopted by the construction industry, as it covers the consistence range of concrete mixes mostly used in practice.

The influence of GGC on the slump of concrete has been reported by many studies, collectively covering several parameters such as glass type (predominantly soda lime), Blaine fineness up to 600 m^2/kg, replacement content up to 50% and water/cement ratio over a range of 0.36–0.67. The results have been sorted into four slump classes (Figure 4.4), based on the slump of the reference concrete, using the slump classification and tolerance of a specified target slump specified in BS EN 206-1 (2003), as summarised in Table 4.2.

It is evident from Figure 4.4 that, for a given water/cement ratio, although the slump of concrete made with GGC may increase, decrease or fluctuate as the GGC content increases, such a variation is generally marginal and still within the same slump class or allowable tolerance. Only three sets of results, in the slump classes of S2 (Dhir et al., 2005c) and S3 (Bajad et al., 2012a; Metwally, 2007), exhibit a decrease in slump beyond the limits when the GGC content is 20% or more. In effect, it can be considered that the slump of GGC concrete remains essentially similar to that of the reference PC concrete. The fineness of the GGC does not seem to have an effect on the slump of concrete (Figure 4.4).

Similarly, when used in mortar, the flow of the mixes remains almost unchanged for GGC content up to 40% as PC replacement (Borosnyoi et al., 2013; Jang et al., 2014, 2015; Kou and Xing, 2012; Lee and Lee, 2016; Matos and Sousa-Coutinho, 2016a; Parghi and Alam, 2016; Pereira-de-Oliveira et al., 2012; Remarque et al., 2001; Sugita et al., 1997; Tognonvi et al., 2015; Wang and Hou, 2011; Wang et al., 2016). However, in only one case the flow of a cement paste showed a rate of 35% increase for every 10% GGC used (Schwarz et al., 2007).

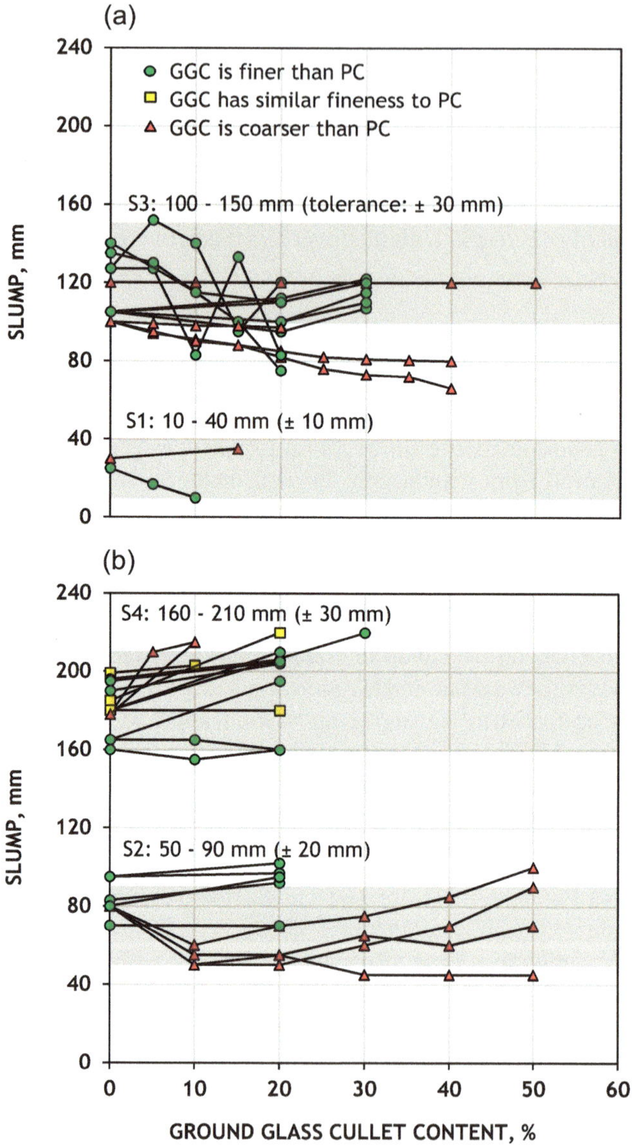

Figure 4.4 Influence of ground glass cullet (GGC) on slump of concrete for (a) S1 and S3 slump classes and (b) S2 and S4 slump classes. *PC*, Portland cement.
Data taken from Bajad et al. (2012a,b), Borosnyoi et al. (2013), Cassar and Camilleri (2012), Cui (2005), Dhir et al. (2005c), Jangid and Saoji (2014), Kamali and Ghahremaninezhad (2015), Kara (2013b), Kara et al. (2012), Kim et al. (2014), Metwally (2007), Nassar and Soroushian (2012a, 2013), Neithalath and Schwarz (2009), Pavoine et al. (2014), Schwarz et al. (2008), Singh et al. (2014), Taha and Nounu (2008a), Tang et al. (2005) and Wattanapornprom and Stitmannaithum (2015).

Table 4.2 Slump classes and tolerances specified in BS EN 206-1 (2003)

Slump Class	Slump, mm	Tolerance, mm
S1	10–40	±10
S2	50–90	±20
S3	100–150	±30
S4	160–210	±30

Overall, the slump (shown in Figure 4.4) and flow results are coherent with the findings mentioned in Section 4.3.1, in that the water required for standard consistence of GGC mortar is generally close to that of the reference PC mortar. This suggests that probably the water content and admixture dosage for a concrete do not have to be modified when GGC is used as a PC replacement.

4.4.2 Stability

The stability of concrete in the fresh state is the key to attaining the desired engineering properties of concrete in its hardened state. A concrete is not stable if it suffers from (i) excessive bleeding, whereby water moves upward to the surface of concrete in an undue manner, or (ii) segregation, whereby aggregates separate from cement paste, resulting in a nonuniform mix.

Table 4.3 summarises the effects of GGC on the stability of a normal concrete (Du and Tan, 2015; Dumitru et al., 2010; Taha and Nounu, 2008a, 2009) and self-compacting concrete (Shi and Wu, 2005). No qualitative data are available in all these studies. However, in general, no signs of bleeding or segregation of concrete containing up to 60% GGC have been reported (Du and Tan, 2015; Taha and Nounu, 2008a, 2009).

Notwithstanding the preceding, it has been reported that concrete made with GGC of Blaine fineness 335 kg^2/m showed higher bleeding than the corresponding reference PC concrete (Dumitru et al., 2010). The same issue has also been reported by Shi and Wu (2005), by whom no bleeding was seen on the self-compacting concrete (SCC) containing either FA or GGC of similar fineness, but SCC made with coarser GGC showed little bleeding. The bleeding observed in these cases may have been caused by the lack of fines in the mix. Thus it is necessary to ensure that the fines content of concrete remains unchanged when PC is replaced by GGC.

4.4.3 Density

Given that the typical specific gravity of PC and GGC is about 3.15 and 2.50, respectively, concrete containing GGC will have a lower density than concrete made with 100% PC. Apart from one case (Kamali and Ghahremaninezhad, 2015), in which the density of concrete was reported to increase with increasing GGC content up to 20%, the reported studies have generally tended to suggest that the density of fresh concrete generally decreases with the use of GGC as a cement component (Dhir et al.,

Table 4.3 Stability of concrete made with ground glass cullet

References	Fineness	w/c	GGC, %	Main Observations
Du and Tan (2015)	PC, 10-μm median diameter GGC, 3.4-μm mean diameter	0.47	0–60	No bleeding or segregation
Dumitru et al. (2010)	PC, not available GGC, 335 kg²/m	–	0–25	Bleeding of GGC concrete was higher than that of PC concrete
Shi and Wu (2005)[a]	FA, 74% passing 45 μm GGC 1, 50% passing 45 μm GGC 2, 80% passing 45 μm	0.34	30	No bleeding in SCC made with FA and GGC 1, but little bleeding was seen for SCC made with GGC 2
Taha and Nounu (2008a, 2009)	PC, not available GGC, 45-μm mean diameter	0.38	0, 20	Both GGC and PC concretes were homogeneous

GGC, ground glass cullet; *PC*, Portland cement; *w/c*, water/cement ratio.
[a]Comparing lightweight self-compacting concrete (SCC) made with fly ash (FA) and two GGCs with different fineness. The fineness data of the materials are taken from Shi et al. (2005).

2005c; Nassar and Soroushian, 2012a, 2013; Nwaubani and Poutos, 2013; Sharifi et al., 2015; Shi and Wu, 2005; Taha and Nounu, 2008a, 2009; Tang et al., 2005). As shown in Figure 4.5, excluding the positive values, which are likely to be due to experimental errors, the density of concrete is calculated to decrease at a rate of about 0.5% on average for every 10% replacement of PC by GGC.

4.4.4 Air Content

There are two types of air that can be found in concrete, namely entrapped air and entrained air. The former is irregularly sized and created inevitably during mixing and compacting of fresh concrete. Entrapped air can be removed by proper compaction techniques. The latter is of a spherical nature and is uniformly distributed in concrete, resulting from the deliberate use of an air-entraining admixture. Entrained air is used to increase the resistance of concrete against cycles of freeze–thaw attack. The increase in cement fineness or decrease in alkali content of cement reduces entrained air in concrete (PCA, 1998).

For a given dosage of air-entraining agent, Nassar and Soroushian (2012a, 2013) show that the air content of concrete made with 20% GGC of 495 m²/kg Blaine fineness is similar to that of reference PC concrete, with a marginal difference of less than 1%. The same observation is also reported by other studies (Kim et al., 2014; Neithalath and Schwarz, 2009; Pavoine et al., 2014), but it is unsure if the dosage of air-entraining agent was kept constant for both the GGC and the reference PC concrete.

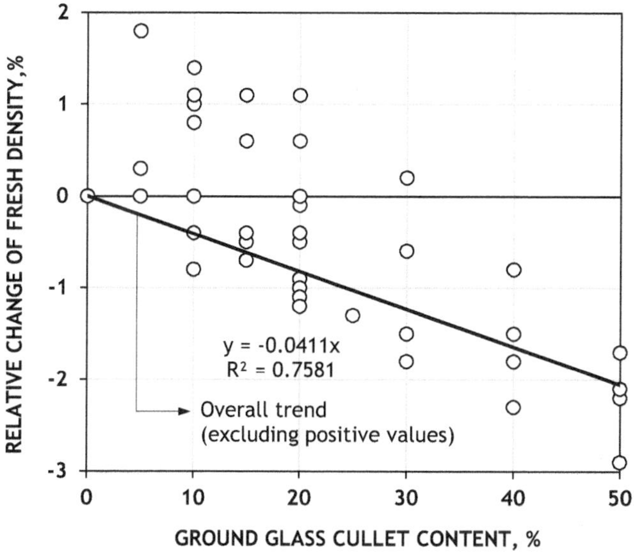

Figure 4.5 Influence of ground glass cullet on fresh density of concrete.
Data taken from Bajad et al. (2012a,b), Bhat and Rao (2014), Cassar and Camilleri (2012), Dhir et al. (2005c), Kadir et al. (2016), Kamali and Ghahremaninezhad (2015), Nassar and Soroushian (2012a, 2013), Nwaubani and Poutos (2013), Oyekan and Oyelade (2011), Schwarz et al. (2008), Singh et al. (2014), Taha and Nounu (2008a, 2009) and Tang et al. (2005).

On the other hand, Lee and Lee (2016) and Sharifi et al. (2015) suggest that the air content decreases as GGC content increases, resulting in a 2% reduction when 20% GGC is used, with a given high-range water-reducing admixture (which is likely to contain an air-entrainment feature). Additionally, in a study undertaken by Shi and Wu (2005), at a fixed air-entraining admixture dosage, lightweight SCC containing 30% FA has an air content of 7%, but when FA is fully replaced by a finer GGC and a coarser GGC, the air content is decreased to around 4%.

It would appear that the effect of GGC on the stability of entrained air has not been clearly established; however, it may be assumed that the use of GGC up to content no greater than 20% does not significantly affect the dosage of air-entraining admixture to achieve a target air content in concrete.

4.4.5 Initial and Final Setting Times

When a concrete sets, it loses its consistence and becomes difficult to handle. The setting time of concrete is important to concrete producers and contractors as it affects the allowable working duration for delivery and casting.

Table 4.4 Initial and final setting of lightweight self-compacting concrete made with fly ash and ground glass cullet (Shi and Wu, 2005)

				Setting Time	
Mix	**Na_2O, %**	**K_2O, %**	**w/c**	**Initial**	**Final**
30% Fly ash	0.72	1.70	0.34	10 h 38 min	12 h 15 min
30% GGC	13.2	0.12	0.34	9 h 3 min	10 h 55 min

GGC, ground glass cullet; *w/c*, water/cement ratio.

The use of GGC up to 25% content can extend the setting times of concrete for up to 1–2 h longer than PC concrete (Dumitru et al., 2010). As to be expected, GGC concrete containing a low dosage of retarding admixture has a shorter initial setting time than reference PC concrete containing a high dosage of retarding admixture (Wattanapornprom and Stitmannaithum, 2015).

In comparing the setting times between concrete made with FA and GGC, the study of Shi and Wu (2005) suggests that, all things being equal, GGC is likely to result in faster initial and final setting times of concrete than FA (Table 4.4). This may be due to GGC containing relatively high Na_2O, which is known to accelerate the setting times of cement (Hewlett, 1998).

4.4.6 Temperature Rise and Heat of Hydration

Hydration of cement is an exothermic process, which generates heat. The heat of hydration can be a problem when it comes to mass concrete and thick section casting, in which a steep temperature gradient between the interior (hotter) and the exterior (cooler) of the concrete may result in thermal cracking, due to different cooling rates and volume change. There are several ways to minimise the excessive temperature differential in concrete, from the material selection to the formwork type and removing time (Concrete Society, 2010), amongst which the use of FA and GGBS as a PC replacement features favourably. This is because the materials reduce the proportions of C_3A and C_3S compounds, which hydrate most rapidly, thus reducing the heat developed and temperature rise in concrete.

Table 4.5 summarises the results of paste and mortar specimens made with up to 60% GGC as regards their temperature rise under semiadiabatic conditions in the first 24 h, and rate of heat evolution as well as cumulative heat during the first 4 days, obtained using an isothermal conduction calorimeter. The main points arising from Table 4.5 are listed below:

(a) Temperature Rise

The use of GGC decreases the peak temperature rise values of specimens (Kamali and Ghahremaninezhad, 2016), and the reduction increases with increasing GGC content (Kara et al., 2016). It shall be mentioned that the peak temperature results reported in these two studies vary greatly, probably due to different experiment setups.

It can be seen from Table 4.5 that the temperature rise of cement containing GGC is not affected by either the glass type (soda lime and aluminosilicate) or the glass colour (green, flint and amber), as shown by Kamali and Ghahremaninezhad (2016) and Kara et al. (2016), respectively. In a parallel study, Kamali and Ghahremaninezhad (2016) also showed that, at 20% PC replacement content, the temperature rise of samples containing GGC and FA is the about the same. This suggests that there is no significant difference between the two materials in terms of temperature rise.

The time corresponding to the peak temperature rise recorded for cement containing GGC may be close to that of reference PC cement (Kamali and Ghahremaninezhad, 2016), but it can be delayed as long as 5 h (Kara et al., 2016).

(b) Rate of Heat Evolution

Table 4.5 shows that the peak of the rate of heat evolution of specimens decreases as the GGC content increases (Dhir et al., 2005a; Dyer and Dhir, 2001a; Du and Tan, 2014; Kara et al., 2016; Tognonvi et al., 2015). On average, an about 8% drop in the peak of the rate of heat evolution is reported for every 10% GGC used, up to 60% content. Such a reduction is to be expected, as the PC content has been reduced, and the pozzolanic reaction of GGC is a slow process, which takes place at later age. No significant change in the time corresponding to the peak of the rate of heat evolution between specimens containing GGC and reference PC specimens has been noted, however (Table 4.5).

(c) Cumulative Heat

The cumulative heat evolved in cement paste containing GGC, over a period of up to four days, has been reported, but the findings were not coherent (Table 4.5). Du and Tan (2014) report that the cumulative heat of cement paste decreases as GGC content is increased; whilst Mirzahosseini and Riding (2014) and Neithalath (2008) suggest the opposite. This is perhaps due to the differences in pozzolanic reactivity of the GGC used. However, it should be noted that the rate of heat evolution is more critical than the cumulative heat evolved in practice (Neville, 1995). In addition, it has been suggested that GGC originating from green-coloured container glass shows higher total heat of hydration than GGC of clear colour (Mirzahosseini and Riding, 2014).

4.5 Mechanical Properties

When a fresh concrete begins to set and harden, it starts to hold its shape as a solid mass and gain in strength owing to cement hydration. The properties of concrete in the hardened state are important, as it must withstand all imposed loads and maintain its volume stability, as well as possessing good durability, to ensure the integrity and longevity of a structure. In this section, the strength and both load-dependent and load-independent deformation properties of concrete made with GGC as a replacement of PC are discussed.

Table 4.5 Temperature rise and heat of hydration of mixes containing ground glass cullet

References	GGC		w/c	Monitoring	Temperature Rise		Rate of Heat Evolution		Cumulative
	Content, %	Type		Duration, h	Peak, °C	Time, h	Peak, W/kg	Time, h	Heat, kJ/kg
Dhir et al. (2005a), Dyer and Dhir (2001a)	0	SL–G	0.50 (paste)	72	–	–	3.0	12.0	–
	10				–	–	2.7	12.0	–
	20				–	–	2.3	12.5	–
	30				–	–	2.1	12.5	–
	40				–	–	1.8	12.5	–
Du and Tan (2014)	0	SL	0.485 (paste)	72	–	–	5.0	6.5	280
	15				–	–	4.3	6.5	235
	30				–	–	4.0	6.0	220
	45				–	–	3.4	6.0	190
	60				–	–	2.5	6.0	150
Kamali and Ghahremaninezhad (2016)[a]	0		0.40 (paste)	24	33.5	8.0	–	–	–
	20	AS			31.5	8.0	–	–	–
	20	SL			30.5	8.0	–	–	–
	20	Fly ash			30.0	8.5	–	–	–
Kara et al. (2016)[b]	0		0.29 (mortar)	Temp, 24 Heat, 120	92	10.0	2.6	11	–
	20	SL–G			80	15.0	2.5	11	–
	30				76	15.5	1.9	11	–
	20	SL–C			81	15.0	2.5	11	–
	30				74	15.5	1.9	11	–
	20	SL–A			80	15.0	2.5	11	–
	30				84	13.0	1.8	11	–
	30	B			–	–	1.5	11	–

References	GGC		w/c	Monitoring	Temperature Rise		Rate of Heat Evolution		Cumulative
	Content, %	Type		Duration, h	Peak, °C	Time, h	Peak, W/kg	Time, h	Heat, kJ/kg
Mirzahosseini and Riding (2014, 2015)	0		0.35 (paste)	96	–	–	–	–	288
	25	SL–G			–	–	–	–	333
	25	SL–C			–	–	–	–	332
Neithalath (2008)	0	SL	0.32 (paste)	72	–	–	–	–	310
	10				–	–	–	–	325
	20				–	–	–	–	340
	30				–	–	–	–	330
	0		0.42 (paste)		–	–	–	–	290
	10				–	–	–	–	340
	20				–	–	–	–	340
	30				–	–	–	–	300
Tognonvi et al. (2015)	0	SL	0.35 (paste)	50	–	–	2.3	9.5	–
	20				–	–	1.8	9.5	–

AS, aluminosilicate; *B*, borosilicate; *GGC*, ground glass cullet; *SL*, soda lime; *SL–A*, amber soda lime; *SL–C*, clear soda lime; *SL–G*, green soda lime; *w/c*, water/cement ratio.
[a]Temperature measured from a sample kept in a semiadiabatic calorimeter.
[b]Temperature measured from an insulated sample using thermocouples.

4.5.1 Compressive Strength

(a) Effect of Ground Glass Cullet Content

The compressive strength of concrete is the property most commonly specified by engineers, as it indicates its ability to resist the intended designed compressive loads and also gives an indirect, albeit approximate, indication of other hardened concrete properties.

The effect of GGC on the strength of concrete for a PC replacement level of up to 60%, at a constant water/cement ratio ranging from 0.28 to 0.64, has been studied by many researchers from different parts of the world, as early as the beginning of the 1980s (Samarin, 1980). Most of the studies have focused on GGC of the soda lime type, such as that derived from containers and flat glasses, though other types of glass, such as borosilicate and aluminosilicate glasses, have been used in some studies.

It is noted that changes in the compressive strength of concrete due to the use of GGC as a replacement for PC are closely associated with three main factors, namely (i) GGC content, (ii) its fineness and (iii) the age of the concrete. Thus, for clarity, the compressive strength of GGC concrete in relation to corresponding reference PC concrete, expressed in a percentage form, is separated into six age groups of strength determination at 3 days (Figure 4.6(a)), 7 days (Figure 4.6(b)), 28 days (Figure 4.6(c)), 56 days (Figure 4.6(d)), 91 days (Figure 4.6(e)) and 365 days (Figure 4.6(f)). Within each strength group, the results are divided into three categories based on the fineness of GGC, being finer than, similar to or coarser than that of the corresponding reference PC (see Table 4.1). The trend lines obtained in each category are based on the mean results at each individual GGC content. For ease of reference, the ranges of FA covered in BS EN 197-1 (2011) cements as CEM II/A (6%–20% FA), CEM II/B (21%–35% FA) and CEM IV (36%–55% FA) are shown in the figure.

In general, the following points have been observed:

- At 3 days (Figure 4.6(a)): The compressive strength of concrete containing GGC is lower than that of PC concrete, and the strength reduction increases as GGC content is increased. As to be expected, for a given GGC content, the use of coarser GGC shows a greater strength reduction compared with finer GGC. Although the data for concrete made with GGC of particle size similar to that of reference PC are unavailable, it is very likely that its trend line would lie between those of concrete made with finer and coarser GGC.
- At 7 days (Figure 4.6(b)): Apart from concrete made with 10% finer GGC, which shows performance similar to that of PC concrete, the compressive strength development of all GGC concrete is lower than that of PC concrete, though the strength reduction in this case is less than that observed at 3 days.
- At 28 days (Figure 4.6(c)): The trend for compressive strength of concrete remains unchanged or shows a slight improvement for concrete made with finer GGC up to 20% content, beyond which it reduces as the GGC content is increased. Concrete made with GGC of particle size similar to or coarser than PC continues to show low compressive strength compared with PC concrete.

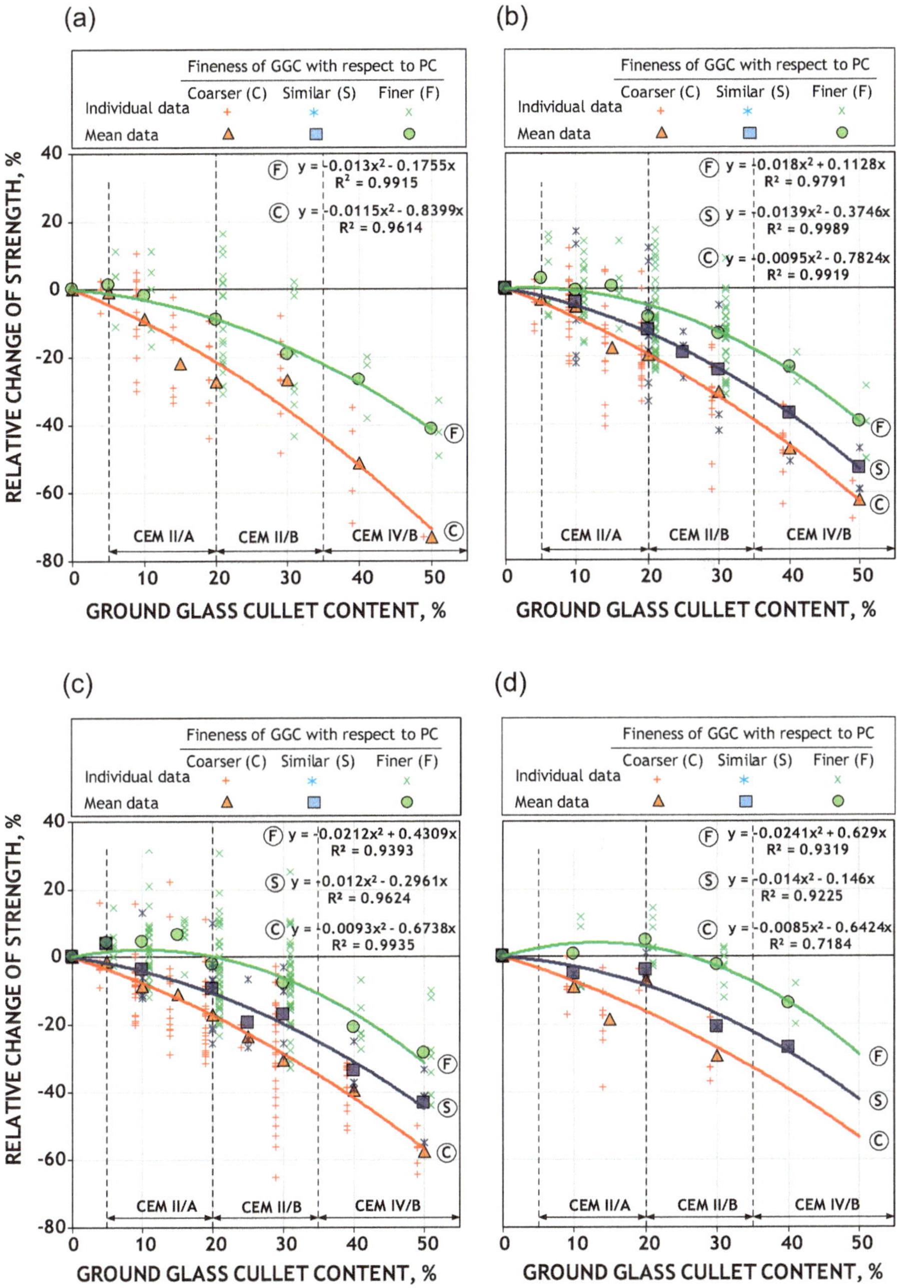

Figure 4.6 cont'd

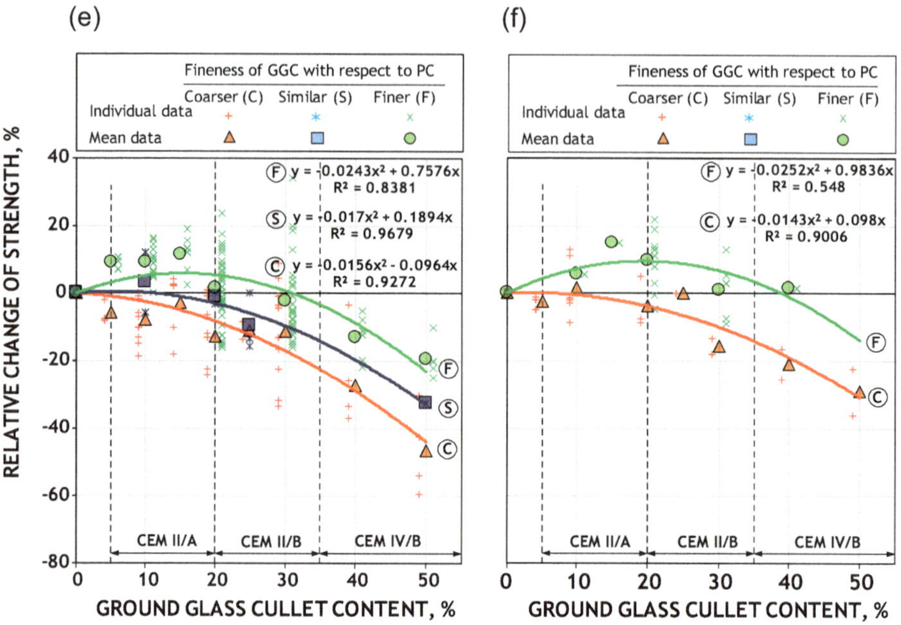

Figure 4.6 Effects of ground glass cullet (GGC) on the compressive strength development of concrete at (a) 3 days and (b) 7 days, (c) 28 days, (d) 56 days, (e) 91 days and (f) 365 days. *PC*, Portland cement.

Data taken from Ali (2015), Altaf et al. (2013), Aly et al. (2011), Bajad and Modhera (2009), Bajad et al. (2011a,b, 2012b, 2014), Bhat and Rao (2014), Byars et al. (2004a), Bignozzi et al. (2015), Borosnyoi et al. (2013), Calmon et al. (2014), Carsana et al. (2014), Cassar and Camilleri (2012), Chaid et al. (2015), Chen et al. (2006), Chidiac and Mihaljevic (2011), Cui (2005), Dahmardeh et al. (2015), Dhir (2005c), Dyer and Dhir (2001a,b), Du and Tan (2014, 2015), Girbes et al. (2004, 2015), Federico (2013), Hussain and Chandak (2015), Jang et al. (2015), Jangid and Saoji (2014), Han et al. (2016), Idir et al. (2009, 2011a,b, 2013), Kadir et al. (2016), Kara (2013a,b, 2015), Kara et al. (2012, 2016), Karamberi and Moutsatsou (2005), Karamberi et al. (2004), Khmiri et al. (2012, 2013), Kim et al. (2015), Kou and Xing (2012), Laldji et al. (2004), Lee and Lee (2016), Lin et al. (2008, 2009), Liu (2011), Liu et al. (2013), Madandoust and Ghavidel (2013), Matos and Sousa-Coutinho (2016a,b), Meena and Singh (2012), Metwally (2007), Miranda et al. (2014b), Mirzahosseini and Riding (2014), Moncea et al. (2013), Nassar and Soroushian (2011, 2012b, 2013), Niang et al. (2015); Nishikawa et al. (1995), Nwaubani and Poutos (2013), Neithalath and Schwarz (2009), Oyekan and Oyelade (2011), Özkan and Yüksel (2008), Pavoine et al. (2014), Peyvandi et al. (2013), Priscilla and Naik (2014), Remarque et al. (2003), Schwarz et al. (2007, 2008), Seju et al. (2015), Shao and Lehoux (2001), Shao et al. (2000), Shayan and Xu (2004, 2006), Shi and Wu (2005), Shi et al. (2005), Singh et al. (2014), Sobolev et al. (2007), Tagnit-Hamou and Bengougam (2012), Taha and Nounu (2008a, 2009), Tamanna et al. (2015), Tang et al. (2005), Tejaswi et al. (2015), Tognonvi et al. (2015), Tuncan et al. (2001), Wang (2011), Wang and Hou (2011), Wang et al. (2009, 2014, 2016), Wattanapornprom and Stitmannaithum (2015), Yilmaz and Degirmenci (2010) and Zheng (2013).

- At 56 days (Figure 4.6(d)): Concrete made with finer GGC up to about 25% shows similar or higher compressive strength compared with PC concrete. No significant improvement in strength is observed for concrete made with GGC of particle size similar to or coarser than PC.
- At 91 days (Figure 4.6(e)): The strength gains in concrete made with finer GGC become pronounced, with the highest compressive strength achieved when 15% GGC is used. This material can be used at up to 30% without compromising the compressive strength of concrete. The same observation is seen with concrete made with GGC of particle size similar to PC, but with its content limited to 10%. The compressive strength of coarser GGC shows some improvement but the strength developed is still less than that of PC concrete.
- At 365 days (Figure 4.6(f)): Concrete made with finer GGC continues to benefit from strength improvement. The peak strength gain is at 20% GGC, beyond which the strength gain reduces until 40% GGC, at which no strength improvement is seen with respect to the corresponding PC concrete. The concrete compressive strength with coarser GGC at this age is comparable to that of PC concrete at 10% replacement. Again, the data for concrete made with GGC of particle size similar to PC are unavailable, but it can be safely assumed that its compressive strength would be similar to or slightly higher than that of PC concrete for a GC content up to 25%.

(b) Water/Cement Ratio and Strength Relationship

In designing a concrete mix, its compressive strength is taken primarily as a function of the water/cement ratio. However, the relationship between water/cement ratio and compressive strength of concrete can be affected by the proportion and properties of its constituent materials, curing conditions and other factors. To provide a general view on how the use of GGC modifies this relationship, the compressive strength data of GGC concrete are plotted against the corresponding water/cement ratio at the age of 3–91 days, as shown in Figure 4.7(a)–(d). The data used therein were based on concrete made with natural aggregate and subjected to a normal moist curing.

It is evident from Figure 4.7(a)–(d) that, whilst the curve profiles of water/cement ratio and compressive strength of GGC concrete are essentially similar to those of PC concrete, different GGC contents and fineness in relation to PC produce different curves at a given concrete age. The development of these curves is important as they allow engineers to estimate the concrete strength that is likely to be achieved for a given GGC content, water/cement ratio and curing age. The curves can also be used to overcome reduction in early age strength in GGC concrete by reducing its water/cement ratio. For example, to obtain similar strength compared to PC concrete of 0.50 water/cement ratio at 3 days, the water/cement ratio of concrete made with 6%–20% and 21%–35% finer GGC would need to be adjusted to 0.45 and 0.40, respectively (Figure 4.7(a)). This can be achieved by increasing its cement content or reducing its water content using a water-reducing admixture, but the latter option would be more attractive in terms of sustainability. Whilst loss of consistency in fresh concrete as a result of the reduction in water content can be compensated for by using a water-reducing admixture, matching the 3-day strength of concrete made with GGC of coarser particle size than PC may not be practically feasible or cost effective, as it requires a huge reduction in water/cement ratio.

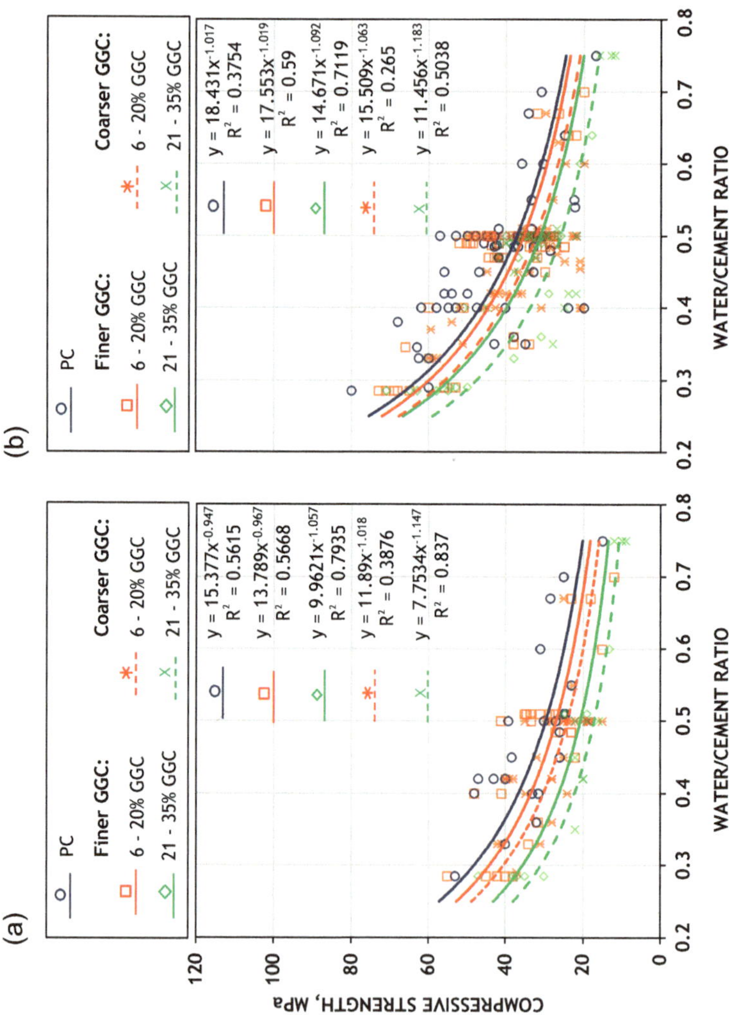
(a)
(b)
PC
Finer GGC:
6 - 20% GGC
21 - 35% GGC
Coarser GGC:
6 - 20% GGC
21 - 35% GGC
y = 15.377x^-0.947 R² = 0.5615
y = 13.789x^-0.967 R² = 0.5668
y = 9.9621x^-1.057 R² = 0.7935
y = 11.89x^-1.018 R² = 0.3876
y = 7.7534x^-1.147 R² = 0.837
y = 18.431x^-1.017 R² = 0.3754
y = 17.553x^-1.019 R² = 0.59
y = 14.671x^-1.092 R² = 0.7119
y = 15.509x^-1.063 R² = 0.265
y = 11.456x^-1.183 R² = 0.5038
COMPRESSIVE STRENGTH, MPa
WATER/CEMENT RATIO

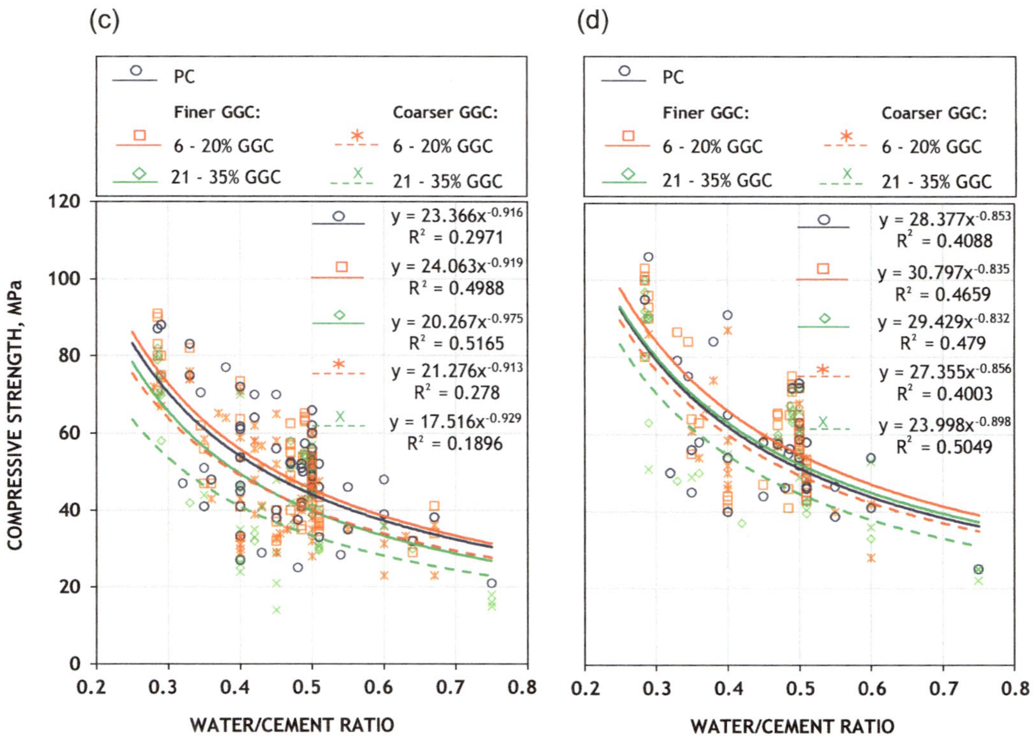

Figure 4.7 Water/cement ratio and strength of concrete containing ground glass cullet (GGC) at (a) 3 days, (b) 7 days, (c) 28 days and (d) 91 days. Data taken from Alidoust et al. (2007), Aly et al. (2011), Bhat and Rao (2014), Borosnyoi et al. (2013), Carsana et al. (2014), Cassar and Camilleri (2012), Cui (2005), Dahmardeh et al. (2015), Dhir et al. (2005c), Du and Tan (2015), Dyer and Dhir (2001a), Girbes et al. (2004), Han et al. (2016), Idir et al. (2011a), Kamali and Ghahremaninezhad (2015), Kara et al. (2012, 2016), Karamberi et al. (2006), Khmiri et al. (2012), Kim et al. (2015), Lin et al. (2008), Liu (2011), Matos and Sousa-Coutinho (2016a,b), Miranda et al. (2014a), Mirzahosseini and Riding (2014), Nassar and Soroushian (2012a, 2013), Neithalath and Schwarz (2009), Özkan and Yüksel (2008), Parghi and Alam (2016), Pavoine et al. (2014), Priscilla and Naik (2014), Remarque et al. (2003), Schwarz et al. (2007, 2008), Shao and Lehoux (2001), Shayan and Xu (2004), Singh et al. (2014), Taha and Naunu (2008a), Tamanna et al. (2015), Tejaswi et al. (2015), Tognonvi et al. (2015), Wang and Hou (2011), Wang et al. (2009, 2014, 2016) and Wattanapornprom and Stitmannaithum (2015).

(c) Strength Development with Age

Similar to other pozzolanic materials, the pozzolanic reaction of GGC with the calcium hydroxide produced by PC hydration essentially changes the rate of strength gain in the concrete. Figure 4.8 shows the compressive strength development of concrete

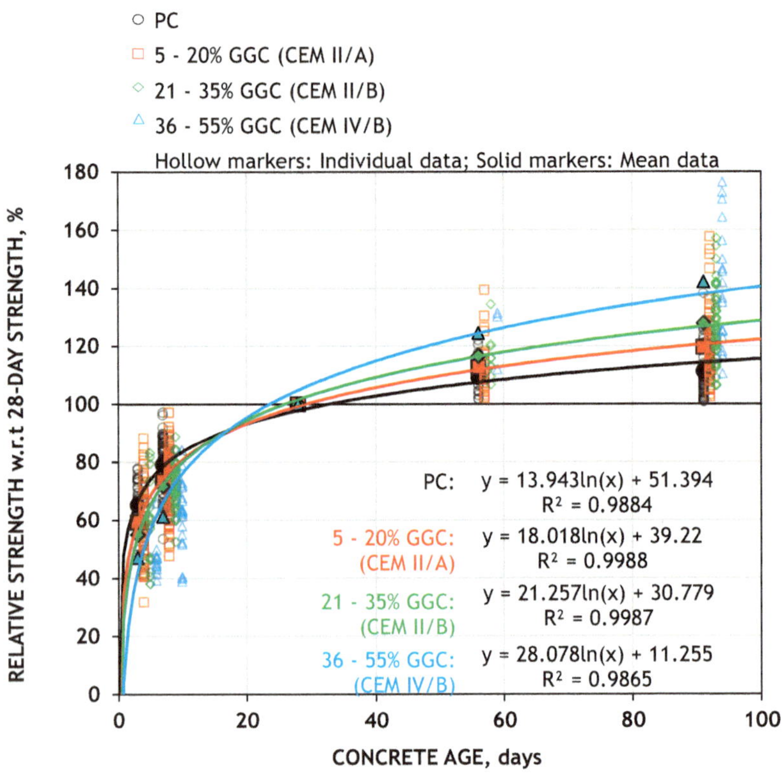

Figure 4.8 Strength development of concrete made with ground glass cullet (GGC). *PC*, Portland cement; *w.r.t*, with respect to.
Data taken from Ali (2015), Bhat and Rao (2014), Borosnyoi et al. (2013), Byars et al. (2004a), Carsana et al. (2014), Corbu et al. (2013), Cui (2005), Dahmardeh et al. (2015), Dhir et al. (2005c), Du and Tan (2015), Dyer and Dhir (2001a), Girbes et al. (2004), Han et al. (2016), Idir et al. (2011a), Jang et al. (2015), Kamali and Ghahremaninezhad (2015), Kara (2013b, 2015), Kara et al. (2012, 2016), Karamberi and Moutsatsou (2005), Karamberi et al. (2006), Khmiri et al. (2012), Kim et al. (2015), Kou and Xing (2012), Lin et al. (2008), Liu (2011), Matos and Sousa-Countinho (2012), Matos and Sousa-Coutinho (2016b), Meena and Singh (2012), Metwally (2007), Mirzahosseini and Riding (2014), Moncea et al. (2013), Nassar and Soroushian (2013), Neithalath and Schwarz (2009), Osmani and Pappu (2010), Özkan and Yüksel (2008), Parghi and Alam (2016), Pavoine et al. (2014), Priscilla and Naik (2014), Remarque et al. (2003), Schwarz et al. (2007, 2008), Seju et al. (2015), Shao and Lehoux (2001), Shi and Wu (2005), Taha and Nounu (2008a), Tejaswi et al. (2015), Tognonvi et al. (2015), Wang and Hou (2011), Wang et al. (2009, 2014, 2016) and Wattanapornprom and Stitmannaithum (2015).

made with PC and GGC for concrete at age up to 91 days, expressed in relation to their corresponding 28-day strength. The data of GGC concrete have been grouped, based on GGC content, into CEM II/A, CEM II/B and CEM IV/B cements in accordance with BS EN 197-1 (2011).

It is evident from Figure 4.8 that the use of GGC as a PC replacement changes the strength development profile of concrete, and its comparison with PC concrete can be viewed at three stages, namely, the first 14 days, between 14 and 28 days and after 28 days. It should be noted that the influence of GGC fineness is not apparent in this case.

Initially, during the first 14 days after casting, the rate of strength gain of PC concrete is significantly higher than that GGC concrete, showing that, on average, 80% of the strength has been gained by the age of 7 days (Figure 4.8). The slow gain in strength due to partial replacement of GGC is a general occurrence at the early age of concrete, and this effect is more pronounced with increasing GGC content.

In the period between 14 and 28 days, differences in the rate of strength gain between PC and GGC concrete decreases to being marginal. After 28 days, the gain in strength is still visible in both PC and GGC concrete, but the corresponding rate, in a descending order, appears to be concrete containing 36%–55% GGC, 21%–35% GGC, 6%–20% GGC and 0% GGC (PC-only concrete). This suggests that the pozzolanic effect of GGC continues to take place, resulting in a significantly higher gain in strength of GGC concrete at the later ages. At 91 days, the average strength gain relative to the corresponding 28-day strength of concrete containing 36%–55% GGC was close to 40%, whilst that of the PC concrete was only 10%.

4.5.2 Tensile Strength

It is known that concrete is strong in compression but weak in tension. For structural design, the tensile strength of concrete is normally ignored, as the tensile resistance is provided by steel reinforcement placed in the tension region within structural concrete elements. However, for road pavement design, the tensile strength of concrete is used to determine its resistance to tensile stresses in flexure. The tensile strength of concrete is also important to control cracking of concrete resulting from tensile stresses caused by restrained concrete movement such as shrinkage and thermal expansion. There are two indirect test methods commonly used to measure the tensile strength of concrete, namely the splitting tensile strength test and the flexural strength test.

Splitting Tensile Strength

Similar to the observation made with compressive strength, the change in splitting tensile strength of GGC concrete in relation to PC concrete is affected by the GGC content and its fineness. For a GGC content up to 20%, an increase in the splitting tensile strength of concrete can be expected when GGC finer than PC is used (Metwally, 2007), and GGC of similar or coarser fineness compared with

PC may maintain or reduce the splitting tensile strength of concrete (Liu, 2011; Priscilla and Naik, 2014; Taha and Nounu, 2008a; Wang, 2011). However, the actual qualitative difference is small, no more than 0.5 MPa in most cases. It can be considered that the effect of GGC, for content up to 20%, on splitting tensile strength of concrete is essentially very small. Concrete containing 20%–50% GGC has relatively low tensile strength compared with corresponding PC concrete (Hussain and Chandak, 2015; Wang, 2011).

Figure 4.9 shows the relationship between characteristic cube strength and tensile strength of concrete made with up to 50% GGC measured at 28 days, together with the relationship obtained from Eurocode 2 (2004) for comparison purposes. Both the characteristic cube strength and the tensile strength are determined using the following equations:

$$f_{ck,\,cube} = f_m\,(1 - 1.64v)$$

where

$f_{ck,cube}$ is the characteristic cube compressive strength;

f_m is the mean cube compressive strength;

1.64 is the constant for 5% of individual cube strength below the design strength;

v is the coefficient of variation, taken as 6% for fair laboratory control (ACI 301, 2005).

$$f_{ct} = 0.9 f_{ct,\,sp} \text{ (based on Eurocode 2, 2004)}$$

where

f_{ct} is the tensile strength;

$f_{ct,sp}$ is the splitting tensile strength.

It can be seen from Figure 4.9 that, excluding two points (both from Taha and Nounu, 2008a), which deviate significantly from the majority of the data, the trend lines obtained for GGC concrete and PC concrete are almost the same, and both are marginally lower than that of Eurocode 2 (2004). This perhaps suggests that Eurocode 2 (2004) is applicable for concrete containing GGC in estimating the tensile strength of concrete without the need of any modification. Additionally, it is evident from Figure 4.9 that for a given compressive strength, the tensile strength of GGC concrete is similar to that of PC concrete.

Flexural Strength

The flexural strength of concrete is determined using a centre-point loading (ASTM C293, 2015) or third-point loading (ASTM C78, 2015 and BS EN 12390-5, 2009) test on a beam specimen. The results obtained from third-point loading are about 15% lower, but less variable, than those obtained from centre-point loading (Lamond and Pielert, 2006).

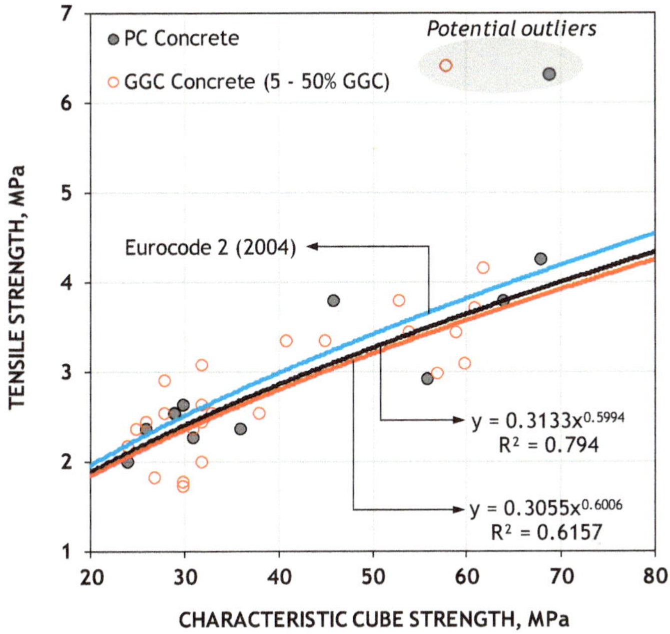

Figure 4.9 Relationship between characteristic compressive and tensile strength of ground glass cullet (GGC) concrete. *PC*, Portland cement.
Data taken from Hussain and Chandak (2015), Liu (2011), Metwally (2007), Mitra et al. (2016), Narayana and Mailar (2015), Priscilla and Naik (2014), Taha and Nounu (2008a, 2009), Tuncan et al. (2001) and Wang (2011).

Figure 4.10 shows the influence of GGC content on the flexural strength of concrete determined at 28 days. No distinction is made for GGC fineness in Figure 4.10, as the effect of GGC fineness is unclear. This is perhaps because the flexural strength test is sensitive to the handling and moisture condition of specimens. Notwithstanding this, it can be seen from Figure 4.10 that, on average, the inclusion of GGC up to 20% content as a cement component does not adversely affect the flexural strength of concrete, but beyond that, the flexural strength decreases as the GGC content is increased.

Owing to the sensitivity of flexural strength measurement, it may be more practical to develop a relationship between the characteristic compressive strength of concrete and its corresponding flexural strength during the design stage, and subsequently, the compressive strength instead of flexural strength can be used for production quality control. The characteristic cube strength of GGC concrete and corresponding reference PC concrete is plotted against their 28-day flexural strength, as shown in Figure 4.11, together with the relationship suggested by RILEM Technical Committee 162 (2000). The calculation of characteristic cube strength was the same as previously described in the splitting tensile strength section.

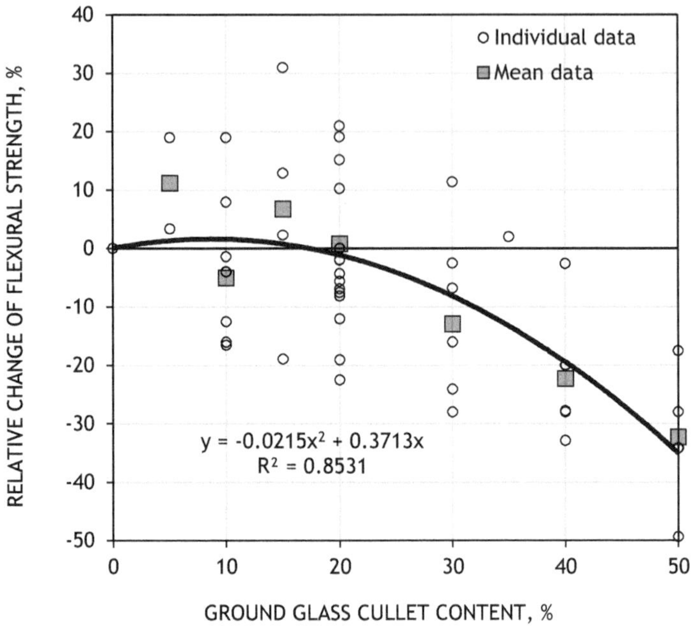

Figure 4.10 Effect of ground glass cullet content on the flexural strength of concrete at 28 days.
Data taken from Ali (2015), Aly et al. (2011), Bajad et al. (2011a), Cassar and Camilleri (2012), Dhir et al. (2005c), Kamali and Ghahremaninezhad (2015), Matos and Sousa-Coutinho (2012, 2016b), Nassar and Soroushian (2013), Parghi and Alam (2016), Taha and Nounu (2008a) and Wang and Hou (2011).

In developing the trend lines for both GGC concrete and PC concrete in Figure 4.11, two sets of results (Taha and Nounu, 2008a; Wang and Hou, 2011) have not been considered, as these data for PC concrete appeared to be deviating from the majority. It is evident from Figure 4.11 that the use of GGC as a cement component is not likely to adversely affect the relationship between compressive strength and flexural strength, with their trend lines being almost identical. However, both trend lines do not align with that obtained from RILEM Technical Committee 162 (2000). At a given compressive strength, the trend lines of PC concrete and GGC concrete suggest a flexural strength value that is 50% greater than that of RILEM Technical Committee 162 (2000) (Figure 4.11).

4.5.3 Elastic Modulus

The elastic behaviour of concrete is usually expressed in terms of modulus of elasticity, calculated from the stress–strain response of a test cylinder subjected to load. As far as the material characteristics are concerned, the modulus of elasticity of concrete is mainly affected by the elastic moduli of aggregate and hardened cement paste, as well as the aggregate–matrix interface.

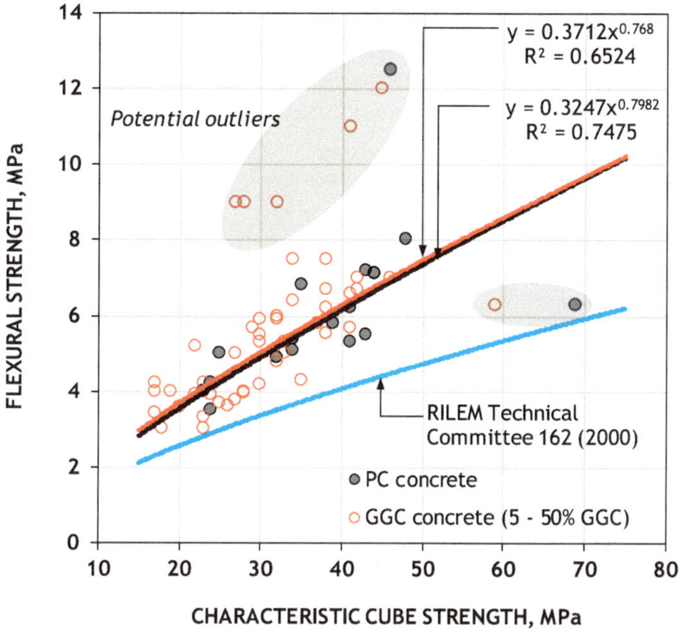

Figure 4.11 Relationship between characteristic compressive and flexural strength of ground glass cullet (GGC) concrete. *PC*, Portland cement.
Data taken from Ali (2015), Aly et al. (2011), Bajad et al. (2011a), Cassar and Camilleri (2012), Dhir et al. (2005c), Kamali and Ghahremaninezhad (2015), Matos and Sousa-Coutinho (2012, 2016b), Nassar and Soroushian (2013), Parghi and Alam (2016), Taha and Nounu (2008a) and Wang and Hou (2011).

The relationship between elastic modulus and compressive strength of GGC concrete and corresponding reference PC concrete, both determined at 28 days, is shown in Figure 4.12. Except for Samarin (1980), all the studies suggest that, for a given water/cement ratio, the use of GGC as a PC replacement reduces both the modulus of elasticity and the compressive strength of concrete, with the reduction in the latter being more prominent than in the former. In the study conducted by Samarin (1980), it was reported that concrete made with 20% GGC as PC cement and 100% GC sand as natural sand (0.57 water/cement ratio) has a modulus of elasticity and compressive strength similar to that of the reference concrete made with PC and natural sand (0.59 water/cement ratio). This could be due to denser cement paste structure in the former concrete, though it is expected that the concrete would show higher elastic modulus owing to the use of GC sand, which is slightly stiffer than natural sand.

Comparing the relationship between elastic modulus and compressive strength for limestone aggregate concrete given in Eurocode 2 (2004) with the results of Madandoust and Ghavidel (2013) and Taha and Nounu (2008a), in which limestone coarse aggregate was used, it can be seen that the former is relatively closer to the trend line in Eurocode 2 (2004), whilst the latter shows a significant deviation (Figure 4.12).

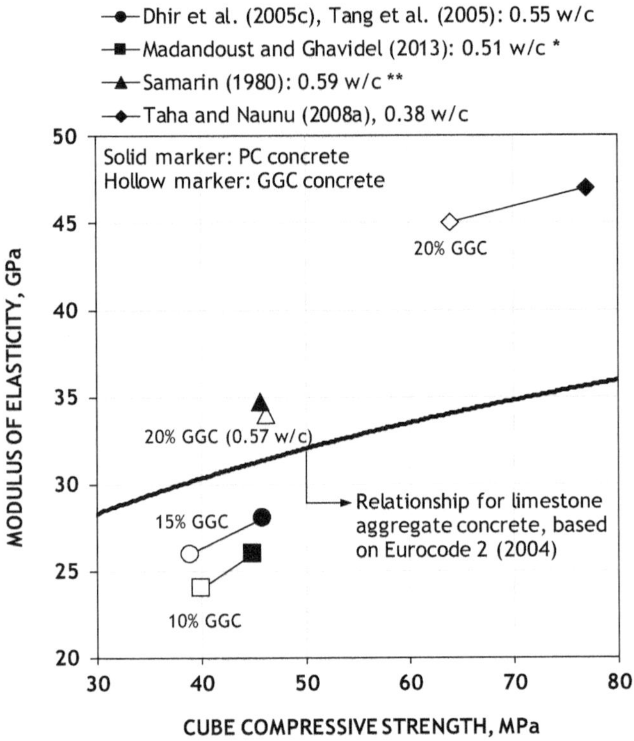

Figure 4.12 Relationship between compressive strength and modulus of elasticity of ground glass cullet (GGC) concrete at 28 days. * Both PC and GGC concrete contained 5% rice husk as cement. ** GGC concrete made with 100% GC as sand. *PC*, Portland cement; *w/c*, water/cement ratio.

4.5.4 Creep

When concrete is subjected to a sustained load, its volume decreases gradually over time, a phenomenon known as creep. Although an important concrete property in designing structures, the study of the creep of GGC concrete is not commonly available in the literature. Given that for the same stress/strength ratio at the time of load application, the long-term hydration and greater relative gain in strength of concrete made with low-heat cements can result in smaller creep than that of concrete made with normal cement (Neville, 1995), it is therefore postulated that GGC may exert a similar effect on the creep of concrete.

4.5.5 Autogenous Shrinkage

Autogenous shrinkage is a result of the self-desiccation process that takes place in cement paste whereby water is consumed for the hydration of the cement. Typically, the autogenous shrinkage is relatively insignificant in normal concrete, but it is pronounced in concrete of low water/cement ratio (for example, high-strength concrete) and concrete with high cement content (self-compacting concrete).

Table 4.6 Autogenous shrinkage of lightweight fly ash and ground glass cullet self-compacting concrete of 0.34 water/cement ratio at 49 days

Type	Fineness[a]	Content, %	28-Day Compressive Strength, MPa	Autogenous Shrinkage, μm
Fly ash	74% passing 45 μm	30	37	18
Soda lime GGC	50% passing 45 μm	30	39	48
Soda lime GGC	80% passing 45 μm	30	50	20

GGC, ground glass cullet.
[a]Based on Shi et al. (2005).
Based on Shi and Wu (2005).

The autogenous shrinkage values of lightweight SCC made with FA and soda lime GGC, studied by Shi and Wu (2005), are given in Table 4.6. Expanded shale was used as lightweight coarse aggregate for the mixes. GGC of 50% and 80% passing 45-μm sieve size and FA of 73.6% passing 45-μm were used as PC replacement. It can be seen from Table 4.6 that, for a given content, coarser GGC results in autogenous shrinkage at least 2 times higher than FA in lightweight SCC, whilst the compressive strength remains unchanged. When the fineness of GGC is kept similar to or slightly finer than that of FA, the autogenous shrinkage of lightweight SCC made with these two materials was about the same, though GGC renders higher compressive strength at 28 days.

4.5.6 Drying Shrinkage

The loss of water from hardened concrete to a nonsaturated environment results in a volume reduction known as drying shrinkage. Except for the case in which shrinkable aggregate is used, normally, shrinkage takes place in cement paste and aggregate provides restraint for such movement. As far as the cement paste is concerned, its quality, judging from the water/cement ratio and degree of hydration, as well as its volume, has a primary effect on the magnitude of drying shrinkage.

The ratio of drying shrinkage of concrete containing up to 40% GGC to that of its corresponding reference PC concrete, expressed in percentage form, is shown in Figure 4.13. The water/cement ratios of both GGC concrete and PC concrete specimens were kept the same, ranging from about 0.30 to 0.67. The specimens were stored in an environment having 50%–65% RH and 20–23°C for different lengths of duration, the shortest being 28 days (Wang et al., 2014) and the longest being close to 2 years (Kara et al., 2014).

It can be seen from Figure 4.13 that, despite some fluctuations in the results, on the whole, the use of GGC as a cement component increases the drying shrinkage of concrete. The effect is likely to become more apparent with increasing GGC content in the cement. Although the shrinkage measurements are not provided, Dumitru

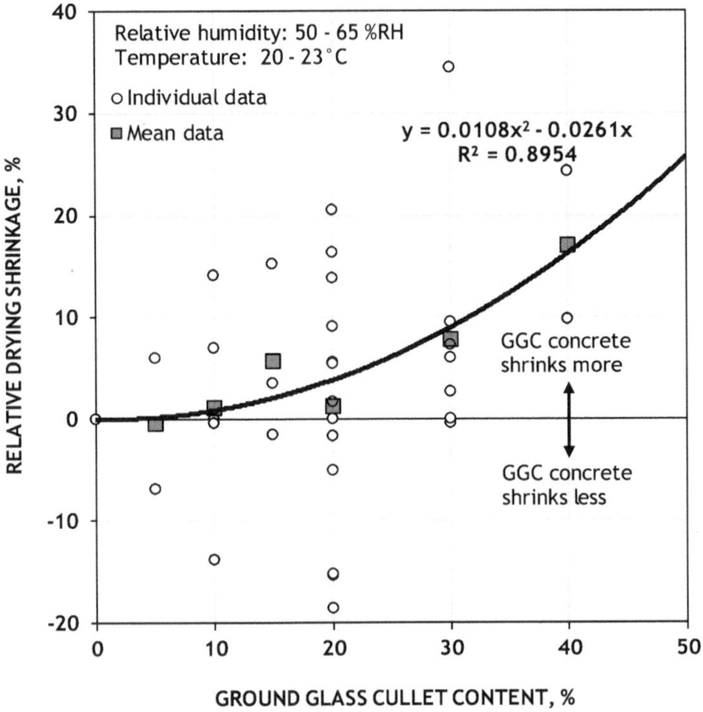

Figure 4.13 Effect of ground glass cullet (GGC) content on the drying shrinkage of concrete. Data taken from Calmon et al. (2014), Dhir et al. (2005c), Dumitru et al. (2010), Kara et al. (2014), Lee and Lee (2016), Nassar and Soroushian (2013), Shayan (2002), Shayan and Xu (2004), Tang et al. (2005), Wang (2011), Wang et al. (2014) and Wattanapornprom and Stitmannaithum (2015).

et al. (2010) also reported that the drying shrinkage of concrete made with 15% and 25% GGC is higher than that of PC concrete. This is perhaps not surprising as other cementitious materials, such as FA, GGBS and silica fume, are known to increase the drying shrinkage of concrete (Neville, 1995).

4.6 Permeation

Water, aggressive ions and gases can enter concrete through three distinct mechanisms, namely absorption, permeability and diffusion. The ease of transportation of fluids into concrete is governed by its porosity in terms of volume, size and interconnectivity of voids, which in turn are essentially influenced by the properties of hardened cement paste and particle packing. Given that the pozzolanic reaction of GGC densifies the cement paste structure, thereby reducing the porosity of concrete, it is expected that the use of GGC as a PC replacement can result in an improvement in the permeation of concrete.

4.6.1 Porosity

There are several types of pores with varying size ranges present in hardened cement paste, such as gel pores, capillary pores and entrained and entrapped air (air voids). These pores have different influences on the various properties of concrete, and only the larger pores, such as capillary pores and air voids, affect its permeation properties (Aligizaki, 2006).

Table 4.7 presents the amount of permeable voids (accessible to water) in cement paste and concrete containing GGC measured at the standard age of 28 days and, in long terms, at 91 and 404 days. The porosity measurements were obtained by determining the weight of saturated specimens in air, after oven drying and immersion in water for a period of time.

At 28 days, the amount of permeable voids of mixes with GGC is similar to that of mixes without GGC (Table 4.7). It should be remembered that although the pozzolanic reaction is normally a slow process and happens at the later age, it appears that the inclusion of GGC up to 60% does not adversely affect the porosity of mixes at the age of 28 days.

At a later age, the amount of permeable voids of GGC concrete mixes is comparable to (Du and Tan, 2015; Pavoine et al., 2014; Shayan and Xu, 2006) or less than that of corresponding reference PC mixes (Kamali and Ghahremaninezhad, 2015, 2016). The reduction in permeable voids is likely to be due to pore refinement in hydrated cement paste resulting from the pozzolanic reaction of GGC. Similar observations are also found in GGC mixes made with coarse recycled concrete aggregate (RCA) (Nassar and Soroushian, 2012a). This is perhaps a good achievement in producing a more sustainable yet durable concrete, as the inclusion of GGC is shown to reduce the negative effect of coarse RCA, that is, its contribution to the porosity of concrete.

It should be noted that the higher amount of permeable voids in concrete containing 30% GGC (Shayan and Xu, 2006) can be attributed to the bleeding reported as being seen during its fresh state, which normally results in a porous structure in hardened concrete.

4.6.2 Absorption

As moist curing ceases, most of the pores in concrete are partially saturated, due to the loss of internal moisture to the surrounding environment and continuation of the hydration process, making the concrete able to absorb any liquids. The absorption capacity of concrete can be determined with a simple water absorption test, which, in principal, measures the weight increase of a dry sample after water immersion.

Figure 4.14(a) and (b) shows the water absorption of concrete containing GGC as a cement additive, with the water/cement ratio ranging from 0.38 to 0.64, measured after 28 to 91 days moist curing, mostly in accordance with the ASTM C642 (2013) method. In general, the changes in the water absorption of concrete due to the inclusion

Table 4.7 Permeable voids of mixes containing ground glass cullet as cement

References	w/c	GGC Content, %	Permeable Voids, %	
			28 days	**91 days**
(a) Cement Paste				
Kamali and Ghahremaninezhad (2015)	0.50	0	26.0	30.0
		20[a]	26.5	25.0
		20[b]	27.5	26.0
Kamali and Ghahremaninezhad (2016)	0.50	0	23.0	22.5
		20[a]	24.0	15.0
		20[b]	20.5	15.0
(b) Concrete Made With Natural Aggregate				
Du and Tan (2015)	0.49	0	11.5	12.2
		15	10.5	11.0
		30	10.0	10.8
		45	9.5	11.8
		60	11.7	12.6
Pavoine et al. (2014)	0.40	0	–	13.0
		20	–	14.0
	0.55	0	–	16.5
		20	–	21.5
Shayan and Xu (2006)	0.49	0	13.2	12.2[c]
		20	14.1	13.8[c]
		30[d]	15.0	16.2[c]
(c) Concrete Made With 100% Coarse RCA				
Nassar and Soroushian (2012a)	0.38	0	–	15.5
		20	–	13.2
	0.50	0	–	15.8
		20	–	13.8

GGC, ground glass cullet; *RCA*, recycled concrete aggregate; *w/c*, water/cement ratio.
[a]Aluminosilicate glass.
[b]Soda lime glass.
[c]Bleeding in fresh state.
[d]Measured at 404 days.

of GGC are considerably small, although the majority of the results tend to suggest that the absorption of concrete decreases with increasing GGC content up to 30%, regardless of the concrete age (Figure 4.14(a)). Thus, this opens up an opportunity for engineers to specify GGC concrete for water-retaining structures such as water dams, reservoirs and drains, in which low absorption capacity of concrete is crucial.

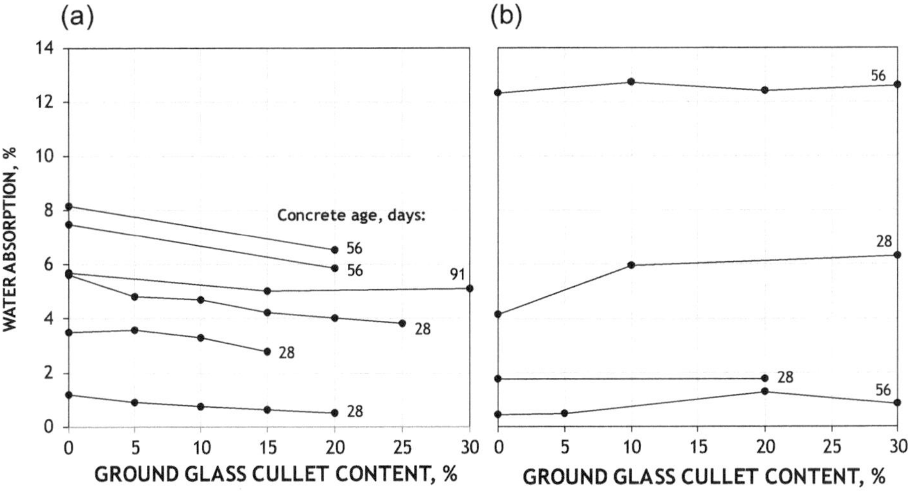

Figure 4.14 Effect of ground glass cullet (GGC) on the water absorption of concrete at different ages showing (a) reduction and (b) increase or no change as GGC is increased. Data taken from Bhat and Rao (2014), Du and Tan (2015), Kadir et al. (2016), Nwaubani and Poutos (2013), Nassar and Soroushian (2012a), Parghi and Alam (2016), Wang et al. (2016), Taha and Nounu (2008a) and Tuncan et al. (2001).

In addition, apart from using GGC alone as a PC replacement, the water absorption of concrete made with a ternary blended cement with GGC (up to 25% content) and FA (up to 30% content) and a quaternary blended cement with GGC (up to 25% content), FA (at 10% content) and silica fume (at 10% content) has been investigated by Parghi and Alam (2016) and Tuncan et al. (2001). The studies suggest that the combination of GGC with FA or a mixture of FA and silica fume offers a slightly better reduction in the water absorption of concrete than the use of GGC alone. This is thought to be due to better packing of cementitious particles, confirming again that GGC is a viable material for use in concrete with low water absorption requirements.

4.6.3 Permeability

Permeability of concrete defines the flow of a liquid or gas into concrete under a pressure gradient. It is controlled mainly by the capillary porosity of hardened cement paste and the interfaces between matrix and aggregate phases. The effect of GGC on the permeability of concrete has been investigated in two different forms, namely oxygen permeability and water penetration, and the results are summarised in Table 4.8. It should be noted that GGC used in these studies is finer than that of corresponding reference PC.

In the oxygen permeability test, specimens having 0.30 water/cement ratio were subjected to three different constant gas pressures of 17, 25 and 50 kPa (Chaid et al., 2015). The findings were positive, as the oxygen permeability of concrete containing 30% GGC is close to zero regardless of the applied gas pressure, whilst that of the reference PC concrete is about 2.25×10^{-16} m^2. This suggests that the pozzolanic reaction of GGC is able to break the connectivity of the capillary pores, making the concrete impermeable to gases. As oxygen gases are responsible for the corrosion of reinforcement, this implies that the use of GGC cement in concrete may be more favourable in an environment conducive to corrosion.

In the water penetration test of concrete, the depth of penetration of the waterfront is measured, after subjecting the test specimen to a fixed water pressure over a period. The results of Du and Tan (2015), which were obtained from 28-day moist-cured specimens under a water pressure of 0.75 MPa, show that the water penetration depth of concrete decreases as GGC content increases up to 60% (Table 4.8). Except for concrete containing 60% GGC, the improvement in the permeability of concrete due to the inclusion of GGC becomes greater at 90 days. This reduction in water penetration of GGC concrete could be attributed to a reduction in the continuity of capillary pores, which are the pathways for water transportation, resulting from pozzolanic reaction of GGC. Such data tend to suggest that GGC is well suited to use in water-retaining structures.

Similar water penetration tests have been carried out (Kara, 2013a) on concrete made using GGC as a PC replacement and RCA as a coarse aggregate replacement. The depth of water penetration was measured on 28-day moist-cured specimens after subjecting them to a water pressure of 5 MPa for 72 h. Table 4.8 suggests that, for a fixed water/cement ratio of 0.44 and GGC content of 20%, the depth of water

Table 4.8 Permeability of ground glass cullet concrete measured in various forms

References	w/c	GGC Content, %	Result
			Oxygen Permeability, $\times 10^{-16}$ m^2
Chaid et al. (2015)	0.30	0	2.25
		30	≈0
			Water Penetration Depth, mm
Du and Tan (2015)	0.49	0	25.5, 21.5[a]
		15	11.5, 6.0[a]
		30	9.0, 3.0[a]
		45	8.0, 4.0[a]
		60	5.5, 7.5[a]
Kara (2013a)	0.44	0	20
		20[b]	18
		20[c]	20

GGC, ground glass cullet; *RCA*, recycled concrete aggregate; *w/c*, water/cement ratio.
[a]Results at 28 days, 90 days.
[b]With 50% coarse RCA.
[c]With 100% coarse RCA.

penetration of concrete containing 50% and 100% coarse RCA is essentially the same as that of the reference normal aggregate concrete. The actual effect of GGC on the permeability of RCA concrete is not known, as the results for RCA concrete made without GGC are not available. However, given that the results are comparable to those of concrete made with 100% PC and 100% natural coarse aggregate (Table 4.8), it may be concluded that the expected high permeability of RCA concrete due to RCA being a relatively a high absorption material is offset by the pore refinement resulting from pozzolanic reaction of GGC.

4.6.4 Diffusion

The transportation of a fluid to concrete under a concentration gradient is referred to as diffusion. The effect of GGC on the moisture diffusion of concrete has been determined by Schwarz et al. (2008) in accordance with the method developed by Neithalath (2006) for concrete made with 0% (reference) and 10% GGC, as well as concrete made with 10% FA for comparison. Two series of concrete were tested: in the first series, the water/cement ratio of all concrete was fixed at 0.40, and in the second series, the water/cement ratio of both 10% GC concrete and 10% FA concrete was reduced to achieve consistency similar to that of the reference concrete. The results for the moisture diffusion coefficient of concrete moist cured for 14 and 90 days, together with their compressive strength determined at 14, 28 and 90 days, are given in Table 4.9.

For concrete with moist-curing age of 14 days, except for 10% FA concrete with 0.40 water/cement ratio, the moisture diffusion coefficient of either 10% GGC or 10% FA concrete, of similar or lower water/cement ratio than reference concrete, is $1.2 \times 10^{-6}\,m^2/h$, which is very close to that of reference concrete at $1.1 \times 10^{-6}\,m^2/h$. The exceptionally high moisture diffusion coefficient of $40 \times 10^{-6}\,m^2/h$ observed in

Table 4.9 Diffusion coefficient of water for concrete containing ground glass cullet and fly ash after curing for 14 and 90 days (Schwarz et al., 2008)

		Compressive Strength, MPa			Diffusion Coefficient, $\times 10^{-6}\,m^2/h$	
Cement	**w/c**	**14 days**	**28 days**	**90 days**	**14 days**	**90 days**
100% PC	0.40	41	45	54	1.1	0.070
90% PC + 10% GGC	0.40	38	43	49	1.2	0.100
	0.36	39	43	50	1.2	0.015
90% PC + 10% FA	0.40	34	40	50	40.0	0.065
	0.39	39	43	52	1.2	0.010

FA, fly ash, 74% passing 45 μm; *GGC*, ground glass cullet, 72% passing 45 μm; *PC*, Portland cement, 95% passing 45 μm; *w/c*, water/cement ratio.

10% FA concrete with 0.40 water/cement ratio could be due to the slow pozzolanic reaction of FA, which is evident in its relatively low 14-day compressive strength compared with other concrete (Table 4.9). Although also a pozzolanic cement, GGC does not seem to significantly affect either the compressive strength or the permeation of concrete at the early age.

For concrete with moist-curing age of 90 days, the moisture diffusion coefficient of all the 0.40 water/cement ratio concrete, regardless of the type of cement used, is essentially similar, varying within 0.065×10^{-6} to $0.100\times10^{-6}\,m^2/h$ (Table 4.9). On the other hand, both 10% GGC and 10% FA concrete with water/cement ratio lower than 0.40 had a lower moisture diffusion coefficient value of 0.015×10^{-6} to $0.010\times10^{-6}\,m^2/h$, respectively, which is likely due to their denser hardened cement paste structure.

Overall, it would appear that when GGC is used at 10% as PC replacement, the diffusion coefficient of the concrete can remain unchanged if its water/cement ratio is unadjusted, or the value can be reduced if the water/cement ratio is reduced as a result of improvement in consistency. Compared with FA, GGC is found to offer higher compressive strength and improved diffusion in concrete at the early age.

4.7 Durability

With the implementation of Eurocode BS EN, 1990:2002+A1, 2005, the indicative design working life of building structures is taken as 50 years and that of monumental building structures, bridges and other civil engineering structures as 100 years. Thus, there is a strong emphasis on ensuring that concrete structures remain robust and stable throughout their service life. The longevity and serviceability of a structure are governed by its durability properties, which are closely related to the hardened cement paste properties. As the pozzolanic reaction of GGC modifies the structure and chemistry of cement paste, this section discusses the durability performance of concrete containing GGC as PC replacement.

4.7.1 Chloride Ingress

The presence of chloride ions in concrete can destroy the protective passive film of the reinforcement and this can lead to its corrosion where oxygen and moisture are present. The presence of chloride ions in concrete can be (i) internal due to the contamination of constituent materials and the use of a chloride-laden admixture, and (ii) external, due to exposure to a chloride-rich environment. The former is generally not a concern as the maximum chloride content of concrete is regulated by standards such as BS EN 206 (2013). Thus, chloride attack on concrete is predominantly due to the ingress of chloride ions from various sources such as seawater, deicing salts and industrial processes.

Commonly, the rapid chloride penetration test (RCPT) is used in accordance with ASTM C1202 (2012) to provide an indication of the resistance of concrete to chloride ingress. Figure 4.15 shows RCPT results for concrete with water/cement ratio mainly varying from 0.40 to 0.50, with GGC content varying over a range up to 60%, and moist cured for 28 and 91 days, together with the chloride ion penetrability classes given in ASTM C1202 (2012) for comparison purposes. It is evident from Figure 4.15 that the inclusion of GGC as PC replacement reduces the charge passed in the concrete at an early age of 28 days. The reduction in charge passed is more pronounced, from high to moderate chloride ion penetrability, or from moderate to low, when the content of GGC is greater than 20% (Figure 4.15). This indicates that concrete made with cement containing GGC can be expected to be less susceptible to chloride ion penetration.

Table 4.10 compares the non-steady-state chloride migration coefficient of GGC concrete with those of PC concrete and FA concrete, at different curing ages, obtained using the NT Build 492 (1999) method. At 7 days, the chloride migration coefficient of concrete appears to be influenced by the fineness of both GGC and PC (Table 4.10), with finer GGC reducing the coefficient value (Du and Tan, 2015), whilst GGC of fineness similar to PC showing the opposite (Wang et al., 2009). At 28 days, the chloride migration coefficient of GGC concrete is similar to or lower than that of PC concrete (Table 4.10). At 90 days, all the results suggest that GGC reduces the chloride migration coefficient of concrete and the effect of GGC fineness is less significant. An average magnitude reduction of 3 times (range: 1–5 times) can be observed for concrete containing up to 20% GGC.

Compared with FA of similar content at 30%, the chloride migration coefficient of GGC concrete, measured at 91 days, is about 2 times lower than that of FA concrete (Table 4.10).

Overall, the results shown in Figure 4.15 and Table 4.10 reinforce the view that the pore refinement of cement structure resulting from the pozzolanic reaction of GGC can increase the resistance of concrete to chloride ingress. Thus, given the data available, it could be concluded that GGC can be specified for structural concrete exposed to a chloride-rich environment.

4.7.2 Carbonation

Carbonation takes place in concrete when carbon dioxide, upon entering the concrete, reacts with calcium hydroxide, producing calcium carbonate. The major concern of carbonation is that it can cause corrosion of steel reinforcements in concrete, as the carbonation reaction lowers the pH of the concrete, which in turn leads to the removal of the protective passive film of the reinforcement.

The carbonation resistance data, expressed in terms of the carbonation depth of GGC concrete and the corresponding reference PC concrete at equal water/cement ratios of 0.50 and 0.55 and subjected to accelerated carbonation at 3%–5% concentration of CO_2, 20–23°C and 65% RH for 2–4 months, are given in Table 4.11. The results

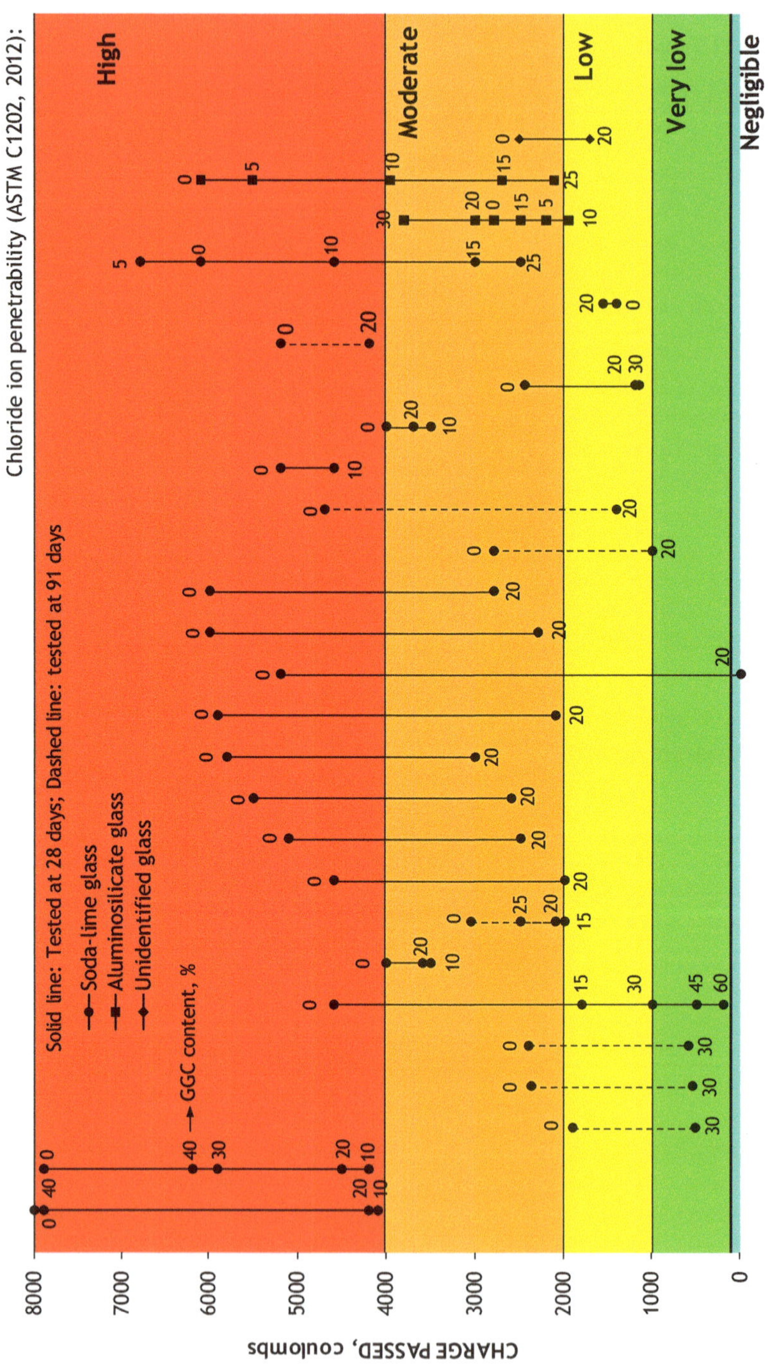

Figure 4.15 Rapid chloride penetration test results of concrete containing ground glass cullet (GGC).
Data taken from Cassar and Camilleri (2012), Chaid et al. (2015), Du and Tan (2014), Jain and Neithalath (2010), Kamali and Ghahremaninezhad (2015), Laldji et al. (2004), Nassar and Soroushian (2011), Omran and Tagnit-Hamou (2016), Shayan and Xu (2006), Tagnit-Hamou and Bengougam (2012) and Zheng (2013).

Table 4.10 Chloride migration coefficient of ground glass cullet concrete measured in the non-steady state at different curing ages

References	Parameter	GGC Content, %	D_{nssm}, $\times 10^{-12}\,m^2/s$		
			7 days	**28 days**	**91 days**
(a) Comparison With PC					
Du and Tan (2015)	PC, 10 μm diam. GGC, 2.4 μm diam. 0.49 w/c	0	21	20	20
		15	20	8	4
		30	17	3	3
		45	6	1	1
		60	5	2	2
Jain and Neithalath (2010)	PC, 13 μm diam. GGC, 20 μm diam. 0.40 w/c	0	–	10.0	8.5
		10	–	10.5	8.0
		20	–	11.0	5.0
Kamali and Ghahremaninezhad (2015)	PC, not available GGC, 8.4 μm diam. 0.50 w/c	0	–	19	17
		10	–	21	10
		20	–	10	3
Wang et al. (2009)	Gradings of PC and GGC are similar[a] 0.48 w/c	0	40	20	19
		25	62	22	10
		50	90	10	8
Matos and Sousa-Coutinho (2012)	Gradings of PC and GGC are similar[a] 0.50 w/c	0	–	16.2	–
		10	–	9.5	–
		20	–	7.8	–
(b) Comparison With Fly Ash					
Shi and Wu (2005)	Fineness not available 0.34 w/c	30 (fly ash)	–	–	7.4[b]
		30	–	–	3.5[b]

GGC, ground glass cullet; *PC*, Portland cement; *w/c*, water/cement ratio.
[a]Gradings given in the study.
[b]At 56 days.

suggest that the carbonation depth of specimens made with GGC up to 20% content was about 2–4 times greater than that of the corresponding PC specimens, suggesting that concrete containing GGC has lower carbonation resistance. This is to be expected, as GGC consumes some calcium hydroxide in the cement paste system in the process of developing the pozzolanic reaction with GGC, consequently making the resultant concrete more susceptible to carbonation. This phenomenon also occurs in concrete containing FA (Lye et al., 2015), GGBS (Lye et al., 2016) and limestone powder (Elgalhud et al., 2017).

Table 4.11 Carbonation depth of concrete made with ground glass cullet

References	w/c	Curing, days	Exposure Conditions	GGC Content %	Carbonation Depth, mm
Dhir et al. (2005a)	0.55	28	3%–4% CO_2, 20°C, 65% RH (for 9 weeks)	0	4
				15	14
Matos and Sousa-Coutinho (2012)	0.50	n.a.	5% CO_2, 23°C, 65% RH (for 16 weeks)	0	3
				10	5
				20	8

GGC, ground glass cullet; *RH*, relative humidity; *w/c*, water/cement ratio.

Compared with FA, whose chemical composition, in some respects, is comparable to that of GGC, it would appear that for a given PC replacement level, GGC concrete is likely to show a higher carbonation effect than FA concrete (Lye et al., 2015).

4.7.3 Acid Attack

Exposure of concrete to acids can cause dissolution of the hydrated cement paste and sometimes the aggregates. Thus, concrete that is prone to acid attack, such as exposure to industrial or agriculture effluent, needs to be properly designed and protected.

Data on the effect of GGC on the resistance of mortar to acid attack, expressed in terms of the change in mass and compressive strength, before and after being subjected to 12 weeks immersion in 5% sulphuric acid solution (Siad et al., 2016) or five drying and wetting cycles in concentrated sulphuric acid solution (Wang, 2011), are given in Table 4.12.

Collectively, the results suggest that the acid resistance of mortar increases or remains unchanged when PC is replaced by GGC up to 15%. However, beyond 15% GGC content, the results are contradictory; Siad et al. (2016) showed that the acid resistance of mortar increases as GGC content is increased up to 45%, but the opposite is observed in Wang (2011) for up to 50%. The reasons for these contradictory results are unknown, but given that calcium hydroxide, which is vulnerable to acid attack (Neville, 1995), will be consumed by GGC during the pozzolanic reaction, the use of GGC as PC replacement is more likely to improve the resistance to acid attack.

In a parallel study (Siad et al., 2016), it was found that the use of 10% GGC in conjunction with either 10% GGBS or 10% FA results in mortar with acid resistance that is on par with that of mortar made with 15% GGC.

Table 4.12 Resistance of mortar made with ground glass cullet to acid attack

References	Parameter	GGC Content, %	Change in Mass[a], %	Change in Strength[a], %
Siad et al. (2016)	w/c 0.45; moist curing 28 days Test: immersed in 5% H_2SO_4 solution for 12 weeks (ASTM C267, 2012)	0	−39	−53
		15	−37	−45
		30	−28	−40
		45	−25	−35
		10% GGC + 10% GGBS	−33	−50
		10% GGC + 10% FA	−32	−48
Wang (2011)	w/c 0.485; moist curing not given Test: five cycles of drying and wetting in concentrated H_2SO_4 solution	0	−7.9	n.a.
		10	−7.8	n.a.
		20	−9.2	n.a.
		30	−9.2	n.a.
		40	−10.0	n.a.
		50	−10.2	n.a.

FA, fly ash; *GGBS*, ground granulated blast furnace slag; *GGC*, ground glass cullet; *w/c*, water/cement ratio.
[a]In conjunction, GGBS and FA.

4.7.4 Sulphate Attack

Sulphate attack in concrete can result from the use of sulphate-contaminated constituent materials or when concrete is in contact with a sulphate-bearing environment such as clayey soils and groundwater. Depending on the type of sulphate, such as calcium, sodium, potassium and magnesium, the associated damage caused by sulphate attack in concrete can vary in its severity.

The results of change in the length of concrete specimens having 0.50 and 0.55 water/cement ratio and incorporating GGC, silica fume and FA and exposed to sodium sulphate solutions are given in Table 4.13. It is evident from Table 4.13 that GGC mortars show less expansion than PC mortars, suggesting that the use of GGC increases the sulphate resistance of the mixes, even at a level of 10% (Matos and Sousa-Coutinho, 2012, 2016b). In addition, increasing fineness of GGC, from 400 to 600 m^2/kg Blaine fineness, resulted in an increase in sulphate resistance of the mix (Carsana et al., 2014).

Comparing the performance of GGC with other pozzolanic cements, it can be seen from Table 4.13 that, for a given PC replacement content at 30%, GGC shows higher sulphate resistance than FA (Matos and Sousa-Coutinho, 2012, 2016b). For silica fume, the same observation is also apparent in the studies conducted by Matos and Sousa-Coutinho (2012, 2016b), although the study of Carsana et al. (2014) suggests that GGC can also result in better sulphate resistance than silica fume but at a higher content.

Table 4.13 Length change of mortar and concrete mixes containing ground glass cullet and other pozzolanic cements immersed in sodium sulphate solution

References	Parameter	Pozzolanic Cement	Change in Length[a], %
Carsana et al. (2014)	**Mortar**, 0.50 w/c; curing duration not known; 5% Na_2SO_4 immersion for 8 weeks (ASTM C1012, 2010)	0	+0.017
		30% GGC[b]	+0.007
		30% GGC[c]	+0.002
		10% SF	+0.005
		30% FA[d]	+0.011
Matos and Sousa-Coutinho (2012)	**Mortar**, 0.50 w/c; 28 days curing; Na_2SO_4 immersion for 26 weeks (LNEC E462, 2004)	0	+0.220
		10% GGC	+0.001
		10% SF	+0.090
Matos and Sousa-Coutinho (2016b)	**Mortar**, 0.50 w/c; 28 days curing; Na_2SO_4 immersion for 26 weeks (LNEC E462, 2004)	0	+0.161
		10% GGC	+0.004
		10% SF	+0.088
Tang et al. (2005)	**Concrete**, 0.55 w/c; 28 days curing; 5% Na_2SO_4 immersion for 20 weeks	0	−0.003
		15% GGC	−0.008

FA, fly ash; *GGC*, ground glass cullet; *SF*, silica fume; *w/c*, water/cement ratio.
[a]Positive and negative values indicate expansion and shrinkage, respectively.
[b]GGC 400 m^2/kg Blaine fineness.
[c]600 m^2/kg Blaine fineness.
[d]Containing 7.4% CaO.

Improvement in sulphate resistance due to the incorporation of GGC with PC as well as other pozzolanic cements is to be expected. This is because it reduces two cement hydrate products in the hydrated cement paste that are more vulnerable to sulphate attack, namely: (i) tricalcium aluminate (C_3A) hydrates by reducing the content of PC and (ii) calcium hydroxide ($Ca(OH)_2$) through its consumption in the pozzolanic reaction. Thus, GGC, like other pozzolanic cements, is also suitable for specification by engineers for use in concrete exposed to sulphate attack, such as foundations, and it is likely to show higher sulphate resistance than FA and probably silica fume also. Surprisingly, the study undertaken by Tang et al. (2005) recorded shrinkage in both PC and GGC concrete, which is not what one would expect from sulphate attack tests, and in the absence of further data it is difficult to draw any sensible conclusion.

It should be mentioned that sulphate resistance of concrete can also be expressed as the change in compressive strength or mass of specimens after immersion in sulphate solutions. However, such tests can be difficult to analyse owing to the complexity of the sulphate reactions involved. Figure 4.16 shows change in compressive strength of concrete and mortar specimens containing GGC after continuous immersion in 4%–8% magnesium sulphate and 4%–5% sodium sulphate solutions for up to

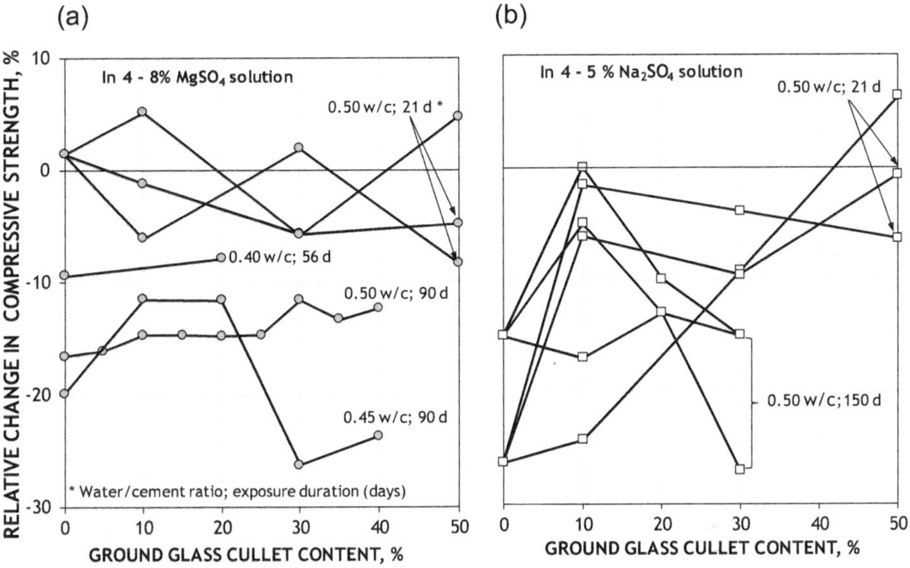

Figure 4.16 Change in strength of specimens containing ground glass cullet in (a) $MgSO_4$ and (b) Na_2SO_4 solutions.
Data taken from Bajad et al. (2011a, 2012a,b), Carsana et al. (2014), Ke et al. (2015), Özkan and Yüksel (2008), Patagundi and Prakash (2015) and Priscilla and Naik (2014).

150 days, conducted by various researchers during 2008–15. The results do not show a consistent change in relation to GGC content. Indeed, surprisingly in some cases, strength gain instead of expected strength loss is reported. Notwithstanding this, overall, the results suggest that the use of GGC can improve the resistance of mortar and concrete mixes to sulphate attack.

4.7.5 Alkali–Silica Reaction

The chemical reaction between alkalis in cement and reactive silica constituents present in aggregate is known as ASR, which produces a gel reaction product that swells when it imbibes water. The expansion of the gel is deleterious, as it increases the internal pressure in concrete, subsequently resulting in cracking of the concrete and finally the loss of integrity of structure.

Commonly, pozzolanic cements such as FA and metakaolin and GGBS are used to mitigate ASR expansion, as these materials lower the pH of the mix through the consumption of $Ca(OH)_2$ in the pozzolanic reaction.

There are a number of test methods that have been developed to assess ASR in concrete by measuring the expansion of specimens subjected to high alkaline conditions. The following three are the most commonly used methods, and their test conditions and the interpretation of the expansion data are described in Table 4.14.

Table 4.14 Interpretation of alkali–silica reaction results obtained using various test methods

	Interpretation of Alkali–Silica Reaction Results			
Test Method	**Duration**	**Expansion, %**	**Description/ Classification**	**Based on**
ASTM C1260 (2014)	14 days	<0.10	Innocuous	ASTM C1260 (2014)
		0.10–0.20	Innocuous or deleterious behaviour in field performance	
		>0.20	Potentially deleterious expansion	
ASTM C1567 (2013)	14 days	>0.10	Potentially deleterious expansion	ASTM C1567 (2013)
BS 812-123 (1999)	52 weeks	≤0.05	Nonexpansive	BRE Digest 330 (2004)
		0.05 to ≤0.10	Probably nonexpansive	
		0.10 to ≤0.20	Possibly expansive	
		>0.20	Expansive	

(i) ASTM C1260 (2014), a mortar-bar method that is used to determine the reactivity of an aggregate

(ii) ASTM C1567 (2013), also a mortar-bar method like ASTM C1260, but used to evaluate the ability of pozzolanic cements and GGBS to control ASR expansion

(iii) BS 812-123 (1999), a concrete prism method, which is used to determine the reactivity of an aggregate

Probably the first ASR study of mixes containing GGC as PC replacement was reported by a group of researchers from Canada in 2000 (Shao et al., 2000), which investigated the effect of the fineness of GGC on ASR expansion (discussed later). Since then, the ASR of mortar and concrete containing GGC, predominantly of the soda lime glass type, with GGC content up to as high as 60% (Dhir et al., 2005a), has been evaluated by different researchers mostly using the three aforementioned test methods.

Collectively, the studies can be separated into two groups based on the types of fine aggregate used, that is (i) natural and unreactive aggregate (or assumed as natural aggregate when the information is unclear) and (ii) reactive aggregate or glass cullet (GC)-sand, as shown in Figures 4.17 and 4.18, respectively. The former group investigates the possible expansion resulting from the silica in GGC, whilst the latter group examines the effectiveness of GGC to nullify the ASR-induced expansion. It should be mentioned that the results presented therein were the expansion readings taken at the recommended test duration (varies depending on the test method used), although the results at later ages were available in some cases.

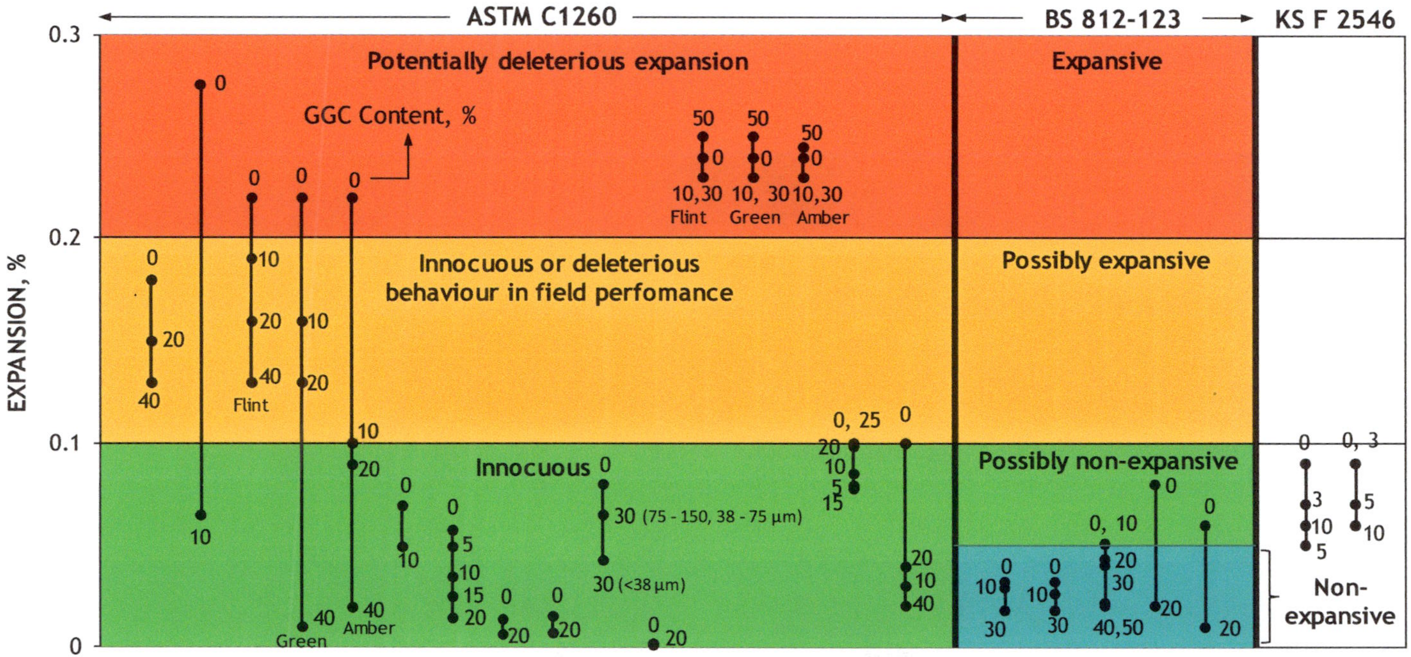

Figure 4.17 Expansion of ground glass cullet (GGC) specimens made with natural and unreactive fine aggregate.
Data taken from Aly et al. (2012), Dhir et al. (2005c), Dyer and Dhir (2001a,b), Jang et al. (2015), Liu (2011), Mandville et al. (2013), Metwally (2007), Nassar and Soroushian (2012a,b, 2013), Özkan and Yüksel (2008), Parghi and Alam (2016), Shao and Lehoux (2001), Shayan and Xu (2004), Taha and Nounu (2008b, 2009) and Zheng (2013).

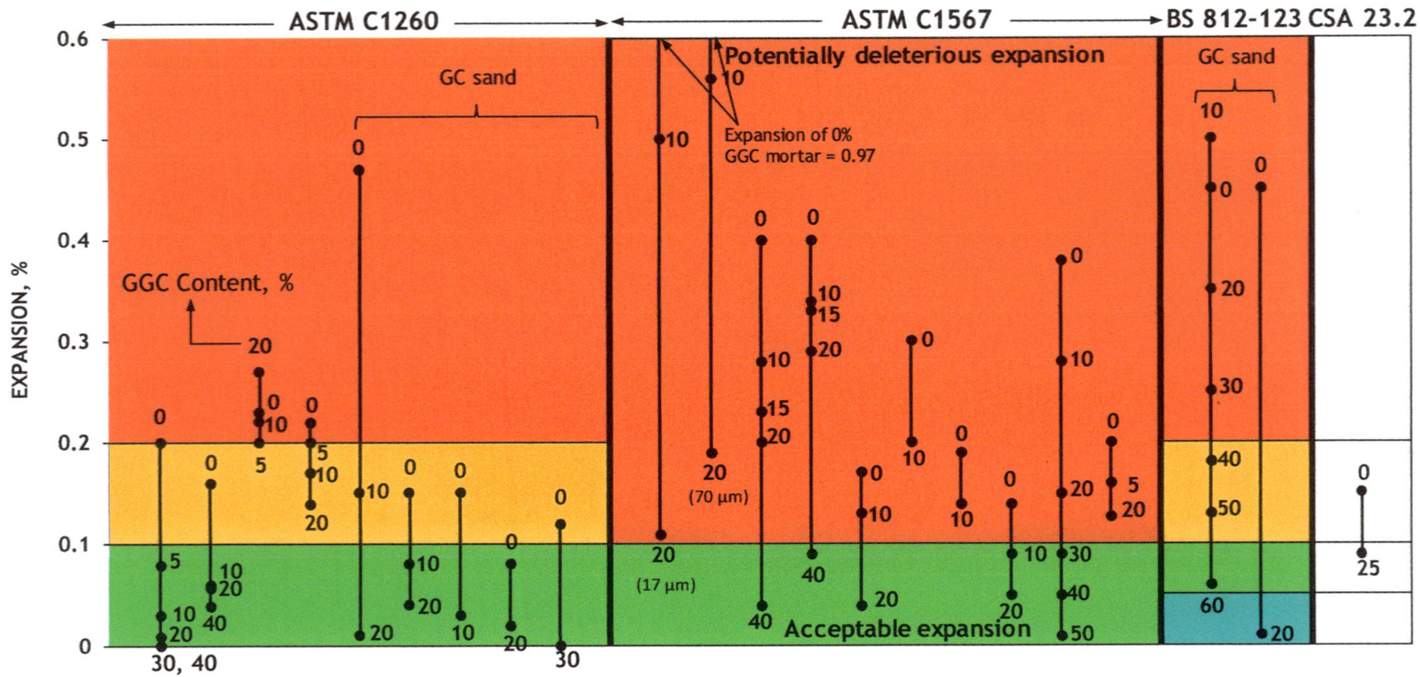

Figure 4.18 Expansion of ground glass cullet (GGC) specimens made with reactive aggregate or glass cullet sand.
Data taken from Afshinnia and Rangaraju (2001), Byars et al. (2004a), Cholleti (2015), Dhir et al. (2005c), Dyer and Dhir (2010), Kim et al. (2015), Matos and Sousa-Coutinho (2012, 2016a,b), Matos et al. (2015), Nassar and Soroushian (2012b), Neithalath and Schwarz (2009), Nunes et al. (2013), Serpa et al. (2013), Shayan and Xu (2004), Schwarz et al. (2008) and Taha and Nounu (2009).

From Figure 4.17, where natural fine aggregate is used, it is evident that GGC is nonexpansive as the expansion of GGC specimens is mostly less than that of reference PC specimens, with the vast majority of the results showing an expansion (i) less than 0.10%, which is the innocuous zone using the ASTM C1260 method, (ii) less than 0.05%, as indicative of nonexpansion using the BS 812-123 method, and (iii) less than 0.10%, known as low risk of deleterious expansion using Korean standard KS F 2546 (Jang et al., 2015). In general, the expansion value decreases as GGC content increases.

Two important observations on the effects of the fineness of GGC and the colour of soda lime GGC on the expansion of specimens should be noted in Figure 4.17. In the fineness study, GGC of three fineness sizes, namely 75–100, 38–75 and less than 38 μm, were used to replace 30% of PC (Shao et al., 2000). The results showed that the expansion of mortar made with the GGC of less than 38 μm is the least. In the colour study, whilst Özkan and Yüksel (2008) show that the colour of soda lime GGC (all having similar fineness) has no influence on the expansion, Dyer and Dhir (2001a,b) suggest that for a given GGC content (all having similar fineness), the expansion of specimens containing flint GGC is the highest compared with that of the green and amber GGC.

From Figure 4.18, where reactive aggregate or GC-sand is used, it can be seen that the expansion of specimens decreases when GGC content increases, suggesting that the ability of GGC to nullify the ASR-induced expansion increases with increasing content of GGC in the mixes. Its effectiveness is more pronounced at higher content, probably at least 30% or more (Figure 4.18).

The influence of fineness of GGC (Afshinnia and Rangaraju, 2015) shown in Figure 4.18 is coherent with the findings in Figure 4.17, which shows that the expansion of mortar made with finer GGC (17-μm average size) is consistently less than that of mortar made with coarser GGC (70 μm). It should be noted from Figure 4.18 that even when GC is used as a fine aggregate, the inclusion of GGC as a PC replacement still effectively suppresses the expansion of mixes to the innocuous zone in the ASTM C1260 method or the nonexpansive zone in the BS 812-123 method. This probably indicates that the ASR-induced expansion resulting from GC in fine aggregate form can be subsided by the same material but in a powder form. Thus, both GC-sand and GGC cement can probably be used concurrently in making concrete without the concern of ASR-induced expansion.

In addition to the findings shown in Figures 4.17 and 4.18, some studies also suggest that the use of GGC in conjunction with FA ranging from 5% to 15% (Kim et al., 2015; Parghi and Alam, 2016; Schwarz et al., 2008) could result in an expansion less than 0.10% (indicative of innocuous performance) using the ASTM C1260 method.

4.7.6 Freeze–Thaw Attack

The presence of moisture in the capillary pores of concrete can be a concern in winter conditions. When water freezes at low temperature, its volume is increased by approximately 9%, which creates an expansion force within concrete. The cumulative cycles of freezing and thawing experienced by concrete can lead to various damages, such as scaling and cracking, and eventually reduce its service life. An air-entraining admixture is normally used to increase the freeze–thaw resistance of concrete.

The freeze–thaw resistances of air-entrained and non-air-entrained concrete containing up to 25% GGC, tested mainly in accordance with ASTM C666 (1992), but in one case with CEN/TS 51-draft (1994), are given in Table 4.15. The ASTM C666 (1992) contains two different procedures: procedure A, in which specimens are both frozen and thawed in water, and procedure B, in which specimens are frozen in air and thawed in water. It can be seen from Table 4.15 that the deterioration of GGC concrete due to freeze–thaw attack has been examined using several methods, commonly measuring the change in the dynamic modulus of elasticity of concrete, and the value can also be used to calculate the durability factor (ASTM C666, 1992) and change in strength.

Only one case, which happens to be an early study conducted by Polley (1996), shows that the freeze–thaw resistance of air-entrained GGC concrete is lower than that of PC concrete (Table 4.15). The possible explanation for this observation is that the entrained air content of PC concrete is higher than that of GGC concrete, i.e., 8.1% in PC concrete compared with 3.4%–7.5% in GGC concrete, although the target entrained air content was 6% ± 2%.

All the other studies, conducted since 2000, show that the freeze–thaw resistance of both the non-air-entrained and the air-entrained concrete remains unchanged or increases when GGC is used (Table 4.15). It should be mentioned that, for air-entrained concrete, when GGC is used as PC replacement, the dosage of air-entraining agent might need to be slightly increased (Laldji and Tagnit-Hamou, 2007) to achieve the same target entrained air content as the corresponding PC concrete.

4.7.7 Abrasion

Abrasion is a deterioration of the concrete surface, in the form of progressive loss of surface material or thinning, which can occur (i) on concrete pavements and floors due to the wear and impact actions of automobiles, (ii) on hydraulic structures due to solid particles carried in flowing water (known as erosion) and (iii) by the formation of vapour bubbles from fast-moving downstream water (cavitation). Whilst the design and composition of concrete can affect its abrasion resistance, proper concrete casting practices, especially finishing and curing, are also important factors as they ensure the quality of the near-surface zone of concrete.

Table 4.15 Resistance to freeze–thaw attack of concrete containing GGC

References	Parameter[a]	Measurement	Resistance to Freeze–Thaw Attack[b]
(a) ASTM C666 Procedure A			
Abendeh et al. (2015a) and Abendeh et al. (2015b)	0%–15% GGC; 0.50 w/c; AEA not given; 28 days curing; from 6°C to −16°C and reverse in 2.5 h; 230 cycles	Weight loss	↑
		Compressive strength loss	≈
		Ultra-pulse velocity	≈
		Dynamic modulus loss	≈
Polley (1996)[c]	0%–20% GGC; 0.43 w/c; AEA for target 6 ± 2% air; 28 days curing; from 4°C to −18°C and reverse in 3 h; 350 cycles	Stiffness loss	↓
Laldji and Tagnit-Hamou (2007)	0%, 25% GGC; 0.45 w/c; AEA for target 5%–8% air; 28 days curing; 300 cycles	Durability factor	≈
(b) ASTM C666 Procedure B			
Al-Akhras (2012)	0%–18% GGC; 0.40, 0.60 w/c; AEA not given; 90 days curing; from 5°C to −18°C and reverse In 3.6 h; 100–300 cycles	Dynamic modulus loss	↑
		Durability factor	↑
Nassar and Soroushian (2012a)[d]	0, 20% GGC; 0.50 w/c; AEA for > 5% target air; 42 days curing; 310 cycles	Weight gain[e]	↑
		Flexural strength loss	↑
Tuncan et al. (2001)	0%–15% GGC; 0.45 w/c; AEA not stated; 28 days curing; from 5°C to −18°C in 4 h; 60 cycles	Compressive strength loss	↑
		Tensile strength loss	↑
		Dynamic modulus loss	↑
(c) CEN/TC 51-draft (1994)			
Dhir et al. (2005c) and Tang et al. (2005)	0, 15% GGC; 0.55 w/c; AEA not used; 7 days curing; from 20°C to −20°C and reverse in 24 h; 56 cycles	Mass scaling	≈

[a] *AEA*, air-entraining agent; *GGC*, ground glass cullet; *w/c*, water/cement ratio.
[b] Comparing the resistance to freeze–thaw attack of GGC concrete to Portland cement concrete. ↑ indicates increase; ↓ indicates reduction; ≈ indicates no significant change with value difference within ±20%.
[c] Containing 20% glass cullet as fine aggregate.
[d] Containing 50% and 100% coarse recycled concrete aggregate.
[e] Weight gain due to water absorption as a result of cracking.

The study of the effect of GGC as a PC replacement on the abrasion resistance of concrete is limited, which is somewhat to be expected, as in such applications the influence of aggregate is more pronounced than that of the cement. Nevertheless, the use of GGC at 15%–37% content in concrete has been reported to result in good abrasion resistance, though detailed information has not been provided (Corbu et al., 2013; Dumitru et al., 2010).

In one of the two more elaborate studies, the abrasion resistance of GGC concrete was determined using the test method proposed by Dhir et al. (1991), in which the abrasion depth of specimens is measured after subjecting the surface of test concrete specimens to rotating steel abrasive wheels for 15 min (Dhir et al., 2005c). The results suggest that, for an equal water/cement ratio at 0.55, the abrasion resistance of 28-day cured 15% GGC content concrete is essentially similar to that of the reference PC concrete, although the 28-day compressive strength of the former concrete is lower than that of the latter (Dhir et al., 2005c).

The other abrasion study (Nassar and Soroushian, 2013) was performed on GGC and PC concrete containing 0%, 50% and 100% coarse RCA at 0.36 and 0.45 water/cement ratio, tested in accordance with ASTM C944 (2012), in which the weight loss of specimens is measured after they are subjected to rotating cutters for 2 min. At the age of 56 days, regardless of water/cement and content of coarse RCA, the results show that the abrasion resistance of GGC concrete is generally higher than that of the corresponding PC concrete.

Overall, it can be concluded that the use of GGC is not likely to significantly change the abrasion resistance of concrete. It should also be mentioned that, as the abrasion resistance of concrete is normally associated with its compressive strength, sufficient curing must be given to concrete containing GGC because of its slower rate of strength gain.

4.8 Environmental Impact

Glass is a stable material, but nonetheless it is necessary to understand the hazardous potential it poses to the environment through leaching of heavy metals from glass. Table 4.16 lists the leached element concentrations of thin-film-transistor liquid-crystal display (TFT-LCD) glass, which was studied by researchers from Taiwan for its suitability as a PC replacement. TFT-LCD is a glass material containing mercury and lead, commonly used in televisions, flat-panel displays, projectors and other electronic products.

Determined in accordance with the toxic characteristic leaching procedure 1311 of the US Environmental Protection Agency [US EPA (1992)], the results given in Table 4.16 suggest that the leached concentrations of arsenic (As), mercury (Hg) and lead (Pb) from TFT-LCD glass are well below the corresponding regulatory level of the US EPA (2016). Other elements such as cadmium (Cd), chromium (Cr) and selenium (Se) are not detected (Table 4.16). TFT-LCD contains small traces of copper (Cu) and zinc (Zn), but the levels of these are not regulated.

Table 4.16 Leached element concentrations of thin-film-transistor liquid-crystal display glass obtained using the toxic characteristic leaching procedure

	Leached Concentration, mg/L		
Element	**US EPA Regulatory Level[a]**	**Lin et al. (2008, 2009)**	**Wang and Hou (2011) and Wang et al. (2014)**
As	5.0	–	0.022
Cd	1.0	N.D.	N.D.
Cr	5.0	N.D.	N.D.
Cu	N.R.	0.25	–
Hg	0.2	–	0.0077
Pb	5.0	N.D.	0.281
Se	1.0	–	N.D.
Zn	N.R.	0.16	–

N.D., not detected; *N.R.*, not regulated.
[a]Based on US EPA (2016).

In addition, some work has been undertaken to investigate the lead leaching potential of cement clinker (Lairaksa et al., 2013) and mortar (Moncea et al., 2013) containing cathode ray tube (CRT) funnel glass as part of the raw feed materials and a replacement for PC, respectively. CRT funnel glass, which is rich in lead, used to be commonly employed in the production of television and computer monitors before the introduction of LCD technology.

Based on the study of five cement clinkers containing 0.1%–0.5% CRT as a raw feed, Lairaksa et al. (2013) found that the encapsulation of lead was the highest in cement clinker containing 0.1% CRT, and the lead encapsulation ability of cement clinker reduces as the C_2S proportion reduces.

When CRT funnel glass (containing less than 5% CRT panel glass) was used as a cementitious material, Moncea et al. (2013) reported that the cumulative lead released after 64 days for mortar made with PC and 28% CRT and mortar made with GGBS cement and 28% CRT was 0.24 and 0.51 mg/m^2, respectively. It is claimed (Moncea et al., 2013) that the lead leaching values were below the limit of 100 mg/m^2 stipulated in the Dutch Building Material Decree.

4.9 Case Studies

Given that GGC possesses pozzolanic properties, there have been some successful field applications yielding positive results in many cases for its use as a cement component and in one case as a raw feed for cement clinker production (see Table 4.17). As summarised in Table 4.17, the material started to spark the interest

Table 4.17 Case studies involving the use of ground glass cullet as clinker raw feed and a cement component in concrete applications

	Application Details				
References	**Location**	**Year**	**Use in**	**GGC Content %**	**Main Observations**
Byars et al. (2003, 2004a)	United Kingdom	2002–04	Precast concrete products	20–30	The use of GGC can minimise the risk of ASR expansion of concrete products containing GC as sand
Byars et al. (2004a) and Zhu and Byars (2005)	United Kingdom	2002–04	Ready-mixed concrete	10–50	Except for concrete made with 50% GGC, the compressive strength of GGC concrete was generally similar to or higher than the reference PC concrete at 90 days
Chen et al. (2002)	Hong Kong	2000	Clinker	–	No significant changes in the chemical and physical properties of clinker containing GGC, as well as the exhaust gases generated during clinker production
Dumitru et al. (2013)	Australia	–	Concrete pavement	15	Target strength was met at either 28 or 90 days; the performance of concrete was satisfactory
Jang et al. (2014)	South Korea	–	Prestressed high-strength concrete	10	No significant changes in mechanical properties of concrete
Nassar and Soroushian (2011), Soroushian (2012)	United States	2008–09	Concrete pavement and curbs	10–23	The long-term strength, permeability and durability of concrete improved the most at an optimum 20% GGC content
Peyvandi et al. (2013)	United States	2012	Exterior flatwork and curbs	n.a.	Nearly 544 tonnes of concrete made with GGC was consumed

References	Application Details				Main Observations
	Location	Year	Use in	GGC Content %	
Omran and Tagnit-Hamou (2016) and Tagnit-Hamou and Bengougam (2012)	Canada	2006–12	Interior and exterior structural elements	10–30	The mechanical properties of concrete were improved at later ages; GGC concrete had better permeability and durability performance than PC concrete
Shayan and Xu (2006)	Australia	2002	Slab	20–30	Target strength was met at 28 days; the performance of concrete after 1 year was satisfactory
Tagnit-Hamou et al. (2015)	Canada	–	Pedestrian bridges	35	GGC was used in conjunction with 17.5% silica fume in the production of ultra-high-performance concrete with superior mechanical, ductility and durability properties

ASR, alkali–silica reaction; *GC*, glass cullet; *GGC*, ground glass cullet; *PC*, Portland cement.

of the construction industry as early as the beginning of the 2000s (Chen et al., 2002), with case studies undertaken in several countries, including Australia, Canada, Hong Kong, South Korea, the United Kingdom and the United States (Table 4.17).

The main points that have emerged from such field applications as listed in Table 4.17 are given below:

- For its use in cement clinker production (Chen et al., 2002), GGC was added into the normal raw feed materials consisting of limestone, silica sand, FA and iron or slag. The exhaust gases generated during the production of clinker containing GGC did not show significant changes in the concentration of nitrogen monoxide (NO), nitrogen dioxide (NO_2), carbon monoxide (CO), carbon dioxide (CO_2) and sulphur dioxide (SO_2) pollutants. The sulphur content (SO_3) and alkali content of the resultant products varied within ±3 times the standard deviation from the past year's statistical record. Both the chemical and the physical properties of cement with GGC addition as a raw feed were reported to be essentially similar to those of cement without GGC addition.
- When used as a cement component in concrete, GGC content as high as 50% has been reported (Byars et al., 2004a), although the common range was between 10% and 30%.
- As the pozzolanic reaction is a slow process, many studies have reported that the compressive strength and other mechanical properties of GGC concrete were lower than those of reference PC concrete initially; it achieved similar or higher values compared with the reference PC concrete at later ages (Byars et al., 2004a; Nassar and Soroushian, 2011; Soroushian, 2012; Omran and Tagnit-Hamou, 2016; Tagnit-Hamou and Bengougam, 2012). However, two cases have been reported in which the target strength of concrete made with 15%–30% GGC was met without modifying its water/cement ratio (Dumitru et al., 2013; Shayan and Xu, 2006).
- The densification of cement microstructure due to the pozzolanic reaction of GGC resulted in an improvement in sorptivity and chloride resistance of concrete (Nassar and Soroushian, 2011; Shayan and Xu, 2006; Soroushian, 2012; Omran and Tagnit-Hamou, 2016; Tagnit-Hamou and Bengougam, 2012). The use of GGC as a PC replacement did not significantly change the abrasion and freeze–thaw resistance of concrete (Nassar and Soroushian, 2011; Soroushian, 2012; Tagnit-Hamou and Bengougam, 2012).

Like FA and metakaolin, the use of GGC as a cement component can minimise the deleterious ASR expansion induced by GC-sand in concrete products (Byars et al., 2004a).

4.10 Conclusions

Owing to its amorphous and silica-rich nature, finely ground GC possesses pozzolanic properties. The influence of GGC as a cement component on the properties of concrete has been researched mainly since the beginning of the 2000s, by individual researchers and established organisations from all over the world. In general, many findings and results confirm that GGC can be used as a replacement for PC, and the material generally enhances the hardened properties of concrete, particularly the compressive strength, permeation and durability properties.

Assessment of its cementitious characteristics suggests that GGC has little influence on the amount of water required for cement paste for the standard consistency test. Cement containing GGC has been confirmed to be sound, and no potential soundness problems have been reported. The increase in the initial setting time of cement resulting from the use of GGC of the soda lime type is considerably insignificant for up to 20% content, beyond which its use extends the initial setting time by an average 25 min for up to 50%. The fineness of GGC affects its strength activity index. When GGC is ground finer than the corresponding PC of the mix, the strength activity index of the resulting cement combination measured at different ages can be expected to meet the requirements stipulated in both ASTM and BS EN standards for FA.

Although the inclusion of GGC modifies the consistence of concrete, such a change is generally within the consistence class and allowable tolerance given in BS EN 206-1 (2003). Concrete containing GGC has no stability issues but it may show more bleeding if the GGC particle size is coarse. For every 10% GGC inclusion, the fresh density of concrete decreases by about 0.5% on average. The entrained air content of concrete is not adversely affected using GGC up to the 20% level. Comparing with FA, GGC is likely to result in shorter initial and final setting times. Both the temperature rise and the rate of heat evolution of concrete are reduced when GGC is used as a replacement for PC.

The compressive strength of concrete is significantly affected by GGC content and its fineness. Although the use of GGC improves the compressive strength of concrete at later ages, for a given water/cement ratio, a similar or slightly higher compressive strength compared with PC concrete at 28 days can be achieved with GGC finer than PC, up to a content of no more than 20% of the total cement used. Due to the pozzolanic activity of GGC, the rate of strength gain of GGC concrete can be higher than that of PC concrete after 28 days. The tensile strength of concrete (both splitting tensile strength and flexural strength) is almost unaffected for a GGC content up to 20%, beyond which it decreases as GGC content is increased. The relationship between compressive strength and tensile strength of concrete is essentially unchanged when GGC is used. The use of GGC can result in a slight decrease in the modulus of elasticity. It is postulated that the effect of GGC on the creep of concrete is similar to that of FA. The shrinkage of concrete tends to increase as the GGC content increases.

When properly designed, the use of GGC as a PC replacement can result in a reduction in porosity, making the concrete less permeable to water, gas and aggressive ions. Indeed, concrete containing GGC has shown to have higher resistance to chloride and sulphate attack. The same is expected in the case of acid attack, though the available information is not conclusive. Similar to other pozzolanic cement, the carbonation resistance of concrete decreases when GGC is used as a PC replacement. Despite having high silica content, the use of GGC does not give rise to ASR concerns, but in fact its inclusion can help to nullify the ASR expansion induced by reactive aggregates. The effectiveness of GGC in this regard increases with increasing GGC content and particle fineness, and may be independent of its colour. The use of GGC is not likely to adversely affect the resistance of concrete to freeze–thaw attack and abrasion.

The toxic elements that may leach from GGC, which is rich in aluminium and lead oxides, can be expected to be well within the regulatory levels set out by the US EPA. The construction industry has used GGC as a raw feed for clinker production and as a cement component in concrete production since the beginning of the 2000s.

References

Abendeh R, Baker M B, Salem Z A and Ahmad H, 2015a. The feasibility of using milled glass wastes in concrete to resist freezing-thawing action. International Science Index, Civil and Environmental Engineering 9 (8), 981–984.

Abendeh R, Ramadan K, Haddad R and Baker M B, 2015b. Freezing thawing resistance of concrete incorporating glass waste. Journal of Materials Science Research 4 (4), 19–25.

ACI 301, 2005. Specifications for Structural Concrete. American Concrete Institute, Farmington Hills, MI, USA. 656 pp.

Afshinnia K and Rangaraju P R, 2015. Influences of fineness of ground recycled glass on mitigation of alkali-silica reaction in mortars. Construction and Building Materials 81, 257–267.

Al-Akhras N, 2012. Performance of glass concrete subjected to freeze-thaw cycling. The Open Construction & Building Technology Journal 6, 392–397.

Ali I, 2015. Behaviour of concrete by using waste glass powder and fly ash as a partial replacement of cement. International Journal of Engineering Research and Technology 4 (5), 1238–1243.

Alidoust O, Sadrinejad I and Ahmadi M A, 2007. A study on cement-based composite containing polypropylene fibres and finely ground glass exposed to elevated temperatures. International Science Index, Civil and Environmental Engineering 1 (4), 253–257.

Aligizaki K K, 2006. Pore Structure of Cement-based Materials. Taylor and Francis, Oxon, London. 432 pp.

Altaf S N, Asmat K M, Seemab G M, Aalam K M and Azim K F, 2013. Suitability of Waste Glass Powder as a Partial Replacement of Cement in Fibre Reinforced Concrete. Civil Engineering Department, Abdul Razzak Kalsekar Polytechnic, Mumbai, India. 142 pp.

Aly M, Hashmi M S J, Olabi A G, Messeiry M and Hussain A I, 2011. Effect of nano clay particles on mechanical, thermal and physical behaviours of waste-glass cement mortars. Materials Science and Engineering A 528, 7991–7998.

Aly M, Hashmi M S J, Olabi A G, Messeiry M and Hussain A I, 2012. Effect of colloidal nano-silica on the mechanical and physical behaviour of waste-glass cement mortar. Materials and Design 33, 127–135.

ASTM 1567, 2013. Standard Test Method for Determining the Potential Alkali-silica Reactivity of Combinations of Cementitious Materials and Aggregate (Accelerated Mortar-bar Method). ASTM International, West Conshohocken, PA, USA.

ASTM C1202, 2012. Standard Test Method for Electrical Indication of Concrete's Ability to Resist Chloride Ion Penetration. American Society for Testing and Materials, West Conshohocken, Pennsylvania, USA.

ASTM C1260, 2014. Standard Test Method for Potential Alkali Reactivity of Aggregates (Mortar-bar Method). ASTM International, West Conshohocken, PA, USA.

ASTM C187, 2011. Standard Test Method for Amount of Water Required for Normal Consistency of Hydraulic Cement Paste. ASTM International, West Conshohocken, PA, USA.

ASTM C191, 2013. Standard Test Methods for Time of Setting of Hydraulic Cement by Vicat Needle. ASTM International, West Conshohocken, PA, USA.

ASTM C293, 2015. Standard Test Method for Flexural Strength of Concrete (Using Simple Beam with Center-Point Loading). ASTM International, West Conshohocken, PA, USA.

ASTM C311, 2013. Standard Test Methods for Sampling and Testing Fly Ash or Natural Pozzolans for Use in Portland-cement Concrete. ASTM International, West Conshohocken, PA, USA.

ASTM C618, 2012. Standard Specification for Coal Fly Ash and Raw or Calcined Natural Pozzolan for Use in Concrete. ASTM International, West Conshohocken, PA, USA.

ASTM C642, 2013. Standard Test Method for Density, Absorption and Voids in Hardened Concrete. ASTM International, West Conshohocken, PA, USA.

ASTM C666, 1992. Standard Test Method for Resistance of Concrete to Rapid Freezing and Thawing. ASTM International, West Conshohocken, PA, USA.

ASTM C78, 2015. Standard Test Method for Flexural Strength of Concrete (Using Simple Beam with Third-point Loading). ASTM International, West Conshohocken, PA, USA.

ASTM C944, 2012. Standard test method for abrasion resistance of concrete or mortar surfaces by the rotating-cutter method. ASTM International, West Conshohocken, PA, USA.

Bajad M N and Modhera C D, 2009. Experimental investigations in developing concrete containing waste glass powder as pozzolana. Journal of Information, Knowledge and Research in Civil Engineering 1 (1), 32–37.

Bajad M N, Modhera C D and Desai A K, 2011a. Effect of glass on strength of concrete subjected to sulphate attack. IJCERD 1 (2), 1–13.

Bajad M N, Modhera C D and Desai A K, 2011b. Influence of a fine glass powder on strength of concrete subjected to chloride attack. International Journal of Civil Engineering & Technology 2 (2), 1–12.

Bajad M N, Modhera C D and Desai A K, 2012a. Effects of sodium chloride and magnesium sulphate on glass concrete. The IUP Journal of Structural Engineering 3 (3), 28–47.

Bajad M N, Modhera C D and Desai A K, 2012b. Resistance of concrete containing waste glass powder against $MgSO_4$ attack. MBMCW (May Issue) 15 pp.

Bajad M N, Modhera C D and Desai A K, 2014. Factors affecting the properties of conglasscrete. IOSR Journal of Mechanical and Civil Engineering 11 (2), 43–48.

Bhat V V and Rao N B, 2014. Influence of glass powder on the properties of concrete. IJETT 16 (5), 196–199.

Bignozzi M C, Saccani A, Barbieri L and Lancellotti I, 2015. Glass waste as supplementary cementing materials, the effects of glass chemical composition. Cement and Concrete Composites 55, 45–52.

Borosnyoi A, Kara P, Mlinarik L and Kase K, 2013. Performance of waste glass powder (WPG) supplementary cementitious material (SCM) – workability and compressive strength. Journal of the Scientific Society of the Silicate Industry 65 (3), 90–94.

BRE Digest 330, 2004. Alkali-silica reaction in concrete. Building Research Establishment, Watford, UK, BRE Digest 330 Part 2.

BS 812-123, 1999. Testing Aggregates – Part 123: Method for Determination of Alkali-silica Reactivity – Concrete Prism Method. BSI, London, UK.

BS EN 12390–12395, 2009. Testing Hardened Concrete. Flexural Strength of Test Specimens. BSI, London, UK.

BS EN 196-1, 2005. Methods of testing cement – Part 1: Determination of strength. BSI, London, UK.

BS EN 196-3:2005+A1, 2008. Methods of Testing Cement. Part 3: Determination of Setting Times and Soundness. BSI, London, UK.

BS EN 196–1, 2016. Methods of Testing Cement. Determination of Strength. BSI, London, UK.

BS EN 197–1, 2011. Cement. Part 1: Composition, Specifications and Conformity Criteria for Common Cements. BSI, London, UK.

BS EN 1990:2002+A1, 2005. Eurocode. Basis of Structural Design. BSI, London, UK.

BS EN 1992, 2004. Eurocode 2. Design of Concrete Structures – Part 1-1: General Rules and Rules for Buildings. BSI, London, UK.

BS EN 206, 2003. Concrete – Specification, Performance, Production and Conformity. BSI, London, UK.

BS EN 450–451, 2012. Fly Ash for Concrete. Part 1: Definition, Specifications and Conformity Criteria. BSI, London, UK.

Byars E A, Zhu H and Meyer C, 2003. Use of waste glass for construction products, legislative and technical issues. In: Dhir R K, Newlands M D and Halliday J E (Eds.), International Symposium on Recycling and Re-use of Waste Materials, Dundee, UK. Thomas Telford, London, UK, pp. 827–838.

Byars E A, Zhu H and Morales B, 2004a. Congrasscrete I. The Waste & Resources Action Programme (WRAP), Oxon, UK Final Report GLA2–006. 54 pp.

Byars E A, Morales B and Zhu H, 2004b. Congrasscrete II. The Waste & Resources Action Programme (WRAP), Oxon, UK Final Report GLA0015–007. 30 pp.

Byars E A, Morales B and Zhu H, 2004c. Waste glass as concrete aggregate and pozzolan. Concrete (January Issue), 41–44.

Calmon J L, Sauer A S, Vieira G L and Teixeira J, 2014. Effects of windshield waste glass on the properties of structural repair mortars. Cement and Concrete Composites 53, 88–96.

Carsana M, Frassoni M and Bertolini L, 2014. Comparison of ground waste glass with other supplementary cementitious materials. Cement and Concrete Composites 45, 39–45.

Cassar J and Camilleri J, 2012. Utilisation of imploded glass in structural concrete. Construction and Building Materials 29, 299–307.

CEN/TC 51, 1994. Draft European Prestandard: Test Method for the Freeze-thaw Resistance of Concrete – Tests with Water or Sodium Chloride Solution, Part 1: Scaling. Comité Européen de Normalisation (CEN), Brussels, Belgium.

Chaid R, Kenai C, Zeroub H and Jauberthie, 2015. Microstructure and permeability of concrete with glass powder addition conserved in the sulphatic environment. European Journal of Environmental and Civil Engineering 19 (2), 219–237.

Chen G, Lee H, Young K, Yue P L, Wong A, Tao T and Choi K K, 2002. Glass recycling in cement production – an innovation approach. Waste Management 22, 747–753.

Chen C H, Huang R, Wu J K and Yang C C, 2006. Waste E-glass particles used in cementitious mixtures. Cement and Concrete Research 36, 449–456.

Chidiac S E and Mihaljevic S N, 2011. Performance of dry cast concrete blocks containing waste glass powder or polyethylene aggregates. Cement and Concrete Composites 33, 855–863.

Cholleti R N R, 2015. Use of Waste Glass as a Supplementary Cementitious Material (Msc thesis). Department of Civil and Environmental Engineering, University of Utah, Salt Lake City, UT, USA. 52 pp.

Concrete Society, 2010. Non-structural Cracks in Concrete Technical Report No. 22. Concrete Society, Surrey, UK. 63 pp.

Corbu O, Chira N, Szilagyi H and Constantinescu H, 2013. Ecological concrete by use of waste glass. In: 13th SGEM GeoConference NANO, BIO and GREEN – Technologies for a Sustainable Future, Albena, Bulgaria, pp. 411–418.

Cui Y, 2005. A Study of the Use of Glass Cullet as Sand, Cement Component and Filler in Concrete Ph.D. thesis. University of Dundee, Dundee, UK.

Dahmardeh S R, Moghaddam M H M, Moghaddam M S S and Mousavi S R, 2015. Assessment the potential use of milled glass as Portland cement in self-compacting concrete. Applied Mathematics in Engineering, Management and Technology 3 (2), 345–353.

GNR Database, 2014. GNR Project Reporting CO_2. Available from: http.//wbcsdcement.org/GNR-2014/index.html.

Dhir R K and Dyer T D, 2001. Glass recycling, combining enhanced recycling methods and alternative routes to close the loop. In: Dhir R K, Limbachiya M C and Dyer T D (Eds.), International Symposium on Recycling and Re-use of Glass Cullet, Dundee, UK. Thomas Telford, London, UK, pp. 277–287.

Dhir R K and Dyer T D, 2004. Maximising opportunities for recycling glass. In: Limbachiya M C and Roberts J J (Eds.), International Conference on Sustainability Waste Management and Recycling – Glass Waste, London, UK. Thomas Telford, London, UK, pp. 2–16.

Dhir R K, Hewlett P C and Chan Y N, 1991. Near-surface characteristics of concrete: abrasion resistance. Materials and Structures 24, 122–128.

Dhir R K, Paine K and Caliskan S, 2005a. Recycled glass. Concrete (June Issue), 44–46.

Dhir R K, Dyer T, Albert T, Cui Y and Wang L, 2005b. Towards maximising the se of glass cullet in concrete. Concrete (August), 32–34.

Dhir R K, Dyer T, Tang M C, Cui Y and Wang L, 2005c. Realising a High-value Sustainable Recycling Solution to the Glass Cullet Surplus Closing the Loop. Concrete Technology Unit. Department of Civil Engineering, University of Dundee, Dundee, UK. 154 pp.

Du H J and Tan K H, 2014. Waste glass powder as cement replacement in concrete. Journal of Advanced Concrete Technology 12, 468–477.

Du H J and Tan K H, 2015. Transport properties of concrete with glass powder as supplementary cementitious material. ACI Materials Journal 112 (3), 429–438.

Dumitru I, Song T, Caprar V, Brooks P and Moss J, 2010. Incorporation of recycled glass for durable concrete. In: Zachar J, Claisse P, Naik T R and Ganjian E (Eds.), Second International Conference on Sustainable Construction Materials and Technologies, Ancona, Italy, pp. 513–521.

Dumitru I, Song T, Bornstein B, Brools P and Moss J, 2013. Field trials using recycled glass as natural sand replacement and powdered glass as cementitious materials replacement in concrete pavement. In: Third International Conference on Sustainable Construction Materials and Technologies, Kyoto, Japan 9 pp.

Dyer T D and Dhir R K, 2001a. Chemical reactions of glass cullet used as cement component. Journal of Materials in Civil Engineering 13 (6), 412–417.

Dyer T D and Dhir R K, 2001b. Use of glass cullet as a cement component in concrete. In: Dhir R K, Limbachiya M and Dyer T D (Eds.), Proceedings of the International Symposium on Recycling and Reuse of Glass Cullet, Dundee, UK. Thomas Telford, London, UK, pp. 157–166.

Dyer T D and Dhir R K, 2010. Evaluation of powdered glass cullet as a means of controlling harmful alkali–silica reaction. Magazine of Concrete Research 62 (10), 749–759.

El-Alami I, 2004. Recycled Glass for Use as Pozzolanic Material. BRE Certification Limited, Watford, UK. Project No. BC2197, 20 pp.

Elgalhud A, Dhir R K and Ghataora G, 2017. Carbonation resistance of concrete: limestone addition effect. Magazine of Concrete Research 69 (2), 84–106.

EN 197-1, 2011. Cement Composition, Specifications and Conformity Criteria for Common Cements. Comité Européen de Normalisation (CEN), Brussels, Belgium.

Federico L M, 2013. Waste Glass – a Supplementary Cementitious Material Ph.D. thesis. Department of Civil Engineering, McMaster University, Ontario, Canada. 88 pp.

Federico L M and Chidiac S E, 2009. Waste glass as a supplementary cementitious material in concrete – critical review of treatment methods. Cement and Concrete Composites 31, 606–610.

Girbes I, Lopez-Tendero M J, Suesta C and Belena I, 2004. Recycled glass cullet as partial Portland cement replacement. In: Limbachiya M C and Roberts J J (Eds.), Proceedings of the International Conference on Sustainable Waste Management and Recycling, Glass Waste, London, UK. Thomas Telford, London, UK, pp. 141–148.

Han W, Sun T, Li X, Sun M and Lu Y, 2016. Using of borosilicate glass waste as a cement additive. Nuclear Instruments and Methods in Physics Research B 381, 11–15.

Hewlett P C (Ed.), 1998. Lea's Chemistry of Cement and Concrete. Arnold, London, UK.

Hussain M V and Chandak R, 2015. Strength properties of concrete containing waste glass powder. IJERA 5 (4), 1–4.

Idir R, Cyr M and Tagnit-Hamou A, 2009. Use of waste glass in cement-based materials. In: 1st International Conference on Sustainable Built Environment Infrastructures in Developing Countries, Oran, Algeria 8 pp.

Idir R, Cyr M and Tagnit-Hamou A, 2011a. Pozzolanic properties of fine and coarse colour-mixed glass cullet. Cement and Concrete Composites 33, 19–29.

Idir R, Cyr M and Tagnit-Hamou, 2011b. Potential pozzolanicity of glass cullet fines and aggregates. Direction régionale et interdépartementale de l'Equipement et de l'Aménagement d'Ile-de-France, Annales Du Bâtiment Et Des Travaux Publics 28–34.

Idir R, Cyr M and Tagnit-Hamou, 2013. Role of the nature of reaction products in the differing behaviours of fine glass powders and coarse glass aggregates used in concrete. Materials and Structures 46, 233–243.

Jain J A and Neithalath N, 2010. Chloride transport in fly ash and glass powder modified concretes – influence of test methods on microstructure. Cement and Concrete Composites 32, 148–156.

Jang H S, So H S, Jeon S H and So S Y, 2014. A study of the possibility of using TFT-LCD waste glass as admixture for steam-cured PHC piles. Magazine of Concrete Research 66 (4), 196–208.

Jang H S, Jeon S H, So H S and So S Y, 2015. Waste glass powder as a cement concrete binder. International Journal of Precision Engineering and Manufacturing 16 (12), 2591–2597.

Jangid J B and Saoji A C, 2014. Experimental investigation of waste glass powder as the partial replacement of cement in concrete production. IOSR-JMCE 55–60.

Kadir A, Ismail M, Abdullah M, Sarani N and Hassan M, 2016. The usage of glass waste as cement replacement. Key Engineering Materials 673, 95–104.

Kamali M and Ghahremaninezhad A, 2015. Effect of glass powders on the mechanical and durability properties of cementitious materials. Construction and Building Materials 98, 407–416.

Kamali M and Ghahremaninezhad A, 2016. An investigation into the hydration and microstructure of cement pastes modified with glass powder. Construction and Building Materials 112, 915–924.

Kara P, 2013a. Application of recycled concrete aggregates containing waste glass powder/ suspension and bottom ash as a cement component in concrete. In: Proceedings of Fib Symposium on Engineering a Concrete Future, Technology, Modelling and Construction, Tel Aviv, Israel, pp. 109–112.

Kara P, 2013b. The influence of waste glass slurry on the properties of concrete. IJAIEM 2 (8), 325–330.

Kara P, 2015. Performance of lamp glass waste powder (LGWP) as supplementary cementitious material (SCM) – viscosity and electrical conductivity. Journal of Silicate Based and Composite Materials 67 (1), 13–18.

Kara P, Korjakins A and Kovalenko K, 2012. The usage of fluorescent waste glass powder in concrete. The Construction Specifier 13, 26–32.

Kara P, Borosnyoi A and Fenyvesi O, 2014. Performance of waste glass powder (WGP) supplementary cementitious material (SCM) – drying shrinkage and early age shrinkage cracking. Journal of Silicate Based and Composite Materials 66 (1), 18–22.

Kara P, Csetenyi L J and Boronsnyoi A, 2016. Performance characteristics of waste glass powder substituting Portland cement in mortar mixtures. In: 3rd International Conference on Competitive Materials and Technology, Misckolc-lillafured, Hungary, pp. 1–7.

Karamberi A and Moutsatsou A, 2005. Participation of coloured glass cullet in cementitious materials. Cement and Concrete Composites 27, 319–327.

Karamberi A, Kerinis M and Moutsatsou A, 2004. Innovative reuse of glass cullet in cement products. In: Limbachiya M C and Roberts J J (Eds.), Proceedings of the International Conference on Sustainable Waste Management and Recycling, Glass Waste, pp. 133–140 London, UK.

Lairaksa N, Moon A R and Makul N, 2013. Utilization of cathode ray tube waste: Encapsulation of PbO-containing funnel glass in Portland cement clinker. Journal of Environmental Management 117, 180–186.

Karamberi A, Chaniotakis E, Papageorgiou D and Moutsatsou A, 2006. Influence of glass cullet in cement pastes. China Particuology 4 (5), 234–237.

Ke G, Zeng D and Zou P, 2015. The study of mortar specimens containing waste glass powder on the resistance to sulfate erosion. Applied Mechanics and Materials 744–746, 1551–1555.

Khmiri A, Samet B and Chaabouni M, 2012. Assessment of the waste glass powder pozzolanic activity by different methods. IJRRAS 10 (2), 322–328.

Khmiri A, Chaabouni M and Samet B, 2013. Chemical behaviour of ground waste glass when used as partial cement replacement in mortars. Construction and Building Materials 44, 74–80.

Kim J, Moon J, Shim J, Sim J, Lee H and Zi G, 2014. Durability properties of a concrete with waste glass sludge exposed to freeze-and-thaw condition and de-icing salt. Construction and Building Materials 66, 398–402.

Kim J, Yi C and Zi G, 2015. Waste glass sludge as a partial cement replacement in mortar. Construction and Building Materials 75, 242–246.

Kou S H and Xing F, 2012. The effect of recycled glass powder and reject fly ash on the mechanical properties of fibre-reinforced ultrahigh performance concrete. Advances in Materials Science and Engineering 2012, 1–8.

Laldji S and Tagnit-Hamou A, 2007. Glass frit for concrete structures, a new, alternative cementitious material. Canadian Journal of Civil Engineering 34 (7), 793–802.

Laldji S, Tagnit-Hamou A and Fares G, 2004. Glass frit in concrete as an alternative cementitious material. In: International RILEM Conference on the Use of Recycled Materials in Buildings and Structures, Barcelona, Spain, pp. 953–962.

Lamond J F and Pielert J H, 2006. Significance of Tests and Properties of Concrete and Concrete-making Materials. ASTM International, West Conshohocken, PA, USA.

Lee S T and Lee J, 2016. Performance of cementitious composites incorporating ground TFT-LCD waste glass. Journal of Testing and Evaluation 44 (1), 213–221.

Lin K L, Wang N F, Shie J L, Lee T C and Lee C, 2008. Elucidating the hydration properties of paste containing thin film transistor liquid crystal display waste glass. Journal of Hazardous Materials 159 (2–3), 471–475.

Lin K L, Huang W J, Shie J L, Lee T C, Wang K S and Lee C H, 2009. The utilization of thin film transistor liquid crystal display waste glass as a pozzolanic material. Journal of Hazardous Materials 163 (2–3), 916–921.

Liu M, 2011. Incorporating ground glass in self-compacting concrete. Construction and Building Materials 25 (2), 919–925.

Liu S, Xie G and Rao M, 2013. Effect of waste glass powder on properties and microstructure of ultrahigh performance cement based materials. Materials Research Innovations 17 (1), 210–214.

LNEC E-462 (Laboratório Nacional de Engenharia Civil), 2004. Resistance of cement to sulfate attack. LNEC, Lisbon, Portugal (in Portuguese).

Lye C Q, Dhir R K and Ghataora G S, 2015. Carbonation resistance of fly ash concrete. Institution of Civil Engineers Journal, Magazine of Concrete Research 67 (21), 1150–1178.

Lye C Q, Dhir R K and Ghataora G S, 2016. Carbonation resistance of GGBS concrete. Institution of Civil Engineers Journal, Magazine of Concrete Research 68 (18), 936–969.

Madandoust R and Ghavidel R, 2013. Mechanical properties of concrete containing waste glass powder and rice husk ash. Biosystems Engineering 116 (2), 113–119.

Mandville A, Jarvis M, Kerrivan V and Fitzgerakd K, 2013. Glass Powder as an Alkali Aggregate Reaction Suppressant in Newfoundland and Labrador. Cretecon Consulting, Newfoundland. 104 pp.

Matos A M and Sousa-Coutinho J, 2012. Durability of mortar using waste glass powder as cement replacement. Construction and Building Materials 36, 205–215.

Matos A M and Sousa-Coutinho J, 2016a. ASR and sulphate performance of mortar containing industrial waste. Structural Concrete 17 (1), 84–95.

Matos A M and Sousa-Coutinho J, 2016b. Waste glass powder in cement, macro and micro scale study. Advances in Cement Research 28 (7), 423–432.

Matos A, Ramos T, Nunes S and Sousa-Coutinho J, 2015. Durability enhancement of SCC with waste glass powder. Materials Research 1–8.

Meena A and Singh R, 2012. Comparative Study of Waste Glass Powder as Pozzolanic Material in Concrete BSc thesis. Department of Civil Engineering, National Institute of Technology, Rourkela, India. 46 pp.

Metwally I M, 2007. Investigations on the performance of concrete made with blended finely milled waste glass. Advances in Structural Engineering 10 (1), 47–53.

Miranda Jr. E J P, Bezerra H J C L, Piliti F S and Paiva A E M, 2014a. Increasing the compressive strength of Portland cement concrete using flat glass powder. Materials Research 17, 45–50.

Miranda Jr. E J P, Paiva A and Rodrigues E, 2014b. Influence of waste glass on the physical properties of Portland cement concrete. Materials Science Forum 798–799, 576–581.

Mirzahosseini M and Riding K A, 2014. Effect of curing temperature and glass type on the pozzolanic reactivity of glass powder. Cement and Concrete Research 58, 103–111.

Mirzahosseini M and Riding K A, 2015. Influence of different particle sizes on reactivity of finely ground glass as supplementary cementitious material (SCM). Cement and Concrete Composites 56, 95–105.

Mitra S, Srimany A and Dey B G, 2016. A practical study on effective utilisation of waste glass in concrete. International Journal of Science, Engineering and Technology Research (IJSETR) 5 (3), 694–696.

Moncea A M, Badanoiu A, Georgescu M and Stoleriu S, 2013. Cementitious composites with glass waste from recycling of cathode ray tubes. Materials and Structures 46 (12), 2135–2144.

Narayana D and Mailar G, 2015. Experimental investigation on the properties and durability of partially replaced recycled aggregates, glass powder concrete for cost effective construction. International Journal of Research in Engineering & Advanced Technology 2 (6), 135–141.

Nassar R and Soroushian P, 2011. Field investigation of concrete incorporating milled waste glass. Journal of Solid Waste Technology and Management 37 (4), 307–319.

Nassar R and Soroushian P, 2012a. Strength and durability of recycled aggregate concrete containing milled glass as partial replacement for cement. Construction and Building Materials 29, 368–377.

Nassar R and Soroushian P, 2012b. Green and durable mortar produced with milled waste glass. Magazine of Concrete Research 64 (7), 605–615.

Nassar R and Soroushian P, 2013. Use of milled waste glass in recycled aggregate concrete. Proceedings of the Institution of Civil Engineers 166 (5), 304–315.

Neithalath N, 2006. Analysis of moisture transport in mortars and concrete using sorption–diffusion approach. ACI Materials Journal 103 (3), 209–217.

Neithalath N, 2008. Quantifying the effects of hydration enhancement and dilution in cement pastes containing coarse glass powder. Journal of Advanced Concrete Technology 6 (3), 397–408.

Neithalath N and Schwarz N, 2009. Properties of cast-in-place concrete and precast concrete blocks incorporating waste glass powder. The Open Construction & Building Technology Journal 3, 42–51.

Neville A M, 1995. Properties of Concrete, fourth ed. Longman, Essex, UK. 844 pp.

Niang A, Roy N and Tagnit-Hamou A, 2015. Structural behavior of concrete incorporating glass powder used in reinforced concrete columns. Journal of Structural Engineering 141 (3), 1–10.

Nishikawa T, Takatsu M and Daimon M, 1995. Fracture behaviour of hardened cement paste incorporating mineral additions. Cement and Concrete Research 25 (6), 1218–1224.

NT Build 492, 1999. Concrete, Mortar and Cement Based Repair Materials: Chloride Migration Coefficient from Non-steady-state Migration Experiment (Nordic Council of Ministers, Nordtest Methods, Espoo, Finland).

Nunes S, Matos A M, Duarte T, Figueiras H and Sousa-Coutinho J, 2013. Mixture design of self- compacting glass mortar. Cement and Concrete Composites 43, 1–11.

Nwaubani S O and Poutos K I, 2013. The influence of waste glass powder fineness on the properties of cement mortars. IJAIEM 2 (2), 110–116.

Omran A and Tagnit-Hamou A, 2016. Performance of glass- powder concrete in field applications. Construction and Building Materials 109, 84–95.

Osmani M and Pappu A, 2010. An assessment of the compressive strength of glass reinforced plastic waste filled concrete for potential applications in construction. Concrete Research Letters 1 (1), 1–5.

Oyekan G L and Oyelade O A, 2011. Crushed waste glass as a partial replacement of cement in normal concrete production with sugar added as an admixture. Journal of Engineering and Applied Sciences 6 (6), 369–372.

Özkan O and Yüksel İ, 2008. Studies on mortars containing waste bottle glass and industrial by- products. Construction and Building Materials 22 (6), 1288–1298.

Parghi A and Alam M S, 2016. Physical and mechanical properties of cementitious composites containing recycled glass powder (RGP) and styrene butadiene rubber (SBR). Construction and Building Materials 104, 34–43.

Patagundi B R and Prakash K B, 2015. Use of glass powder as mineral admixture in steel fibre reinforced concrete. IJREST 1 (1), 25–30.

Pattengill M and Shutt T C, 1973. Use of ground glass as pozzolan. In: Symposium of Utilization of Waste Glass in Secondary Products, Albuquerque, Mexico, pp. 156–166.

Pavoine A, Darbec D, Chaussadent T, Tagnit-Hamou A and Diver L, 2014. Impact of alternative cementitious material on mechanical and transfer properties of concrete. ACI Materials 111 (3), 251–261.

PCA, 1998. Control of air content in concrete. Concrete Technology Today 19 (1), 1–3.

Pereira-de-Oliveira L A, Castro-Gomes J P and Santos P M S, 2007. Optimization of pozzolanic reaction of ground waste glass incorporated in cement mortars. In: Conference on Sustainable Construction, Materials and Practices: Challenge of the Industry for the New Millennium, Lisbon, Portugal 7 pp.

Pereira-de-Oliveira L A, Castro-Gomes J P and Santos P M S, 2012. The potential pozzolanic activity of glass and red-clay ceramic waste as cement mortars components. Construction and Building Materials 31, 197–203.

Peyvandi A, Soroushian P and Nassar R, 2013. Recycled glass concrete. Concrete International (January), 29–32.

Polley C, 1996. The effects of waste glass aggregate on the strength and durability of Portland cement concrete. MSc thesis. Department of Civil and Environmental Engineering, University of Wisconsin-Madison, WI, USA. 174 pp.

Priscilla M and Naik P A, 2014. Strength and durability study on recycled aggregate concrete using glass powder. IJETT 11 (5), 259–264.

Remarque W, Helnz D and Schleusser C, 2001. Glass powder as a reactive addition for blast furnace cements. In: Dhir R K, Limbachiya M C and Dyer T D (Eds.), International Symposium on Recycling Ad Re-use of Glass Cullet, Dundee, UK. Thomas Telford, London, UK, pp. 229–238.

Remarque W, Urbonas L and Helnz D, 2003. About the behaviour of glass powder in pastes and mortar with cement. In: 11th International Congress on Chemistry of Cement, Durban, SA, USA, pp. 3051–3065.

RILEM Technical Committee 162, 2000. Test and design methods for steel fibre reinforced concrete. Materials and Structures 33 (249), 75–81.

Samarin A, 1980. Use of fine crushed bottle glass sand and partial cement replacement in concrete. In: School of Civil Engineering, University of New South Wales, Australia, UNICIV Report No. R-193 16 pp.

Schwarz N, DuBois M and Neithalath N, 2007. Electrical conductivity based characterization of plain and coarse glass powder modified cement pastes. Cement and Concrete Composites 29, 656–666.

Schwarz N, Cam H and Neithalath N, 2008. Influence of a fine glass powder on the durability characteristics. Cement and Concrete Composites 30, 486–496.

Seju D, Makwana A, Vankar D, Patel B and Patel V, 2015. Effect of Glass Powder on Compressive Strength of Concrete BSc thesis. Gujarat Technology University, Ahmedabad, India. 31 pp.

Serpa D, Santos Silva A, de Brito J, Pontes J and Soares D, 2013. ASR of mortars containing glass. Construction and Building Materials 47, 489–495.

Shao Y and Lehoux P, 2001. Feasibility of using ground waste glass as a cementitious material. In: Dhir R K, Limbachiya M C and Dyer T D (Eds.), International Symposium on Recycling and Re-use of Glass Cullet, Dundee, UK. Thomas Telford, London, UK, pp. 209–219.

Shao Y, Lefort T, Moras S and Rodriguez D, 2000. Studies on concrete containing ground waste glass. Cement and Concrete Research 30, 91–100.

Sharifi Y, Afshoon I and Firoozjaie Z, 2015. Fresh properties of self-compacting concrete containing ground waste glass microparticles as cementing material. Journal of Advanced Concrete Technology 13, 50–66.

Shayan A, 2002. Value-added utilisation of waste glass in concrete. In: IASBSE Symposium, Melbourne, Australia, pp. 1–11.

Shayan A and Xu A, 2004. Value-added utilisation of waste glass in concrete. Cement and Concrete Research 34, 81–89.

Shayan A and Xu A, 2006. Performance of glass powder as a pozzolanic material in concrete: a field trial on concrete slabs. Cement and Concrete Research 36, 457–468.

Shekhawat B S and Aggarwal V, 2014. Utilisation of waste glass powder in concrete – a literature review. IJIRSET 3 (7), 822–826.

Shi C and Wu Y, 2005. Mixture proportioning and properties of self-consolidating lightweight concrete containing glass powder. ACI Materials Journal 102 (5), 355–363.

Shi C and Zheng K, 2007. A review on the use of waste glasses in the production of cement and concrete. Resources, Conservation and Recycling 52, 234–247.

Shi C, Wu Y, Riefler C and Wang H, 2005. Characteristics and pozzolanic reactivity of glass powders. Cement and Concrete Research 35, 987–993.

Siad H, Lachemi M, Sahmaran M and Hossain K, 2016. Effect of glass powder on sulfuric acid resistance of cementitious materials. Construction and Building Materials 113, 163–173.

Singh G, Singh A K, Bhaskar A and Attree A S, 2014. A critical study of effectiveness of waste glass powder in cocnrete. International Archive of Applied Sciences and Technology 5 (3), 31–35.

Sobolev K, Turker P, Soboleva S and Iscioglu G, 2007. Utilization of waste glass in ECO-cement: Strength properties and microstructural observations. Waste Management 27, 971–976.

Soroushian P, 2012. Towards Broad Use of Recycled Glass Concrete on MSU Campus (Innovation in Sustainability Seed Grant). Michigan State University, East Lansing, MI, USA. 48 pp.

Sugita S, Yu Q, Shoya M, Tsukinaga Y and Isojima Y, 1997. The effect of waste glass powder on the properties and pore structure of ordinary Portland cement hydrates. In: 10th International Congress on the Chemistry of Cement, Goteburg, Sweden 8 pp.

Tagnit-Hamou A and Bengougam A, 2012. The use of glass powder as supplementary cementitious material. Concrete International (March), 56–61.

Tagnit-Hamou A, Soliman N, Omran A F, Mousa M T, Gauvreau N and Provencher M, 2015. Novel ultra-high-performance glass concrete. Concrete International 37 (3), 41–47.

Taha B and Nounu G, 2008a. Properties of concrete contains mixed colour waste recycled glass as sand and cement replacement. Construction and Building Materials 22, 713–720.

Taha B and Nounu G, 2008b. Using lithium nitrate and pozzolanic glass powder in concrete as ASR suppressors. Cement and Concrete Composites 30, 497–505.

Taha B and Nounu G, 2009. Utilizing waste recycled glass as sand/cement replacement in concrete. Journal of Materials in Civil Engineering 21 (12), 709–721.

Tamanna N, Sutan N M, Yakub I, Lee D and Ahmad E, 2015. Influence of mortar incorporating silica based material on the formation of C-S-H and mechanical strength properties. Applied Mechanics and Materials 695, 647–650.

Tang M C, Wang L and Dhir R K, 2005. Engineering and durability properties of cocnrete containing waste glass. In: Dhir R K, Dyer T D and Newlands M D (Eds.), Proceedings of the International Conference on Achieving Sustainability in Construction, Dundee, UK. Thomas Telford, London, UK, pp. 219–228.

Tejaswi S S, Rao R C, Vidya B and Renuka J, 2015. Experimental investigation of waste glass powder as partial replacement of cement and sand in concrete. The IUP Journal of Structural Engineering 8 (4), 14–22.

Tognonvi M, Zidol A, Aitcin P and Tagnit-Hamou A, 2015. Aging of glass powder surface. Journal of Non-crystalline Solids 427, 175–183.

Tuncan M, Karasu B and Yalcin M, 2001. The suitability for using glass and fly ash in Portland cement concrete. In: Proceedings of the 11th International Offshore and Polar Engineering Conference, Stavanger, Norway, pp. 146–152.

US EPA, 1992. Method 1311, Toxicity Characteristics Leaching Procedure (TCLP). US Environmental Protection Agency, Washington, DC, USA.

US EPA, 2016. Code of Federal Regulations, Identification and Listing of Hazardous Waste, Subpart C – Characteristics of Hazardous Waste. 40 CFR 261.24. US Environmental Protection Agency, Washington, DC, USA. Available on http//www.ecfr.gov/cgi-bin/text-idx?SID=2d31612016bdbd3a1b9c99af2a3649dc&mc=true&node=sp40.28.261.c&rgn=div6#se40.28.261_124.

Wang H Y, 2011. The effect of the proportion of thin film transistor–liquid crystal display (TFT–LCD) optical waste glass as a partial substitute for cement in cement mortar. Construction and Building Materials 25, 791–797.

Wang H Y and Hou T C, 2011. A study of elevated temperatures on the strength properties of LCD glass powder cement mortars. Integrated Waste Management I, 391–404.

Wang Z, Shi C and Song J, 2009. Effect of glass powder on chloride ion transport and alkali-aggregate reaction expansion of lightweight aggregate concrete. Journal of Wuhan University of Technology-Material 24 (2), 312–317.

Wang H Y, Zeng H H and Wu J Y, 2014. A study on the marco and micro properties of concrete with LCD glass. Construction and Building Materials 50, 664–670.

Wang H Y, Wang C, Chang S and Lin J, 2016. A study of engineering and electricity properties of cement mortar added with recycled materials and piezoelectric powders. Construction and Building Materials 113, 297–305.

Wattanapornprom R and Stitmannaithum B, 2015. Comparison of properties of fresh and hardened concrete containing finely ground glass powder, fly ash, or silica fume. Engineering Journal 19 (3), 35–47.

WBCSD, 2012. The Cement Sustainability Initiative. World Business Council for Sustainable Development, Geneva, Switzerland. 30 pp.

WBCSD, 2016. Low Carbon Technology Partnerships Initiative. World Business Council for Sustainable Development, Geneva, Switzerland. 23 pp.

Yilmaz A and Degirmenci N, 2010. The combined usage of ground waste glass with an industrial by-product in manufacturing Portland cement mortar. Advances in Cement Research 22 (2), 73–80.

Zheng K, 2013. Recycled glass concrete. In: Parcheco-Torgal F, Jalali S, Labrincha J and John V M (Eds.), Eco-efficient Concrete. Woodhead Publishing, Cambridge, UK, pp. 241–270.

Zhu H and Byars E, 2005. Potential for use of waste glass in concrete. Concrete (February), 41–45.

Use of Glass Cullet as a Sand Component

5

Main Headings

- Fresh properties
- Strength properties
- Deformation
- Permeation
- Durability
- Environmental impact
- Case studies

Synopsis

This section evaluates the performance of concrete made with glass cullet (GC) sand. The grading, particle shape and specific gravity of GC-sand can affect the consistence, stability and density of fresh concrete. The strength properties, modulus of elasticity and shrinkage of concrete are also affected by the grading of GC-sand. It is shown that the material can be used up to 100% without compromising the mechanical properties of concrete. The permeation of concrete could be reduced due to the impervious nature of GC-sand. The durability of concrete is mostly improved apart from the resistance to carbonation, freeze–thaw attack and abrasion. The alkali–silica reaction of GC-sand is affected by its type, colour, size and content; and the expansion can be nullified with the use of pozzolanic cements. GC is safe for the environment and it has been used in the concrete construction industry since the 2000s.

Keywords: Glass cullet, Sand, Concrete, Fresh properties, Strength, Deformation, Permeation, Durability, Environmental impact, Case studies.

Sustainable Construction Materials: Glass Cullet. https://doi.org/10.1016/B978-0-08-100984-0.00005-9

5.1 Introduction

Aggregate is one of the three fundamental ingredients, the other two being cement and water, in making concrete, which is the most widely used human-made material in the construction industry the world over. The annual global production of concrete is estimated to be 25 billion tonnes (WBCSD, 2009), and this figure is likely to rise due to the rapid expansion of the human population and the continuous improvements in social welfare and living standards, particularly in the developing countries. The consumption of aggregates in concrete-related usage represents about 45% of the total aggregate production (UEPG, 2016), which is forecast to reach 35.1 billion tonnes to meet the global construction demand by 2020 (Dhir et al., 2016).

The production of concrete relies heavily on the natural sources of aggregate, such as sand and gravel deposits and crushed-rock quarries. The excavation of these materials intensifies the pressures on the environment. Although a slow transition process, the use of aggregates derived from alternative sources, as replacement for natural aggregates, is becoming increasingly important in promoting sustainability in the concrete construction sector. Commonly, these alternative materials are divided into two groups: (i) recycled aggregates, which are the processed natural aggregates recovered from construction, demolition and excavation waste, and (ii) secondary aggregates, which are mainly sourced from industrial by-products. A notable example of the latter is blast furnace slag, which is generated from iron manufacturing. Both recycled and secondary aggregates are recognised in BS EN 12620: 2002+A1 (2008) for use in concrete.

Glass cullet (GC), a processed waste glass sourced from industrial and household wastes, is considered to have potential for use as a secondary aggregate. When crushed and sieved to a particle size similar to that of natural sand, the near-zero water absorption and relatively high hardness characteristics of GC can improve the performance and serviceability of structural concrete. This is achievable only if the concrete mix is designed properly, considering the characteristics of GC, such as its angularity and particle size distribution.

In this chapter, the effects of GC as a replacement for natural sand on both the fresh and the hardened properties of concrete, as well as the associated environmental impact and case studies, are discussed using the database developed from over 200 publications originating from 25 countries, over the period of 1972–2017, with about 95% published since 2000.

5.2 General Information

Owing to its flakiness and elongated shape, the use of GC as a coarse aggregate (of size greater than 4 mm) in concrete would be unsuitable (CCANZ, 2011; Dhir et al., 2006). When processed into a granular form of size less than 4 mm, GC can be used as a manufactured sand in the same way as natural sand in making concrete. This section provides general information on some of the important properties of concrete made with GC-sand, covered in review articles.

(a) Fresh Properties

The smooth surface texture and zero water absorption characteristics of GC-sand could improve the consistence of concrete (Meyer et al., 2001; Rashad, 2014, 2015), but some studies have reported the opposite and related it to the angularity of GC particles (Shi and Zheng, 2007; Rashad, 2014; Zheng, 2013). GC-sand concrete has lower fresh density than natural sand concrete, attributed to its lower specific gravity (Rashad, 2014, 2015; Shi and Zheng, 2007). Prior to its use as a fine aggregate, the material should be washed, as the presence of sugar residue (in container glass) can extend the setting time of concrete (CWC, 1996).

(b) Strength and Deformation

The use of GC-sand up to 20% increases the compressive, flexural and splitting tensile strengths of concrete (CCANZ, 2011). However, the literature tends to suggest that the use of GC-sand could reduce the strength of concrete, and this is thought to be due to its smooth surface texture, which results in a weaker bond between GC-sand and cement paste (CCAA, 2008; Rashad, 2014, 2015; Shi and Zheng, 2007; Zheng, 2013). The resistance of concrete to deformation increases when GC-sand is used (CCAA, 2008; Rashad, 2014, 2015; Zheng, 2013).

(c) Permeation and Durability

The use of GC-sand as a natural sand replacement reduces the permeation of concrete due to its near-zero water absorption property (Rashad, 2014, 2015). Most of the studies agreed that GC-sand could improve the durability of concrete in terms of (i) chemical aspects, such as chloride ingress, carbonation, sulphate attack and acid attack, and (ii) physical aspects, such as abrasion and exposure to elevated temperature (Cree et al., 2013; Meyer et al., 2001; Rashad, 2014, 2015; Richardson, 2013; Zheng, 2013).

Perhaps the major concern of using GC-sand in concrete is its high silica content, which can potentially cause alkali–silica reaction (ASR) and lead to deleterious expansion. A review article prepared by Figg (1981) for a conference held in Cape Town, South Africa, in 1981, documented laboratory and field studies on the ASR of GC concrete in several countries, such as Canada, Germany and the United Kingdom, during the 1960s and 1970s. The article concluded that:

(i) All types of GC could undergo an expansion reaction with cements, with GC containing high boron and/or alkali metal content being the most reactive.

(ii) The ASR of GC in concrete is qualitatively different from that of reactive aggregates.

(iii) GC could result in deleterious expansions even when low-alkali Portland cement is used.

Some attempts were made to describe the ASR behaviour of mixes containing GC in studies undertaken during the 2000s. These include two mathematical models developed based on different aspects, namely, fracture mechanics (Bazant et al., 1998, 2000) and the kinetics of the chemical reactions and water diffusion involved in the ASR process (Bazant and Steffens, 2000), as well as the dissolution of GC in an alkaline environment and ASR expansion mechanisms of GC concrete (Shi, 2009).

Nevertheless, it has been suggested that the ASR expansion in GC-sand concrete can be minimised by several methods, such as:

- Use of fly ash (FA) (CCAA, 2008; CCANZ, 2011; Dhir et al., 2006; Rashad, 2014; Shi and Zheng, 2007; TriVitro, 2004; Zheng, 2013), ground granulated blast furnace slag (GGBS) (Dhir et al., 2006; Kumar and Naik, 2010; Rashad, 2014; Richardson, 2013; TriVitro, 2004), metakaolin (Dhir et al., 2006; Meyer, 2003a,b; Meyer et al., 2001; Rashad, 2014; Shi and Zheng, 2007) and even finely ground GC itself (Kumar and Naik, 2010; TriVitro, 2004).
- Use of low alkali cement with Na_2O_{eq} less than 0.60% (CWC, 1996; Copland et al., 2009).
- Use of lithium compounds (Kumar and Naik, 2010; Rashad, 2014; Richardson, 2013).

Overall, the literature presents some contradictory views on the influence of GC on the properties of concrete, probably due to the differences in the coverage and depth of analysis. Notwithstanding that, GC is considered to be a viable material as a replacement for natural sand in concrete, although additional care needs be taken to prevent the risk of ASR.

5.3 Fresh Properties

Without proper attention given to a freshly mixed concrete, between its manufacturing and its final setting, the expected hardened properties of concrete may never be fully realised. Although admixtures are used to modify the fresh properties of concrete, aggregates can also play a part through their characteristics, such as surface texture and shape. This section discusses the effects of using GC-sand as a replacement for natural sand on the consistence, stability, density and air content of fresh concrete. The initial and final setting times of concrete are not covered, as it is thought that, unless contaminated with sugar residue, the use of GC as a natural sand should have no influence in this respect.

5.3.1 Consistence

Consistence, which used to be known as workability, is one of the most specified fresh concrete properties. It provides a measure of the ease with which a freshly mixed concrete can be handled during mixing, transporting, placing, compacting and finally finishing, without the loss of its uniformity. Consistence can also give an indication of the energy required to compact fresh concrete.

To evaluate the effect of GC-sand on the consistence of concrete, particular attention should be paid to the particle size distribution of both the GC-sand and the corresponding reference natural sand used. This is because even for the same material, different particle gradings can give rise to different particle packing, void size and distribution and other dissimilarities, all of which in turn will affect the properties of concrete. Although important, this factor has often been overlooked by researchers when it comes to comparing the performance of GC-sand and natural sand. Thus, the effect of GC-sand on the consistence of concrete is discussed here based on the following three groups:

- Group 1: Gradings of GC-sand and reference natural sand are different.
- Group 2: Gradings of GC-sand and/or reference natural sand are unavailable.
- Group 3: Gradings of GC-sand and reference natural sand are the same.

The consistence of concrete made with GC-sand and natural sand as per groups 1 and 2, at equal water/cement ratios, measured in terms of slump, is shown in Figure 5.1. Based on the slump of reference concrete, the results shown in Figure 5.1 are separated into four slump classes, namely, S1 (for 10–40 mm slump), S2 (50–90 mm), S3 (100–150 mm) and S4 (160–210 mm), in accordance with BS EN 206 (2013). In addition, the tolerances for different target slumps given in the same standard are also shown in the figure for reference purposes.

The effect of GC-sand on the slump of concrete changes, depending on the slump class of the reference natural sand concrete, as described below:

- For the S1 slump class (Figure 5.1(a)), the slump of concrete increases as GC content is increased.
- For the S2 slump class (Figure 5.1(b)), the effect of GC becomes less apparent, and most of the results show that the slump of GC concrete fluctuates mostly within the S2 slump class. However, concrete containing more than 50% GC-sand tends have a slump value beyond the range of the S2 slump class.
- For the S3 slump class (Figure 5.1(c)), the inclusion of GC-sand does not significantly change the slump of concrete if the initial value is close to the lower limit of the S3 slump class (at 100 mm). However, the slump of concrete tends to decrease as GC-sand content increases if the initial value is near to the upper limit (at 150 mm). Similar to the observation seen in Figure 5.1(b), the slump of concrete made with more than 50% GC-sand is likely to fall below the lower limit of the S3 class.
- For the S4 slump class (Figure 5.1(d)), the results are limited and contradicting, with one showing a tendency to increase the slump when GC is used, whilst another shows a drastic reduction in slump. The observation in the latter might be expected, as the natural sand with a fineness modulus of 2.78 was replaced by a very fine GC-sand with average diameter of 12.4 μm (Chen et al., 2006), which could increase the water demand due to the increase in surface area.

Thus far, only one group of researchers (Dhir et al., 2006), from the Concrete Technology Unit of the University of Dundee in the United Kingdom, has made an effort to study the slump of concrete made with GC-sand and natural sand based on an equal particle size distribution. In the study, three grading ranges (coarse, medium and fine), with each grading

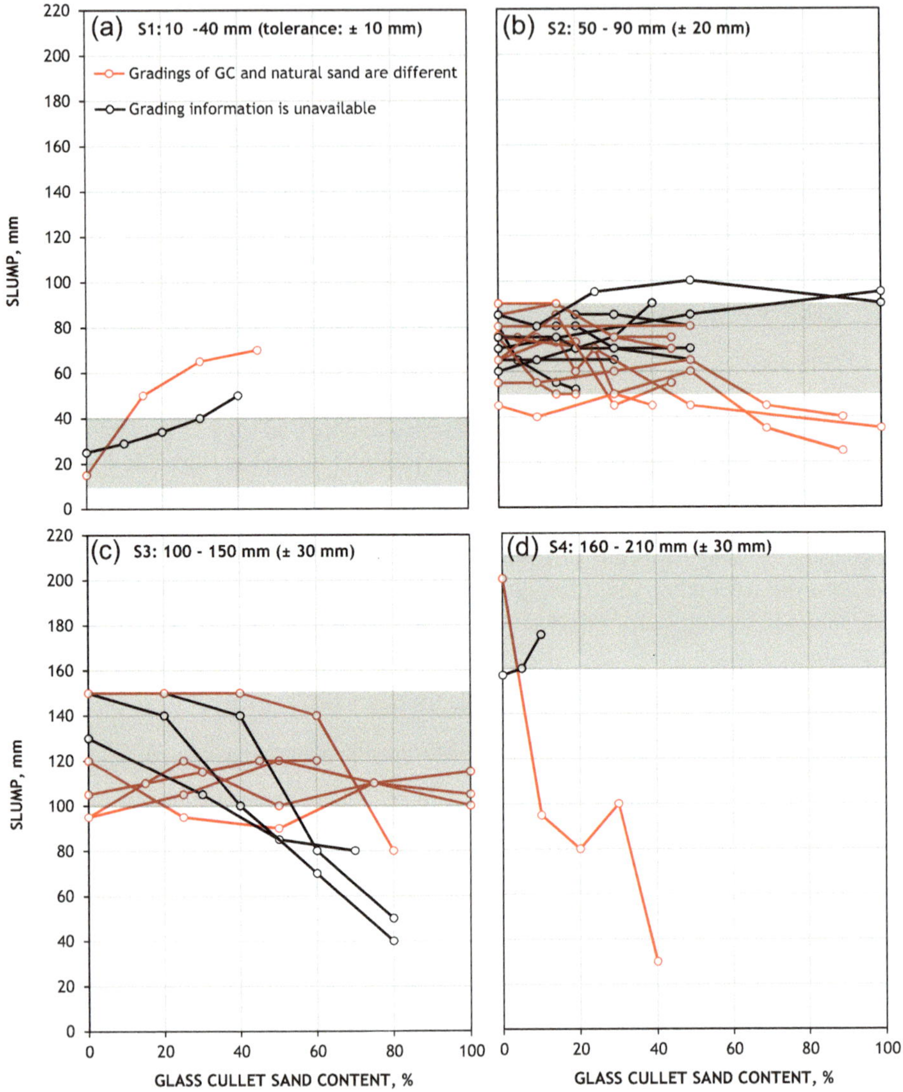

Figure 5.1 Influence of glass cullet (GC) sand on slump of concrete for (a) S1 slump class, (b) S2 slump class, (c) S3 slump class and (d) S4 slump class.
Data taken from Abdallah and Fan (2014), Adaway and Wang (2015), Batayneh et al. (2007), Chen et al. (2006), Du (2011), Du and Tan (2014), Dumitru et al. (2010), Huang et al. (2015), Ismail and Al-Hashmi (2009), Jia et al. (2015), Koh (2015), Limbachiya (2009), Malik et al. (2013, 2014), Mardani-Aghabaglou et al. (2015), Morrison (2004), Naik and Wu (2001), Park et al. (2004), Perkins (2008), Rajabipour et al. (2012), Samarin (1980), Serpa et al. (2015), Taha and Nounu (2008a, 2009), Terro (2006), Topcu and Canbaz (2004), Wang (2009a), Wang et al. (2014), Wright (2012), Wright et al. (2014), Zammit (1990) and Zammit et al. (2004).

range further divided into three curves (lower, medium and upper), were developed by combining the grading classifications given in BS 882 (1992) and BS EN 12620: 2002+A1 (2008). In total, nine grading curves with different fineness moduli, ranging from 0.95 (very fine) to 4.20 (very coarse), for GC-sand and natural sand were studied.

Except for concrete made with a natural sand of 4.20 fineness modulus for which the result was not available, Table 5.1 shows that the slump decreases as the sand grading becomes finer (or fineness modulus value becomes smaller) for both GC and natural sand concrete. However, for a given grading curve, the use of GC-sand as a replacement for natural sand tends to result in a slight reduction in the slump of concrete. In addition, it would appear that the type of GC-sand has no significant influence on the slump of concrete.

As the gradings of the two sands are the same, which suggests similar aggregate packing is likely to be achieved, the reduction in slump observed in GC-sand concrete may be attributed to the aggregate shape. The angular shape of GC-sand increases the interlocking effect and thus makes the aggregate skeleton more resistant to movement.

In addition, the effect of GC-sand on the flow diameter of mortar has also been studied, but the findings have not been conclusive. When the gradings of both GC-sand and natural sand are the same, the flow diameter shows an increase not greater than 20 mm for mortar containing up to 100% GC-sand (Penacho et al., 2014; Sikora et al., 2016). However, when the gradings of the two materials are different, conflicting trends have been observed (Hui and Sun, 2011; Ling and Poon, 2011a,b, 2012a,d, 2013; Takata et al., 2004; Yang et al., 2009; Zhao et al., 2013a).

5.3.2 Stability

As the specific gravities of the materials used in making concrete are different, it is important to ensure that concrete remains stable, with a homogeneous distribution of constituent materials, during its fresh state.

When fine-graded natural sand (55%–100% passing a 0.5-mm sieve based on BS EN 12620:2002+A1, 2008) was replaced by coarse-graded GC-sand (5%–45% passing a 0.5-mm sieve) at 50% and 100% contents, the resultant concrete was harsh and showed signs of bleeding and segregation (Taha and Nounu, 2008a, 2009). Other studies have also reported the bleeding of concrete containing GC-sand, but the effect of the sand could not be assessed as (i) its use was in conjunction with crumb rubber particles as sand (Chen et al., 2013), (ii) the water/cement ratio of the GC-sand concrete was dissimilar to that of the reference concrete (Dumitru et al., 2010) or (iii) the reference natural sand concrete was inhomogeneous (Zhao et al., 2013b).

Perhaps the study undertaken by Dhir et al. (2005a) is more useful in evaluating the effect of GC-sand on the stability of concrete as the gradings of GC-sand and natural sand used were the same. The stability of concrete, in the form of cohesiveness of concrete, was assessed in accordance with BS 1881-102 (1983) by the response of the slumped concrete (after the completion of slump test) to gentle tapping with a tamping bar. Table 5.2 shows the cohesiveness and appearance of concrete containing GC-sand and natural sand of different fineness moduli, together with the description of cohesiveness given in Dhir et al. (2005a).

Table 5.1 Effect of glass cullet (GC) sand on the slump of concrete when the gradings of GC and natural sand are the same

Reference	Glass type	Fineness Modulus*	Glass Cullet, %	Slump, mm
Cui (2005), Dhir et al. (2004)	Container	4.20	0	Not available
			100	10
		3.55	0	60
			100	40
		3.43	0	30
			100	35
		2.75	0	50
			100	50
		2.70	0	50
			100	40
		2.66	0	45
			100	50
		1.96	0	30
			100	20
		1.83	0	30
			100	5
		0.95	0	10
			100	0
Dhir et al. (2005a)	Natural sand	2.75	0	65
	Container (amber)		100	40
	Container (green)		100	65
	Container (flint)		100	55
	Flat (building)		100	40
	Flat (automobile)		100	50
	Borosilicate		100	60
	Mixed		100	50
Tang et al. (2005)	Container	2.75	0	50
			50	45

w/c ratio, water/cement ratio.
* Larger value indicates coarser particles.

It can be seen from Table 5.2 that, as the fineness modulus of sand decreases from 4.20 (very coarse) to 0.95 (very fine), GC-sand concrete changes from being 'cohesive' to 'very cohesive,' whilst natural sand concrete changes from being 'not cohesive' to 'very cohesive.' As for the appearance of concrete, which was based on visual evaluation, both coarse GC-sand and natural sand made the mix harsh, whilst finer sands made the mix dry (Table 5.2).

Table 5.2 Cohesion and appearance of concrete made with glass cullet (GC) and natural sands of different fineness moduli

Fineness Modulus	Sand Type	Cohesiveness[a]	Appearance
4.20	Natural	Not cohesive	Harsh
	GC	Cohesive	Very harsh
3.55	Natural	Slightly cohesive	Slightly harsh
	GC	Cohesive	Harsh
3.43	Natural	Slightly cohesive	Slightly harsh
	GC	Cohesive	Harsh
2.75	Natural	Very cohesive	Good
	GC	Very cohesive	Good
2.70	Natural	Very cohesive	Good
	GC	Very cohesive	Good
2.66	Natural	Cohesive	Good
	GC	Very cohesive	Good
1.96	Natural	Very cohesive	Slightly dry
	GC	Very cohesive	Dry
1.83	Natural	Very cohesive	Slightly dry
	GC	Very cohesive	Dry
0.95	Natural	Very cohesive	Very dry
	GC	Very cohesive	Very dry

[a]Very cohesive: gradually slumps further, no shearing. Cohesive: gradually slumps further, some shearing. Little cohesive: gradually slumps further, then partial collapse. Not cohesive: slumped concrete shears.
Based on Dhir et al. (2005a).

Overall, it would appear that both GC-sand and natural sand of fineness modulus from 2.66 to 2.75 would result in concrete with acceptable cohesiveness and visual appearance. The use of GC-sand or natural sand with fineness modulus falling beyond that range is likely to affect the stability of the concrete, though this can be overcome by adjusting its cement paste content.

5.3.3 Density

The density of concrete depends on the specific gravity of its constituent materials. As shown in Section 3.6.6 of Chapter 3, the specific gravity of natural sand is typically about 2.60, whilst that of GC varies depending on its chemical composition, with soda lime GC at 2.46, alumina silicate at 2.54 and lead glass at 3.04.

Thus, as to be expected, the use of the soda lime and aluminate GC would result in a marginal decrease in the density of concrete (Adaway and Wang, 2015; Batayneh et al., 2007; Borhan, 2012; Camilleri et al., 2004; De Castro and De Brito, 2013; Dhir et al., 2005a; Du, 2011; Du and Tan, 2014; Dumitru et al., 2010; Hui and Sun, 2011;

Ismail and Al-Hashmi, 2009; Mardani-Aghabaglou et al., 2015; Naik and Wu, 2001; Oliveira et al., 2013, 2015; Penacho et al., 2014; Samarin, 1980; Topcu and Canbaz (2004); Taha and Nounu, 2008a, 2009; Tan and Du, 2013; Wang and Huang, 2010b; Wright, 2012; Wright et al., 2014).

On the other hand, the use of lead GC-sand, which is derived from cathode ray tubes (CRTs), has been shown to increase the density of concrete at about 1% for every 10% addition of lead GC-sand (Hui and Sun, 2011; Romero, 2013; Romero et al., 2013; Zhao et al., 2013a). However, the material might not be suitable for use in concrete because of its lead leaching potential (see Section 5.9).

5.3.4 Air Content

The placement of concrete is always followed by compaction to expel the air entrapped in the loose concrete during mixing and handling. A fully and properly compacted concrete is dense and with few air voids. The presence of air voids in concrete reduces its strength and increases its permeation properties, which can facilitate the transportation of fluid into the concrete.

Figure 5.2 shows the difference in the air content between GC-sand concrete mixes and the corresponding natural sand concrete mixes, made with and without an air-entraining agent, at equal water/cement ratios. For non-air-entrained concrete, the replacement of natural sand by GC-sand of the same grading leads to a slight increase in air content (no more than 0.5% on average), which can be considered as within the experimental error. A similar observation is also seen when natural sand is replaced by GC of different gradings in non-air-entrained concrete. For non-air-entrained mortar, the replacement of natural sand by GC-sand of either different or the same grading is found to increase the air content (Figure 5.2).

For air-entrained concrete, the air is deliberately introduced into the mix to provide resistance to freeze–thaw attack. The results shown in Figure 5.2 suggest that GC-sand tends to increase the air content of air-entrained concrete for a given dosage of air-entraining agent. This might be due to the difference in the gradings of GC-sand and natural sand used, but such information was unavailable in the studies.

Considering the sensitivity and accuracy of the air content test, it can be assumed that the use of GC-sand as a replacement for natural sand does not significantly affect the air content of non-air-entrained and air-entrained concrete mixes.

5.4 Strength Properties

The strength of concrete determines its ability to carry loads and is the most commonly specified property of hardened concrete in structural applications. It can be measured in various forms, namely, compressive strength, splitting tensile strength and flexural strength.

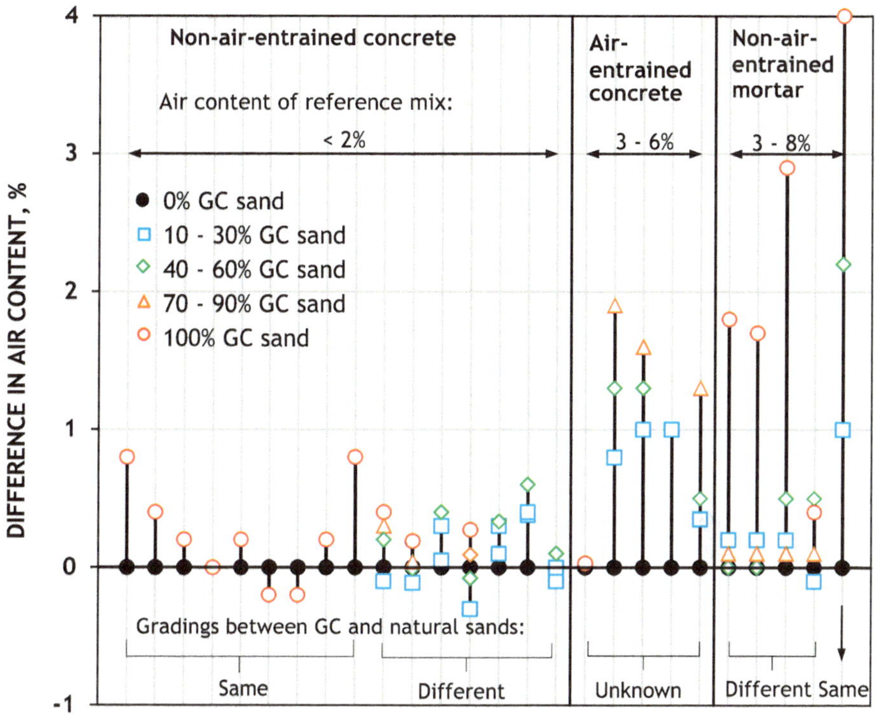

Figure 5.2 Effects of glass cullet (GC) sand on air content of concrete and mortar. Data taken from Cui (2005), Dhir et al. (2005a), Du (2011), Du and Tan (2014), Dumitru et al. (2010), Naik and Wu (2001), Park et al. (2004), Penacho et al. (2014), Rajabipour et al. (2012), Tan and Du (2013), Wright (2012) and Wright et al. (2014).

5.4.1 Compressive Strength

The compressive strength of concrete is affected by its water/cement ratio and aggregate properties, as well as the bond between cement paste and aggregate. The compressive strength of concrete made with GC-sand content up to 100% of the total fine aggregate component has been investigated since 1980, over a wide range of water/cement ratios from 0.15 to 1.20, but with 80% of the studies undertaken with a water/cement range of 0.35–0.65.

The effect of GC-sand as a replacement for natural sand on the compressive strength of concrete has been analysed by comparing the 28-day compressive strength of GC concrete with that of the corresponding reference natural sand concrete at equal water/cement ratios, expressed in percentage. Given that particle packing can affect the compressive strength of concrete, the results are separated based on the gradings of GC-sand and natural sand, in a manner similar to that adopted previously in Section 5.3.1, described as follows.

(i) Gradings of GC-Sand and Reference Natural Sand are the Same (Figure 5.3(a))

Only a small number of studies fall into this group (Cui, 2005; Dhir et al., 2005a; Panchakaria and Hall, 1996; Penacho et al., 2014; Poutos and Nwaubani, 2013; Serpa et al., 2015; Sikora et al., 2016; Tang et al., 2005). As shown in Figure 5.3(a), when GC is used up to 100% as a natural sand replacement, the change in compressive strength of each individual data point fluctuates from −15% to 20%, giving an overall average 0.5% increase. Thus, the results suggest that the use of GC-sand as a natural sand replacement does not adversely affect the compressive strength of concrete, provided that the gradings of the two sands are the same.

(ii) Gradings of GC-Sand and Reference Natural Sand are Different (Figure 5.3(b))

Replacing natural sand with GC-sand of a different grading can result in dissimilar aggregate packing that leads to a different strength response when the concrete is loaded. In this case, Figure 5.3(b) shows that the change in strength is mostly within the +20% to −20% range for concrete containing up to 90% GC-sand, though at 10% and 20% GC-sand

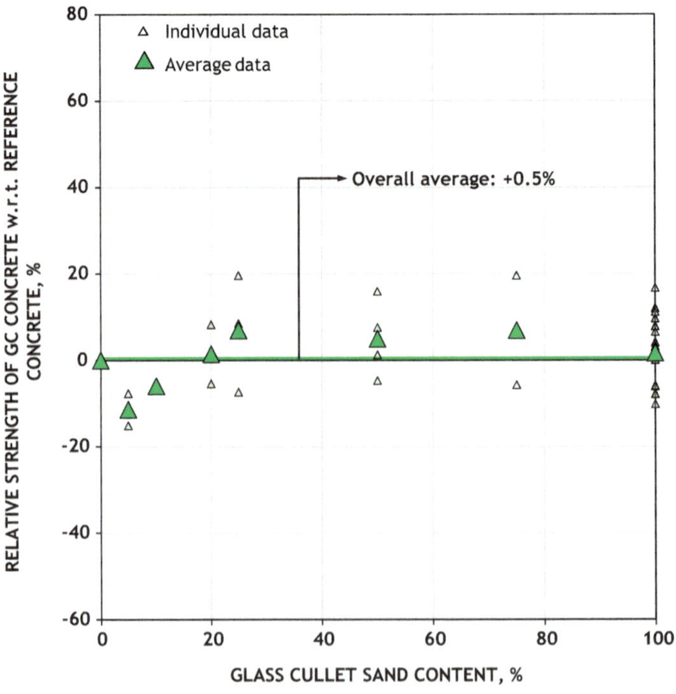

Figure 5.3(a) Relative compressive strength of glass cullet sand concrete with respect to natural sand concrete, where the sand gradings are the same.

content, the majority of the data fall on the positive side (indicating increases in strength). The spread of data at 100% GC-sand concrete is almost double, ranging from +40% to −40%. The overall average of these results is −1%, suggesting that the combination of two sands of different gradings is likely to reduce the compressive strength of concrete.

In general, it would appear that, when the gradings of GC-sand and natural sand are different, the compressive strength of the concrete may remain essentially unchanged up to 20% GC content, beyond which the strength tends to be reduced as the GC-sand content is increased up to 100%.

(iii) Gradings of GC-Sand and/or Reference Natural Sand are Unavailable (Figure 5.3(c))

For the studies in which the grading information of either GC-sand or natural sand or both is not reported, the laboratory control is considered to be doubtful. It is evident from Figure 5.3(c) that a much higher fluctuation is observed in these results than in those in Figure 5.3(a) and (b). The data distribution in Figure 5.3(c) also shows that the values are predominantly negative (reduction in strength), particularly for concrete containing more than 20% GC-sand. The overall average of 12.5% reduction in strength shown in Figure 5.3(c) might prove the point that the laboratory control for these studies is possibly unsatisfactory.

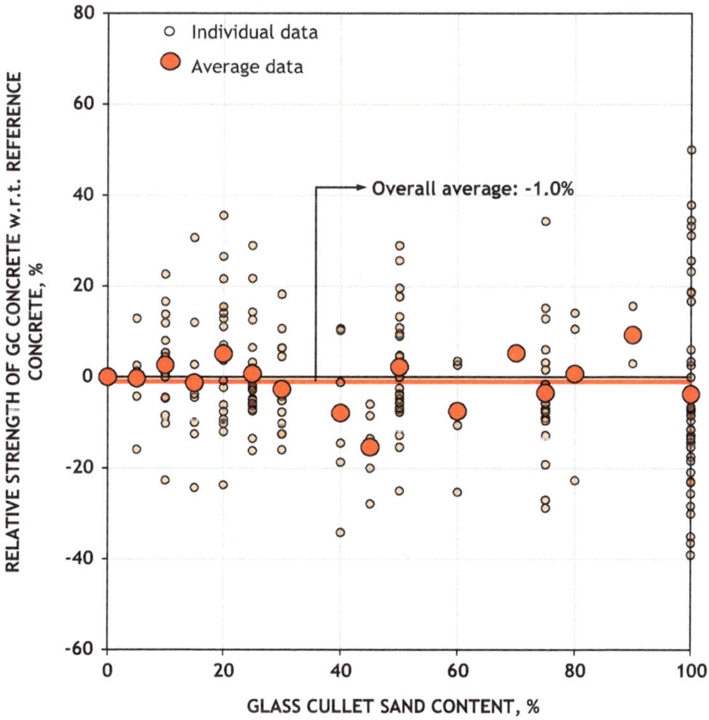

Figure 5.3 (b) Relative compressive strength of glass cullet (GC) sand concrete with respect to natural sand concrete, where the sand gradings are different.

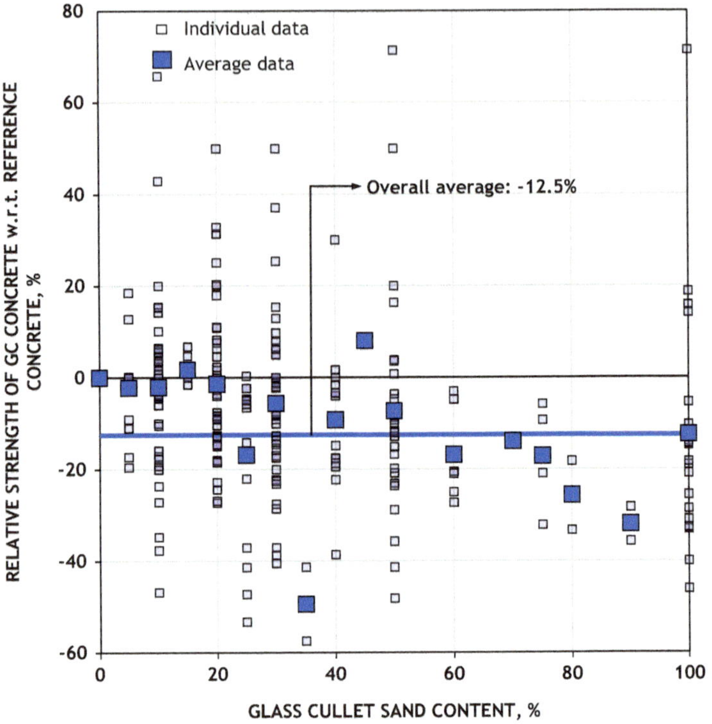

Figure 5.3 (c) Relative compressive strength of glass cullet (GC) sand concrete with respect to natural sand concrete, where sand gradings are unavailable.

Data of Figures 5.3 (a)–(c) taken from Abdallah and Fan (2014), Abendeh et al. (2015), Adaway and Wang (2015), Ali and Al-Tersawy (2012), Anna (2013), Ammash et al. (2009), Batayneh et al. (2007), Bhandari and Tajne (2013), Bhandari et al. (2014), Bignozzi and Saccani (2012), Borhan (2012), Borhan and Bailey (2014), Camilleri et al. (2004), Chen et al. (2006, 2011), Corinaldesi et al. (2016), Cui (2005), De Castro and De Brito (2013), Degirmenci et al. (2011), Dhir et al. (2004, 2005a), Du (2011), Du and Tan (2014), Dumitru et al. (2013), Ganiron (2013), Guo et al. (2015), Huang et al. (2015), Hui and Sun (2011), Ismail and Al-Hashmi (2009), Isler (2012), Jones et al. (2005), Ke et al. (2011), Kim and Soh (2002), Koh (2015), Kou and Poon (2009), Kulkarni et al. (2015), Lee (2011), Lee et al. (2008), Lim (2014), Limbachiya (2009), Ling and Poon (2011a,b, 2012a,c,d, 2013, 2014d), Ling et al. (2011a,b, 2012a,b), Malik et al. (2013, 2014), Mardani-Aghabaglou et al. (2015), Maschio et al. (2013), Mavroulidou et al. (2011), Miranda et al. (2014), Mitra et al. (2016), Morrison (2004), Naik and Wu (2001), Noruziaan and Buskell (2010), Oliveira et al. (2008), Panchakaria and Hall (1996), Park and Lee (2004), Park et al. (2004), Penacho et al. (2014), Perkins (2008), Polley (1996), Polley et al. (1998), Poutos and Nwaubani (2013), Rajabipour et al. (2012), Ram et al. (2015), Ramana and Samdani (2013), Richardson et al. (2002), Romero (2013), Romero et al. (2013), Rose and Saji (2015), Sagoe-Crentsil et al. (2001), Samarin (1980), Sangha et al. (2004), Serpa et al. (2015), Sharif et al. (2014), Sharifi et al. (2013), Shehata et al. (1996), Shklyan (2009), Sidharthan et al. (2016), Sikora et al. (2016), Soutsos et al. (2006), Sua-iam and Makul (2013), Suresh et al. (2015), Taha and Nounu (2008a, 2009), Tan and Du (2013), Tang et al. (2005), Terro (2006), Walczak et al. (2015), Wang (2009a), Wang and Huang (2010a), Wang et al. (2014, 2015), Wright (2012), Wright et al. (2014), Yang et al. (2009), Zammit (1990), Zammit et al. (2004) and Zhao et al. (2013a).

Apart from these, additional analysis on the variability of the 28-day compressive strength of concrete used in developing Figure 5.3 has been carried out, but it is limited to the mixes made with 100% Portland cement and of 0.45, 0.50 and 0.55 water/cement ratio. The numbers of samples for natural sand concrete and GC-sand concrete (regardless of GC content) used in this analysis were 13 and 48, respectively. The results showed that, for natural sand concrete, the mean standard deviation and coefficient of variance values were 8.2 MPa (with a range of 6.3–10.5 MPa) and 19.7% (with a range of 16.1%–25.1%), respectively. The corresponding values for GC-sand concrete were 6.3 MPa (with a range of 4.5–7.6 MPa) and 17.2% (with a range of 13.8%–20.5%).

The relatively low values of variability found in GC concrete could be associated with the inherent quality of the GC-sand material itself. As the original form of GC-sand is glass, which is produced under controlled conditions, it would have a more consistent quality (less variable) compared with naturally occurring sand. Thus, concrete made containing GC-sand is likely to show less variability in its strength.

5.4.2 Tensile Strength

Although weak in tension, the tensile strength of concrete is important, as the fracture of concrete, caused by compression or other loading, is a result of cracking due to tensile failure (Popovics, 1998). Cracks propagate when tensile stresses exceed the tensile strength of concrete. This section discusses the effect of GC-sand on the splitting tensile strength and flexural strength of concrete.

Splitting Tensile Strength

In the splitting tensile strength test, a load is applied along the length of a horizontally placed concrete cylinder, which normally fails by splitting into two halves.

Although relatively less reported compared with compressive strength, the splitting tensile strength of concrete containing up to 100% GC-sand has been investigated since 1996 and mostly in the range of 0.38–0.55 water/cement ratio (Ali and Al-Tersawy, 2012; Ammash et al., 2009; Anna, 2013; Batayneh et al., 2007; Borhan, 2012; Du and Tan, 2014; Ke et al., 2011; Kou and Poon, 2009; Kulkarni et al., 2015; Lee et al., 2008; Ling and Poon, 2012b; Malik et al., 2013; Mardani-Aghabaglou et al., 2015; Mavroulidou et al., 2011; Naik and Wu, 2001; Noruziaan and Buskell, 2010; Park et al., 2004; Ramana and Samdani, 2013; Romero, 2013; Serpa et al., 2015; Sharif et al., 2014; Sharifi et al., 2013; Shehata et al., 1996; Taha and Nounu, 2008a; Tan and Du, 2013; Wang, 2009a).

Using the same method as adopted previously in Section 5.4.1, the tensile strength results are separated based on the grading information of GC-sand and natural sand. In general, it is found that the spread of data in tensile strength results for each sand grading group is similar to that of compressive strength results. This suggests that the influence of GC-sand on both the tensile and the compressive strength is essentially the same.

Other studies have also reported the splitting tensile strength of GC concrete, but because of the lack of natural sand concrete, its effect on the tensile strength of concrete could not be assessed (Al-Sibahy and Edwards, 2012a,b; Berry et al., 2011; Meyer and Baxter, 1998b; Meyer et al., 2002; Shayan and Xu, 2004).

The relationship between tensile strength and characteristic cube strength of GC concrete and its corresponding reference natural sand concrete at equal water/cement ratio, determined at 28 days, is shown in Figure 5.4. Both the characteristic cube strength and the tensile strength have been determined using the following equations:

$$f_{ck,cube} = f_m \ (1 - 1.64v)$$

where $f_{ck,cube}$ is the characteristic cube compressive strength; f_m is the mean cube compressive strength;1.64 is the constant for 5% of individual cube strength below the design strength;v is the coefficient of variation, taken as 6% for fair laboratory control (ACI 301, 2005).

$$f_{ct} = 0.9 f_{ct,sp} \ \text{(based on Eurocode 2, 2004)}$$

where f_{ct} is the tensile strength; $f_{ct,sp}$ is the splitting tensile strength (Oliveira et al., 2013, 2015; Penacho et al., 2014).

The obtained relationship is also compared with that of natural sand concrete given in Eurocode 2 (2004) and copper slag (CS) sand concrete (Dhir et al., 2016), as shown in Figure 5.4.

It can be seen from Figure 5.4 that the relationship between tensile strength and characteristic cube strength of GC-sand concrete is essentially similar to that of natural sand concrete. Comparing it with CS-sand, which also has almost zero water absorption, the tensile strength of CS-sand concrete is higher than that of GC-sand concrete for a given characteristic cube strength.

Figure 5.4 also shows that, when the characteristic cube strength is below 45 MPa, the tensile strengths of both GC and natural sand concrete are similar to or slightly lower than the estimated tensile strength from Eurocode 2 (2004). When the characteristic cube strength is above 45 MPa, the tensile strength of GC-sand and natural sand concrete tends to be higher than the estimated value. This observation is more significant for CS-sand concrete (Figure 5.4).

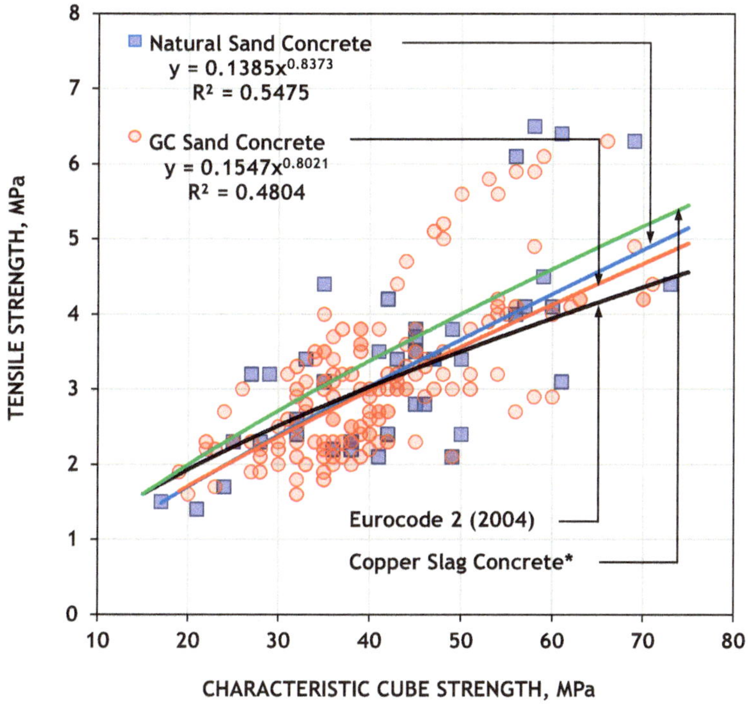

Figure 5.4 Relationship between characteristic cube strength and tensile strength of glass cullet (GC) sand concrete.
Data taken from Ali and Al-Tersawy (2012), Ammash et al. (2009), Batayneh et al. (2007), Borhan (2012), Du (2011), Du and Tan (2014), Dumitru et al. (2010), Ke et al. (2011), Kou and Poon (2009), Kulkarni et al. (2015), Lee et al. (2008), Ling and Poon (2012b), Malik et al. (2013), Mardani-Aghabaglou et al. (2015), Mavroulidou et al. (2011), Naik and Wu (2001), Noruziaan and Buskell (2010), Park et al. (2004), Ramana and Samdani (2013), Romero (2013), Serpa et al. (2015), Sharif et al. (2014), Sharifi et al. (2013), Shehata et al. (1996), Taha and Nounu (2008a, 2009), Tan and Du (2013) and Wang (2009a). *Modified from Dhir et al. (2016).

Notwithstanding this, it can be concluded that the use of GC-sand, and probably other secondary materials of similar characteristics to GC, does not adversely affect the relationship between tensile strength and characteristic cube strength and the use of Eurocode 2 (2004) in estimating the tensile strength of concrete.

The tensile strength of concrete made with these materials can, for all practical purposes, be estimated using Eurocode 2 (2004) without any modification.

Flexural Strength

The flexural strength of concrete is normally specified for the design of road pavements, airport pavements and floor slabs. For the test, a beam is subjected to a third-point loading until failure occurs in the tensile region of the specimen.

Again, a similar data separation method, based on sand grading, is used to compare the flexural strength of GC-sand concrete with respect to natural sand concrete. The data are taken from the literature published since 1996, for GC-sand content up to 100%, covering a wide range of water/cement ratio from 0.15 to 0.67. Flexural strength results were found to vary similar to the compressive strength results in Section 5.4.1, that is:

(i) Replacing natural sand with GC-sand of similar grading resulted in no significant change in flexural strength.

(ii) Replacing natural sand with GC-sand of different grading resulted in a slight reduction in flexural strength as an overall effect.

(iii) When the grading information of GC-sand and/or natural sand is unavailable, the overall flexural strength reduction in GC-sand concrete appears to be the greatest.

Figure 5.5 shows the 28-day characteristic cube strength of GC-sand concrete and reference natural sand concrete plotted against their corresponding 28-day flexural strength. The characteristic cube strength of concrete was determined using the method previously discussed in Section 5.4.2.1.

In developing the trend lines for GC-sand and natural sand concrete in Figure 5.5, some of the data were not considered, because of either (i) unusually high flexural strength, which is taken as a potential outlier, shown as circled in the figure, or (ii) the strength data of the reference natural sand concrete being unavailable (Ling and Poon, 2014b; Meyer and Baxter, 1998b; Meyer et al., 2002).

The scattered distribution of data points seen in Figure 5.5 is somewhat expected, as the flexural strength test is known to be sensitive to the specimen condition. Nevertheless, Figure 5.5 shows that the relationship between the characteristic cube strength and the flexural strength of GC-sand concrete is close to that of natural sand concrete up to about 60 MPa, beyond which, for a given compressive strength, the flexural strength of GC-sand concrete tends to be higher than that of natural sand concrete.

The same relationship for CS-sand concrete, modified from Dhir et al. (2016), is also shown in Figure 5.5 for comparison purposes. It can be seen from Figure 5.5 that, at a given compressive strength, the flexural strength of CS-sand concrete is generally higher than that of GC-sand concrete and natural sand concrete. In addition, the relationship between characteristic cube strength and flexural strength of concrete proposed by RILEM Technical Committee 162 (2000) is also shown in Figure 5.5, and the results suggest that, though conservative, the RILEM model can also be used to estimate the flexural strength of concrete containing GC-sand.

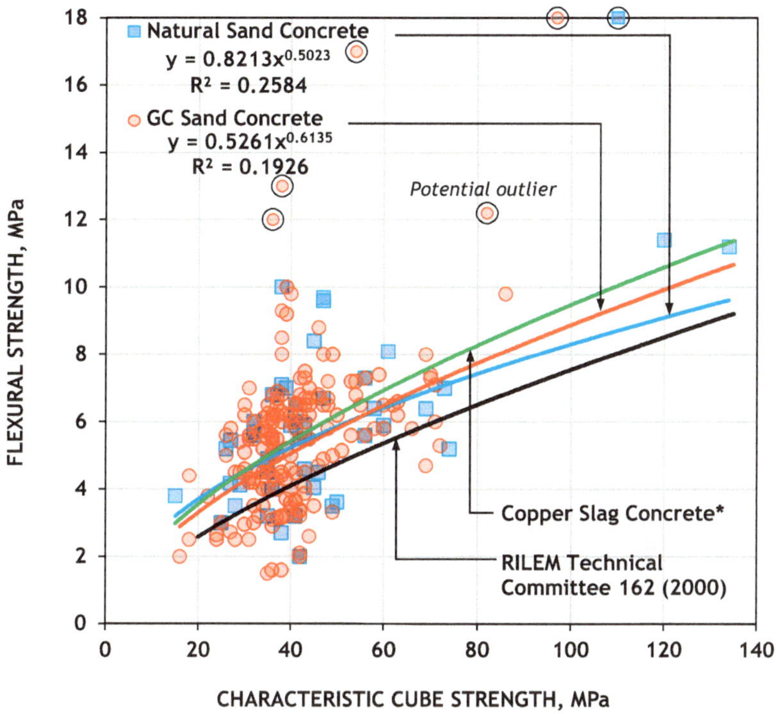

Figure 5.5 Relationship between characteristic cube strength and flexural strength of glass cullet (GC) sand concrete.
Data taken from Ali and Al-Tersawy (2012), Ammash et al. (2009), Batayneh et al. (2007), Degirmenci et al. (2011), Dhir et al. (2005a), Du (2014), Du and Tan (2014), Dumitru et al. (2010), Hui and Sun (2011), Ismail and Al-Hashmi (2009), Kim and Soh (2002), Kulkarni et al. (2015), Limbachiya (2009), Ling and Poon (2011a,b,c, 2013), Ling et al. (2011a, 2012b, 2013), Mavroulidou et al. (2011), Park et al. (2004), Ram et al. (2015), Ramana and Samdani (2013), Serpa et al. (2015), Sharif et al. (2014), Sharifi et al. (2013), Sikora et al. (2016), Soutsos et al. (2006), Taha and Nounu (2008a, 2009), Tan and Du (2013), Tang et al. (2005), Walczak et al. (2015), Wang (2009a), Wang and Huang (2010a), Yang et al. (2009) and Zhao et al. (2013a). *Modified from Dhir et al. (2016).

5.5 Deformation

In the hardened state, concrete undergoes various types of deformation, which occur due to internal factors, such as chemical reaction of cement particles and hydration products, and external factors, such as moisture condition, temperature and applied loads. The deformation of concrete is important to the design engineer, as it can cause deflections and cracking in concrete and losses in prestressed concrete structural elements, leading to potential serviceability failures. Deformation normally takes place in hardened cement paste, with aggregates in concrete providing restraint against volume changes. The following sections discuss the influence of GC-sand on the three primary deformation properties of concrete, namely, modulus of elasticity, creep and shrinkage.

5.5.1 Elastic Modulus

The elastic property of concrete is often expressed as the elastic modulus, which is the slope of the stress–strain response of concrete under uniaxial compression. Figure 5.6 shows the ratio of the 28-day elastic modulus of GC-sand concrete to that of its corresponding reference natural sand concrete, expressed as a percentage. In developing Figure 5.6, each data point has been slightly displaced to avoid overlapping. For each GC-sand level, a box-and-whisker plot was created to show the distribution of data and to identify potential outliers.

In general, the individual average relative elastic modulus at each GC-sand level varies between +10% and −5%, giving an overall average of 2% increase. The variation in individual average value is likely to be due to the influence of the vast majority of the studies in which the gradings of GC-sand and natural sand used are different, giving rise to different particle packing and thus behaviour that is somewhat different under loads.

When the grading of the two sands is the same, the results, shown as triangle symbols in Figure 5.6, suggest that the elastic modulus of concrete can increase by 10.5% on average when 100% GC sand is used (Dhir et al., 2004, 2005a; Wright et al., 2014). However, slight reductions in the values at 5%, 10% and 20% GC-sand contents are most probably affected by their higher water/cement ratio for consistence similar to reference concrete (Serpa et al., 2015). The increase in elastic modulus could be associated with the higher stiffness of GC-sand compared with natural sand, which in return provides a greater restraining effect under load. Notwithstanding this, it would safe to assume that the elastic modulus of GC-sand concrete can be higher than, or at least similar to, that of natural sand concrete, providing that the grading of the sand is unchanged.

Figure 5.7 shows the relationship between the characteristic cube strength and the modulus of elasticity of GC-sand concrete and its corresponding reference natural sand concrete, which is based on the research undertaken by Lye et al. (2017). The relationship is also compared with that of (i) concrete made with aggregate of rock types within the range of basalt to sandstone adopted in Eurocode 2 (2004) and (ii) normal-weight concrete obtained from ACI 318-11 (2011) and (iii) CS-sand concrete, modified from Dhir et al. (2016).

For a given characteristic cube strength, below 70 MPa, the use of near-zero porosity materials such as GC-sand and CS-sand results in higher elastic modulus than the corresponding concrete made with natural sand. However, this benefit becomes somewhat insignificant at a characteristic strength above 70 MPa. Additionally, it can be seen from Figure 5.7 that the elastic modulus–compressive strength relationships of GC-sand, CS-sand and natural sand concrete are closer to that of ACI 318-11 (2011), but different to any of those of natural rock concrete given in Eurocode 2 (2004).

5.5.2 Creep

When a load is applied on a concrete for a longer period, the concrete will deform gradually, and this is known as creep. In other words, creep is a time- and load-dependent deformation. The influence of GC-sand on the creep of concrete has not been commonly investigated. This is perhaps to be expected as the test is both time and cost consuming.

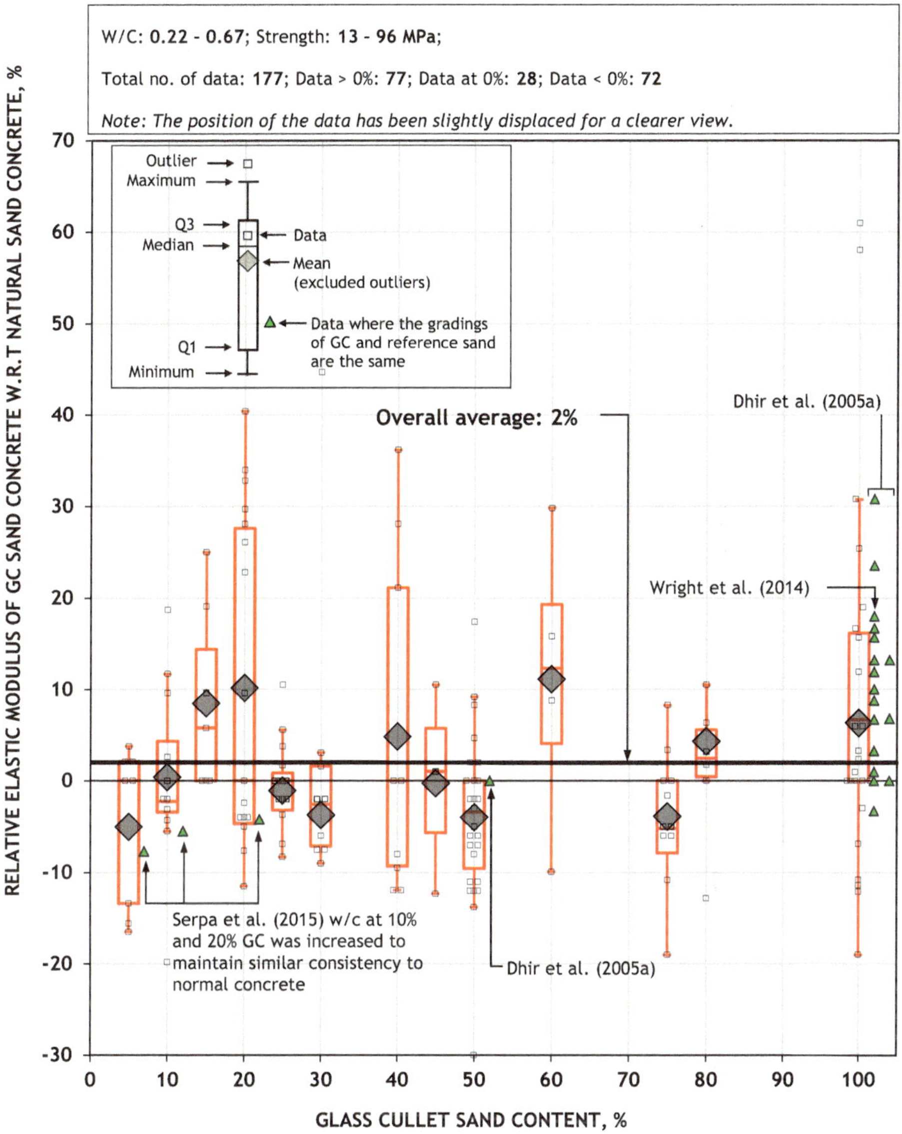

Figure 5.6 Relative modulus of elasticity of glass cullet (GC) sand concrete with respect to corresponding natural sand concrete. *w/c*, water/cement ratio.
Data taken from Abdallah and Fan (2014), Ali and Al-Tersawy (2012), Al-Sibahy and Edwards (2012a), Dhir et al. (2004, 2005a,b), Du and Tan (2014), Dumitru et al. (2010), Ganiron (2013), Guo et al. (2015), Hui and Sun (2011), Jia et al. (2015), Kulkarni et al. (2015), Lee et al. (2008), Limbachiya (2009), Ling and Poon (2012b,c), Mavroulidou et al. (2011), Noruziaan and Buskell (2010), Roskos et al. (2015), Samarin (1980), Serpa et al. (2015), Shehata et al. (1996), Taha and Nounu (2008a), Tan and Du (2013), Tang et al. (2005), Wang (2009a), Wright et al. (2014) and Zhao et al. (2013a,b).

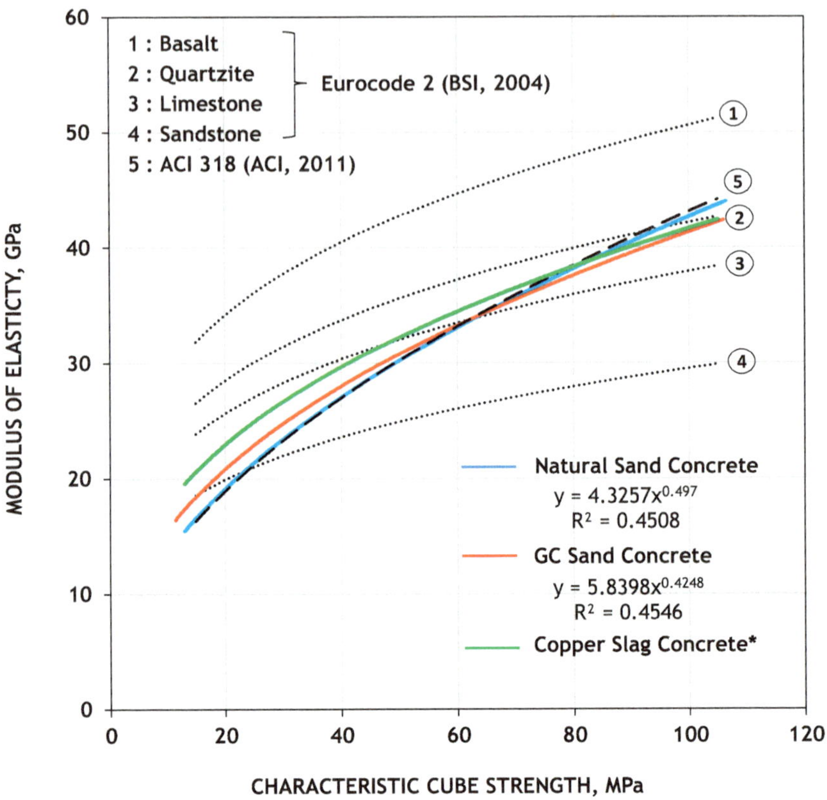

Figure 5.7 Relationship between characteristic cube strength and modulus of elasticity of glass cullet (GC) sand concrete.
*Modified from Dhir et al. (2016).

Nevertheless, it has been suggested that the creep of concrete is not affected significantly by the modulus of elasticity of aggregate for a value greater than 70 GPa (Neville et al., 1983). As shown previously in Section 3.7.8, the moduli of GC-sand and natural sand are close to 70 GPa, thus implying that the use GC-sand in concrete is likely to result in a creep strain similar to that of concrete made with natural sand.

5.5.3 Shrinkage

Shrinkage of concrete is a time-dependent and load-independent deformation, which can occur as soon as concrete is freshly placed and thereafter continues throughout its service life. This volume reduction in concrete takes several forms, such as plastic shrinkage, autogenous shrinkage, drying shrinkage and carbonation shrinkage.

Figure 5.8 shows the relative shrinkage of GC-sand concrete with respect to that of the corresponding reference natural sand concrete, as analysed and evaluated by Lye et al. (2017) based on 31 studies published during 1972–2015. It should be noted that the generic term 'shrinkage' is used here, as the test methods adopted in the studies researched did not generally distinguished between the type of shrinkage experienced by the test specimens.

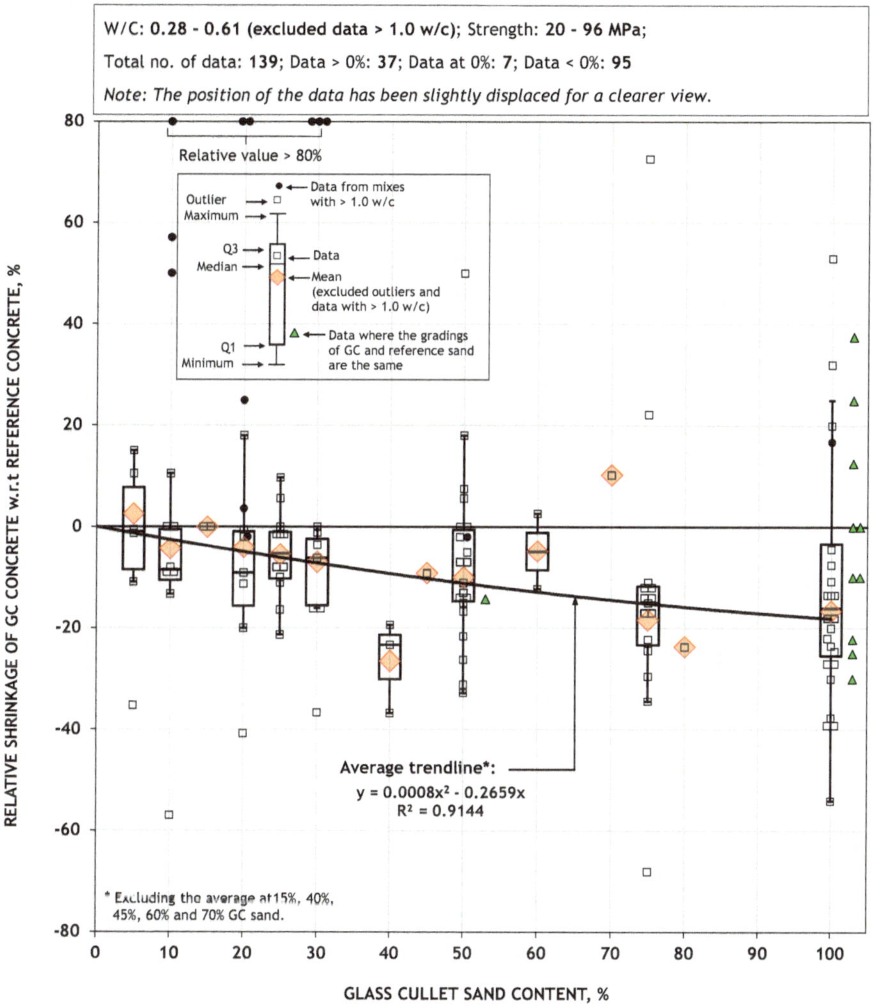

Figure 5.8 Relative shrinkage of glass cullet (GC) sand concrete with respect to corresponding natural sand concrete.
Data taken from Boniface (2006), De Castro and De Brito (2013), Dhir et al. (2005a), Du and Tan (2014), Dumitru et al. (2010), Huang et al. (2015), Hui and Sun (2011), Limbachiya (2009), Ling and Poon (2011a,b, 2012a,b, 2013), Ling et al. (2011a), Oliveira et al. (2013, 2015), Penacho et al. (2014), Phillips et al. (1972), Poon and Ling (2010), Sagoe-Crentsil et al. (2001), Sharif et al. (2014), Shayan (2002), Shayan and Xu (2006), Tan and Du (2013), Wang and Chen (2010), Wang and Huang (2010a), Wang et al. (2014), Wright et al. (2014) and Zhao et al. (2013a,b).

The polynomial regression obtained from the average shrinkage values at each GC-sand content shown in Figure 5.8 suggests that as the GC-sand content increases, the relative shrinkage of GC-sand concrete to natural sand concrete decreases at a decreasing rate, giving an average shrinkage reduction of 16% when 100% GC-sand is used. This benefit is most likely to be associated with the near-zero porosity and water absorption properties of GC-sand, which restrict the ease of moisture movement from concrete to the surroundings. However, it should be noted that the results obtained by Dhir et al. (2005a) show that, on average, replacing natural sand with GC of similar grading may not significantly influence the shrinkage of concrete.

Although based on limited data, the preliminary findings of Lye et al. (2017) suggest that the shrinkage of concrete made with GC-sand may be underestimated using Eurocode 2 (2004).

5.6 Permeation

Permeation of concrete plays an important role in determining the durability of structures, as it can affect the ingress of harmful liquids and gases, which may deteriorate the performance of concrete, for example, ingress of chlorides resulting in corrosion of reinforcement. There are three main transport mechanisms whereby a fluid can enter concrete, namely, absorption, permeability and diffusion. The ease with which fluids can enter concrete ultimately depends on the porosity of (i) cement paste, (ii) aggregate and (iii) the interface between the two, as well as, of course, the interconnectivity of the pores.

5.6.1 Porosity

Measurements of permeable porosity of concrete, with water/cement ratio varying within a narrow range of 0.40–0.58, made with different GC-sand replacements are presented in Table 5.3. Though showing some variable trends, the overall impression from the limited data available appears to be that the use of GC-sand up to a 100% replacement level is unlikely to significantly change the porosity of concrete. In this context, it is interesting to note that the GC-sand used, in all cases, was coarser than the natural sand, with the modulus of fineness in the ratio of approximately 2:1.

5.6.2 Absorption

As concrete is a porous material, a liquid can be absorbed into empty or partially saturated pores through capillarity suction. The effect of GC-sand on the absorption of concrete has been examined by many researchers since 2005, using the three commonly known methods, namely, sorptivity, initial surface absorption and water absorption. The results of each individual test are discussed below.

Table 5.3 Volume of permeable voids in concrete containing glass cullet (GC)

References	Fineness Modulus of Sand		w/c	GC Content, %	Permeable Voids, %
	GC	Natural			
(a) 28 Days Curing					
Shayan and Xu (2006)	3.25[a] 1.67[b]	NA	0.49	0	13.2
				50	14.2
Miranda et al. (2014)	3.88	1.64	0.50	0	9.2
				5	8.0
				10	8.5
				20	13.8
			0.55	0	11.0
				5	8.5
				10	8.5
				20	8.2
			0.58	0	14.5
				5	9.9
				10	9.5
				20	9.3
(b) 90 Days Curing					
Ling and Poon (2011a)	4.67[c] 3.82[d]	1.94	0.40	0	19.0
				100	17.8
Ling et al. (2011a,b)	4.19	1.92	0.40	0	19.0
				25	20.0
				50	20.5
				75	19.5
				100	19.0

w/c, water/cement ratio.
[a]For 0.60–2.36 mm.
[b]For 0.15–0.30 mm.
[c]For 2.35–5 mm.
[d]For 0–2.36 mm.

Table 5.4 Sorptivity of concrete made with glass cullet (GC) sand having fineness modulus similar to that of natural sand

References	Experimental Variable	GC Content, %	Sorptivity, $\times 10^{-4}$ mm/s$^{0.5}$
Mardani-Aghabaglou et al. (2015)	0.45 w/c; Conditions: Cured at 20 ± 2°C and 95% RH for 28 days and then dried at 105°C	0	0.58
		15	0.58
		30	0.57
		45	0.56
		60	0.55
Rajabipour et al. (2012)	0.48 w/c; Conditions: stored at 50°C and 80% RH for 3 days and at 23°C for 20 days	0	15.8
		100	10.0

w/c, water/cement ratio.

(a) Sorptivity

The sorptivity test determines the rate of uptake of water by a partially submerged unsaturated specimen. Table 5.4 compares the sorptivity of concrete containing GC-sand with that of natural sand, reported by Mardani-Aghabaglou et al. (2015) and Rajabipour et al. (2012), where in each case the fineness moduli of the two sands were kept similar. Although subjected to different conditioning prior to testing, it can be seen from Table 5.4 that the sorptivity value of concrete is reduced or remains essentially unchanged when GC-sand is used up to 100%.

The results of other studies, by and large, suggest that the use of GC-sand would tend to reduce the sorptivity of concrete (De Castro and De Brito, 2013; Oliveira et al., 2008, 2013, 2015). However, the actual results may be affected by the dissimilarity in porosity arising from differences in the gradings of the sands used, differences in water/cement ratio of the mixes or the combination of the two.

(b) Initial Surface Absorption

The initial surface absorption test (ISAT) measures the rate of flow of water into concrete under a constant head. Figure 5.9 shows the effect of GC-sand on the initial surface absorption at 10 min (ISAT-10) of concrete samples having water/cement ratios of 0.52, 0.57 and 0.67, tested by Limbachiya (2009) in accordance with BS 1881-208 (1996). In the study, natural sand was replaced directly by GC-sand up to 50% content and the proportions of the aggregates were adjusted to maintain the yield of mix. The gradings of the two sands used were not provided.

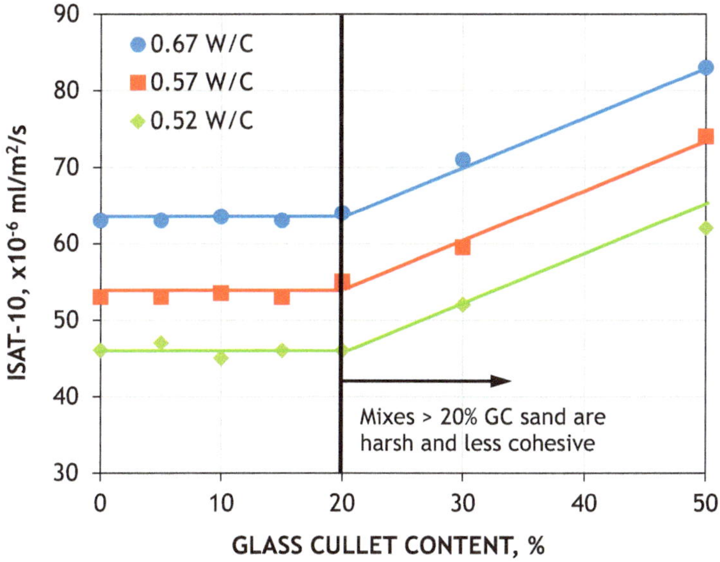

Figure 5.9 Initial surface absorption test at 10 min (ISAT-10) of glass cullet (GC) concrete at different water/cement ratios (w/c).

It can be seen from Figure 5.9 that regardless of the water/cement ratio of concrete, there is no significant change in the ISAT-10 value of concrete containing up to 20% GC-sand, beyond which the value was found to increase linearly as GC-sand content is increased, up to 50%. The increase in ISAT-10 value is most likely to be associated with the harshness and lack of cohesiveness observed in the fresh state of concrete containing more than 20% GC-sand during testing.

The most likely and viable conclusion that can be drawn from these tests is that the use of GC-sand as a direct replacement for natural sand in concrete cannot be adopted on a like-for-like basis without considering the characteristics of GC-sand, such as its angularity and particle size distribution, in designing the test mixes.

On the other hand, Dhir et al. (2005a) also measured the ISAT of concrete containing 100% GC-sand and natural sand of similar fineness modulus, ranging from 0.95 (very fine) to 4.20 (very coarse), using the BS 1881-208 (1996) test method with measurements taken at 10, 30 and 60 min. The replacement with GC-sand was made on a volume basis. Whilst the ISAT values were found to decrease with increasing fineness of both the sands, the ISAT value of GC-sand concrete was always lower than that of natural sand concrete for a given fineness modulus (Dhir et al., 2015). This is most likely to be a correct effect, and it sends the right message that GC-sand must be carefully manufactured, taking full account of the resulting fineness and particle size distribution of the sand produced.

(c) Water Absorption

The water absorption test measures the weight gain of an oven-dried specimen after its immersion in water for a period. Perhaps due to the relatively easy experimental setup compared with the former two tests, many studies have tended to opt for this test in assessing the effect of GC-sand on the absorption of concrete. For a more meaningful comparison, only the results of studies (i) in which the water/cement ratios of the GC-sand and natural sand concrete mixes were kept the same and (ii) that contain more than one GC replacement level have been used to develop Figure 5.10.

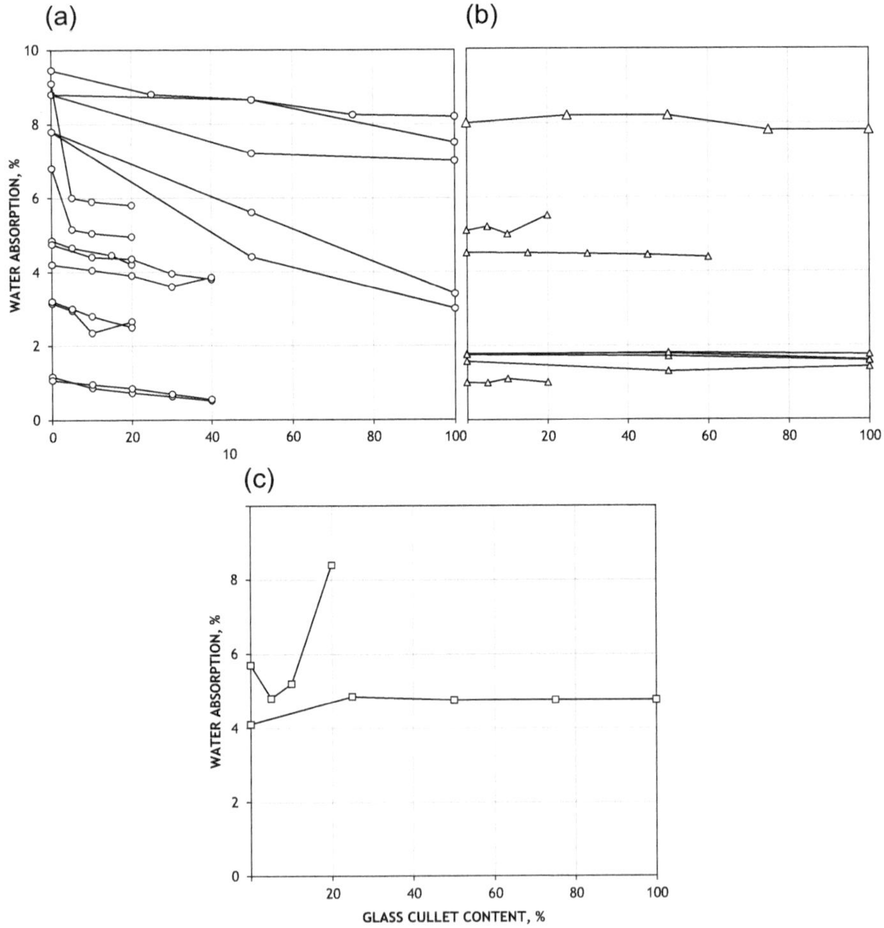

Figure 5.10 Influence of glass cullet sand on water absorption of concrete. The results show (a) a reduction, (b) no significant change and (c) an increase, when glass cullet sand is used. Data taken from Abdallah and Fan (2014), Mardani-Aghabaglou et al. (2015), Lee et al. (2013), Ling and Poon (2011a, 2014a,b), Malik et al. (2013, 2014), Maschio et al. (2013), Mirandra et al. (2014), Sharif et al. (2014) and Taha and Nounu (2008a).

It can be seen from Figure 5.10 that:

- The water absorption of concrete decreases when GC-sand content is increased. This is supported by the majority of the results (Figure 5.10(a)).
- The water absorption of concrete remains relatively unchanged when GC-sand content is increased (Figure 5.10(b)).
- The water absorption of GC-sand concrete is higher than that of natural sand concrete, as shown by only two set of results (Figure 5.10(c)), which can be considered either to be within experimental error (for the one with up to 100% GC-sand content) or as an isolated case (for the one with up to 20% GC-sand).

Thus, in general, the results obtained from water absorption tests are coherent with the findings observed in the sorptivity test and ISAT, which show that the near-zero water absorption characteristics of GC-sand would help to reduce the absorption of property of concrete. This suggests that GC-sand could also be specified for use in water-retaining structures, but due care must be exercised in developing the correct procedure for the manufacture of GC-sand.

5.6.5 Permeability

The transportation of a fluid through concrete under a pressure gradient is known as permeability. The effects of GC-sand on the permeability of concrete to (i) gases, measured in the form of oxygen permeability, and (ii) water, measured in the form of water permeability, water penetration and penetration ratio, are given in Table 5.5. Although in some instances the oxygen permeability of GC-sand concrete appears to be higher than that of natural sand concrete (e.g., Cota et al., 2015), overall, for a constant water/cement ratio, the oxygen permeability of concrete can generally be expected to reduce when natural sand is replaced by GC-sand up to 100% content.

Similarly, the water permeability on controlled low-strength materials (Wang, 2009b; Wang and Chen, 2010) and rendering mortar (Oliveira et al., 2015; Penacho et al., 2014) containing GC-sand up to 100% content has been evaluated using Chinese standard CNS 3763 (2009) and European standard EN 1015-21 (2002), respectively. In most cases, the water permeability of GC-sand mixes is lower than that of the corresponding reference natural sand mixes.

Overall, given that GC is an impervious material, all things being equal, its use as a fine aggregate in concrete should reduce both the air and the water permeability of concrete.

5.6.6 Diffusion

Diffusion of a fluid through concrete takes place when there is a difference in ion concentration on the two sides of the concrete. Although the relevant studies are not available, it is believed that the diffusion of concrete should decrease or remain unchanged when GC-sand is used, because of its characteristic.

Table 5.5 Permeability of glass cullet (GC) sand concrete expressed in different forms

References	Experimental Variable	GC Content, %	Measurement
(a) Oxygen Permeability, $\times 10^{-16}$ m^2			
Oliveira et al. (2008)	0.60 w/c; 28 days moist curing; based on proposed method by Gomes et al. (2002)	0	2.10
		25	2.55
		50	2.80
		100	1.70
Cota et al. (2015)	0.42 w/c; cured at 50°C for 6 h followed by 28 days moist curing; proposed method by Cabrera and Lynsdale (1988)	0	7.9[a]
		7.5	10.0[a]
		15	13.2[a]
(b) Water Permeability, $\times 10^{-18}$ m^2			
Oliveira et al. (2008)	0.60 w/c; 28 days moist curing; proposed method by Gomes et al. (2002)	0	5.2
		25	6.5
		50	6.8
		100	4.1
(c) Water Penetration, mm			
Mardani-Aghabaglou et al. (2015)	0.45 w/c; 28 days moist curing; EN 12390-8 (2009)	0	19.5
		15	18.5
		30	18.0
		45	17.5
		60	16.0
(d) Permeability Ratio			
Wang and Huang (2010a)	0.28 w/c; 28 days moist curing; CNS 3763 (2001)	0	0.24
		10	0.22
		20	0.20
		30	0.14

[a]Average results of concrete containing 0%–15% metakaolin. Unit: $\times 10^{-5}$ m^2.

5.7 Durability - Chemical Attack

The deterioration of concrete in terms of chemical attack can result from various sources such as seawater, carbon dioxide or deicing salts, as well as the constituent materials of concrete themselves such as reactive aggregate. The mechanism of this type of degradation in concrete mainly involves the transport of fluid and dissolution of hydration products, leading to the formation of detrimental compounds or disturbance of the passive film of reinforcement.

5.7.1 Chloride Ingress

The rapid chloride penetration test (RCPT), as described in ASTM C1202 (2012), is a popular test used to determine the chloride resistance of concrete by measuring the penetrability of chloride ions through a 50 mm diameter x 100 mm long specimen under a potential difference of 60 V for a duration of 6 h.

Figure 5.11 shows the results of the RCPT for concrete made with different types of GC-sand up to 100% content, together with the five chloride ion penetrability categories provided in ASTM C1202 (2012). Except for Chen et al. (2006), the water/cement ratio of the specimens in all the studies was no more than 0.55, which is within the requirements for the exposure classes XD1 and XD2 for chloride-induced corrosion other than from seawater given in BS EN 206 (2013).

It can be seen from Figure 5.11 that the chloride ion penetrability in concrete generally decreases, suggesting increase in chloride resistance, as the GC-sand content increases, up to 100%. This improvement is visible in all GC specimens, regardless of:

(i) Glass type: soda lime GC, aluminosilicate GC and lead GC.

(ii) Colour of GC: amber, green, clear and mixed.

(iii) Age of the concrete: 28 and 90 days.

(vi) Use of pozzolanic cement: FA and GGBS.

In addition, Dumitru et al. (2010) determined the chloride penetration of GC concrete using the NT Build 443 (1995) method. Natural sand was replaced by 30%, 45% and 60% GC-sand in concrete, with a slight reduction in water/cement ratio to maintain its consistence. After immersion in chloride solution for 35 days, the chloride diffusion coefficient was found to reduce slightly from $1.74 \times 10^{-11}\,m^2/s$ for natural sand concrete to $1.32–1.53 \times 10^{-11}\,m^2/s$ for GC-sand concrete.

In summary, although pozzolanic cement, rather than sand, is commonly used to increase the chloride resistance of concrete, the improvement in chloride resistance seen with GC-sand concrete suggests that GC-sand is an ideal material for use in concrete exposed to chlorides.

5.7.2 Carbonation

Carbonation is a chemical reaction between atmospheric carbon dioxide and calcium hydroxide in cement paste, which can lower the pH of the concrete and subsequently destroy the passive oxide film of steel reinforcement. Depassivation of steel makes the reinforcement more prone to corrosion.

The effect of GC-sand on the carbonation resistance of concrete subjected to accelerated carbonation has been assessed by a few studies using different methods: the LNEC E 391 (1993) method in De Castro and De Brito (2013), the Dhir et al. (1985) method in Dhir et al. (2005a) and the BRE Digest 405 (1995) method in Ling and Poon (2012b). The carbonation and compressive strength results, expressed in relative forms, together with the experimental variables in each study, are given in Table 5.6.

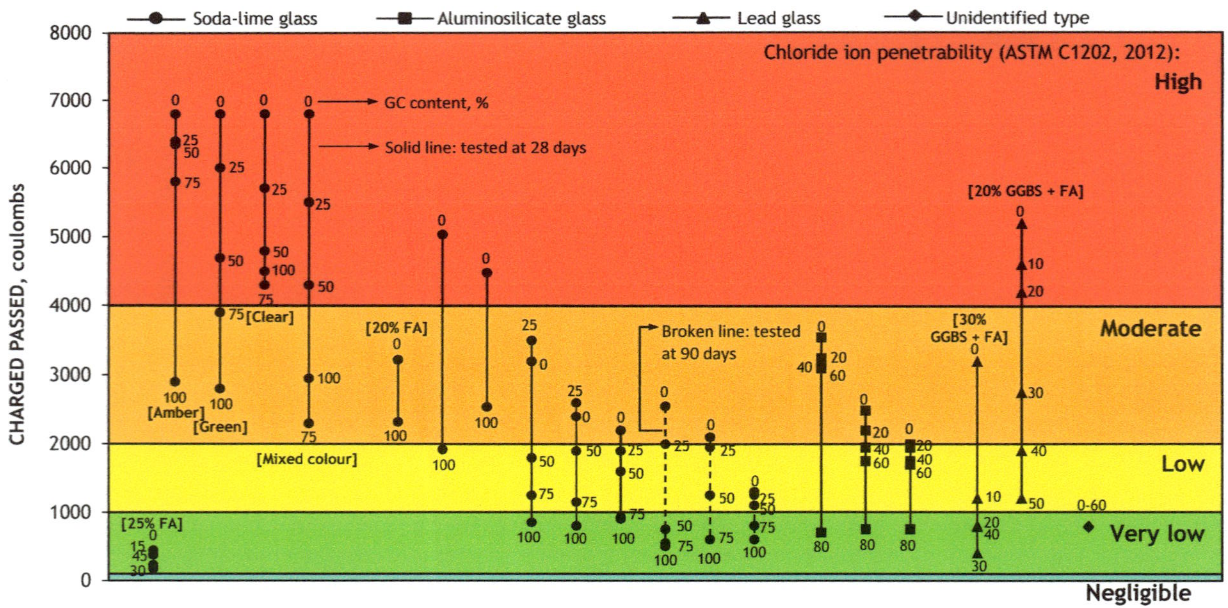

Figure 5.11 Rapid chloride penetration test results of concrete containing glass cullet (GC) sand. *FA*, fly ash; *GGBS*, ground granulated blast furnace slag.

Data taken from Mardani-Aghabaglou et al. (2015), Chen et al., 2006, Du, 2011, Du and Tan, 2014, Kou and Poon, 2009, Tan and Du, 2013, Wang, 2009a, Wang et al., 2014, Wright, 2012 and Wright et al., 2014.

Table 5.6 Carbonation resistance of concrete containing glass cullet (GC) sand subjected to accelerated carbonation

			Relative Change	
References	**Experimental Variable**	**GC, %**	**Strength[a], %**	**Carb. Depth[b], %**
De Castro and De Brito (2013)	0.55 w/c; curing and exposure condition not given; measurement taken at the age of 91 days	0	–	0
		5	–	+10
		10[c]	–	0
		20[d]	–	+7
Dhir et al. (2005a)	0.61 w/c; 28-day-cured specimens; 3%–4% CO_2; 20°C; 65% RH for 63 days	0	+6	0
		50	+19	+100
Ling and Poon (2012b)	0.48 w/c; 28-day-cured specimens; 4% CO_2; 25°C; 60% RH for 90 days	0	+5.0	0
		25[e]	+7.5	+33
		50[e]	+8.0	+50
		75[e]	+6.0	+38
		100[e]	+6.5	+56

RH, relative humidity.
[a]Relative compressive strength at 28 days with respect to that at 90 days. The higher the value, the higher the increase in strength.
[b]Relative carbonation (Carb.) depth of GC concrete with respect to that of natural sand concrete. Positive value indicates increase in carbonation depth.
[c]0.57 water/cement ratio (w/c).
[d]0.58 w/c.
[e]Average of treated and untreated GC samples.

Except for the data of De Castro and De Brito (2013) for 10% and 20% GC-sand content, it can be seen from Table 5.6 that, for a given water/cement ratio, the use of GC-sand increases the carbonation depth in concrete, suggesting a reduction in carbonation resistance. The available data do not suggest a clear relationship between the GC content and the relative change in carbonation depth.

The possible explanation for the reduction in carbonation resistance when GC-sand is used is that some of the calcium hydroxide in the concrete is consumed by the fine GC particles for pozzolanic reaction. This is possibly supported by the level of strength increase from 28 to 90 days for concrete containing GC-sand being greater than that of natural sand concrete (Table 5.6). Thus, lower calcium hydroxide in GC-sand concrete suggests that it is easier for carbonation to take place. Further research is required to substantiate this explanation.

5.7.3 Acid Attack

Acid attack causes decomposition of hydration products, particularly calcium hydroxide, and eventually affects the integrity of concrete. As far as aggregate is concerned, calcareous aggregates are known to be susceptible to acid attack, whilst siliceous aggregates are acid resistant (Alexander and Mindess, 2005).

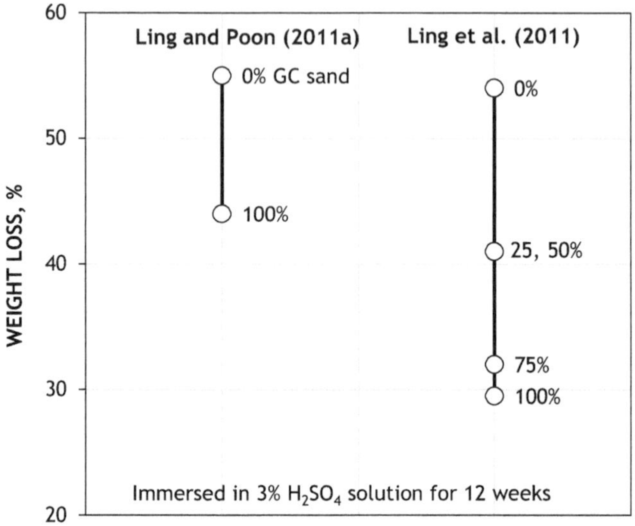

Figure 5.12 Weight loss of glass cullet (GC) sand specimens subjected to sulphuric acid solution.

The acid resistance of GC-sand mortar has been investigated by Ling and Poon (2011a) and Ling et al. (2011a,b) by immersing 28-day-cured specimens in 3% sulphuric acid solution in accordance with ASTM C267-01 (2006). The water/cement ratio of the mortar mix was 0.4 and it contained 10% metakaolin as a Portland cement replacement and up to 100% GC-sand as a natural sand replacement. As shown in Figure 5.12, after 12 weeks of acid immersion, the weight loss in GC-sand specimens was less than that of the natural sand specimens, indicating an improvement in acid resistance when GC-sand is used. The results of Ling et al. (2011a,b) suggest that the acid resistance of concrete generally increases as the GC-sand content is increased.

5.7.4 Sulphate Attack

Certain cement hydration products are vulnerable to sulphate attack. Depending on the type of sulphate, the reaction can form either expansive products that cause cracking in concrete or products with little cementing value that disintegrate cement paste.

The effect of GC-sand as a natural sand replacement on sulphate resistance is commonly studied in the form of loss of weight of mortar specimens subjected to cyclic drying and wetting in sulphate solution, the results of which are shown plotted in Figure 5.13. Except for Du (2011) and Tan and Du (2013), in which the measurement was made on 28-day cured specimens in accordance with ASTM C267-01 (2006), all other studies performed the test on 3 or 7-day cured specimens using the ASTM C1012 (2013) method. The water/cement ratio of the mixes ranged from 0.28 to 0.55.

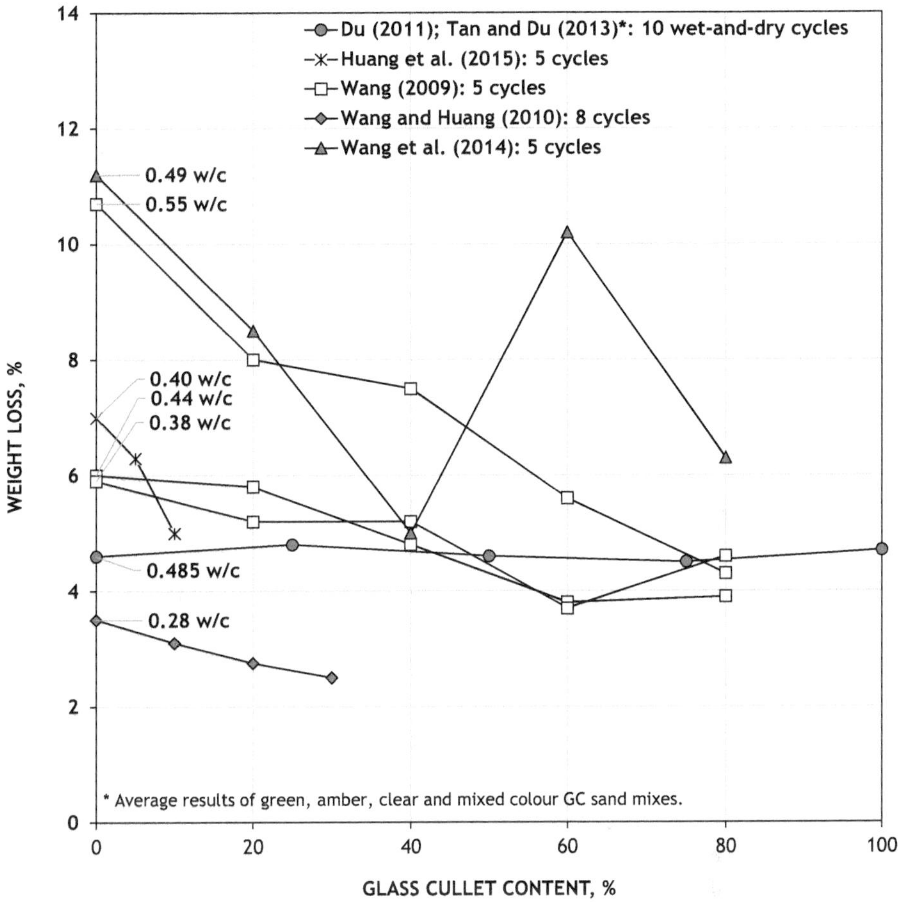

Figure 5.13 Weight loss of glass cullet (GC) sand specimens subjected to different cycles of wet-and-dry sulphate attack. *w/c*, water/cement ratio.

Figure 5.13 shows that the sulphate resistance of GC-sand mortar is similar to or better than that of the corresponding reference natural sand mortar. This is to be expected, as sulphates attack only the hydration products; thus aggregates have little influence in this respect, though they can affect the permeation of mixes owing to aggregate packing. It should be mentioned that Du (2011) and Tan and Du (2013) also examined the effect of GC-sand colour (green, amber, clear and mixed) on the sulphate resistance of concrete, and the results suggest that the weight loss of specimens was comparable regardless of the colour of GC-sand used.

Other types of sulphate-resistance tests, such as change in length (Dhir et al., 2005a; Tang et al., 2005) and change in compressive and flexural strengths (Du, 2011; Tan and Du, 2013), of GC-sand concrete specimens after sulphate exposure have also been performed. However, the expected results of expansion and loss in strength were not observed in either GC-sand or natural sand concrete specimens, and instead shrinkage and gain in strength were recorded.

5.7.5 Alkali–Silica Reaction

The ASR occurs when the siliceous minerals in certain types of aggregate react with the alkali pore fluid in the cement matrix to form alkali–silicate gels, which can swell and expand in the presence of moisture, leading to cracking and disintegration of concrete. The use of GC-sand, especially of the soda lime type, often raises a concern over ASR, as it contains a reactive form of silica and also a potential source of alkali (Dhir et al., 2003). In this section, the effects of the type, colour, aggregate size and content of GC-sand on ASR, as well as the effect of pozzolanic cements in minimising the ASR expansion induced by GC-sand, are discussed.

Types of Glass Cullet

The alkali–silica reactivity of GC-sand made of different types of glass, namely soda lime (normally used as container glass and flat glass), borosilicate (typically Pyrex glass), fused glass and lead-silicate (used in tableware), have been compared, and the results are given in Table 5.7. Of all the types of glass tested, fused glass and lead-silicate glass are the most reactive, followed by borosilicate glass, and soda lime glass has been consistently shown to be least reactive. This suggests that soda lime glass, which is abundantly available, is relatively more suitable for use as a sand component in concrete compared with other types of glass.

Colours of Glass Cullet

Colourant additives are used to impart specific colours to the glass. For soda lime GC, especially those derived from container glass, the main colour types are green, amber and clear, as well as mixed colours (which is a mixture of green, amber and/or clear). The ASR expansion of mortar specimens tested in accordance with ASTM C1260 (2014), RILEM TC 106-AAR (2000) and JIS A 1146 (2000), and concrete specimens in accordance with BS 812-123 (1999), made with different colour types of GC-sand as a replacement for natural sand for up to 100% is given in Table 5.8.

Table 5.7 Effect of type of glass cullet sand on alkali–silica reaction

References	Standard	Testing Age	Expansion
Dhir et al. (2005a)	BS 812-123	78 weeks	Borosilicate > soda lime
Jin et al. (2000)	ASTM C1260	14 days	Fused silica[a] > borosilicate > soda lime
Saccani and Bignozzi (2010)	ASTM C1260	14 days	Lead silicate > borosilicate > soda lime

[a]Containing 99.97% SiO_2.

Table 5.8 Effect of colour of glass cullet sand on alkali–silica reaction

Reference	GC Content, %	Expansion, % Highest	⟶	Lowest
(a) ASTM C1260				
Byars et al (2004a)	100	0.050	0.040 0.040	-
Du (2011), Tan and Du (2013)	25	0.049	0.038 0.038	0.032
	50	0.065	0.022	0.020 0.020
	75	0.128	0.012	0.009 0.009
	100	0.137	0.010 0.010	0.005
Jin et al. (2000)	10	0.410	0.280	0.100
Liu (2011)	10	0.060	0.050	-
Park and Lee (2004)	30	0.370	0.180	
	50	0.480	0.210	
	100	0.650	0.250	
Saccani and Bignozzi (2010)	10	0.120	0.105	-
	25	0.119	0.100	-
	35	0.150	0.090	-
Sagoe-Crentsil et al. (2001)	100	0.170	0.025	0.020
Topcu et al. (2008)	25	0.450	0.390	0.350
	50	0.600	0.550	0.510
	75	0.750	0.620	0.600
	100	0.830	0.740	0.730
(b) ASTM C1293				
Yuksel et al. (2013)	10	0.018	0.012	-
	20	0.022	0.015	-
	40	0.023	0.020	-
	50	0.024	0.021	-
(c) RILEM AAR-2				
Vegt et al. (2004)	100	0.450	0.029	0.024
Yuksel et al. (2013)	25, 50, 75, 100	0 0 0	-	-
(d) BS 812-123				
Dhir et al. (2008)	50	0.070	0.065	0.050
	100	0.060 0.060	0.045	-
(e) JIS A 1146				
Yamada et al. (2004)	25	0.880	0.820	0.080

Note:

Clear	Green	Amber	Mixed

It appears that no clear trend was observed in describing the effect of the colour of GC-sand on the ASR expansion. However, many results tend to show that the ASR expansion of specimens containing clear GC-sand is the highest, whilst that of specimens containing green GC-sand is the least. Indeed, it has been suggested that the alkali–silica reactivity of green GC-sand is affected by its Cr_2O_3 content, which is used in making green container glass, as the ASR expansion was found to decrease with increasing Cr_2O_3 content in green GC-sand (Jin et al., 2000; Meyer and Baxter, 1998b).

Glass Cullet Aggregate Size

The effects of container GC of different colours on ASR expansion at various aggregate sizes, ranging from 6 to 0.04 mm, has been assessed to investigate the pessimum size at which the expansion is maximum. As shown in Figure 5.14, regardless of the colour, the pessimum size of GC-sand is generally in between 1 and 3 mm, and the expansion decreases as the GC aggregate size decreases. In addition, it has also been found that concrete or mortar made with very fine GC-sand, with size less than 0.15 mm, has an expansion less than that of a reference made with natural aggregate, indicating the possible occurrence of a pozzolanic reaction of very fine GC-sand, which is able to control the ASR expansion (Byars et al., 2004c; Dhir et al., 2005a; Jin et al., 2000; Meyer, 2001, 2003a; Zhu and Byars, 2005). This is line with the findings observed in Chapter 5 on the ASR expansion of mixes containing ground GC as a cement component.

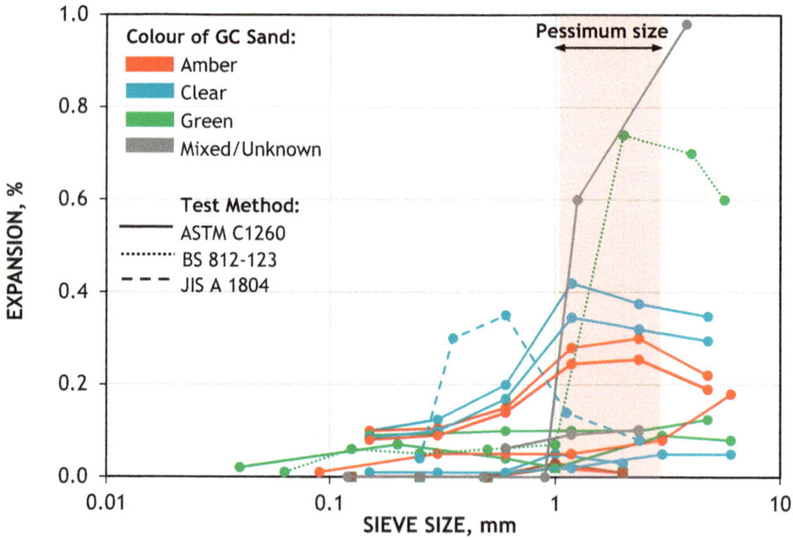

Figure 5.14 Pessimum size of glass cullet (GC) sand in the alkali–silica reaction.
Data taken from Byars et al. (2004a), Dhir et al. (2005a), Idir et al. (2009, 2010), Jin et al. (2000), Lee et al. (2011), Meyer (2003a, 2003b), Meyer and Baxter (1998a, 1998b), Meyer et al. (1996b, 1996b, 1998), Rajabipour et al. (2010), Yamada and Ishiyama (2005), Yamada et al. (2004), Yuksel et al. (2013) and Zhu and Byars (2005).

Content of Glass Cullet Sand

The effect of GC-sand (mostly soda lime) as a replacement for natural sand, up to 100%, on the ASR expansion has been assessed by many researchers, commonly using the ASTM C1260 (2014) method (see Figure 5.15), as well as other methods, such as BS 812-123 (1999), RILEM TC 106-AAR (2000), ASTM C1293-8 (2015), ASTM C227 (2010) and ASTM C1567 (2013) (Figure 5.16). It should be mentioned that the results presented therein were based on the recommended test duration of each test method (e.g., 14 days in ASTM C1260 and 52 weeks in BS 812-123), though there are studies that continued the test beyond the recommended duration. For reference, the guidance for interpretation of the expansion results of each method is also shown in the figures.

It can be seen from Figures 5.16 and 5.17 that, regardless of the test method used, the expansion of mortar or concrete increases with increasing GC-sand content, although a few results showed the opposite (Abdallah and Fan, 2014; Chen et al., 2006; Ismail and Al-Hashmi, 2009; Sagoe-Crentsil et al., 2001). In general, when GC-sand is used for no more than about 40% as a replacement for natural sand, the induced ASR expansion is considered to be innocuous or acceptable, depending on the test method used. At 100% GC content, the expansion is deleterious.

Effect of Pozzolanic Cements

It is widely recognised that the potential risk of ASR expansion induced by GC fine aggregate can be minimised with the use of pozzolanic cements, though the effectiveness of lithium compound additives has also been explored (Meyer and Baxter, 1998b; Meyer et al., 1996a; Taha and Nounu, 2008a,b). Figure 5.17 shows the ASR expansion of mortar made with GC-sand, up to 100%, as a replacement for natural sand, and different types of pozzolanic cements as a replacement for Portland cement, which is determined in accordance with the ASTM C1260 (2014) method. The tested pozzolanic cements include those covered in BS EN 197-1 (2011), such as FA (5%–60%), metakaolin (5%–30%), GGBS (25% and 60%), as well as ground GC itself (20%–30%).

It is evident from Figure 5.17 that pozzolanic cements can be used to minimise/nullify the ASR expansion of mortar induced by GC-sand, even at 100% GC-sand content. The effect of pozzolanic cements in minimising the expansion becomes more pronounced as their content increases. Depending on its type, the effectiveness level of each pozzolanic cement at which the expansion is considered as innocuous is different; for FA this is likely to be at least 25%; for metakaolin, 10%; for GGBS, 60% and for ground GC itself, 20%–30%. Similar findings were also observed (Dhir et al., 2005a, 2009; Taha and Nounu, 2009) when tested using BS 812-123 (1999) methods.

5.8 Durability - Physical Attack

Physical attack on concrete causes either (i) cracking and spalling such as that due to freeze–thaw cycles and exposure to high temperature or (ii) thinning of the concrete surface due to abrasion.

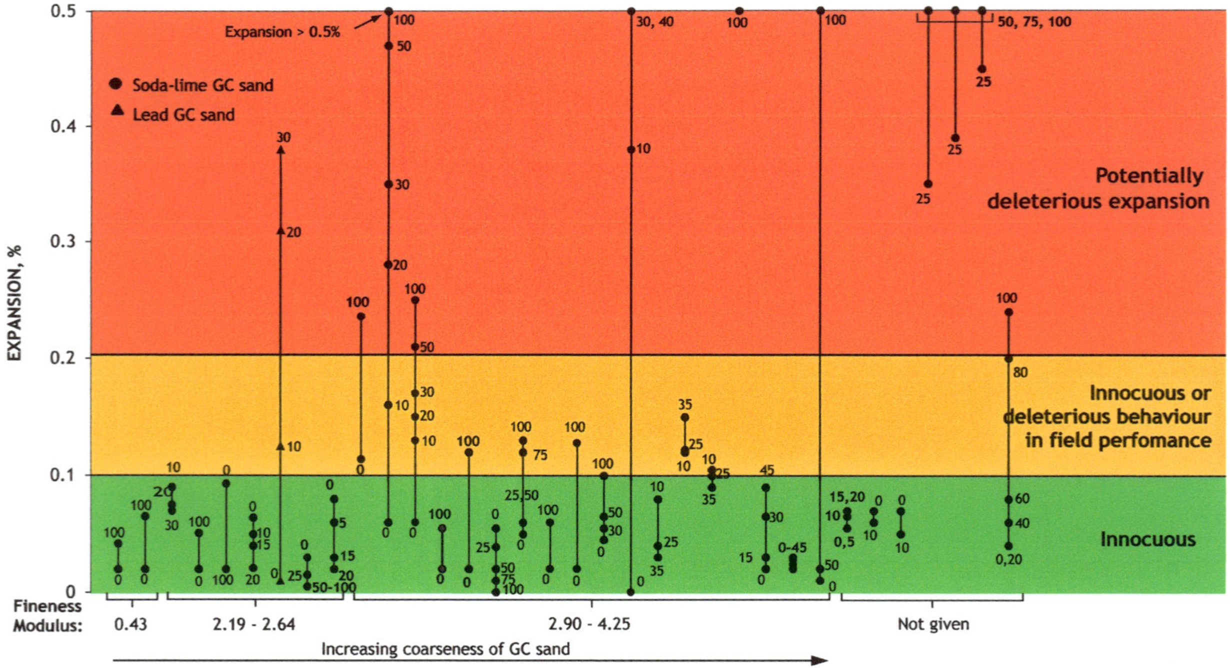

Figure 5.15 Influence of glass cullet (GC) sand content on the alkali–silica reaction expansion tested using the ASTM C1260 method.
Data taken from Abdallah and Fan (2014), Dhir and Tang (2008), Du (2011), Du and Tan (2014), Hudec and Ghamari (2000), Ismail and Al-Hashmi (2009), Kozlova et al. (2003), Kou and Poon (2009), Lam et al. (2007), Lee et al. (2011), Limbachiya (2009), Limbachiya et al. (2012), Liu (2011), Naik and Wu (2001), Park and Lee (2004), Park et al. (2004), Romero et al. (2013), Saccani and Bignozzi (2010), Shayan and Xu (2004), Tan and Du (2013), Topcu et al. (2008) and Zhu and Byars (2004).

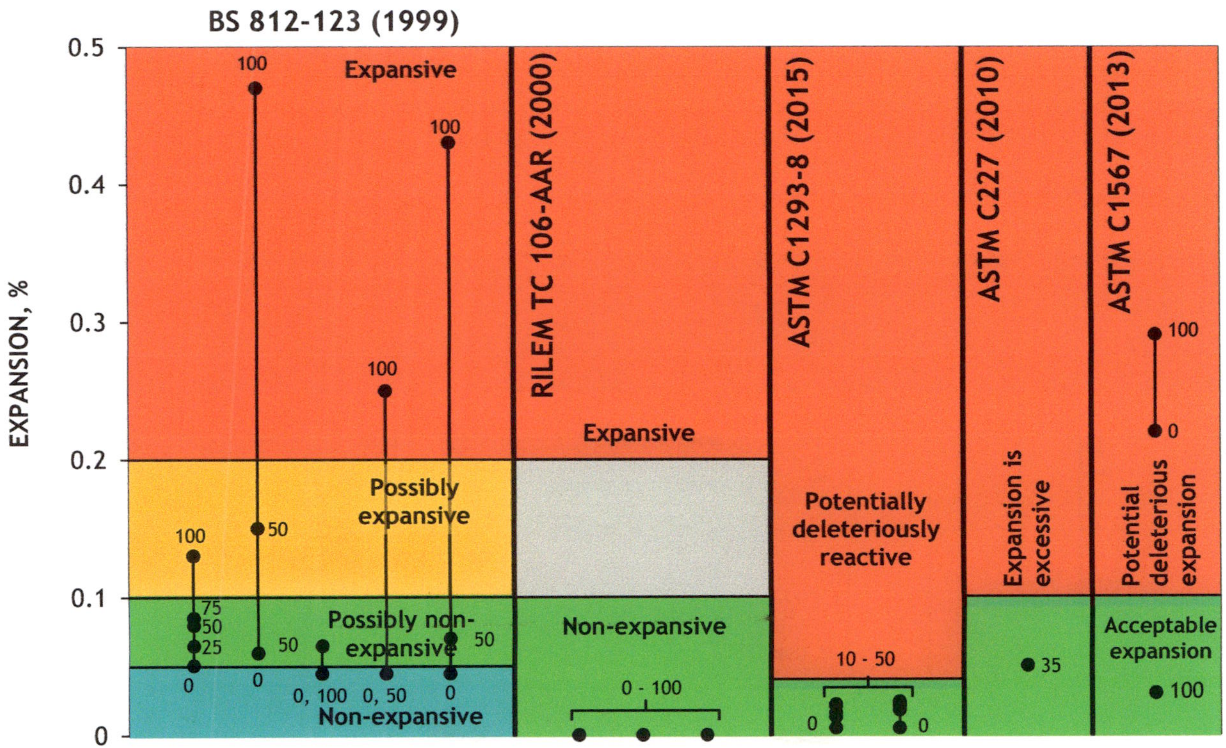

Figure 5.16 Influence of glass cullet sand content on the alkali–silica reaction expansion tested using other methods.
Data taken from Dhir et al. (2005a, 2009), Maier and Durham (2012), Phillips et al. (1972), Serpa et al. (2013), Taha and Nounu (2008b, 2009) and Yuksel et al. (2013).

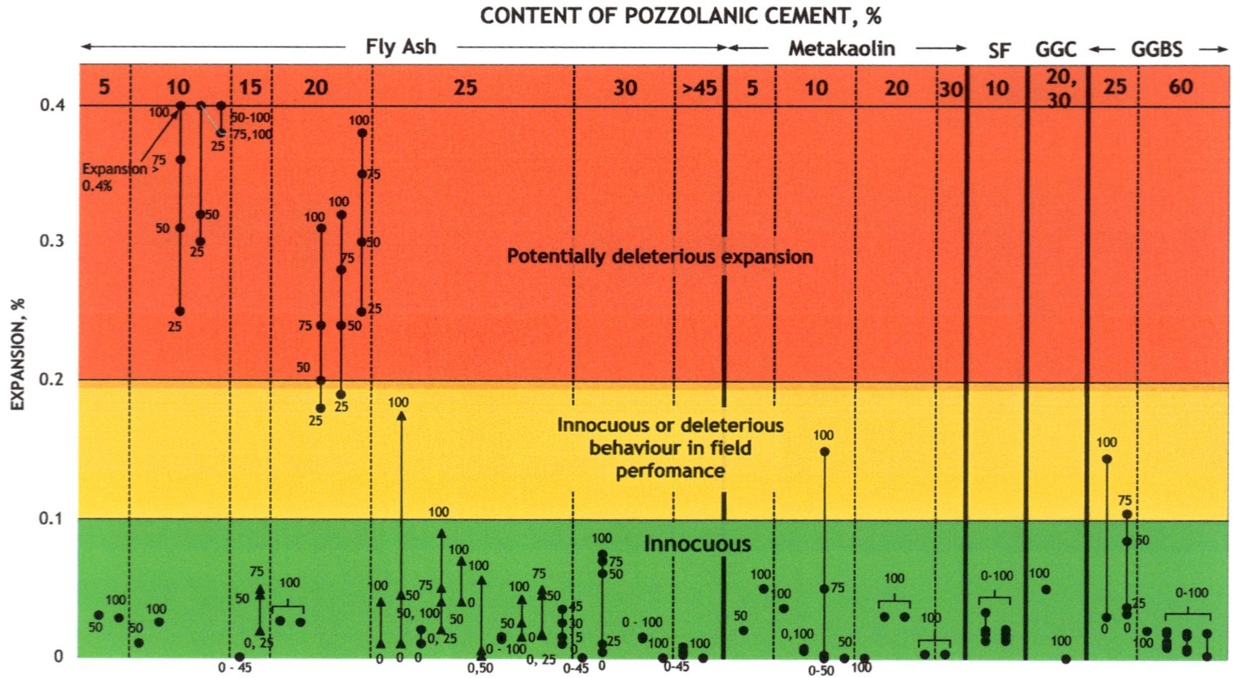

Figure 5.17 Effect of pozzolanic cement to minimise/ nullify the alkali–silica reaction expansion of glass cullet sand using the ASTM C1260 (2014) method. *GGBS*, ground granulated blast furnace slag; *GGC*, ground glass cullet; *SF*, silica fume.
Data taken from Du and Tan (2013, 2014), Hui and Sun (2011), Kou and Poon (2009), Lam et al. (2007), Lee et al. (2011), Ling and Poon (2011a,b, 2012c, 2013, 2014a,c), Ling et al. (2011a,b), Maier and Durham (2012), Naik and Wu (2001), Sagoe-Crentsil et al. (2001), Topcu et al. (2008) and Zhao et al. (2013a, 2013b).

5.8.1 Freeze–Thaw Attack

Water expands by 9% when it turns into ice. This volume increase can be detrimental to saturated concrete when the expansive pressure generated during freezing of water exceeds the tensile strength of the concrete. The freeze–thaw resistance of air-entrained and non-air-entrained concrete containing up to 50% GC-sand, subjected to various freeze–thaw exposures in accordance with ASTM C666 (2015) and CEN/TC 51-draft (1994) test procedures, is given in Table 5.9.

The results, measured as change in performance of concrete in several ways, suggest that the use of GC-sand up to 20% content does not significantly influence the freeze–thaw resistance of air-entrained and non-air-entrained concrete (Abendeh et al., 2015; Romero, 2013). However, at 50% GC-sand content, it would appear that GC-sand concrete has lower resistance to freeze–thaw attack than natural sand concrete (Dhir et al., 2005a). Given the limited information, further tests would be necessary to substantiate any conclusion.

5.8.2 Abrasion

The abrasion resistance of concrete is affected by the quality of hardened cement paste, properties of aggregates and surface treatment. Figure 5.18 shows the abrasion resistance of GC-sand concrete and mortar, measured using various standard methods, such as: (i) the BS 6717 (2001) method in Ling and Poon (2011a) and Ling et al. (2013), (ii) ASTM C944 (1999) in Rajabipour et al. (2012), (iii) DIN 52108 (2002) in Serpa et al. (2015) and (iv) a self-developed method by Dhir et al. (1991) in Tang et al. (2005). Except for Serpa et al. (2015) and Dumitru et al. (2010), in which the water/cement ratio of GC-sand concrete mixes was adjusted to keep their consistence similar to that of normal sand concrete mixes, all other studies kept the water/cement ratio of GC-sand concrete mixes, which varied from 0.40 to 0.61, the same as that of their corresponding reference natural sand concrete mixes.

The following main points can be observed from Figure 5.18:

- For a given water/cement ratio, the abrasion resistance of mortar/concrete mixes containing GC-sand, up to 100%, is lower than that of the corresponding reference mixes (Ling and Poon, 2011a; Ling et al. (2011a,b); Rajabipour et al., 2012; Tang et al., 2005). It is claimed that the observed reduction in abrasion resistance is likely to be due to weaker bonding between GC-sand and cement paste (Rajabipour et al., 2012).
- Similar results were also observed in the Dumitru et al. (2010) study, although the water/cement ratio of the concrete with up to 60% GC-sand was slightly lower than that of the reference concrete.
- One study (Serpa et al., 2015), however, although having a higher water/cement ratio, shows that the abrasion resistance of concrete remains essentially unchanged with the use of GC-sand up to 20% content.

Table 5.9 Resistance to freeze–thaw attack of glass cullet sand concrete

References	Experimental Variables[a]	Measurement	Resistance to Freeze–Thaw Attack[b]
(a) ASTM C666 Procedure A (2015)			
Abendeh et al. (2015)	0%–20% GC; 0.50 w/c; non-air-entrained; 28 days curing; from −16 to 6°C and reverse in 2.5 h; 230 cycles	Weight loss	↑
		Compressive strength loss	≈
		Ultra-pulse velocity	≈
		Resonance frequency	≈
Romero (2013)	0%–20% GC; 0.47 w/c; air-entrained; 210 days curing in 19°C and 50°C; from −18°C to 4°C and reverse in 2.75 h; 300 cycles	Weight loss	≈
		Durability factor	≈
(b) CEN/TC 51-Draft (1994)			
Dhir et al. (2005a)	0% and 50% GC; 0.61 w/c; non-air-entrained; 7 days curing; from −20°C to 20°C and reverse in 24 h; 56 cycles	Mass scaling	↓

[a]*GC*, glass cullet; *w/c*, water/cement ratio.
[b]Comparing the resistance to freeze–thaw attack of ground GC concrete to natural sand concrete: ↑ indicates increase; ↓ indicates decrease; ≈ indicates no significant change with value difference within ±20%.

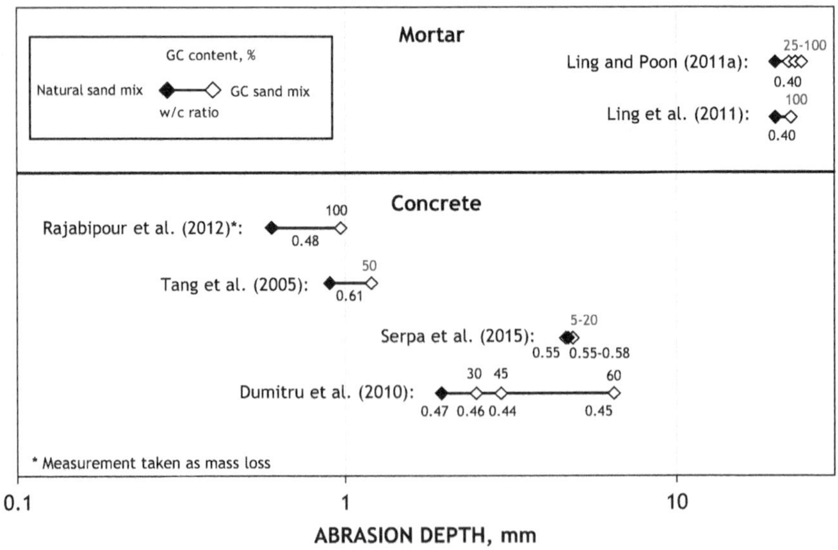

Figure 5.18 Abrasion resistance of mixes made with glass cullet (GC) sand.

Overall, it would appear that the angularity and high stiffness of GC-sand do not increase the abrasion resistance of concrete. Notwithstanding this, the available data show that where concrete is likely to be exposed to traffic wear, GC-sand can be used up to about 20% as a natural sand replacement without compromising its abrasion resistance.

5.8.3 Exposure to Elevated Temperature

When concrete is exposed to elevated temperature, changes in its hardened properties occur due to the loss of water, decomposition of hydration products, buildup of thermal stresses and several other factors that can affect its structural integrity. Studies have been undertaken to evaluate the effects of GC-sand on the mechanical properties of mortar (Guo et al., 2015) and concrete (Ling and Poon, 2012c,d; Ling et al., 2012a; Terro, 2006), with water/cement ratios of the test mixes ranging from 0.35 to 0.48 and the test specimens subjected to elevated temperatures up to 800°C. In relative terms (ratio of high temperature to room temperature), the loss of compressive strength and the modulus of elasticity in GC-sand concrete mixes were either similar to or less than those of natural sand mixes.

In addition, at high temperatures, the use of GC-sand has been shown to generally improve the performance of structural fibre-reinforced concrete (Borhan and Bailey, 2014) and structural lightweight concrete (Al-Sibahy and Edwards, 2012a,b).

5.9 Environmental Impact

The major environmental problem associated with the use of GC-sand is the potential leaching of heavy metals that may be released from the concrete. This is of particular concern when the GC contains high levels of lead, barium and other heavy metals. Table 5.10 gives the experimental variables and main observations of leaching studies of concrete and mortar and of concrete mixes having water/cement ratio ranging from 0.38 to 0.60 and containing up to 100% GC-sand. The toxicity characteristic leaching procedure, method 1311, developed by the US Environmental Protection Agency (US EPA, 1992), which is used to identify potentially hazardous materials, has been most commonly adopted for the assessment of GC. Other methods such as the synthetic precipitation leaching procedure, method 1312 (US EPA, 1994), and leaching test by BS EN 12457 Part 1–4 (2002) have also been used.

The majority of the leaching tests were performed on concrete and mortar mixes made with CRT GC-sand, and only two studies were on mortar containing soda lime GC-sand (Ling and Poon, 2012a, 2013). This is because CRT, which is derived from television sets and computer monitors, has a high lead content in the funnel and neck components and high barium content in the panel component, whereas soda lime glass, which is abundantly available, does not normally contain appreciable amounts of toxic metals (refer to Section 3.5.1).

Table 5.10 Leaching studies of concrete and mortar containing glass cullet sand

References	Experimental Variables	Main Observations
Kim et al. (2009)	**GC**: CRT, 0.149–0.42 mm **Mortar**: 0% and 10% GC, 0.4 w/c, 7 days curing **Addition**: two types of biopolymer solutions **Method**: modified TCLP	The use of biopolymer solutions reduced the leached concentration of lead
Ling and Poon (2011b) and Ling and Poon (2012b)	**GC**: acid-treated CRT funnel, <5 mm **Mortar**: 0%–100% GC, 25% fly ash, 0.45 w/c, 28 days curing **Method**: TCLP	The leaching of lead was below the detection limit of 0.06 mg/L
Ling and Poon (2012a)	**GC**: acid-treated and untreated CRT funnel, soda lime glass, <5 mm **Mortar**: 0%–100% GC, 25% fly ash, 0.45 w/c, 28 days curing **Method**: TCLP	Except for untreated CRT GC, the leaching of lead from mixes containing treated CRT GC and soda lime GC was well within the permissible limit of 5 mg/L
Ling and Poon (2012b)	**GC**: acid-treated and untreated CRT funnel, <5 mm **Concrete**: 0%–100% GC, 15% fly ash, 0.48 w/c, 28 days curing **Method**: TCLP	The leaching of barium was low for all concrete; only untreated CRT GC concrete had lead concentrations that exceeded the permissible limit
Ling and Poon (2013)	**GC**: acid-treated CRT funnel, different sizes, soda lime glass, <5 mm **Mortar**: 0% and 100% GC, 25% fly ash, 0.45 w/c, 28 days curing **Method**: TCLP	The leaching of lead increased with decreasing particle size, but all within the permissible limit
Maschio et al. (2013)	**GC**: mixture of CRT, <500 μm **Concrete**: 20% GC, 0.44 w/c, 28 days curing **Method**: TCLP	The leaching of barium and lead was below the permissible limits
Morrison (2004)	**GC**: CRT, <5 mm **Concrete**: 40% GC, 0.6 w/c, 35% fly ash, 50% GGBS, curing age not given **Method**: BS EN 12457 (2002) Part 1–4	The use of fly ash and GGBS reduced the leaching of barium and lead to an acceptable level
Romero (2013) and Romero et al. (2013)	**GC**: CRT, <5 mm **Mortar**: 10%–30% GC, 0.47 and 0.49 w/c, 28 days curing **Addition**: biopolymer solution **Method**: SPLP	Mortar containing biopolymer solution showed the lowest leached concentrations of lead
Sua-iam and Makul (2013)	**GC**: CRT funnel, 0.69 fineness modulus **Concrete**: 0%–40% GC, 0.38 w/c, 14 days curing **Addition**: limestone powder **Method**: TCLP	The leaching of lead of all concretes was below the permissible limit, although concrete containing limestone powder showed slightly higher lead concentration
Zhao et al. (2013a)	**GC**: untreated CRT funnel, <5 mm **Mortar**: 0%–75% GC, 25% fly ash, 25% GGBS, 0.45 w/c, 28 days curing **Method**: TCLP	The leaching of lead exceeded the permissible limit

CRT, cathode ray tube; *GC*, glass cullet; *GGBS*, ground granulated blast furnace slag; *SPLP*, synthetic precipitation leaching procedure; *TCLP*, toxicity characteristic leaching procedure; *w/c*, water/cement ratio.

The results of Maschio et al. (2013) and Sua-iam and Makul (2013) suggest that the leached concentrations of lead and barium from mixes containing CRT GC-sand were below the US EPA regulatory limits of 5 and 100 mg/L, respectively. However, the results of Ling and Poon (2012a,b) and Zhao et al. (2013a) reported the opposite. These conflicting findings may be due to the size of the CRT used. In the former studies, the CRT used was very fine, with fineness modulus values of 0.72 (Maschio et al., 2013) and 0.69 (Sua-iam and Makul, 2013). In the latter cases, the CRT was crushed to the size less than 5 mm (Ling and Poon, 2012a,b; Zhao et al., 2013a).

Nevertheless, attempts have been made to reduce the leaching of lead and barium from mixes made with CRT GC-sand to an acceptable level. These include:

(i) Immersing crushed CRT in 5% nitric acid for 3 h (Ling and Poon, 2011b,a, 2012b,d, 2013).

(ii) Adding biopolymer solutions to concrete and mortar mixes, such as a mixture of xanthan and guar gum (Kim et al., 2009) or boric acid and guar gum (Kim et al., 2009; Romero, 2013; Romero et al., 2013).

(iii) Using FA and GGBS as Portland cement replacement (Morrison, 2004).

On the other hand, for the soda lime GC, its leached lead concentration is within the US EPA regulatory limit of 5.0 mg/L (Ling and Poon, 2012a, 2013). Thus, unless contaminated, the use of soda lime GC-sand in concrete is not likely to release any harmful elements to a threatening level.

5.10 Case Studies

Although the laboratory findings have been encouraging, case studies with GC-sand are needed to share the experiences of the use of the material in real conditions. The case studies involving the use of GC-sand in concrete and concrete-related applications are briefly described below, though a more detailed assessment of case study findings is presented in Chapter 10. It should be mentioned that most of the studies were undertaken during 2000–06, after which no further work has been reported.

From the reported case studies, the following main points can be observed:

- All the projects from the United Kingdom were of a collaborative nature, between the Waste and Resources Action Programme, which is a UK-based government body aimed at reducing waste, and the industry. Most of the field trials were conducted on concrete-related products containing GC-sand, such as semidry cast blocks, wet-pressing kerbs, glass aggregate exposed concrete products, paint-grade blocks and masonry-grade blocks. No ASR expansion was observed in the work undertaken (Bell, 2006; Byars et al., 2004a,b).
- The performance of mortar products containing GC-sand was generally reported to be similar to that of reference mortar containing silica sand, although its consistencce was affected in a few cases (Boniface, 2006).

- The use of GC-sand as a replacement for natural sand up to 75% in conjunction with ground glass cullet as a Portland cement replacement up to 30% in making concrete has been undertaken, and the compressive strength and other hardened concrete properties were found to be satisfactory (Byars et al., 2004a; Shayan and Xu, 2006).
- For concrete pavement applications, Polley et al. (1998) claimed that the abrasion resistance of GC-sand concrete was good, but Dumitru et al. (2013) reported otherwise. In the former case, GC-sand concrete also showed good durability during the exposure of the test specimens to three winters.

5.11 Conclusions

When GC is crushed and sieved to the size of fine aggregate, conforming to standards such as BS EN 12620:2002+A1 (2008), it can be used as an alternative to natural sand in making concrete. However, when evaluating the effect of GC on the performance of concrete, the biggest omission seen in the studies currently reported in the literature is that little effort has been made to ensure that the particle size distribution of GC-sand is the same as that of the reference natural sand. Thus, the changes in the performance of concrete in many cases may have been be due to the differences in aggregate packing rather than the properties of the material itself. Since the grading of GC-sand used in the current literature tends to be coarse (see Section 3.6.2), if this factor and the angularity of GC-sand were considered in the design of the test mixes, one would expect to produce a concrete with better performance than what the current studies let one believe.

In the fresh state, the use of GC-sand as a replacement for natural sand tends to reduce the consistence of concrete, which is due to its angular shape. Like natural sand, the stability of concrete is affected by the particle size distribution of GC-sand, and sand with a fineness modulus of 2.66–2.75 is found to produce concrete with acceptable cohesiveness and visual appearance. Depending on the type of GC-sand used, its inclusion could result in a marginal change in the density of concrete. The air content of mixes is not significantly affected when GC sand is used.

For the strength and deformation properties of concrete, the influence of the particle size distribution of GC-sand can be significant. It is evident that the compressive strength of concrete remains essentially unaffected when GC of (i) similar and (ii) different grading compared with natural sand is used, although the former tends to show a slight improvement. However, concrete made with GC-sand has less variability in its compressive strength compared with concrete made with natural sand. The same procedures that are used for natural sand concrete can be adopted in estimating the tensile and flexural strength of GC-sand concrete based on its characteristic cube strength. Concrete made with GC-sand tends to show higher modulus of elasticity and lower shrinkage strain than concrete made with natural sand, especially when the two materials have similar grading. Although no information is available, the use of GC-sand should result in creep strain similar to that of natural sand concrete.

Given that water absorption and porosity of GC-sand are almost zero, all things being equal, its use as a fine aggregate in concrete would reduce the permeation properties of concrete. The improvement in permeation also suggests that the use of GC-sand should not adversely affect the durability of concrete. Except for carbonation, the durability of GC-sand concrete in terms of its resistance to chloride ingress, acid attack and sulphate attack is similar to or better than that of natural sand concrete. The ASR of GC-sand is shown to be affected by the glass type, GC aggregate size and its content. It has been shown that soda lime GC is the least reactive glass type, green among the soda lime GC colours tends to show the least expansion, the pessimum size of GC-sand is between 1 and 3 mm regardless of its colour and the expansion increases with increasing GC-sand content. Although ASR expansion can be a concern, a great deal of literature suggests that the expansion can be nullified with the use of FA, ground blast furnace slag, metakaolin and even GC itself in a ground form. In terms of the resistance to physical attack, the available information in the literature suggests that GC-sand can be used up to about 20% without compromising the resistance to freeze–thaw attack and abrasion. The inclusion of GC-sand could maintain or improve the performance of concrete exposed to elevated temperature.

When high-lead-bearing or high-barium-bearing glass is used as sand in concrete, some successful attempts have been made to reduce the leaching of heavy metals to nonhazardous limits. The use of GC-sand of the soda lime type in concrete is not likely to cause environmental concern. Case studies undertaken with GC-sand in concrete, mortar and concrete-related applications generally suggest that GC-sand is a viable material for use in concrete/mortar mixes as a natural sand replacement.

References

Abdallah S and Fan M, 2014. Characteristics of concrete with waste glass as fine aggregate replacement. International Journal of Engineering and Technical Research 2 (6), 11–17.

Abendeh R, Ramadan K, Haddad R and Baker M, 2015. Freezing thawing resistance of concrete incorporating glass waste. Journal of Materials Science Research 4 (4), 19–25.

ACI 301, 2005. Specifications for Structural Concrete. American Concrete Institute, Farmington Hills, MI, USA. 656 pp.

ACI 318-11, 2011. Building Code Requirements for Structural Concrete and Commentary. American Concrete Institute, Farmington Hills, MI, USA.

Adaway M and Wang Y, 2015. Recycled glass as a replacement for fine aggregate in structural concrete – effects on compressive strength. Electronic Journal of Structural Engineering 14 (1), 116–122.

Alexander M and Mindess S, 2005. Aggregates in Concrete. Taylor & Francis, Oxon, UK. 448 pp.

Ali E E and Al-Tersawy S, 2012. Recycled glass as partial replacement for fine aggregate in self-compacting concrete. Construction and Building Materials 35, 785–791.

Al-Sibahy A and Edwards R, 2012a. Mechanical and thermal properties of novel lightweight concrete mixtures containing recycled glass and metakaolin. Construction and Building Materials 31, 157–167.

Al-Sibahy A and Edwards R, 2012b. Thermal behaviour of novel lightweight concrete at ambient and elevated temperatures: experimental, modelling and parametric studies. Construction and Building Materials 31, 174–187.

Ammash H K, Muhammed M S and Nahhab A H, 2009. Using of waste glass fine aggregate in concrete. Al-Qadisiya Journal for Engineering Sciences 2 (2), 206–214.

Anna A S, 2013. Characteristics of Concrete Building Units Containing Crushed Waste Glass (MSc thesis). Department of Construction Management and Engineering, North Dakota State University of Agriculture and Applied Science, Fargo, ND, USA.

ASTM C227, 2010. Standard Test Method for Potential Alkali Reactivity of Cement-Aggregate Combinations (Mortar-Bar Method). ASTM International, West Conshohocken, PA, USA.

ASTM C267-01, 2006. Standard Test Methods for Chemical Resistance of Mortars, Grouts, and Monolithic Surfacings and Polymer Concretes. ASTM International, West Conshohocken, PA, USA.

ASTM C666, 2015. Standard Test Method for Resistance of Concrete to Rapid Freezing and Thawing. ASTM International, West Conshohocken, PA, USA.

ASTM C944, 1999. Standard Test Method for Abrasion Resistance of Concrete or Mortar Surfaces by the Rotating-Cutter Method. ASTM International, West Conshohocken, PA, USA.

ASTM C1012, 2013. Standard Test Method for Length Change of Hydraulic-Cement Mortars Exposed to a Sulfate Solution. ASTM International, West Conshohocken, PA, USA.

ASTM C1202, 2012. Standard Test Method for Electrical Indication of Concrete's Ability to Resist Chloride Ion Penetration. ASTM International, West Conshohocken, PA, USA.

ASTM C1260, 2014. Standard Test Method for Potential Alkali Reactivity of Aggregates (Mortar-Bar Method). ASTM International, West Conshohocken, PA, USA.

ASTM C1293-8, 2015. Standard Test Method for Determination of Length Change of Concrete Due to Alkali-Silica Reaction. ASTM International, West Conshohocken, PA, USA.

ASTM 1567, 2013. Standard Test Method for Determining the Potential Alkali-Silica Reactivity of Combinations of Cementitious Materials and Aggregate (Accelerated Mortar-Bar Method). ASTM International, West Conshohocken, PA, USA.

Batayneh M, Marie I and Asi I, 2007. Use of selected waste materials in concrete mixes. Waste Management 27, 1870–1876.

Bazant Z P and Steffens A, 2000. Mathematical model for kinetics of alkali-silica reaction in concrete. Cement and Concrete Research 30, 419–428.

Bazant Z P, Jin W and Meyer C, 1998. Micromechanical fracture analysis of ASR-induced damage in concrete with waste glass. In: Proceedings of the 12th Engineering Mechanics Conference, Engineering Mechanics: A Force for the 21st Century, La Jolla, CA, USA, pp. 48–51.

Bazant Z P, Zi G and Meyer C, 2000. Fracture mechanics of ASR in concretes with waste glass particles of different sizes. Journal of Engineering Mechanics 126 (3), 226–232.

Bell G, 2006. Full Scale Operational Trials Involving the Use of Recycled Glass in the Manufacture of Concrete Blocks. The Waste and Resources Action Programme, Oxon, UK. 14 pp.

Berry M, Stephens J and Cross D, 2011. Performance of 100% fly ash concrete with recycled glass aggregate. ACI Materials Journal 108 (4), 378–384.

Bhandari P S and Tajne K M, 2013. Use of waste glass in cement mortar. International Journal of Civil and Structural Engineering 3 (3), 704–711.

Bhandari P S, Dhale S A, Ghutke V S and Pathan V G, 2014. Influence of fine glass aggregate on cement mortar. IJECS 3 (1), 3607–3610.

Bignozzi M C and Saccani A, 2012. Ceramic waste as aggregate and supplementary cementing material: a combined action to contrast alkali silica reaction (ASR). Cement and Concrete Composites 34 (10), 1141–1148.

Boniface T, 2006. Full Scale Operational Trials Using Recycled Glass in Selected End Markets: Using Recycled Glass in Cementitious Products. The Waste & Resources Action Programme (WRAP), Oxon, UK. Project GLA44-014, 24 pp.

Borhan T, 2012. Properties of glass concrete reinforced with short basalt fibre. Materials and Design 42, 265–271.

Borhan T and Bailey C, 2014. Structural behaviour of basalt fibre reinforced glass concrete slabs. Materials and Structures 47 (1), 77–87.

BRE Digest 405, 1995. Carbonation of Concrete and Its Effect on Durability. Building Research Establishment, Garston, UK.

BS 812-123, 1999. Testing Aggregates. Method for Determination of Alkali-Silica Reactivity. Concrete Prism Method. BSI, London, UK.

BS 882, 1992. Specification for Aggregates from Natural Sources for Concrete. BSI, London, UK.

BS 1881-102, 1983. Testing Concrete. Method for Determination of Slump. BSI, London, UK.

BS 1881-208, 1996. Testing Concrete. Recommendations for the Determination of the Initial Surface Absorption of Concrete. BSI, London, UK.

BS 6717, 2001. Precast, Unreinforced Concrete Paving Blocks. Requirements and Test Methods. BSI, London, UK.

BS EN 1992, 2004. Eurocode 2. Design of Concrete Structures Part 1-1: General Rules and Rules for Buildings. BSI, London, UK.

BS EN 197-1, 2011. Cement. Part 1: Composition, Specifications and Conformity Criteria for Common Cements. BSI, London, UK.

BS EN 206, 2013. Concrete. Specification, Performance, Production and Conformity. BSI, London, UK.

BS EN 12457-1, 2002. Characterisation of Waste. Leaching. Compliance Test for Leaching of Granular Waste Materials and Sludges. One Stage Batch Test at a Liquid to Solid Ratio of 2 L/kg for Materials with High Solid Content and with Particle Size below 4 mm (Without or with Size Reduction). BSI, London, UK.

BS EN 12457-2, 2002. Characterisation of Waste. Leaching. Compliance Test for Leaching of Granular Waste Materials and Sludges. One Stage Batch Test at a Liquid to Solid Ratio of 10 L/kg for Materials with Particle Size below 4 mm (without or with Size Reduction). BSI, London, UK.

BS EN 12457-3, 2002. Characterisation of Waste. Leaching. Compliance Test for Leaching of Granular Waste Materials and Sludges. Two Stage Batch Test at a Liquid to Solid Ratio of 2 L/kg and 8 L/kg for Materials with a High Solid Content and with a Particle Size below 4 mm (without or with Size Reduction). BSI, London, UK.

BS EN 12457-4, 2002. Characterisation of Waste. Leaching. Compliance Test for Leaching of Granular Waste Materials and Sludges. One Stage Batch Test at a Liquid to Solid Ratio of 10 L/kg for Materials with Particle Size below 10 mm (without or with Size Reduction). BSI, London, UK.

BS EN 12620:2002+A1, 2008. Aggregate for Concrete. BSI, London, UK.

Byars E A, Morales B and Zhu H Y, 2004a. Conglasscrete I. In: The Waste and Resources Action Programme, Oxon, UK Project code: GLA2-006, 54 pp.

Byars E A, Morales B and Zhu H Y, 2004b. Conglasscrete II. In: The Waste and Resources Action Programme, Oxon, UK Project code: GLA0015-007, 30 pp.

Byars E A, Morales B and Zhu H Y, 2004c. Waste glass as concrete aggregate and pozzolan, laboratory and industrial projects. Concrete (January) 41–44.

Cabrera J G and Lynsdale C J, 1988. A new gas permeameter for measuring the permeability of mortar and concrete. Magazine of Concrete Research 40, 177–182.

Camilleri J, Montesin F E and Sammut M, 2004. The use of waste glass and pulverized fuel ash in concrete construction. In: Limbachiya M C and Roberts J J (Eds.), International Conference on Sustainable Waste Management and Recycling: Glass Waste, London, UK. Thomas Telford, London, UK, pp. 83–90.

CCAA, 2008. Use of Recycled Aggregates in Construction. Cement, Concrete and Aggregates Australia (CCAA), Australia. 25 pp.

CCANZ, 2011. Best Practice Guide for the Use of Recycled Aggregates in New Concrete. Cement and Concrete Association of New Zealand (CCANZ), New Zealand. TR14, 49 pp.

CEN/TC 51, 1994. Draft European Prestandard: Test Method for the Freeze-Thaw Resistance of Concrete – Tests with Water or Sodium Chloride Solution, Part 1: Scaling. Comité Européen de Normalisation (CEN), Brussels, Belgium.

Chen C H, Huang R, Wu J K and Yang C C, 2006. Waste E-glass particles used in cementitious mixtures. Cement and Concrete Research 36 (3), 449–456.

Chen S, Chang C, Wang H and Huang W, 2011. Mixture design of high performance recycled liquid crystal glasses concrete (HPGC). Construction and Building Materials 25 (10), 3886–3892.

Chen S H, Wang H Y and Jhou J W, 2013. Investigating the properties of lightweight concrete containing high contents of recycled green building materials. Construction and Building Materials 48, 98–103.

CNS 3763, 2001. Waterproof Agent of Cement for Concrete Construction. Chinese National Standards, Taiwan.

CNS 3763, 2009. Waterproof Agent of Cement for Concrete Construction. Chinese National Standards, Taiwan.

Copland R, Robertson C and Slaughter G, 2009. A brewery built from millions of beer bottles: use of glass aggregate concrete. In: WasteMINZ Conference, New Zealand 6 pp.

Corinaldesi V, Nardinocchhi A and Donnini J, 2016. Reuse of recycled glass in mortar manufacturing. European Journal of Environmental and Civil Engineering 140–151.

Cota F P, Melo C C D, Panzera T H, Araújo A G, Borges P H R and Scarpa F, 2015. Mechanical properties and ASR evaluation of concrete tiles with waste glass aggregate. Sustainable Cities and Society 16, 49–56.

Cree D, Green M and Noumowe A, 2013. Residual strength of concrete containing recycled materials after exposure to fire: a review. Construction and Building Materials 45, 208–223.

Cui Y, 2005. A Study of the Use of Glass Cullet as Sand, Cement Component and Filler in Concrete (Ph.D. thesis). University of Dundee, Dundee, UK.

CWC, 1996. Recycled Glass in Portland Cement Concrete. Clean Washington Center, WA, USA. BP-GL4-05-01, 2 pp.

De Castro S and De Brito J, 2013. Evaluation of the durability of concrete made with crushed glass aggregates. Journal of Cleaner Production 41, 7–15.

Degirmenci N, Yilmaz A and Cakir O A, 2011. Utilization of waste glass as sand replacement in cement mortar. Indian Journal of Engineering and Materials Sciences 18, 303–308.

Dhir R K and Tang M C, 2008. ASR Testing on Concrete Blocks. The Waste and Resources Action Programme, Oxon, UK. 13 pp.

Dhir R K, Jones M R and Munday J G L, 1985. A practical approach to studying carbonation of concrete. Concrete 19 (20), 32–34.

Dhir R K, Hewlett P C and Chan Y N, 1991. Near-surface characteristics of concrete: abrasion resistance. Materials and Structures 24 (2), 122–128.

Dhir R K, Dyer T, Tang A and Cui Y, 2004. Towards maximising the value and sustainable use of glass. Concrete (January) 38–40.

Dhir R K, Dyer T D, Tang M C, Cui Y and Wang L, 2005a. Realising a High-Value Sustainable Recycling Solution to the Glass Cullet Surplus Closing the Loop. Concrete Technology Unit, Department of Civil Engineering, University of Dundee, Dundee, UK. 155 pp.

Dhir R K, Dyer T D, Tang A, Cui Y and Wang L, 2005b. Towards maximising the use of glass cullet in concrete. Concrete (August) 32–34.

Dhir R K, Dyer T D and Paine K, 2006. Appropriate use of sustainable construction materials. Concrete (October) 20–23.

Dhir R K, Dyer T and Tang M, 2003. Expansion due to alkali-silica reaction (ASR) of glass cullet used in concrete. In: Dhir R K, Newlands M D and Halliday J E (Eds.), Proceedings of the International Symposium on Recycling and Reuse of Waste Materials. Thomas Telford, London, pp. 751–760.

Dhir R K, Dyer T and Tang M, 2009. Alkali-silica reaction in concrete containing glass. Materials and Structures 42 (10), 1451–1462.

Dhir R K, Dyer T D, Tang A, Cui Y and Wang L, 2015. Towards maximising the use of glass cullet in concrete. Concrete 39 (8), 32–34.

Dhir R K, de Brito J, Mangabhai R and Lye C Q, 2016. Sustainable Construction Materials: Copper Slag. Woodhead Publishing, Cambridge, UK. 336 pp.

DIN 52108, 2002. Testing of Inorganic Non-metallic Materials—Wear Test Using the Grinding Wheel According to Boehme—Grinding Wheel Method (in German). Deutsches Institut für Normung, Berlin, Germany.

Du H, 2011. Glass Sand Concrete and "Sandless Concrete" (Ph.D. thesis). Department of Civil and Environmental Engineering, National University of Singapore, Singapore. 234 pp.

Du H and Tan K H, 2013. Use of waste glass as sand in mortar: part II – alkali– silica reaction and mitigation methods. Cement and Concrete Composites 35 (1), 118–126.

Du H and Tan K H, 2014. Concrete with recycled glass as fine aggregates. ACI Materials Journal 111 (1), 47–57.

Dumitru I, Song T, Caprar V, Brooks P and Moss J, 2010. Incorporation of recycled glass for durable concrete. In: Second International Conference on Sustainable Construction Materials and Technologies, Ancona, Italy 9 pp.

Dumitru I, Song T, Bornstein B, Brooks P and Moss J, 2013. Field trials using recycled glass as natural sand replacement and powdered glass as cementitious materials replacement in concrete pavement. In: Third International Conference on Sustainable Construction Materials and Technologies, Kyoto, Japan 9 pp.

EN 1015-21, 2002. Methods of Test for Mortar for Masonry – Part 21: Determination of the Compatibility of One-Coat Rendering Mortars with Substrates. Comité Européen de Normalisation (CEN), Brussels, Belgium.

EN 12390-8, 2009. Testing Hardened Concrete—Part 8: Depth of Penetration of Water Under Pressure. European Standard, Brussels.

Figg J W, 1981. Reaction between cement and artificial glass in concrete. In: Conference on Alkali-Aggregate Reaction in Concrete, Cape Town, South Africa, pp. 1–18.

Ganiron Jr. T, 2013. Use of recycled glass bottles as fine aggregates in concrete mixture. International Journal of Advanced Science and Technology 61, 17–28.

Gomes J P C, Oliveira L A P, Gonilho C N P and Torgal F M A S P, 2002. Discussion of Aggregate and Concrete Water Absorption and Permeability Testing Methodology. In: Oktay U, Abrantes V and Tadeu A (Eds.), XXX IAHS World Congress on Housing, Coimbra, Portugal, pp. 1761–1768.

Guo M, Chen Z, Ling T C and Poon C S, 2015. Effects of recycled glass on properties of architectural mortar before and after exposure to elevated temperatures. Journal of Cleaner Production 101, 158–165.

Huang L, Wang H and Wang S, 2015. A study of the durability of recycled green building materials in lightweight aggregate concrete. Construction and Building Materials 96, 353–359.

Hudec P P and Ghamari R C, 2000. Ground waste glass as an alkali-silica reactivity inhibitor. In: 11th International Conference on Alkali-Aggregate Reaction, Quebec, Canada, pp. 663–672.

Hui Z and Sun W, 2011. Study of properties of mortar containing cathode ray tubes (CRT) glass as replacement for river sand fine aggregate. Construction and Building Materials 25, 4059–4064.

Idir R, Cyr M and Tagnit-Hamou A, 2009. Use of waste glass as powder and aggregate in cement-based materials. In: 1st International Conference on Sustainable Built Environment Infrastructures in Developing Countries, Oran, Algeria, pp. 109–116.

Idir R, Cyr M and Tagnit-Hamou A, 2010. Use of fine glass as ASR inhibitor in glass aggregate mortars. Construction and Building Materials 24 (7), 1309–1312.

Isler J W, 2012. Assessment of Concrete Masonry Units Containing Aggregate Replacements of Waste Glass and Rubber Tire Particles (MSc thesis). University of Colorado Denver, USA. 108 pp.

Ismail Z Z and Al-Hashmi E A, 2009. Recycling of waste glass as a partial replacement for fine aggregate in concrete. Waste Management 29 (2), 655–660.

Jia S, Gong H, Zhang B, Zhang C and Li Y, 2015. Research on waste glass concrete elasticity modulus. Applied Mechanics and Materials 744–746, 1477–1480.

Jin W, Meyer C and Baxter S, 2000. 'Glasscrete' – concrete with glass aggregate. ACI Materials Journal 97 (2), 208–213.

JIS A 1146, 2000. Method of Test for Alkali-Silica Reactivity of Aggregates by Mortar-Bar Method. JSA, Tokyo, Japan.

Jones M R, McCarthy A and Dhir R K, 2005. Recycled and secondary aggregates in foamed concrete. In: DTI/WRAP Aggregates Research Programme STBF 13/13C. The Waste and Resources Action Programme, Oxon, UK. 66 pp.

Ke G, Bai J, Tann D and Abu-Tair A, 2011. The effect of waste glass on concrete performance under high temperatures. In: 2nd International Conference on Waste Engineering and Management, Shanghai, China, pp. 324–331.

Kim W and Soh Y, 2002. Properties of unsaturated polyester mortars using crushed waste glass. Journal of Asian Architecture and Building Engineering 1 (1), 7–12.

Kim D, Quinlan M and Yen T F, 2009. Encapsulation of lead from hazardous CRT glass wastes using biopolymer cross-linked concrete systems. Waste Management 29, 321–328.

Koh C J, 2015. Characterisation of Shape of Fine Recycled Crushed Coloured Glass and the Effect on the Properties of Structural Concrete when Used as a Fine Aggregate Replacement (Ph.D. thesis). University of Wolverhampton, UK.

Kou S C and Poon C S, 2009. Properties of self- compacting concrete prepared with recycled glass aggregate. Cement and Concrete Composites 31 (2), 107–113.

Kozlova S, Millrath K, Meyer C and Shimanovich S, 2003. A suggested screening test for ASR in cement-bound composites containing glass aggregate based on autoclaving. Cement and Concrete Composites 26, 827–835.

Kulkarni P, Shah R, Mantri M, Patidar M and Nahar S, 2015. Experimental investigation on properties of concrete made using glass cullet as a partial replacement to fine aggregates. IJMTER 2 (7), 942–948.

Kumar R and Naik T R, 2010. Sustainable concrete with industrial and post-consumer by-products. In: Second International Conference on Sustainable Construction Materials and Technologies, Ancona, Italy, pp. 1899–1910.

Lam C S, Poon C S and Chan D, 2007. Enhancing the performance of pre- cast concrete blocks by incorporating waste glass – ASR consideration. Cement and Concrete Composites 29 (8), 616–625.

Lee G, 2011. Effects of Using Waste Glass with Different Particle Sizes to Replace Sand in Pre-cast Concrete Blocks (M Phil thesis). Department of Civil and Structural Engineering, The Hong Kong Polytechnic University, Hong Kong. 113 pp.

Lee C, Chen C, Huang R and Wu J, 2008. Properties of concrete made with recycled epoxy/glass particles. Journal of the Chinese Institute of Engineers 31 (6), 1061–1067.

Lee G, Ling T, Wong Y and Poon C, 2011. Effects of crushed glass cullet sizes, casting methods and pozzolanic materials on ASR of concrete blocks. Construction and Building Materials 25 (5), 2611–2618.

Lee G, Poon C S, Wong Y L and Ling T C, 2013. Effects of recycled fine glass aggregates on the properties of dry- mixed concrete blocks. Construction and Building Materials 38, 638–643.

Lim A, 2014. The Use of Waste Glass as a Partial Substitution for Fine Aggregate in Concrete Design (BSc thesis). School of Engineering and IT, Charles Darwin University. 44 pp.

Limbachiya M C, 2009. Bulk engineering and durability properties of washed glass sand concrete. Construction and Building Materials 23 (2), 1078–1083.

Limbachiya M, Meddah M S and Fotiadou S, 2012. Performance of granulated foam glass concrete. Construction and Building Materials 28, 758–768.

Ling T and Poon C, 2011a. Properties of architectural mortar prepared with recycled glass with different particle sizes. Materials and Design 32 (5), 2675–2684.

Ling T and Poon C, 2011b. Utilization of recycled glass derived from cathode ray tube glass as fine aggregate in cement mortar. Journal of Hazardous Materials 192 (2), 451–456.

Ling T and Poon C, 2012a. A comparative study on the feasible use of recycled beverage and CRT funnel glass as fine aggregate in cement mortar. Journal of Cleaner Production 29–30, 46–52.

Ling T and Poon C, 2012b. Feasible use of recycled CRT funnel glass as heavyweight fine aggregate in barite concrete. Journal of Cleaner Production 33, 42–50.

Ling T and Poon C, 2012c. Stress– strain behaviour of fire exposed self- compacting glass concrete. Fire and Materials 37 (4), 297–310.

Ling T and Poon C, 2012d. Development of a method for recycling of CRT funnel glass. Environmental Technology 33 (22), 2531–2537.

Ling T and Poon C, 2013. Effects of particle size of treated CRT funnel glass on properties of cement mortar. Materials and Structures 46 (1), 25–34.

Ling T and Poon C, 2014a. Use of recycled CRT funnel glass as fine aggregate in dry- mixed concrete paving blocks. Journal of Cleaner Production 68, 209–216.

Ling T and Poon C, 2014b. Feasible use of large volumes of GGBS in 100% recycled glass architectural mortar. Cement and Concrete Composites 53, 350–356.

Ling T and Poon C, 2014c. Use of CRT funnel glass in concrete blocks prepared with different aggregate-to- cement ratios. Green Materials 2 (1), 43–51.

Ling T and Poon C, 2014d. High temperatures properties of barite concrete with cathode ray tube funnel glass. Fire and Materials 38, 279–289.

Ling T, Poon C and Kou S, 2011a. Feasibility of using recycled glass in architectural cement mortars. Cement and Concrete Composites 33, 848–854.

Ling T, Poon C and Kou S, 2011b. Utilization of recycled glass derived from cathode ray tube glass as fine aggregate in cement mortar. Journal of Hazardous Materials 192 (2), 451–456.

Ling T, Poon C and Kou S, 2012a. Influence of recycled glass content and curing conditions on the properties of self-compacting concrete after exposure to elevated temperatures. Cement and Concrete Composites 34, 265–272.

Ling T, Poon C, Lam W, Chan T and Fung K, 2012b. Utilization of recycled cathode ray tubes glass in cement mortar for X-ray radiation-shielding applications. Journal of Hazardous Materials 199–200, 321–327.

Ling T, Poon C, Lam W, Chan T and Fung K, 2013. X-ray radiation shielding properties of cement mortars prepared with different types of aggregates. Materials and Structures 46, 1133–1141.

Liu M, 2011. Incorporating ground glass in self- compacting concrete. Construction and Building Materials 25 (2), 919–925.

LNEC E 391, 1993. Concrete: Determination of the Carbonation Resistance (in Portuguese). Laborat6rio Nacional De Engenharia Civil, Lisbon, Portugal.

Lye C Q, Dhir R K and Ghataora G S, 2017. Deformation properties of concrete made with glass cullet. Proceedings of Institution of Civil Engineers Proceedings, Structures and Buildings 170 (5), 321–335.

Maier P L and Durham S A, 2012. Beneficial use of recycled materials in concrete mixtures. Construction and Building Materials 29, 428–437.

Malik M I, Bashir M, Ahmad S, Tariq T and Chowdhary U, 2013. Study of concrete involving use of waste glass as partial replacement of fine aggregates. IOSR Journal of Engineering 3 (7), 8–13.

Malik M I, Manzoor A, Ahmad B, Asima S, Ali R and Bashir M, 2014. Positive potential of partial replacement of fine aggregates by waste glass (<600 micron) in concrete. International Journal of Civil Engineering & Technology 5 (11), 146–153.

Mardani-Aghabaglou A, Tuyan M and Ramyar K, 2015. Mechanical and durability performance of concrete incorporating fine recycled concrete and glass aggregates. Materials and Structures 48, 2629–2640.

Maschio S, Tonello G and Furlani E, 2013. Recycling glass cullet from waste CRTs for the production of high strength mortars. Journal of Waste Management 2013, 1–8.

Mavroulidou M, Candengue D and Ramachandran S, 2011. Use of glass cullet aggregate in concrete with binders counteracting alkali-silica reaction. In: Proceedings of the 12th International Conference on Environmental Science and Technology, Rhodes, Greece, pp. 1202–1209.

Meyer C, 2001. Recycled glass – from waste material to valuable resource. In: Dhir R K, Limbachiya M C and Dyer T D (Eds.), Symposium on Recycling and Reuse of Glass Cullet, Dundee, Scotland. Thomas Telford, London, UK, pp. 1–10.

Meyer C, 2003a. Glass Concrete. Columbia University, New York, USA. 3 pp.

Meyer C, 2003b. Glass concrete: putting sustainability to work in a decorative way. Concrete International (June) 55–58.

Meyer C and Baxter S, 1998a. Use of Recycled Glass for Concrete Masonry Block. New York State Energy Research and Development Authority, New York, USA. 154 pp.

Meyer C and Baxter S, 1998b. Use of Recycled Glass and Fly Ash for Precast Concrete. New York State Energy Research and Development Authority, New York, USA. 144 pp.

Meyer C, Baxter S and Jin W, 1996a. Potential of waste glass for concrete masonry block. In: Fourth Materials Engineering Conference on Materials for the New Millennium, Washington DC, USA, pp. 666–673.

Meyer C, Baxter S and Jin W, 1996b. Alkali-silica reaction in concrete with waste glass as aggregate. In: Fourth Materials Engineering Conference on Materials for the New Millennium, Washington DC, USA, pp. 1388–1397.

Meyer C, Jin W and Baxter S, 1998. Concrete with glass as aggregate. In: Proceedings of the 13th FIP Congress on Challenges for Concrete in the Next Millennium, Amsterdam, the Netherland, pp. 573–576.

Meyer C, Egosi N and Andela C, 2001. Concrete with waste glass as aggregate. In: Dhir R K, Limbachiya M C and Dyer T D (Eds.), International Symposium on Recycling and Re-use of Glass Cullet, Dundee, UK. Thomas Telford, London, UK, pp. 179–180.

Meyer C, Shimanovich S and Vilkner G, 2002. Precast Concrete Wall Panels with Glass Concrete. Department of Civil Engineering Mechanics, Columbia University, New York, USA. NYSERDA Report-02, 45 pp.

Miranda Jr. E J P, Paiva A and Rodrigues E, 2014. Influence of waste glass on the physical properties of Portland cement concrete. Materials Science Forum 798–799, 576–581.

Mitra S, Srimany A and Dey B G, 2016. A practical study on effective utilisation of waste glass in concrete. International Journal of Science, Engineering and Technology Research 5 (3), 694–696.

Morrison C, 2004. Reuse of CRT glass as aggregate in concrete. In: Limbachiya M C and Roberts J J (Eds.), International Conference on Sustainable Waste Management and Recycling: Glass Waste, London, UK. Thomas Telford, London, UK, pp. 91–98.

Naik T R and Wu Z, 2001. Crushed post-consumer glass as a partial replacement of sand in concrete. In: 5th CANMET/ACI International Conference on Recent Advance of Concrete Technology, Singapore 20 pp.

Neville A M, Dilger W H and Brooks J J, 1983. Creep of Plain and Structural Concrete. Construction Press, London, UK.

Noruziaan B and Buskell C, 2010. Investigating the feasibility of using post-consumer aggregates in concrete mix. In: 2nd International Conference on Waste Engineering and Management, Shanghai, China, pp. 368–377.

NT Build 443, 1995. Concrete, Hardened: Accelerated Chloride Penetration. Nordic Council of Ministers, Nordtest Methods, Espoo, Finland.

Oliveira L A P, Gomes J P C and Santos P, 2008. Mechanical and durability properties of concrete with ground waste glass sand. In: 11 DBMC International Conference on Durability of Building Materials and Components, Istanbul, Turkey, pp. 1–8.

Oliveira R, de Brito J and Veiga R, 2013. Incorporating of fine glass aggregate in renderings. Construction and Building Materials 44, 329–341.

Oliveira R, de Brito J and Veiga R, 2015. Reduction of the cement content in rendering mortars with fine glass aggregates. Journal of Cleaner Production 95, 75–89.

Panchakaria V S and Hall M W, 1996. Glasscrete – disposing of non-recyclable glass. In: Proceedings of 4th ASCE Materials Engineering Conference on Material for a New Millennium, Washington DC, USA, pp. 509–518.

Park S and Lee B, 2004. Studies on expansion properties in mortar containing waste glass and fibers. Cement and Concrete Research 34 (7), 1145–1152.

Park S B, Lee B C and Kim J H, 2004. Studies on mechanical properties of concrete containing waste glass aggregate. Cement and Concrete Research 34, 2181–2189.

Penacho P, De Brito J and Veiga R, 2014. Physico-mechanical and performance characterization of mortars incorporating fine glass waste aggregate. Cement and Concrete Composites 50, 47–59.

Perkins, 2008. Development of concrete containing waste glass. In: Plassmann P and Roach P (Eds.), Proceedings of the 3rd Research Student Workshop, pp. 78–80 Pontypridd, UK.

Phillips J C, Cahn D and Keller G, 1972. Refuse glass aggregate in Portland cement concrete. In: Proceeding of the Third Mineral Waste Utilization Symposium, Chicago, IL, USA, pp. 385–390.

Polley C, 1996. The Effects of Waste Glass Aggregate on the Strength and Durability of Portland Cement Concrete (MSc thesis). Department of Civil and Environmental Engineering, University of Wisconsin, Madison, USA. 174 pp.

Polley C, Cramer S M and Cruz D L, 1998. Potential for using waste glass in Portland cement concrete. Journal of Materials in Civil Engineering 10 (4), 210–219.

Poon C and Ling T, 2010. Use of recycled glass in architectural mortars. Journal of Wuhan University of Technology 32 (7), 231–234.

Popovics S, 1998. Strength and Related Properties of Concrete: A Qualitative Approach. John Wiley & Son, New York, USA. 535 pp.

Poutos K I and Nwaubani S O, 2013. Strength development of concrete made with recycled glass aggregates subjected to frost curing conditions. IJAIEM 2 (2), 19–28.

Rajabipour F, Maraghechi H and Fischer G, 2010. Investigating the alkali silica reaction of recycled glass aggregates in concrete materials. Journal of Materials in Civil Engineering 22 (12), 1201–1208.

Rajabipour F, Shafaatian S and Wright J, 2012. Utilizing Coal Fly Ash and Recycled Glass in Developing Green Concrete Materials. The Pennsylvania State University, PA, USA. PUS-2010-07, 51 pp.

Ram B R, Rajeswari A and Vinay K M, 2015. Recycling of glass in concrete. IJERT 4 (11), 251–255.

Ramana K V and Samdani S S, 2013. Study of influence of crushed waste glass on properties of concrete. IJSR 4 (9), 1034–1039.

Rashad A M, 2014. Recycled waste glass as fine aggregate replacement in cementitious materials based on Portland cement. Construction and Building Materials 72, 340–357.

Rashad A M, 2015. Recycled cathode ray tube and liquid crystal display glass as fine aggregate replacement in cementitious materials. Construction and Building Materials 93, 1236–1248.

Richardson A E, 2013. Strength and durability of concrete using recycled aggregates (RAs). In: Handbook of Recycled Concrete and Demolition Waste. Woodhead Publishing, Cambridge, UK, pp. 330–348.

Richardson A E, Coventry K and Graham S, 2002. Concrete manufacture with un-graded recycled aggregates. Structural Survey 27 (1), 62–70.

RILEM TC 106-AAR, 2000. Recommendations: a – TC 106-2 – detection of potential alkali-reactivity of aggregates – the ultra-accelerated mortar-bar test B – TC 106-3 – detection of potential alkali-reactivity of aggregates – method for aggregate combinations using concrete prisms. Materials and Structures 33, 283–293.

RILEM TC 162, 2000. Test and design methods for steel fibre reinforced concrete. Materials and Structures 33 (249), 75–81.

Romero D F, 2013. Concrete Durability and Environmental Performance of Mixtures Containing Recycled Hazardous Waste Aggregates (Ph.D. thesis). University of Miami, Florida, USA. 230 pp.

Romero D, James J, Mora R and Hays C D, 2013. Study on the mechanical and environmental properties of concrete containing cathode ray tube glass aggregate. Waste Management 33 (7), 1659–1666.

Rose A L and Saji M S, 2015. Using glass waste as a partial replacement for fine aggregate on the strength of concrete. International Journal of Engineering Research-Online 3 (1), 96–99.

Roskos C, White T and Berry M, 2015. Structural performance of self-cementitious fly ash concretes with glass aggregates. Journal of Structural Engineering 141 (3), 3–10.

Saccani A and Bignozzi M C, 2010. ASR expansion behavior of recycled glass fine aggregates in concrete. Cement and Concrete Research 40 (4), 531–536.

Sagoe-Crentsil K, Brown T and Taylor A, 2001. Guide Specification: Recycled Glass as Sand Replacement in Premix Concrete. CSIRO, Victoria, Australia. 44 pp.

Samarin A, 1980. Use of Fine Crushed Bottle Glass Sand and Partial Cement Replacement in Concrete. School of Civil Engineering, University of New South Wales, Australia. UNICIV Report No. R-193, 16 pp.

Sangha M C, Alani A M and Walden P J, 2004. Relative strength of green glass cullet concrete. Magazine of Concrete Research 56 (5), 293–297.

Serpa D, Silva A S, de Brito J, Pontes J and Soares D, 2013. ASR of mortars containing glass. Construction and Building Materials 47, 489–495.

Serpa D, de Brito J and Pontes J, 2015. Concrete made with recycled glass aggregates: mechanical performance. ACI Materials Journal 112 (1), 29–38.

Sharif M, Mehmood A, Yousaf M, Ghaffar A and Goraya R, 2014. Performance of concrete containing waste powdered glass as partial replacement of fine aggregate. NED University Journal of Research 11 (4), 43–49.

Sharifi Y, Houshiar M and Aghebati B, 2013. Recycled glass replacement as fine aggregate in self-compacting concrete. Frontiers of Structural and Civil Engineering 7 (4), 419–428.

Shayan A, 2002. Value-added utilisation of waste glass in concrete. In: IASBSE Symposium, Melbourne, Australia, pp. 1–11.

Shayan A and Xu A, 2004. Value-added utilisation of waste glass in concrete. Cement and Concrete Research 34 (1), 81–89.

Shayan A and Xu A, 2006. Performance of glass powder as a pozzolanic material in concrete: a field trial on concrete slabs. Cement and Concrete Research 36 (3), 457–468.

Shehata I, Varzavand S, El-Sawy A and Fahmy M, 1996. The use of solid waste materials as fine aggregate substitutes in cementitious concrete composites. In: Semi Sesquicentennial Transportation Conference Proceedings, Ames, IA, USA, pp. 42–45.

Shi C, 2009. Corrosion of glasses and expansion mechanism of concrete containing waste glasses as aggregates. Journal of Materials in Civil Engineering 21 (10), 529–534.

Shi C and Zheng K, 2007. A review on the use of waste glasses in the production of cement and concrete. Resources, Conservation & Recycling 52 (2), 234–247.

Shklyan R, 2009. Enhancing Concrete Barrier Reflectivity with a Focus on Recycled Glass Aggregate Replacement (MSc thesis). Civil and Environmental Engineering Department, University of Massachusetts Amherst, Amherst, MA, USA. 114 pp.

Sidharthan O, Vijayaraghavan C and Sivaraman S, 2016. A study on compressive strength of concrete using crushed glass as fine aggregate. IJRESTs 2 (2), 73–77.

Sikora P, Augustyniak A, Cendrowski K, Horszczaruk E, Rucinska T, Nawrotek P and Mijowska E, 2016. Characterization of mechanical and bactericidal properties of cement mortars containing waste glass aggregate and nanomaterials. Materials 9 (8) 16 pp.

Soutsos M N, Millard S G and Karaiskos K, 2006. Mix design, mechanical properties, and impact resistance of reactive powder concrete (RPC). In: Fischer G and Li V C (Eds.), International RILEM Workshop on High Performance Fiber Reinforced Cementitious Composites (HPFRCC) in Structural Applications, Honolulu, Hawaii. RILEM Publications, pp. 549–559.

Sua-iam G and Makul N, 2013. Use of limestone powder during incorporation of Pb-containing cathode ray tube waste in self-compacting concrete. Journal of Environmental Management 128, 931–940.

Suresh D, Narasimhulu K and Nagaraju K, 2015. Study on compressive strength of M20 grade of concrete with partial replacement of cement by fly ash, fine aggregate by waste glass powder and coarse aggregate by e-plastic waste. IJIRSET 4 (9), 9123–9129.

Taha B and Nounu G, 2008a. Properties of concrete contains mixed colour waste recycled glass as sand and cement replacement. Construction and Building Materials 22 (5), 713–720.

Taha B and Nounu G, 2008b. Using lithium nitrate and pozzolanic glass powder in concrete as ASR suppressors. Cement and Concrete Composites 30 (6), 497–505.

Taha B and Nounu G, 2009. Utilizing waste recycled glass as sand/cement replacement in concrete. Journal of Materials in Civil Engineering 21 (12), 709–721.

Takata R, Sato S, Nonaka T, Ogata H and Hattori K, 2004. Investigation on alkali-silica reaction utilizing waste glass in concrete and suppression effect by natural zeolite. In: 29th Conference on Our World in Concrete and Structures, Singapore, pp. 523–528.

Tan K H and Du H, 2013. Use of waste glass as sand in mortar: part I – fresh, mechanical and durability properties. Cement and Concrete Composites 35 (1), 109–117.

Tang M C, Wang L and Dhir R K, 2005. Engineering and durability of concrete containing waste glass. In: Dhir R K, Dyer T D and Newlands M D (Eds.), Proceedings of the International Conference on Achieving Sustainability in Construction, Dundee, UK. Thomas Telford, London, UK, pp. 219–228.

Terro M J, 2006. Properties of concrete made with recycled crushed glass at elevated temperatures. Building and Environment 41 (5), 633–640.

Topcu I B and Canbaz M, 2004. Properties of concrete containing waste glass. Cement and Concrete Research 34, 267–274.

Topcu I B, Boga A R and Bilir T, 2008. Alkali– silica reactions of mortars produced by using waste glass as fine aggregate and admixtures such as fly ash and Li_2CO_3. Waste Management 28 (5), 878–884.

TriVitro, 2004. Using Recycled Glass with Portland Cement. Vitro Hue Glass Aggregate, Technical Bulletin VH-002. 2 pp.

UEPG, 2016. Estimates of Aggregates Production Data 2014. European Aggregates Association, Brussels, Belgium. Available from: http://www.uepg.eu/statistics/estimates-of-production-data/data-2014.

US EPA, 1992. Method 1311: Toxicity Characteristics Leaching Procedure (TCLP). US Environmental Protection Agency, Washington DC, USA.

US EPA, 1994. Method 1312: Synthetic Precipitation Leaching Procedure (SPLP). US Environmental Protection Agency, Washington DC, USA. 30 pp.

Walczak P, Malolepszy J, Reben M and Rzepa K, 2015. Mechanical properties of concrete mortar based on mixture of CRT glass cullet and fluidized fly ash. Procedia Engineering 108, 453–458.

Wang H Y, 2009a. A study of the effects of LCD glass sand on the properties of concrete. Waste Management 29 (1), 335–341.

Wang H Y, 2009b. A study of the engineering properties of waste LCD glass applied to controlled low strength materials concrete. Construction and Building Materials 23 (6), 2127–2131.

Wang H and Chen J S, 2010. Mix proportions and properties of CLSC made from thin film transition liquid crystal display optical waste glass. Journal of Environmental Management 91 (3), 638–646.

Wang H Y and Huang W L, 2010a. Durability of self-consolidating concrete using waste LCD glass. Construction and Building Materials 24 (6), 1008–1013.

Wang H Y and Huang W L, 2010b. A study on the properties of fresh self-consolidating glass concrete (SCGC). Construction and Building Materials 24 (4), 619–624.

Wang H, Zeng H and Wu J, 2014. A study on the macro and micro properties of concrete with LCD glass. Construction and Building Materials 50, 664–670.

Wang F, Gong H, Jia S, Zhang B and Zhang C, 2015. Research on waste glass concrete basic mechanical property. Applied Mechanics and Materials 711, 406–409.

WBCSD, 2009. Recycling Concrete. World Business Council for Sustainable Development. Geneva, Switzerland, 40pp.

Wright J R, 2012. Examining Concrete Properties Containing Recycled Glass Cullet as a 100% Fine Aggregate Replacement (MSc thesis). Department of Civil and Environmental Engineering, The Pennsylvania State University, Stage College, PA, USA. 119 pp.

Wright J R, Cartwright C, Fura D and Rajabipour F, 2014. Fresh and hardened properties of concrete incorporating recycled glass as 100% sand replacement. Journal of Materials in Civil Engineering 26 (10) 11 pp.

Yamada K and Ishiyama S, 2005. Maximum dosage of glass cullet as fine aggregate in mortar. In: Dhir R K, Dyer T D and Newlands M D (Eds.), Proceedings of the International Conference on Achieving Sustainability in Construction, Dundee, UK. Thomas Telford, London, UK, pp. 185–192.

Yamada K, Abe Y and Ishiyama S, 2004. ASR expansion caused by glass cullet under rapid test method. In: Limbachiya M C and Roberts J J (Eds.), International Conference on Sustainable Waste Management and Recycling: Glass Waste, London, UK. Thomas Telford, London, UK, pp. 109–116.

Yang S L, Millard S G, Soutsos M N, Barnett S J and Le T T, 2009. Influence of aggregate and curing regime on the mechanical properties of ultra-high performance fibre reinforced concrete (UHPFRC). Construction and Building Materials 23 (6), 2291–2298.

Yuksel C, Ahari R S, Ahari B A and Ramyar K, 2013. Evaluation of three test methods for determining the alkali- silica reactivity of glass aggregate. Cement and Concrete Composites 38, 57–64.

Zammit L, 1990. The Use of Waste Materials as Aggregates in Concrete Construction: Crushed Waste Glass Fines in Concrete. University of Malta, Malta. 15 pp.

Zammit L, Montesin F E and Torpiano A, 2004. The use of crushed glass waste as fines in concrete construction. In: Limbachiya M C and Roberts J J (Eds.), International Conference on Sustainable Waste Management and Recycling: Glass Waste, London, UK. Thomas Telford, London, UK, pp. 125–131.

Zhao H, Poon C S and Ling T C, 2013a. Utilizing recycled cathode ray tube funnel glass sand as river sand replacement in the high-density concrete. Journal of Cleaner Production 51, 184–191.

Zhao H, Poon C S and Ling T C, 2013b. Properties of mortar prepared with recycled cathode ray tube funnel glass sand at different mineral admixture. Construction and Building Materials 40, 951–960.

Zheng K, 2013. Recycled glass concrete. In: Pacheco-Torgal E, Jalali S, Labrincha J and John V M (Eds.), Eco-efficient Concrete. Woodhead Publishing, Cambridge, UK, pp. 241–270.

Zhu H and Byars E, 2004. Post-consumer glass in concrete: alkali-silica reaction and case studies. In: Limbachiya M C and Roberts J J (Eds.), International Conference on Sustainable Waste Management and Recycling: Glass Waste, London, UK. Thomas Telford, London, UK, pp. 99–108.

Zhu H and Byars E, 2005. Potential for use of waste glass in concrete. Concrete 39 (2), 41–45.

Use of Glass Cullet as Filler Aggregate

6

Main Headings

- Fresh properties
- Strength properties
- Deformation
- Permeation
- Durability

Synopsis

The performance of glass cullet (GC) filler as a fine aggregate replacement in concrete and as a filler component in self-compacting concrete (SCC) is examined in this chapter. The fresh properties of concrete and SCC are largely unaffected when GC filler is used, except for consistence, in which high contents of GC filler can lead to a reduction. At a fixed water/cement ratio, the use of GC filler tends to increase the compressive strength of concrete and SCC. The pozzolanic effect of GC filler can be beneficial to the long-term strength. Although limited, other data show that the impact of GC filler on flexural strength, elastic modulus and shrinkage is generally positive. The use of GC filler as a fine aggregate in concrete or a filler component in SCC results in either an improvement or no significant change in the permeation and durability properties, except for the resistance to carbonation, which could be affected if the pozzolanic effect of the GC filler is pronounced. The use of GC filler at low content is not likely to give rise to alkali–silica reaction concern. No relevant information has been found on the environmental impact or in case studies related to the use of GC filler.

Keywords: Glass cullet, Filler, Concrete, Self-compacting concrete, Fresh properties, Strength, Deformation, Permeation, Durability.

Sustainable Construction Materials: Glass Cullet. https://doi.org/10.1016/B978-0-08-100984-0.00006-0

6.1 Introduction

Aggregates occupy about 75% of the total volume of concrete. More than acting as a low-cost bulk inert filler, aggregates are now also considered to play a major role in the development of concrete properties, in both the fresh state, such as consistence and stability, and the hardened state, such as strength and volume stability associated with load-dependent and load-independent deformation that concrete experiences during its lifetime. In addition, as they are relatively harder, more durable and less porous than hardened cement paste, both the permeation and the long-term durability of concrete can be affected greatly by the characteristics of the aggregate used.

The aggregate used in concrete has to be graded so that the end product is dense, solid and homogeneous, which can ensure satisfactory performance. In terms of size, aggregates, in accordance with BS EN 12620:2002+A1 (2008), can be classified in three categories, namely, coarse aggregate, fine aggregate and filler aggregate. Coarse aggregate has particles greater than 4 mm, whilst fine aggregate has particles less than or equal to 4 mm. Filler aggregate, the focus of this chapter, has a size fraction that mostly passes a 0.063-mm sieve (BS EN 12620:2002+A1, 2008).

The role of filler aggregate in concrete is different to that of coarse and fine aggregates. The material can be used in small quantities, to control the tendency of concrete to bleed in the fresh state (Alexander and Mindess, 2005). In self-compacting concrete (SCC), fillers are commonly used in the mix design owing to the increased fines content requirement of the mix to achieve the required rheological properties (Gaimster and Dixon, 2003; Concrete Society, 2005).

The assessment of glass cullet (GC) used as a filler in concrete applications, undertaken in this chapter, covers broadly all types of concrete mixes, but in the main, (i) normal concrete mixes, in which a small proportion of filler may be used as a component of fine aggregate, cement, or both, and (ii) SCC mixes, in which a filler is usually used in large proportions to produce concrete with special rheological properties in the fresh state.

6.2 Fresh Properties

6.2.1 Consistence (Workability)

Consistence (workability) determines the ease with which a freshly mixed mortar or concrete can be handled without losing its homogeneity. The usual measurements are the flow test for mortar, slump test for concrete and slump flow test (along with V-funnel, L-box and J-ring tests) for SCC. The effects of GC filler on the consistence of mortar, concrete and SCC have been studied, as a fine aggregate replacement up to 20% or a filler replacement up to 100% in SCC and mortar mixes. The experimental details and findings of the studies undertaken by the various researchers are given in Table 6.1.

Table 6.1 Effect of glass cullet filler as a fine aggregate and a filler replacement on the consistence of mortar, concrete and self-compacting concrete

References	Main Points Emerging
(a) Mortar	
Karamberi and Moutsatsou (2005)	Soda lime GC; size: n.a.; content: 100% filler replacement; w/c: n.a. (similar water content) The flow of mortar remained unchanged
(b) Concrete	
Dhir et al. (2005)	Soda lime GC; size: <63 μm; content: 0%–20% as fine aggregate component; w/c: 0.51 The slump of concrete was reduced when GC filler was used
Dhir et al. (2005); Tang et al. (2005)	Soda lime GC; size: <63 μm; content: 0% and 5% as fine aggregate component; w/c: 0.65 The slump of concrete remained unchanged
Klevbo (1998)	Experimental details not given GC filler improved the slump of concrete
Korjakins et al. (2009)	Borosilicate GC; size: <70 μm; content: 0%–30% as fine aggregate component; w/c: 0.57 (0.62 at 30% GC content) The water content of concrete with 30% GC filler was increased to achieve slump similar to that of reference concrete
Persson (2000)	GC type unknown; size: n.a.; content: 21–275 kg/m^3; w/c: 0.37–0.80 The slump of GC filler concrete was between 80 and 115 mm (note: data for the corresponding reference concretes were unavailable)
(c) SCC	
Anagnostopoulos et al. (2009)	GC type unknown; size: n.a.; content: 100% as limestone filler; w/c: 0.50 and 0.55 The slump flow of GC-filler SCC was similar to that of reference SCC, but with a higher water content and lower admixture dosage
Bignozzi et al. (2009)	Soda lime GC; size: <70 μm; content: 100% as limestone filler; w/c: 0.53 The slump flow of GC-filler SCC was similar to that of reference SCC, but the admixture dosage was slightly increased
Georgiadis et al. (2007)	GC type unknown; size: n.a.; content: 100% as limestone filler; w/c: 0.48–0.60 The slump flow of GC-filler SCC was similar to that of reference SCC, but with a higher water content and lower admixture dosage
Matos et al. (2016)	Soda lime GC; size: <90 μm; content: 50% as limestone filler; w/c: 0.41 The slump flow of GC-filler SCC was similar to that of reference SCC
Shakhmenko et al. (2010)	Borosilicate GC; size: <70 μm and <30 μm; content: 100% as dolomite filler; w/c: 0.46–0.50 The slump flow of GC-filler SCC was similar to that of reference SCC, but with a higher water content

GC, glass cullet; *n.a.*, not available; *SCC*, self-compacting concrete; *w/c*, water/cement ratio.

When GC was used as a filler aggregate, at 5% fine aggregate content, the slump of concrete was found to remain unchanged (Dhir et al., 2005; Tang et al., 2005); at higher filler contents, up to 30%, the slump of concrete showed a noticeable reduction (Dhir et al., 2005) and the water content of such concrete mixes had to be increased to achieve slump similar to that of the reference concrete (Korjakins et al., 2009). In only one case GC filler was observed to increase the consistence of concrete, but detailed information on the mix composition has not been made available (Klevbo, 1998).

When used as a filler aggregate, at the same water content, the flow of mortar mixes made with GC filler has been reported to be similar to that of the reference mortar (Karamberi and Moutsatsou, 2005). However, in the case of SCC, the admixture dosage and/or water content used tends to be considerably higher to maintain the slump flow when limestone filler is fully replaced by GC filler (Anagnostopoulos et al., 2009; Bignozzi et al., 2009; Georgiadis et al., 2007; Shakhmenko et al., 2010).

Overall, although the available results are showing varying effects, in one sense it would appear that the use of GC filler, at the 5% content level with respect to fine aggregate content, does not adversely affect the consistence of the mix. However, its use as a filler aggregate tends to reduce the consistence of mortar and that of SCC, but this effect can be overcome by using a water-reducing admixture.

6.2.2 Stability

A freshly mixed concrete is deemed to be unstable if it shows signs of excessive bleeding, usually observed as an upward migration and accumulation of water at the concrete surface, and/or segregation, separation of coarse and fine aggregates or that of aggregate and cement paste. Although important, the stability of concrete is not measured with any specific type of test method, either in the laboratory or in the field, apart from SCC, for which its segregation can be tested in accordance with BS EN 12350-11 (2010).

The research data on the bleeding of concrete containing GC filler are limited, but given that bleeding is affected by the fines content of a mix, the use of GC filler as fine aggregate replacement is most likely to reduce the risk of a concrete mix experiencing bleeding in its fresh state. Through visual inspection, no segregation issues have been observed in SCC containing GC filler that was used as a full replacement for limestone filler (Bignozzi et al., 2009; Matos et al., 2016). In addition, the use of GC filler was found to improve the stability of concrete at the same consistence level (Klevbo, 1998).

6.2.3 Fresh Density

When fully compacted, the density of fresh concrete varies with the specific gravity and content of its constituent materials. As shown previously in Chapter 3, the average specific gravity of soda lime GC is about 2.46, which is slightly lighter than the natural fine aggregates normally used in concrete construction. The changes in fresh density in concrete made with GC filler as a fine aggregate replacement of up to 20% (Dhir et al., 2005; Tang et al., 2005), or mortar and SCC made with GC filler as a full filler aggregate replacement (Bostrom, 2002; Karamberi and Moutsatsou, 2005), have been shown to be in the ±40 kg/m^3 range, which is within acceptable fluctuations of density measurement of normal concrete mixes in practice.

6.2.4 Air Content

The presence of air is undesirable in most concretes, apart from air-entrained concrete, into which air is introduced intentionally, using an air-entraining chemical admixture, to increase its resistance to freeze–thaw attack. Typically, the air content of normal concrete is less than 1.5%, whilst that of air-entrained concrete is about 5%–7%.

For the same water/cement ratio, the air content of mortar (Karamberi and Moutsatsou, 2005) and SCC (Bostrom, 2002; Persson, 2004; Anagnostopoulos et al., 2009) containing GC filler as a replacement for filler aggregate, made with or without an air-entraining agent, is close to that of the corresponding reference mixes. However, it should be mentioned that, in all these cases, the aggregate contents and/or air-entraining agent dosage was not kept constant between the two set mixes.

6.2.5 Plastic Shrinkage

Before hardening, concrete may experience plastic shrinkage, if the rate of evaporation at the surface exceeds the bleeding rate, which can lead to cracking. A limited amount of testing has been undertaken to examine the plastic shrinkage properties of normal concrete and SCC, both made with and without GC filler. The shrinkage measurements in this case were taken during the 24-h period that followed after four hours of the casting of the test mixes (Persson, 2000). Part of results are tabulated in Table 6.2, but they cannot be compared, as the mix proportions and types

Table 6.2 Plastic shrinkage of normal concrete and self-compacting concrete containing glass cullet filler

Concrete Type	Filler Type	Content, kg/m^3		w/c	24-h Plastic Shrinkage, μm
		Cement	Filler		
Normal concrete	No filler	430	0	0.38	–0.23
	Glass cullet	340	21	0.52	–0.10
Self-compacting concrete	Limestone	400	150	0.39	–0.27
	Glass cullet	450	80	0.38	–0.30
		330	190	0.55	–0.47
		251	275	0.8	–0.23

w/c, water/cement ratio.
Based on Persson (2000).

and sizes of aggregates and cement used in each mix were different and may suggest different bleeding characteristics. Notwithstanding this, no abnormalities in the plastic shrinkage behaviour of concrete were observed, and the 24-h plastic shrinkage value of all the concrete specimens made with or without GC filler were broadly the same.

6.3 Strength

6.3.1 Compressive Strength

Compressive strength is one of the most commonly specified properties of hardened concrete. Except for road pavements, concrete structures are normally designed based on its compressive strength, and the tensile stresses are resisted by reinforcement of concrete. The compressive strength is also used widely to correlate with other mechanical properties and durability of concrete, as well as concrete quality during manufacturing. In this section, the effect of GC filler on the compressive strength of concrete is discussed in terms of its use both in normal concrete and in SCC.

Glass Cullet Filler Use in Normal Concrete

(a) Effect of Glass Cullet Filler Content on Compressive Strength

Results on the effect of using GC filler up to 30% replacement of fine aggregate, obtained in two separate studies (Dhir et al., 2005; Korjakins et al., 2009), are shown plotted in Figure 6.1. Although the strength results are not conclusive, some trends are visible:

(i) Five percent is the most effective filler use, with the relative improvement in strength decreasing with age.

(ii) The effectiveness of filler content decreases at higher contents, with the lowest performance being recorded at the dose of 30%.

In addition, the research reported by Korjakins et al. (2010, 2011) has shown that increasing the grinding time by an additional 15 and 60 min, respectively, could improve the performance of GC filler. However, this would need to be weighed against energy consumption cost.

(b) Effect of Glass Cullet Filler on Strength Development With Age

It is evident in Chapters 3 and 4 that finely ground GC with fineness close to or higher than that of a typical Portland cement exhibits pozzolanic properties. This is because the chemical composition of glass is high in SiO_2 content and contains small quantities of major constituents of Portland cement such as CaO, Al_2O_3 and Fe_2O_3. Even though used in a filler size range that is coarser than Portland cement particles, GC filler has shown to improve the long-term compressive strength of concrete when used at 5% as a fine aggregate replacement, at a water/cement ratio of 0.65 (Dhir et al., 2005), as illustrated in Figure 6.2. The results suggest that the

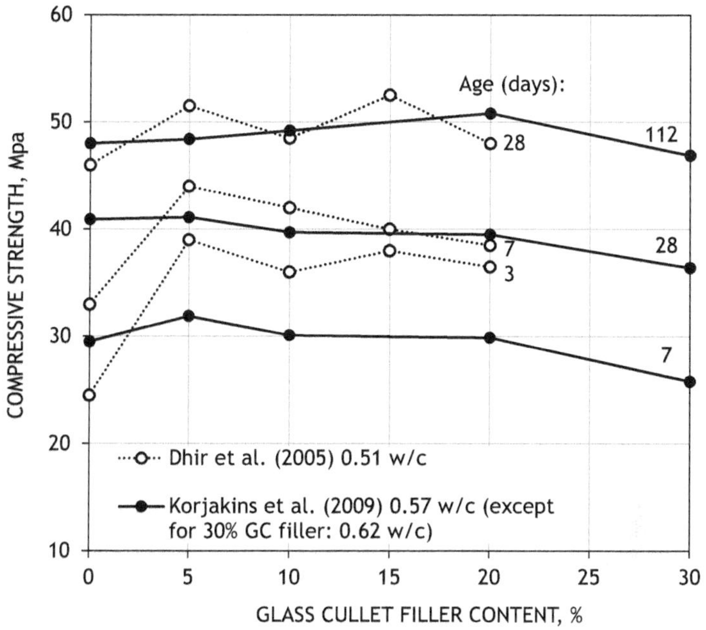

Figure 6.1 Effect of glass cullet filler as fine aggregate replacement on the compressive strength of concrete. *GC*, glass cullet; *w/c*, water/cement ratio.

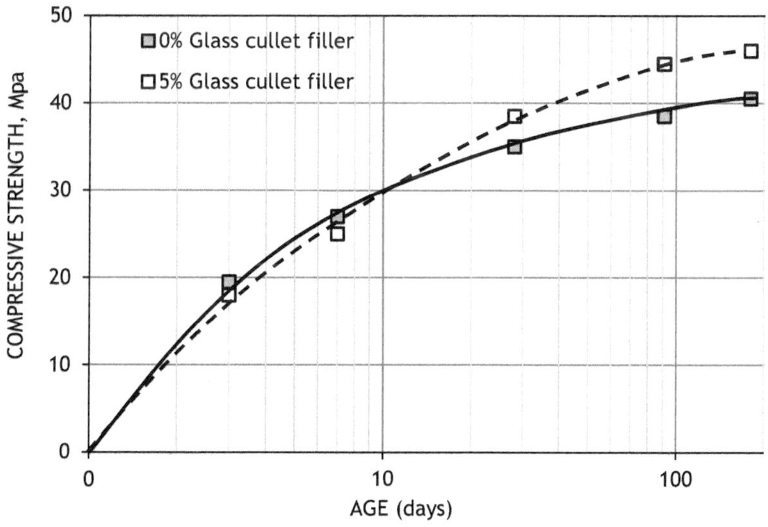

Figure 6.2 Strength development of concrete made with glass cullet filler as fine aggregate. Based on Dhir et al. (2005).

pozzolanic properties of GC filler become pronounced as early as 28 days. The improvement in long-term concrete strength due to the use of GC filler has also been observed by Klevbo (1998), where strength development in concrete was monitored for up to 1 year.

(c) Effect of Glass Cullet Filler on Water/Cement Ratio–Compressive Strength Relationship

The relationship between compressive strength and the water/cement ratio of concrete offers engineers guidance on designing concrete mixes. This relationship has been developed by Dhir et al. (2005) for concrete made with and without 5% GC filler, for water/cement ratio ranging from 0.40 to 0.72, using CEM I 42.5 cement and uncrushed aggregate of 20-mm maximum size and having a target slump of 60 mm. The results are plotted in Figure 6.3. For a given water/cement ratio, the 7-day strengths of concrete made with or without 5% GC filler are essentially similar. At 28 days, owing to the pozzolanic effect, the compressive strength of concrete made with 5% GC filler is higher than that of concrete made without GC filler, showing an average increase of 10%. This effect continues and becomes more significant at ages beyond 28 days, suggesting that with GC filler, a target strength of concrete can be achieved with reduced cement contents, thereby potentially offering some scope for improving the carbon footprint of concrete.

Glass Cullet Filler Aggregate Use in Self-Compacting Concrete

The 28-day compressive strength results for SCC containing GC filler aggregate in place of limestone and dolomite filler aggregates are given in Table 6.3, with data separated in terms of (a) equal water/cement ratio and (b) equal consistence, with respect to reference SCC. The filler aggregate content, calculated as the percentage of the total fine aggregate (sum of fine aggregate and filler aggregate), in all the SCC mixes was in the range from 5% to 25%.

At equal water/cement ratio, the use of GC filler aggregate as a full or partial replacement of limestone filler aggregate at a content less than 8.5% of the total fine aggregate was found to result in an increase in compressive strength (Table 6.3). The improvement in strength might be associated with the pozzolanic effect of GC filler aggregate, as shown previously.

At equal consistence, the use of additional water, with corresponding increase in water/cement ratio, to maintain constant consistence with ground GC filler, as to be expected, resulted in a reduction in the strength of concrete of up to 6% (Anagnostopoulos et al., 2009, see Table 6.3). However, it has been shown that grinding of GC filler aggregate for an additional 60 min could increase the strength of SCC, despite having a slightly higher water/cement ratio than the reference SCC (Shakhmenko et al., 2010). The data of Georgiadis et al. (2007) also suggest higher compressive strength in SCC containing GC filler aggregate, but these are considered to be outliers, as no special modification was made to the material.

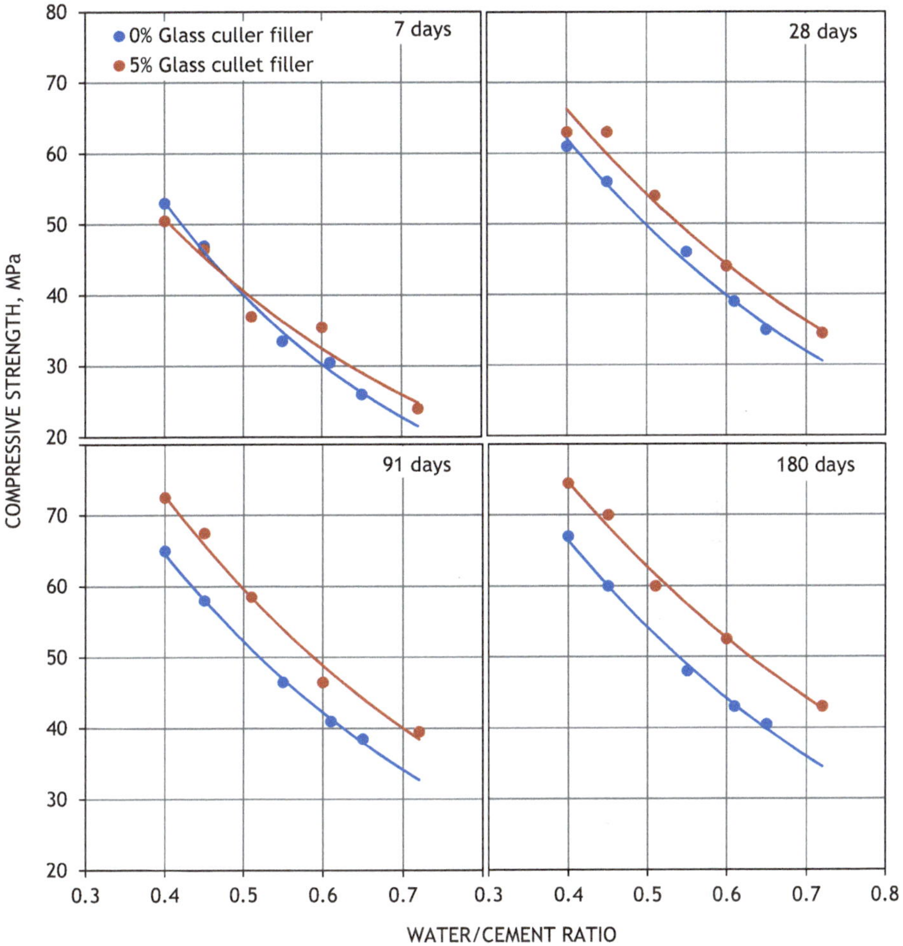

Figure 6.3 Relationship between compressive strength and water/cement ratio of concrete containing glass cullet filler as fine aggregate at different ages.
Based on Dhir et al. (2005).

Overall, the ideal content of GC filler aggregate in SCC is about 8.5%, at which it does not affect the water demand of the concrete to achieve a required consistence and at the same time enhances the strength gain. Higher GC filler aggregate content, however, can result in an increase in the water demand of SCC, thus reducing its compressive strength.

6.3.2 Tensile Strength

The tensile strength of concrete is normally determined using indirect methods and expressed in the forms of splitting tensile strength and flexural strength.

Table 6.3 Effect of glass cullet filler as replacement for limestone/dolomite filler on the compressive strength of self-compacting concrete

References	Filler Aggregate[a]	w/c	28-Day Strength, MPa	Strength Difference With Respect to Reference, %
(a) At Equal Water/Cement Ratio				
Bignozzi et al. (2009)	5% LS	0.53	35.1	0
	5% GC	0.53	41.4	+17.9
Matos et al. (2016)	19% LS	0.41	61.1	0
	8.5% LS°+°8.5% GC	0.41	63.0	+3.1
(b) At Equal Consistence				
Anagnostopoulos et al. (2009)	10% LS	0.50	55.0	0
	10% GC	0.51	53.2	3.3
	13% LS	0.55	42.0	0
	13% GC	0.56	39.5	−6.0
Georgiadis et al. (2007)	10% LS	0.48	54.0	0
	10% GC	0.51	49.0	−9.3
	13% LS	0.52	37.1	0
	13% GC	0.59	38.3	+3.2
Shakhmenko et al. (2010)	25% DL	0.46	67.9	0
	25% GC	0.50	60.2	−11.3
	25% GC[b]	0.48	75.4	+11.0

DL, dolomite filler; *GC*, glass cullet; *LS*, limestone filler; *w/c*, water/cement ratio.
[a]Percentage of sum of (fine aggregate + filler aggregate).
[b]Additional grinding for 60 min.

Information on the splitting tensile strength of concrete containing GC filler aggregate is limited. The only available data suggest that the splitting tensile strength of SCC made with GC filler aggregate is lower than that of SCC made with limestone filler aggregate (Georgiadis et al., 2007). However, the comparison could be unfair, since the former has a higher water/cement ratio.

Similar to splitting tensile strength, information on the effect of GC filler on the flexural strength of concrete is thin. However, the available data, as given in Table 6.4, are positive. They show that, for a given water/cement ratio, the use of GC filler either as a 5% fine aggregate replacement in concrete or as a full filler aggregate replacement in SCC results in an increase in flexural strength. This might open some scope for the material to be used in pavement concrete applications where the flexural strength is a design factor.

Table 6.4 Flexural strength of concrete and self-compacting concrete containing glass cullet filler aggregate

References	w/c	Content	Flexural Strength, MPa	
			28 days	**90 or 180 days**
(a) As Fine Aggregate Replacement				
Dhir et al. (2005)	0.65	0% GC filler	4.1	4.7[a]
	0.65	5% GC filler	5.0	5.3[a]
(b) As Filler Aggregate Replacement				
Bignozzi and Sandrolini (2004)	0.53	17% LS filler	12.0	12.5[b]
	0.53	17% GC filler	15.0	15.5[b]

GC, glass cullet; *LS*, limestone; *w/c*, water/cement ratio.
[a]At 180 days.
[b]At 90 days.

6.4 Deformation Properties

Three deformation properties of concrete that are closely related to deflections and cracking in structure, namely, elastic modulus, creep and shrinkage, are discussed in this section.

6.4.1 Elastic Modulus

The elastic modulus of concrete is much affected by the stiffness of its aggregates and hardened cement paste, as well as their volume fractions. It can be determined from the secant slope of a stress–strain curve of concrete when subjected to a stress equal to one-third of its strength (BS EN 12350-13, 2013).

The limited information available on elastic modulus results for concrete containing GC filler suggests that, for a given water/cement ratio, the use of GC filler as 5% fine aggregate replacement in concrete (Tang et al., 2005) or as 50% limestone filler replacement in SCC (Matos et al., 2016) results in an increase in the elastic modulus. The opposite effect is reported in the study undertaken by Georgiadis et al. (2007), in which GC filler was used as 100% replacement of limestone filler in SCC. However, this is to be expected, as the water content of SCC containing GC filler was increased to keep its flow properties similar to those of reference SCC, thus resulting in higher water/cement ratio and more porous cement paste structure.

From the available data, the compressive strength of concrete containing GC filler has been plotted against its elastic modulus, both determined at 28 days, in Figure 6.4, together with the compressive strength–elastic modulus relationship for basalt, quartzite, limestone and sandstone concrete obtained from Eurocode 2 (2004). However, the information on the type of coarse aggregate used in these concretes was

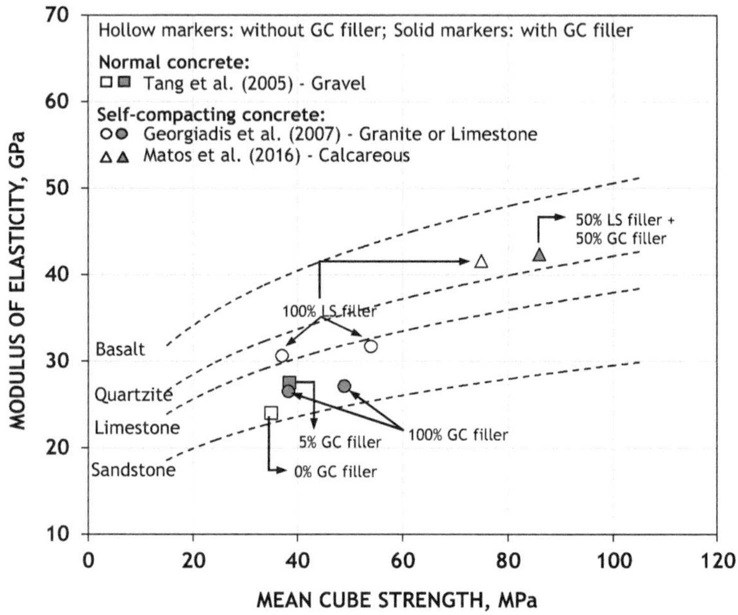

Figure 6.4 Relationship between elastic modulus and compressive strength of concrete containing glass cullet filler. *GC*, glass cullet; *LS*, limestone filler.

not clear and the limited data for concrete made with or without GC filler could not suggest whether they follow any of the trend lines. Notwithstanding this, the results are somewhat coherent with those observed by Lye et al. (2016) for natural aggregate concrete, in that the data tended to fall within the sandstone and limestone concrete trend lines when the strength was less than 60 MPa.

6.4.2 Creep

The increase in strain of concrete under a sustained load over time is known as creep. The research on the creep of concrete made with GC filler is limited; thus rigorous experimental investigations on this subject would not be possible. Notwithstanding that, owing to its void-filling effect as a very fine material, the use of GC filler as a partial replacement for fine aggregate may lead to a small reduction in the creep of concrete.

6.4.3 Shrinkage

Concrete undergoes different forms of load-independent deformation throughout its service life, the major cause being the loss of internal moisture to the surrounding environment, commonly known as drying shrinkage. The shrinkage development of concrete mixes made with and without 5% GC filler as a replacement for fine

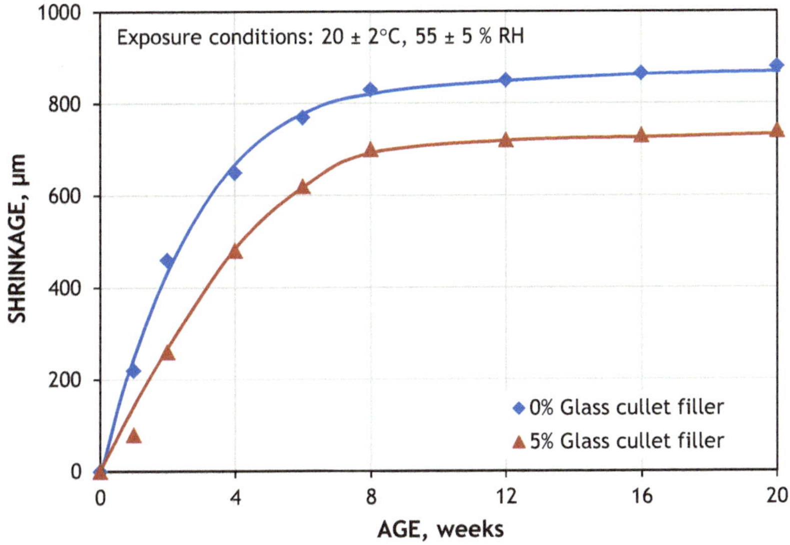

Figure 6.5 Shrinkage of concrete made with and without 5% glass cullet filler (Dhir et al., 2005; Tang et al., 2005). *RH*, relative humidity.

aggregate, with a water/cement ratio of 0.65, and stored at 20±2°C and 55±5% relative humidity (RH) for 20 weeks is presented in Figure 6.5. It is evident that, at a given age, the shrinkage of concrete incorporating GC filler is lower than that of the reference concrete made without GC filler. The shrinkage values for both concretes become almost constant after exposure to drying conditions for 8 weeks.

Limited testing has been undertaken on the shrinkage of SCC containing GC filler, having different water/cement ratios and filler contents and subjected to 20°C and 60% RH storage conditions (Persson, 2000). The shrinkage measurements up to 120 days are shown plotted in Figure 6.6. Whilst the effect of water/cement ratio on the shrinkage of concrete is clearly visible, with shrinkage increasing with water/cement ratio, no clear trend is shown for the filler effect. Notwithstanding this, it can be seen that, in each case, the shrinkage values become almost stable after 60 days, which is similar to that observed in Figure 6.5 for normal concrete.

6.5 Permeation

Concrete is a porous material, and thus it allows the ingress of fluids and gases from the surrounding environment. Commonly, the permeation properties of concrete involve three distinct transport mechanisms, namely, absorption, permeability and diffusion (Jackson and Dhir, 1996). The research undertaken on the permeation properties of concrete containing GC filler has been limited, and the available information deals with only the first two transport mechanisms.

6.5.1 Absorption

The effect of GC filler on absorption, in the form of sorptivity, has been determined in accordance with RILEM TC 116 (1999) for SCC containing GC filler as a replacement for limestone filler (Anagnostopoulos et al., 2009; Matos et al., 2016). The test parameters and sorptivity results are given in Table 6.5. It is evident that the use of GC filler as a partial or full replacement of limestone filler in SCC did not adversely affect its sorptivity, despite the fact that the GC filler used was coarser (Matos et al., 2016) or the mixes had slightly higher water/cement ratio (Anagnostopoulos et al., 2009).

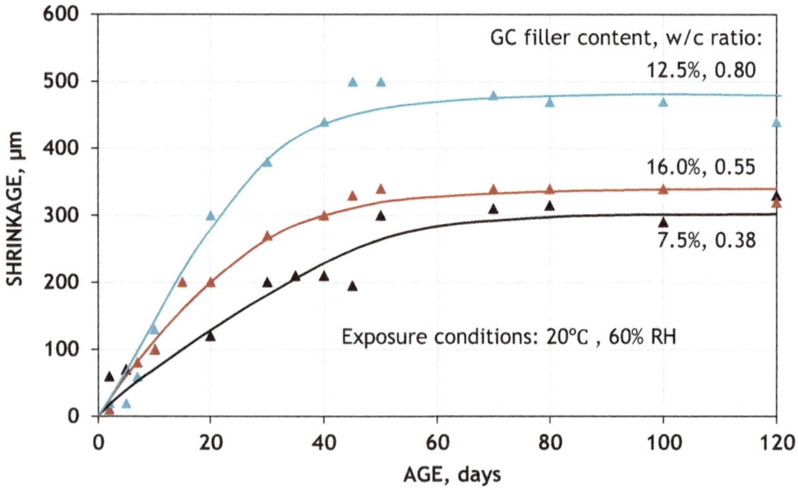

Figure 6.6 Shrinkage of self-compacting concrete containing glass cullet filler (Persson, 2000). *GC*, glass cullet; *RH*, relative humidity; *w/c*, water/cement ratio.

Table 6.5 Sorptivity of self-compacting concrete containing glass cullet filler

References	Duration	w/c	Filler	Sorptivity, g/cm²
Anagnostopoulos et al. (2009)[a]	Curing: 28 days Testing: 24 h	0.55	100% LS	0.5133
		0.56	100% GC	0.5180
		0.50	100% LS	0.4133
		0.51	100% GC	0.4533
Matos et al. (2016)[b]	Curing: 2 months Testing: 4.5 h	0.41	100% LS	0.0031
		0.41	50% LS + 50% GC	0.0026

GC, glass cullet; *LS*, limestone; *w/c*, water/cement ratio.
[a]The size of fillers was unavailable.
[b]GC filler was coarser than LS filler.

Similarly, in another study (Shamenko et al., 2014), it has been shown that replacing quartz powder with a combination of GC filler and silica fume in high-performance concrete does not result in a significant change in its sorptivity profile, monitored for about 15 days.

6.5.2 Permeability

Oxygen permeability testing has been undertaken on SCC containing GC filler as 50% replacement for limestone filler, cured for 90 days at 20°C and 100% RH and then stored for 28 days under laboratory conditions at 20°C and 55%–65% RH (Matos et al., 2016). Although the details of the results are not available, it was claimed that oxygen was unable to penetrate through both concretes made with and without GC filler.

The effect of GC filler fineness, using additional grinding time from 0 to 60 min, on the water permeability of concrete has been studied by Korjakins et al. (2011). The specimens were treated in water under a pressure of 0.5 MPa for 72 h, and the depth of water penetration was determined by splitting the test specimens. The relationship between the water penetration depth and the 28-day compressive strength for concrete made with GC filler with different grinding times (i.e., fineness) is shown in Figure 6.7.

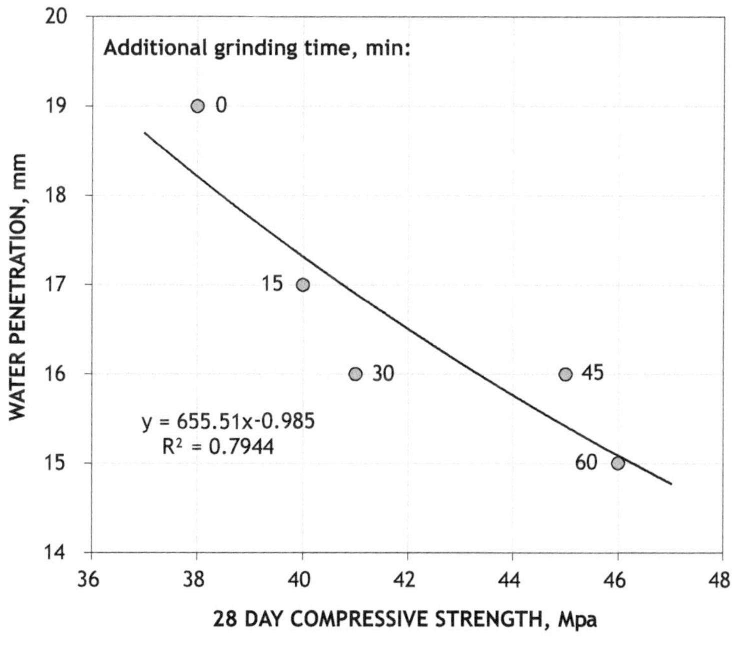

Figure 6.7 Effect of grinding time of glass cullet filler on water penetration and compressive strength of concrete.

As to be expected, increasing the grinding time of GC filler leads to a reduction in water penetration depth and an increase in compressive strength of concrete. However, the improvement in both properties, due to the increase in the fineness of GC filler, possibly becomes less pronounced after grinding for 45 min, even though the trend line does not bring this out (Figure 6.7).

6.6 Durability

The anticipated design life of a concrete structure is related to its durability. Concrete needs to be durable and strong to resist chemical attack, in the form of chloride ingress, carbonation, sulphate and acid attacks and alkali–silica reaction (ASR), and physical attack, in the form of cracking, freeze–thaw attack, and abrasion.

6.6.1 Chloride Ingress

Chloride ingress is a concern to engineers in structural concrete primarily because it destroys the protective passivity layer of embedded steels, subsequently leading to corrosion. The mix proportions and chloride migration coefficient of SCC containing GC filler are given in Table 6.6. It would appear that, at a constant water/cement ratio, replacing half of the content of limestone filler with GC filler results in about 60% reduction in chloride migration coefficient (Matos et al., 2016). It should be of interest to note that the fineness of the GC filler used in this case was coarser than that of limestone filler. In another study (Persson, 2004), the chloride migration coefficient of SCC made with GC filler was essentially similar to that of SCC made with fly ash, although the mix proportions of the mixes were grossly different. Overall, the limited test data available suggest that the use of GC filler could improve, or at the minimum maintain, the original chloride ingress resistance of concrete.

Table 6.6 Chloride migration coefficient of self-compacting concrete containing glass cullet filler

References	w/c	C:FA[a]:CA	Filler/Total Fine Aggregate[a] Ratio	D, $\times 10^{-12}$ m^2/s
Matos et al. (2016)	0.41	1:2.4:2.1	18.5% LS	9.58
	0.41		9.25% LS + 9.25% GC	3.72
Persson (2004)	0.38	1:2.72:1.07	4.4% fly ash	6.0
	0.37	1:2.45:1.32	5.5% GC	5.0

C, cement; *CA*, coarse aggregate; *D*, chloride migration coefficient; *FA*, fine aggregate; *GC*, glass cullet; *LS*, limestone; *w/c*, water/cement ratio.
[a]Including filler.

6.6.2 Carbonation

In addition to chloride-induced corrosion, carbonation is another major cause of corrosion of reinforced concrete. The reaction of carbon dioxide with calcium hydroxide produced during cement hydration lowers the pH of concrete and can destroy the protective passivity layer of the steel reinforcement. The research undertaken on the effect of GC filler on carbonation is limited, and the available information obtained from the accelerated carbonation test, as given in Table 6.7, appears to be contradictory.

When GC filler was used to replace 50% of limestone filler in SCC, in which the filler/total fine aggregate ratio was 18.5%, the carbonation depth of SCC was increased by about 2-fold (Matos et al., 2016). However, when GC filler was used to replace 5% of fine aggregate in concrete, the carbonation depth of the concrete made with GC filler was about half of that of concrete made without GC filler (Tang et al., 2005).

These conflicting results are probably associated with the pozzolanic reactivity of GC filler. In the study of Matos et al. (2016), the compressive strength of GC-filler SCC at 83 days was 11.65 MPa higher than that at 28 days, indicating that some of the calcium hydroxide in the hardened concrete might have been consumed for the pozzolanic reaction induced by the GC filler. In the study of Tang et al. (2005), the compressive strength of GC-filler concrete showed only about 5 MPa increase in strength at 180 days, suggesting that the pozzolanic reaction of the GC filler in this case was not similarly significant.

6.6.3 Sulphate Attack

The reaction of sulphates that may be present in groundwater and soils or within the concrete constituents with the hydrated cement paste leads to the formation of expansive chemical products and subsequently the deterioration of concrete. The sulphate resistance of concrete is governed by its permeability properties and the chemical composition of the hydrated cement paste (which is affected by the type of cement used).

Table 6.7 Effect of glass cullet filler on carbonation of concrete

References	Conditions	w/c	Filler/Total Fine Aggregate[a] Ratio	Carbonation Depth, mm
Matos et al. (2016)	28 days in a 5% CO_2, 60% RH and 23°C chamber	0.41	18.5% LS	0.93
		0.41	9.25% LS+9.25% GC	1.82
Tang et al. (2005)	63 days in a 3%–4% CO_2, 65% RH and 20°C chamber	0.65	No filler	16.0
		0.65	5% GC as FA	8.5

FA, fine aggregate; *GC*, glass cullet; *LS*, limestone; *RH*, relative humidity; *w/c*, water/cement ratio.
[a]Including filler.

The effect of GC filler on the sulphate resistance of concrete has been assessed by Tang et al. (2005), in which the length change of concrete prisms made with and without 5% GC filler as a natural sand replacement, exposed to 5% sodium sulphate solution for up to 20 weeks, was determined in accordance with BRE Digest 363 Part 1 (2003a) and Part 2 (2003b). The water/cement ratio of the two concretes was kept at 0.65.

Although the test specimens were expected to expand from sulphate attack, they were found to display shrinkage instead. The reasons leading to this were unclear, but the results suggested that GC filler concrete had more shrinkage (less expansion) compared with reference concrete. It was concluded (Tang et al., 2005) that the use of GC filler increases the sulphate resistance of concrete owing to the improvement in particle packing of the concrete, making the concrete less permeable to sulphates.

6.6.4 Acid Attack

In general, the acid resistance of concrete is considered weak, owing to its high-alkaline nature. Information on the effect of GC filler on the resistance of concrete to acid attack has been limited. When used as a fine aggregate replacement, GC filler is not likely to cause any significant change in the resistance of concrete to acid attack. However, in the case of SCC, the use of GC filler may offer some advantages over limestone filler in this regard, as the latter material is calcareous, which is known to be vulnerable to acid attack (Neville, 1995).

6.6.5 Alkali–Silica Reaction

Owing to the high silica content of GC filler, no doubt its use in concrete can always lead to concerns over ASR, which causes expansion and subsequently cracking in concrete. The ASR expansion of concrete made with nonreactive and active natural aggregates in which GC filler (soda lime type, different colours) was used as a fine aggregate replacement at 20% has been assessed by a single group of researchers in the United Kingdom (Dhir et al., 2005, 2009), carried out in accordance with the method defined in BS 812-123 (1999).

The types of aggregate used and the alkali content, in terms of sodium oxide equivalents, of the test concrete mixes are given in Table 6.8, and the corresponding ASR measurement results at 52 weeks and 3 years are shown in Figure 6.8. It should be mentioned that the minimum measurement period is 52 weeks, and the test should continue if the expansion between 39 and 52 weeks is greater than 0.01% of the initial measurement. For reference, the guidance for interpretation of the results at 52 weeks given in BRE Digest 330 (2004) is also shown in Figure 6.8.

Table 6.8 Details of concrete mixes used for the alkali–silica reaction test

Aggregate Type	Series	Cement Content, kg/m^3	Na_2O_{eq}, kg/m^3
Low-Reactivity Aggregate			
Natural aggregate	(1), (11), (21), (22)	697	7.0
High-Reactivity Aggregate			
North Yorkshire limestone coarse aggregate and Thames Valley chert sand	(2), (12)	400	4.8
	(3), (13)	426	5.1
	(4), (14)	452	5.4
Trent Valley coarse and fine aggregate	(5), (15)	390	4.2
	(6), (16)	400	4.8
	(7), (17)	452	5.4
Scottish Borders greywacke coarse and fine aggregate	(8), (19)	327	3.6
	(9), (19)	400	4.8
	(10), (20)	503	6.0

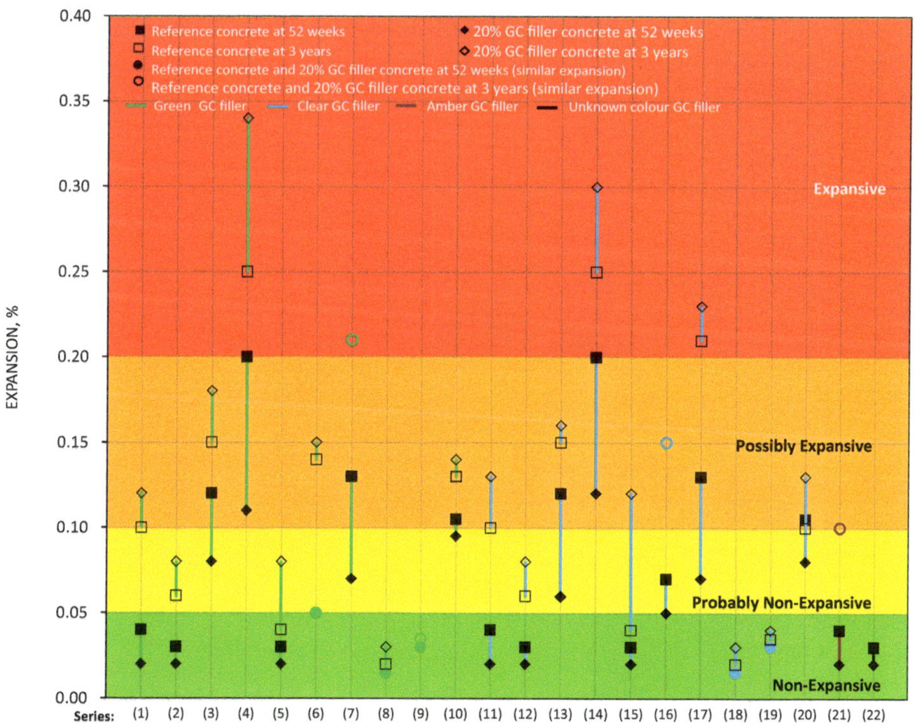

Figure 6.8 Effect of glass cullet (GC) filler on alkali–silica reaction expansion of concrete. Data taken from Dhir et al. (2003) and Dhir et al. (2005, 2009).

It can be seen from Figure 6.8 that, for concrete made with low-reactivity natural aggregates, the 52-week ASR expansion of concrete containing 20% GC filler, regardless of its colour, is lower than that of reference concrete (without GC filler), although both concretes are considered as 'nonexpansive.' However, after 3 years, the length change of both concrete suggests that they are 'possibly expansive,' and the expansion of concrete made with either green or clear GC filler is higher than that of the reference.

For concrete made with high-reactivity aggregate, as to be expected, its ASR expansion increases with increasing alkali content. It has been found that the expansion of concrete containing 20% GC filler is similar to or lower than that of reference concrete at 52 weeks, but the reverse is observed at 3 years. There is no distinguishable difference between the ASR expansion of concrete made with green GC filler and that made with clear GC filler. In a few cases, in which the alkali content of concrete made with or without GC filler is high (greater than 5.4 kg/m^3 Na_2O_{eq}), the expansion results suggest the concrete is 'expansive.'

Overall, it would appear that the use of 20% GC filler as a fine aggregate replacement in concrete results in a reduction in expansion of concrete at early ages, but a higher expansion than that of reference concrete at later ages. This observation is independent of the colours of GC filler and is the same for concrete made with both low- and high-reactivity aggregate. Notwithstanding this, it should be noted that the content of the GC filler used in this case could be too high, which might affect the overall aggregate packing and subsequently the hardened properties of concrete. Indeed, it has been claimed that the optimum level of GC filler for use as a fine aggregate replacement is 5%, at which the compressive strength and ASR expansion of the concrete are acceptable (Tang et al., 2005).

6.6.6 Freeze–Thaw Resistance

A saturated concrete can be damaged when exposed to repeated cycles of freeze–thaw, which can led to expansion (formation of ice) and cracking of concrete. The freeze–thaw resistance of non-air-entrained concrete made with and without 5% GC filler has been assessed by Tang et al. (2005) in accordance with the method described in CEN/TC 51 (1994). The test specimens were immersed in water and subjected to a temperature cycle of −20°C to 20°C in 24 h 56 times. The freeze–thaw resistance of the concrete was expressed in terms of the quantity of scaled material per unit area of the test surface, as shown in Figure 6.9.

After 56 freeze–thaw cycles, the cumulative scaling of concrete made with 5% GC filler as a replacement for fine aggregate was less than 0.01 kg/m^2, whilst that of the reference concrete was 0.03 kg/m^2. This clearly suggests that the use of GC filler increases the freeze–thaw resistance of concrete. A similar finding has also been observed by Klevbo (1998).

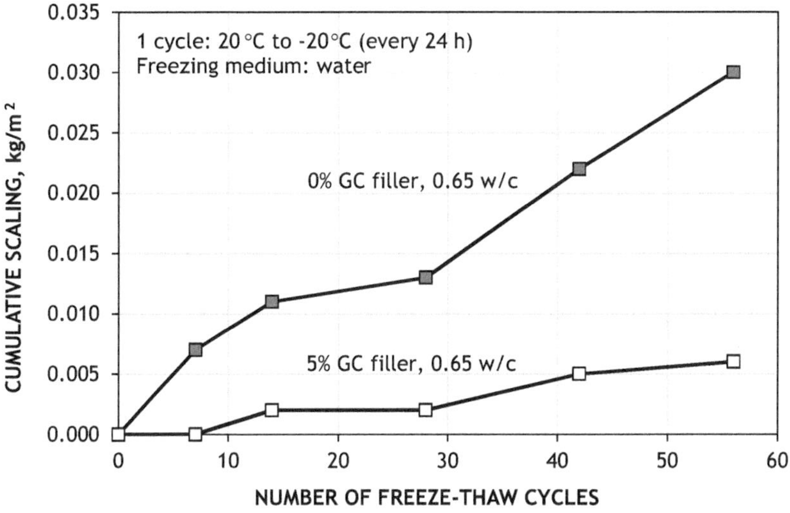

Figure 6.9 Effect of glass cullet filler as fine aggregate on the freeze–thaw resistance of concrete. *GC*, glass cullet; *w/c*, water/cement ratio.
Based on Tang et al. (2005).

6.6.7 Abrasion

Abrasion resistance is particularly important for concrete used in pavements, owing to surface wearing from moving vehicles, and hydraulic structures, owing to the action of solid particles in water. In a study on the abrasion resistance of concrete containing 5% GC filler (Tang et al., 2005), the concrete specimens were subjected to rotating steel abrasive wheels under pressure for 15 min, in accordance with the test method described in Dhir et al. (1991). It was found that the abrasion depth of GC filler concrete was three times smaller than that of the reference concrete, suggesting that the use of GC filler improves the abrasion resistance of concrete.

6.7 Environmental Impact and Case Studies

Information on the environmental impact of concrete containing GC filler has been limited. However, given that GC filler originates from glass, which is an inert material, its use in concrete as part of the constituents should not give rise to environmental concerns in terms of releasing heavy metals, provided that it has been properly processed.

Similarly, case studies associated with the use of GC filler in concrete are lacking. Perhaps this is an area worth considering in the future, to enrich the relevant laboratory findings and facilitate industrial applications, particularly how to achieve optimum blending when GC filler is mixed with fine aggregate during the production of concrete on sites.

6.8 Conclusions

This chapter assesses the performance of GC filler as a fine aggregate replacement, up to 20%, in normal concrete and as a partial or full replacement for filler aggregate, mainly limestone, in SCC. In general, the use of GC in a filler aggregate form in concrete applications, especially in SCC, has not been commonly studied, compared with its use in a ground form as a cement component (Chapter 4) or fine aggregate form (Chapter 5).

When GC filler is used as a replacement for fine aggregate, the consistence of concrete remains unchanged at 5% GC filler content, beyond which it decreases as GC filler content increases up to 30%. No stability issue has been observed when GC filler is used, although it is thought that its small particle size might help to reduce excessive bleeding in concrete. The inclusion of GC filler is not likely to result in significant change in the density, air content or plastic shrinkage of concrete.

In the hardened state, the compressive strength of concrete is found to increase or remain almost unchanged when GC filler is used as a fine aggregate replacement up to 20%. The use of GC filler has shown to improve the long-term strength of concrete, suggesting that the material possesses pozzolanic properties. At a given water/cement ratio, the compressive strength of concrete made with GC filler at 28 days and later is higher than that of concrete made without GC filler. Information on the effect of GC filler on the tensile strength of concrete has been limited, though the available data suggest its positive impact on flexural strength.

At a fixed water/cement ratio, the use of GC filler results in an improvement in the deformation properties of concrete, in the form of elastic modulus and shrinkage. Although information is lacking, concrete containing GC filler is likely to creep less, owing to its void-filling effect.

Increasing the fineness of GC filler by increasing the grinding time could be beneficial to both the permeability and the compressive strength of concrete, but this method needs to be weighed against the energy consumption and cost.

In terms of durability, the use of 5% GC filler as a fine aggregate replacement is found to have a positive effect on the resistance of concrete to carbonation, sulphate attack, freeze–thaw attack and abrasion. No information was available for its performance on the resistance to chloride ingress and acid attack, but it can be safely assumed that these properties are likely to be improved or unaffected when GC filler is used. As far as ASR is concerned, the use of 20% GC filler results in a reduction in expansion at early ages, but a higher expansion than that of reference concrete at later ages. The effect of the colour of GC filler on the ASR expansion of concrete is unclear. It has been suggested that the use of 5% GC filler results in an acceptable ASR expansion.

When GC filler is used as a filler component in SCC, it is found that its inclusion tends to decrease the slump flow of SCC. This negative effect of GC filler on the consistence of concrete might be overcome with the use of a water-reducing admixture. No adverse effect has been found on the stability, density, air content or plastic shrinkage of SCC when GC filler is used.

The use of GC filler at a content no greater than 8.5% (of total content of fine aggregate + filler aggregate) results in an increase in the compressive strength of SCC. Higher content of GC filler can compromise the compressive strength if the water content used is increased to maintain the consistence of SCC. At a fixed water/cement ratio, the use of GC filler increases the flexural strength of SCC. Information on the effect of GC filler on the deformation properties of SCC is thin, although it would be expected that these properties are not likely to be significantly affected.

Similarly, research on the permeation and durability of SCC has been limited. However, where the information is available, it has been shown that the use GC filler in SCC does not adversely affect the sorptivity, oxygen permeability or resistance to chloride ingress. However, its inclusion reduces the resistance to carbonation, which is likely due to its pozzolanic effect.

The environmental impact and case studies associated with the use of GC filler in concrete applications are badly lacking. Notwithstanding this, it can be safely assumed that the material is inert and not likely to release heavy metals, unless it is contaminated. Case studies on this subject are needed to enrich the relevant laboratory findings and facilitate industrial applications.

References

Alexander M and Mindess S, 2005. Aggregates in Concrete. Taylor & Francis, Oxon, UK. 448 pp.

Anagnostopoulos N, Sideris K and Georgiadis A, 2009. Mechanical characteristics of self-compacting concretes with different filler materials, exposed to elevated temperatures. Materials and Structures 42 (10), 1393–1405.

Bignozzi M, Saccani A and Sandrolini F, 2009. Matt waste from glass separated collection: an eco-sustainable addition for new building materials. Waste Management 29 (1), 329–334.

Bignozzi M C and Sandrolini F, 2004. Waste by glass separated collection: a feasible use in cement mortar and concrete. In: Limbachiya M C and Roberts J J (Eds.), Proceedings of the International Conference on Sustainable Waste Management and Recycling: Glass Waste, London, UK. Thomas Telford, London, UK, pp. 117–124.

Bostrom L, 2002. The Performance of Some Self Compacting Concretes When Exposed to Fire. SP Report 2002:23. SP Swedish National Testing and Research Institute, Boras, Sweden. 121 pp.

BRE Digest 330, 2004. Alkali–silica Reaction in Concrete. Part 2-Detailed Guidance for New Construction. BRE, Watford, UK.

BRE Digest 363, 2003a. Concrete in Aggressive Ground. Part 1: Assessing the Aggressive Chemical Environment. BRE, Watford, UK.

BRE Digest 363, 2003b. Concrete in Aggressive Ground. Part 2: Specifying Concrete and Additional Protective Measures. BRE, Watford, UK.

BS 812-123, 1999. Testing Aggregates – Part 123: Method for Determination of Alkali–silica Reactivity – Concrete Prism Method. BSI, London, UK.

BS EN 1992-1-1, 2004. Eurocode 2: Design of Concrete Structures – Part 1-1: General Rules and Rules for Buildings. BSI, London, UK.

BS EN 12350-11, 2010. Testing Fresh Concrete. Part 11: Self-compacting Concrete – Sieve Segregation Test. BSI, London, UK.

BS EN 12620:2002+A1, 2008. Aggregates for Concrete. BSI, London, UK.

BS EN 12350-13, 2013. Testing Hardened Concrete. Part 13: Determination of Secant Modulus of Elasticity in Compression. BSI, London, UK.

CEN/TC 51, 1994. Draft European Prestandard: Test Method for the Freeze-thaw Resistance of Concrete – Tests with Water or Sodium Chloride Solution – Part 1: Scaling – Reference Method B (Cube Test).

Concrete Society, 2005. Self-compacting Concrete – a Review Technical Report No. 62. Concrete Society, Surrey, UK.

Dhir R K, Hewlett P C and Chan Y N, 1991. Near-surface characteristics of concrete: abrasion resistance. Materials and Structures 24, 122–128.

Dhir R K, Dyer T D and Tang M C, 2003. Expansion due to alkali-silica reaction (ASR) of glass cullet sued concrete. In: Dhir R K, Newlands M D and Halliday J E (Eds.), Proceedings of the International Symposium on Recycling and Reuse of Waste Materials, pp. 751–760.

Dhir R K, Dyer T D, Tang M C, Cui Y and Wang L, 2005. Realising a High-value Sustainable Recycling Solution to the Glass Cullet Surplus Closing the Loop. Concrete Technology Unit. Department of Civil Engineering, University of Dundee, Dundee, UK. 155 pp.

Dhir R K, Dyer T D and Tang M C, 2009. Alkali-silica reaction in concrete containing glass. Materials and Structures 42, 1451–1462.

Gaimster R and Dixon N, 2003. Self-compacting concrete. In: Newman J and Choo B S (Eds.), Advanced Concrete Technology. Butterworth-Heinemann, Oxford, UK.

Georgiadis A, Anagnostopoulos N, Sideris K, De Schutter G and Boel V, 2007. Mechanical characteristics of self-compacting concretes produced with different filler materials. In: De Schutter G and Boel V (Eds.), 5th International RILEM Symposium on Self-compacting Concrete, Ghent, Belgium. RILEM Publications SARL, Cachan, France, pp. 611–618.

Jackson N and Dhir R K, 1996. Civil Engineering Materials, fifth ed. Palgrave Macmillan, London, UK. 534 pp.

Karamberi A and Moutsatsou A, 2005. Participation of coloured glass cullet in cementitious materials. Cement and Concrete Composites 27, 319–327.

Klevbo G, 1998. Recycled glass as filler in ready-mixed concrete. In: Proceedings of the 12th European Ready Mixed Concrete Congress, Lisbon, Portugal, pp. 673–680.

Korjakins A, Šahmenko G and Būmanis G, 2009. Boro-silicate glass waste of lamp as a micro-filler for concrete. Scientific Journal of Riga Technical University. Construction Science 10 (10), 131–138.

Korjakins A, Šahmenko G and Būmanis Ģ, 2010. Recycling of boro-silicate glass waste by utilization as filler in concrete. In: Proceedings of 10th International Scientific Conference VSU'2010, Sofia, Bulgaria, pp. 13–18.

Korjakins A, Shakhmenko G, Bajare D and Bumanis G, 2011. Effect of ground glass fineness on physical and mechanical properties of concrete. In: Broekmans M A T M (Ed.), Proceedings of the 10th International Congress for Applied Mineralogy (ICAM), Trondheim, Norway. Springer-Verlag, pp. 395–402.

Lye C Q, Dhir R K and Ghataora G S, 2016. Elastic modulus of concrete made with recycled aggregates. Structures and Buildings 169 (5), 314–339.

Matos A M, Ramos T, Nunes S and Sousa-Coutinho J, 2016. Durability enhancement of SCC with waste glass powder. Materials Research 19 (1), 67–74.

Neville A M, 1994. Properties of Concrete, fourth ed. Longman, Essex, UK. 844 pp.

Persson B, 2000. Shrinkage and strength of self-compacting concrete with different kinds of filler. In: Baroghel-Bouny V and Aïtcin P C (Eds.), International RILEM Workshop on Shrinkage of Concrete, Paris, France. RILEM Publications SARL, Cachan, France, pp. 191–213.

Persson B, 2004. Fire resistance of self-compacting concrete, SCC. Materials and Structures 37 (9), 575–584.

RILEM TC 116, 1999. Technical recommendations: determination of the capillary absorption of water of hardened concrete. Materials and Structures 32 (4), 174–179.

Shakhmenko G, Korjakins A and Bumanis G, 2010. Concrete with microfiller obtained from recycled lamp glass. In: Proceedings of the 10th International Conference on Modern Building Materials, Structures and Techniques, Vilnius, Lithuania, pp. 280–284.

Tang M C, Wang L and Dhir R K, 2005. Engineering and durability properties of concrete containing waste glass. In: Dhir R K, Dyer R D and Newlands M D (Eds.), Proceedings of the International Conference on Achieving Sustainability in Construction, Dundee, UK. Thomas Telford, London, UK, pp. 219–228.

Use of Glass Cullet in Geotechnical Applications

7

Main Headings

- Characteristics of glass cullet
- Geotechnical properties
- Durability
- Environmental impact
- Guidance on using glass cullet

Synopsis

This chapter assesses the performance of glass in geotechnical applications, dealing with the material characteristics, geotechnical properties, durability, environmental impact, case studies and guidance relating to the use of this material. Glass cullet (GC) tends to be used in the coarse sand-to-medium gravel size range and its grading does not change significantly after compaction. The material is almost impermeable and is slightly lighter than natural sand and gravel. The compaction characteristics and permeability of GC are similar to those of natural material. Depending on the size, the frictional angle of GC is about 40–50 degrees. GC is inert and safe to use provided that its debris content is low. GC has been successfully used as a full and partial replacement for sand and gravel in various geotechnical applications such as trench backfill, structural fill and artificial beach/dune fill.

Keywords: Glass cullet, Geotechnical properties, Geotechnical applications, Environmental impact, Case studies.

Sustainable Construction Materials: Glass Cullet. https://doi.org/10.1016/B978-0-08-100984-0.00007-2

7.1 Introduction

Glass cullet (GC), a processed waste glass, can be used as a part of the raw materials in making new glass. This process is known as closed-loop recycling and is normally practiced in the glass manufacturing industry as it reduces the energy consumption required for melting virgin materials. However, such a practice is often limited for several reasons, such as the presence of non-glass materials, commingling of different glass types and mixing of glass colours. Thus, the development of alternative markets for GC is necessary in dealing with this valuable source of construction material.

GC has been used in geotechnical engineering fields since about the 1980s. Much of the early research done in the United States in the 1990s suggested its suitability for use in various applications, including granular layers in road construction, drainage media and structural fills (Dames and Moore, 1993; Nash et al., 1995a,b; TxDOT, 1999). Thereafter, further developments have progressed to materialise a greater potential use of GC in geotechnical applications (Cosentino et al., 1995; PennDOT, 2002; DECC, 2007a). In general, GC may be used at as high as 100% content in the replacement of natural granular materials, for example, pipe bedding and non-load-bearing applications, or as a fill in the construction of embankments. However, in applications in which a structural load is to be supported, the use of GC is limited to a maximum level of about 30%, though a higher content is probably achievable with careful planning and proper design.

To provide a holistic view of the use of GC as a granular material in geotechnical fields, this chapter presents a systematic analysis and evaluation of the relevant published global information dealing with its material characteristics and properties in the context of engineering performance, durability, environmental impact, case studies and guidance relating to this use.

7.2 General Information

Geotechnical-related research on using GC as a granular material dates back to 1984, and has covered a spectrum of work on characterising and measuring the engineering properties of GC for its use in various geotechnical applications. However, the vast majority of this research has been undertaken in the United States and Australia since 2007. Depending on its intended use, GC can be crushed into any size, ranging from dust to medium gravel, though larger-sized GC is likely to cause aggregate packing issues due to its flaky and elongated shape. GC used in research before the year 2000 had a maximum size greater than 19 mm, but after 2000, its maximum size has generally been no more than 9.5 mm (Amlashi et al., 2015). The GC used has tended to be well graded, with a uniformity coefficient (C_u) greater than 4.4.

The maximum dry density (MDD) of GC, determined by both standard and modified compaction methods, has ranged between 15 and 17 kN/m^3 and from 18.0 to 19.5 kN/m^3, respectively (Lee, 2007; Nash et al., 1995a ; Ooi et al., 2008; Soil and Environment Engineers, 1998). It has been found that an increase in compaction effort (as in the modified compaction test) resulted in an insignificant change in both its optimum moisture content (OMC) and its MDD.

GC exhibits a high drained angle of internal friction (φ'), ranging between about 48 and 63 degrees for confining pressures of less than about 100 kPa. A small reduction in φ' was observed as confining pressure was increased (Amlashi et al., 2015). The angle of internal friction of GC can be affected by its grading; for well-graded GC, its values are in the range of 37–43 degrees, whilst for uniformly graded GC, this can be between 31 and 36 degrees (Ooi et al., 2008). Clean GC is not cohesive, but it may exhibit some degree of cohesion if debris and contamination, such as labels, are present (TRB, 2013).

The California bearing ratio (CBR) of GC, with a size similar to that of coarse sand or fine gravel (maximum size of 9.5 mm), is in the range of 40%–80%, which is comparable to that of a typical compacted granular material (Amlashi et al., 2015; Ooi et al., 2008; Soil and Environment Engineers, 1998). However, the CBR value of GC in the form of poorly graded gravel or sand can be very low and no more than 5% (TRB, 2013).

Typically, the permeability of GC ranges from 1×10^{-2} to about 9×10^{-6} m/s depending on its particle size and degree of packing (Amlashi et al., 2015; Lee, 2007; Nash et al., 1995a,b; Soil and Environmental Engineers, 2008). It has been noted that an increase in the largest size of GC has no significant effect on its permeability (Nash et al., 1995a ; Amlashi et al., 2015).

Overall, the engineering properties of GC are comparable to those of natural granular material, and provided the particle size and distribution are properly considered, GC can be used in many geotechnical applications without compromising the technical performance.

7.3 Geotechnical Characteristics of Glass Cullet

One of the most important aspects of using a material in geotechnical engineering is its initial characterisation, such as particle shape, particle size and distribution, specific gravity and consistence. This gives the engineer an initial impression of its suitability for the envisaged applications. If the material is marginal or unsuitable in its current state, the initial characterisation may give an indication of possible options that may be explored to ensure the safe use and compliance of the material with regulatory requirements.

7.3.1 Particle Size Distribution

'Soil' in geotechnical engineering terms describes a material ranging in a size from less than 0.002 mm (clay) to greater than 200 mm (boulder). Between the two extremes lie materials, in an ascending order, described as silt, sand, gravel and cobbles. Each of these terms is associated with a particular size range given in soil classification systems, such as those based on AASHTO M145-91 (2012) and Eurocode 7 (BS EN 1997-2, 2007). The size fractions of soils up to gravel in these two systems are given in Table 7.1, as they are more relevant to the study of GC in geotechnical applications.

The particle size distribution of a soil describes the range and proportion of various sizes of particles, and is typically represented by a graph of cumulative percentage passing a set of sieve sizes. Figure 7.1 shows examples of the particle size distribution of uniformly graded, gap-graded and well-graded soils. Two nondimensional coefficients, namely the C_u and the coefficient of curvature (C_c), as defined below, are commonly used to compare the distribution curves of soils. For well-graded granular soil, its coefficients C_u and C_c are greater than 4 and between 1 and 3, respectively:

$$\text{Coefficient of Uniformity, } C_u = \frac{D_{60}}{D_{10}}$$

$$\text{Cofficient of Curvature, } C_c = \frac{(D_{30})^2}{(D_{60} \times D_{10})}$$

where D_{60}, D_{30} and D_{10} correspond to particle sizes passing at 60%, 30% and 10%, respectively.

Particle Size Distribution of Glass Cullet

The particle size distributions of over 30 GC samples reported from 1993 to 2013 are shown in Figure 7.2, in which the shaded area represents about 80% of the total data. The bulk of the material has a maximum size between 6 and 13 mm and the overall grading lies in the sand and gravel range. In most cases, approximately 2%–7% of the GC is finer than 0.075 mm, with 80% having a D_{50} value between 0.2 and 8.5 mm with

Table 7.1 Description of soils in terms of size

	Range (mm)	
Soil Type	**AASHTO (2004)**	**Eurocode 7 (2007)**
Gravel	2–76.2	2–63
Sand	2–0.075	2–0.063
Silt	0.002–0.075	0.002–0.063
Clay	≤0.002	≤0.002

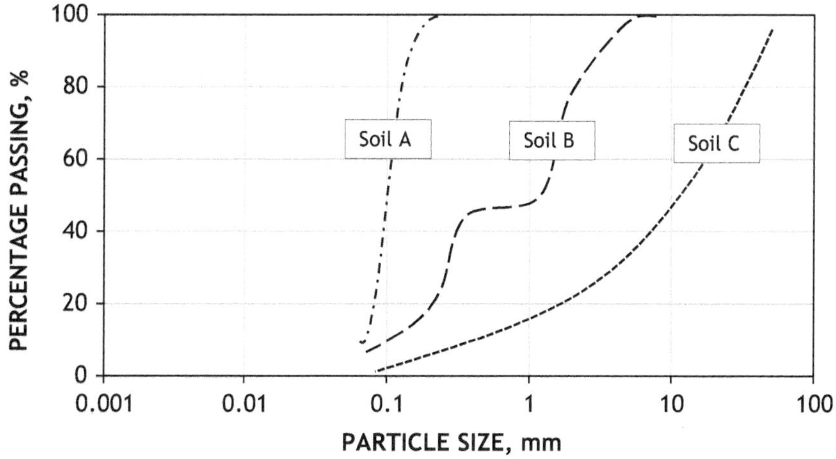

Figure 7.1 Particle size distribution of various soils. Soil A, uniformly graded, where a majority of the particles lie in a narrow range of sizes; soil B, gap graded, where a certain size fraction is missing; soil C, well graded, where a large range of particle sizes is present in good proportions.

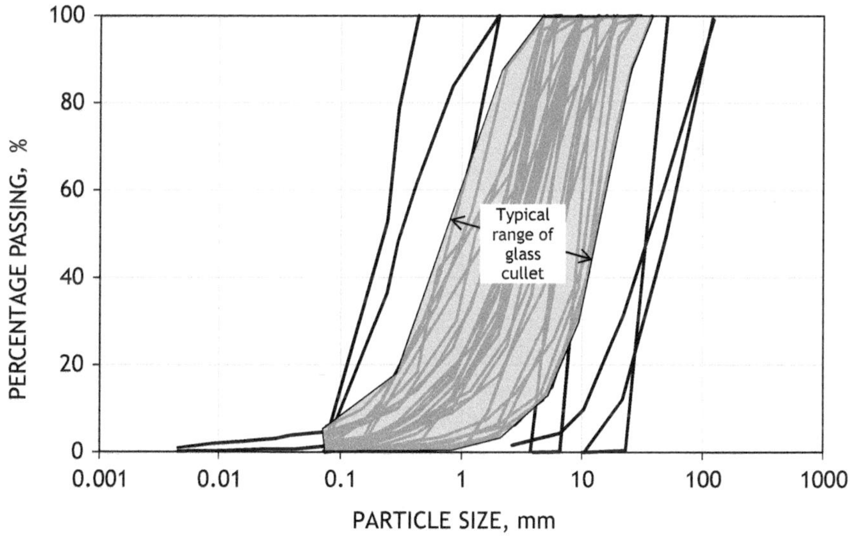

Figure 7.2 Particle size distribution of glass cullet.
Data taken from Arulrajah et al. (2013), Arnold et al. (2008), Basari (2012), Dames and Moore (1993), Cosentino et al. (1995), Eberemu et al. (2013), Finkle and Ksaibati (2007), Grubb et al. (2006b), Henry and Morin (1997), Mavroulidou and Ahmed (2011), Mikami et al. (2009), Nash et al. (1995b), PennDOT (2001a), Piratheepan et al. (2013), SCE Recycling (2013) and Wartman et al. (2004a).

a mean of 2.6 mm. On average, the C_u of the material is 6.5, with each sample normally having a C_u value greater than 4, although in a few cases a value of less than 3 was noted (Basari, 2012; Dames and Moore, 1993). The C_c values for particle size distribution of the GC samples confirm that they had a broader range of particles in the main; the mean value is about 1.57, with less than 10% of the data falling outside the range of 1–3.

It appears that it is possible to produce GC to a range of particle size distributions that may be uniformly graded, well graded or any size range less than about 15 mm, comprising silt, silty sand, sand and gravel. Larger GC particles may be produced but they are more likely to be flat or elongated. Thus, they are not suitable for use as they would have poor compaction properties and, consequently, affect the strength, permeability and other engineering properties of soils.

Particle Size Distribution of Glass Cullet Mixed With Other Materials

A number of researchers have explored the use of GC to modify the engineering properties of other materials such as kaolin clay (PennDOT, 2001b), kaolin and bentonite mix (Malasavage et al., 2007), natural sandy gravel (PennDOT, 2001b), river dredged material (Grubb et al., 2006a,b), crushed limestone (Nash et al., 1995b) and both quarry fines and quarry screenings (PennDOT, 2001b). The host materials to which GC has been added can be divided into two groups: fines, in which a significant proportion of the soil is finer than 0.075 mm (such as silts and clays), and granular materials, in which a significant proportion is larger than 0.075 mm (sand and gravel). In the case of the former, the addition of GC simply adds to the coarser fraction, extending the range of sizes of the soil.

The effect of adding GC to coarse-grained materials on particle size distribution is given in Table 7.2. In general, when the particle size of the host soil is not too dissimilar to that of GC, the effect of adding GC is minimal. However, when the particle sizes of the two materials are different, the addition of GC would fill the missing size fraction of the host soil, thus amending the particle size distribution of the resulting soil.

The changes in the C_u and C_c of sand and gravel materials due to the use of GC, up to 50% content, are shown in Figures 7.3 and 7.4, respectively. The results in Figure 7.3 suggest that as the GC content increases, C_u tends to decrease. This is expected as the C_u of GC ranges from 4.5 to 10, whilst that of the host soils is in the range of 5–88. On the other hand, the C_c of the soils does not show significant change when GC is used (Figure 7.4), and bulk of the results lie in the range 1–3, indicating that the resulting material is well graded.

7.3.2 Effect of Compaction on Particle Size Distribution

Compaction is a process of improving the packing of a material by applying a compaction effort. In a laboratory, this may be done by packing material into a fixed-size mould and applying compaction either by dropping a weight at a given height or by applying vibration and compaction force for a given period. In general, there are two types of compaction tests, namely standard compaction and modified compaction. The latter test equates to application of about 4.5 times the energy per unit volume compared with the former test.

Table 7.2 Particle size distribution of mixes containing glass cullet (GC) and other materials

Reference	Materials	Effect on Particle Size Distribution
Ali and Arulrajah (2012)	Host: recycled concrete aggregate GC (%): 10, 15, 20, 30, 40, 50	Minimal effect as both materials have similar particle size distributions
Ali et al. (2011a)	Host: crushed rock GC (%): 10, 15, 20, 30, 40, 50	Minimal effect as both materials have similar particle size distributions
Eberemu et al. (2013)	Host: laterite (60% finer than 0.075 mm) GC (%): 5, 10, 15, 20	The addition of 20% GC reduces the percentage passing at 0.075 mm to 50%
Grubb et al. (2006a,b)	Host: river dredged material (about 93% finer than 0.075 mm) GC (%): 20, 40, 50, 60, 80	Reduction in the fraction smaller than 0.075 mm; at 80% GC content, particles passing 0.075 mm were reduced to 15%
Malasavage et al. (2007)	Host: kaolinite:bentonite (3:1) mixture (100% finer than 0.075 mm) GC (%): 20, 40, 50, 60, 80	Linear increase in percentage passing 0.075 mm as GC content increases
Nash et al. (1995b)	Host: crushed limestone (gravel) GC (%): 5, 10, 20, 50	Reduction in the fraction larger than 4.75 mm
PennDOT (2001b)	Host: kaolin (clayey silt), King of Prussia soil (sandy gravel), quarry fines (sandy silt) and quarry screenings (sand and gravel) GC (%): 10, 20, 35, 50	Negligible effect on sandy gravel, but increase in the sand fraction of sandy silt and clayey silt

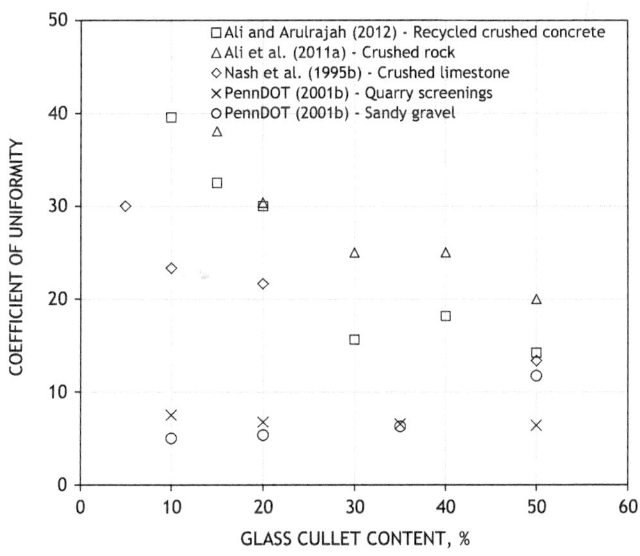

Figure 7.3 Coefficient of uniformity of sand- and gravel-sized materials containing glass cullet.

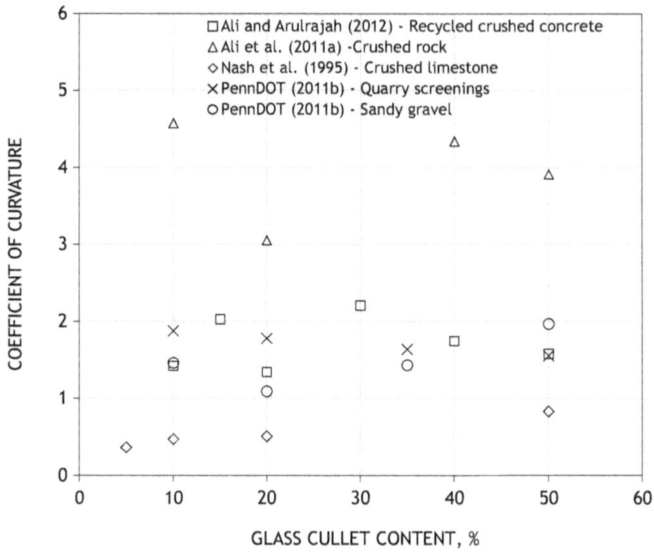

Figure 7.4 Coefficient of curvature of sand- and gravel-sized materials containing glass cullet.

As compaction energy is applied to GC, some breakage of glass particles may be expected. The effect of particle breakage of GC due to compaction energy has been assessed by comparing its particle size distribution curves before and after compaction (Dames and Moore, 1993; PennDOT, 2001a; Disfani et al., 2011b; Ali and Arulrajah, 2012). In general, the results suggest that there is a typical 2.5%–3.5% increase in fines, mainly in the sand-size fraction; and most of the breakage was in the coarser fraction (gravel size) of the material. More breakage was observed in the samples prepared using modified compaction.

Overall, there was little change in either the C_u or the C_c value, except for in two cases (PennDOT, 2001a; Wartman et al., 2004a), in which the breakage of GC particles was up to about 10%. Notwithstanding this, in cases in which the structural performance or the permeability of a soil is critical, the effect of breakage of the GC should be carefully examined.

7.3.3 Shape

BS EN ISO 14688-1 (2002) gives six descriptive terms for the shape of particles, namely, very angular, angular, subangular, subrounded, rounded and well rounded. Since GC is produced by crushing and processing of waste glass, the shape of GC is expected to be angular (Dames and Moore, 1993; So et al., 2015). However, since glass is brittle, its sharp edges might break due to particle attrition, resulting in a change in shape from angular to subrounded.

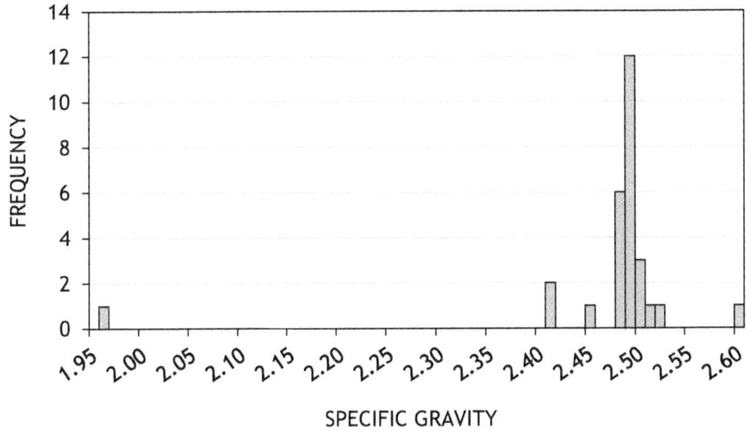

Figure 7.5 Specific gravity of glass cullet samples used in geotechnical studies.
Data taken from Arulrajah et al. (2013, 2014), Ali et al. (2011a,b), Amlashi et al., 2015, Basari (2012), Cosentino et al. (1995), CWC (1996a), Dames and Moore (1993), Disfani et al. (2011a), Eberemu et al. (2013), Grubb et al. (2007), Hagerty et al. (1993), Kang et al. (2009), Kunishima et al. (2000), Li et al. (2013), Malasavage et al. (2007), Mavroulidou and Ahmed (2011), PennDOT (2001a, 2002, 2006), Shin and Sonntag (1994) and Wartman et al. (2004a,b).

7.3.4 Flakiness Index

A particle is defined as flaky if the smallest dimension (i.e., thickness) is less than 0.6 times the mean size of the other two dimensions. Flakiness index measures the percentage of flaky particles present in a sample (BS 812-105.1, 1989). The flakiness index of an aggregate is normally examined for its use in concrete and asphalt mixes, in which a high content of flaky particles can affect the workability of the mixes. Since flakiness index is not applicable to aggregates of less than 6.3 mm and GC is more suited to be used as a fine aggregate, the flakiness index of GC is unlikely to be a concern.

7.3.5 Specific Gravity

Specific gravity is defined as the ratio of the density of the solid part of a material to the density of water at 20°C. Typically, the specific gravity of soils is in the range 2.60 to about 2.80. The specific gravity of GC samples used in geotechnical studies has been found to range from 1.96 to 2.54, with most values between 2.48 and 2.50 (Figure 7.5). The variation in the value is probably due to the presence of debris or organic matter in the GC samples. Notwithstanding this, excluding the two extreme data points, the mean specific gravity of GC is about 2.49, which is 6% lower than that of quartzite sand (2.65).

7.3.6 Water Absorption

The water absorption of an aggregate can be used as a screening test for its freeze–thaw resistance. Aggregates with water absorption less than 0.5% are generally considered to be resistant to freeze–thaw action in practice, though in BS EN 13242:2002+A1 (2007) this value is set to be no greater than 2%.

As glass is an impermeable material, its water absorption values are very low, ranging from 0.2% to 1% (ARRB, 2010; Arulrajah et al., 2013; Ooi et al., 2008), although a value close to 0% may be more appropriate. The presence of debris or organic content in GC may result in higher water absorption value.

7.3.7 Los Angeles Abrasion Value

The Los Angeles abrasion test measures the resistance of an aggregate to abrasion, which can be considered as a durability test. This test is essentially applicable to coarse aggregate used in road pavement. Depending on the intended use, typical acceptable Los Angeles values of aggregate are in the range of 30%–50% (Highways Agency, 2016; INDOT, 2012), though this limit may be as high as 60% (SCDOT, 2011). Materials with lower Los Angeles values are normally used in the upper layers of a road pavement owing to the high stress level.

Figure 7.6 shows the Los Angeles abrasion test results for GC samples used in geotechnical studies. Most of the values fall within the range of 24%–42%, except for one, which is at 52%. Although GC appears to have an acceptable Los Angeles value, as mentioned previously, the material is not suitable for use in a coarse form owing to the potential flakiness and elongation issues.

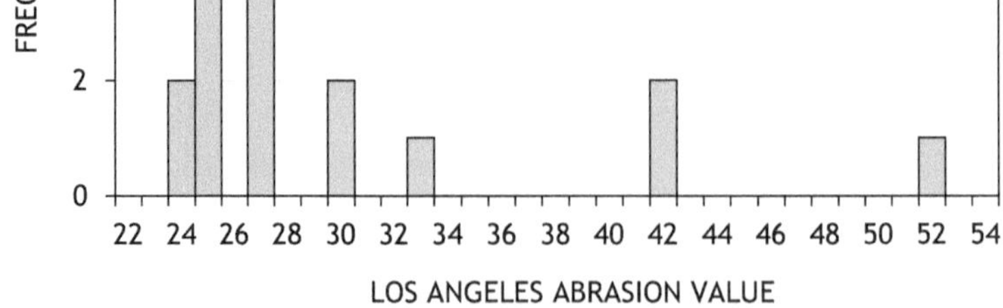

Figure 7.6 Los Angeles abrasion values for glass cullet samples used in geotechnical studies. Data taken from Ali and Arulrajah (2012), Ali et al. (2011a,b), Arulrajah et al. (2013, 2014), Disfani et al. (2011a,b, 2012), Henry and Morin (1997), PennDOT (2001a,b) and Wartman et al. (2004a).

7.3.8 Soundness

The soundness test is a measure of the resistance of both coarse and fine aggregates to weathering by immersing the aggregates in either sodium sulphate or magnesium sulphate solution for a specific period of time (ASTM C88, 2013). In the case of the sodium sulphate soundness test, the minimum acceptable soundness value is 12%; for magnesium sulphate, this may be 18% (Prowell et al., 2005). The soundness test is commonly regarded as a proxy for the freeze–thaw resistance test. The results of soundness tests for GC, using a sodium sulphate solution, have been not greater than 7.5% (SCE Recycling, 2013; Wartman, 2004a), which is well within the acceptable limits, indicating that the material has good freeze–thaw resistance.

7.3.9 pH

The pH of a soil can be affected by its mineralogy and solubility as well as the presence of organic content. Standards such as ASTM D4972 (2001) and BS 1377-3 (1990) provide specific guidance on measuring the pH of soil. Although glass is considered to be an inert material, the sodium component of soda lime GC may leach from the glass and elevate its alkalinity. Limited information on the pH of GC suggests that its pH may be as high as about 10 (Disfani et al., 2011a,b, 2012). The pH of GC should be checked from a leachate point of view, particularly if it is used as a fill material adjacent to galvanised steel or steel structures in earthworks. In the United Kingdom, the pH of fill materials in such an application is limited to within 5–10 (Highways Agency, 2016).

7.3.10 Organic Content

High organic matter in soils makes them more susceptible to settlement under loading and may also result in a reduction in strength. Unless properly cleaned during the manufacturing process, GC will inevitably contain some organic matter arising from paper and food residues. The organic content of GC has been reported to be as low as 0.5% (Ali and Arulrajah, 2012) to as high as 3.1% (Grubb et al., 2008a), and typically it is less than about 1.3% (Ali and Arulrajah, 2012; Ali et al., 2011a,b; Arulrajah et al., 2013, 2014; Disfani et al., 2011a,b, 2012).

It appears that the organic content of GC is less than that of the most productive agricultural soils, which have a typical value ranging from 3% to 6%. In many states in the United States, the maximum values of organic matter in road subgrade and backfills are limited to 2%–7% (Huang et al., 2009). Thus, the utilisation of GC in such applications would be acceptable.

7.3.11 Atterberg Limits

In terms of soil characterisation for geotechnical purposes, the liquid limit (LL), i.e., the moisture content at which a soil changes from liquid to plastic, and the plastic limit (PL), i.e., from plastic to semisolid, are the key indices. They may be used as a rough proxy for the strength of a soil, as the change from LL to PL can give rise to an ~100-fold increase in strength. The difference in these limits is known as the plastic index (PI=LL−PL).

A number of researchers (Basari, 2012; Eberemu et al., 2013; Fauzi et al., 2016; Grubb et al., 2006b; Malasavage et al., 2007) have used up to 80% GC to modify the engineering properties of cohesive soils. The changes in LL, PL and PI of soils due to the use of GC are shown in Figures 7.7–7.9, respectively. In general, an increase in GC content results in a 10%–20% reduction in LL, PL and PI.

The significance of these changes can be seen in a Casagrande chart plotted in Figure 7.10, in which the results for the maximum GC used in the relevant studies are shown. It can be seen that the addition of GC tends to reduce the plasticity of soils. Soils treated with GC are a little coarser and have higher permeability, and for a given moisture content and density, their strength may be higher than that of untreated soils. It should be mentioned that the response of the soils observed in Figure 7.10 is similar to that when lime is added.

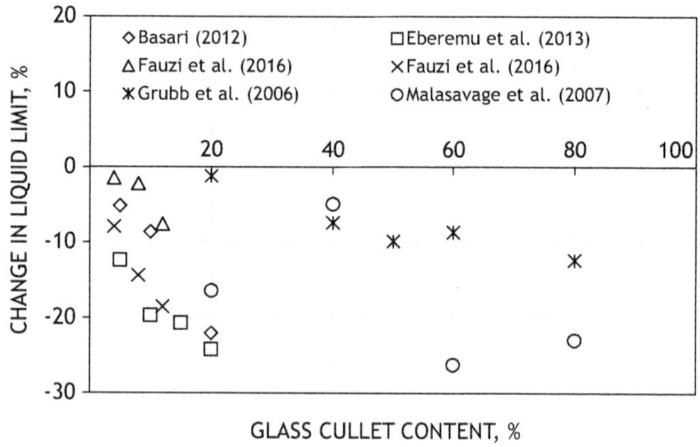

Figure 7.7 Change in liquid limit of cohesive soils with glass cullet content.

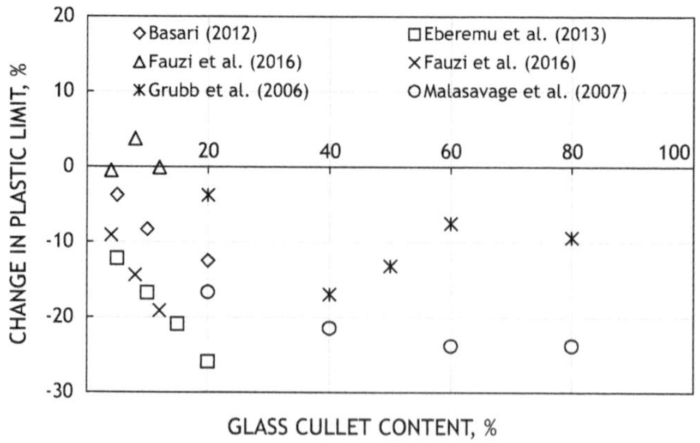

Figure 7.8 Change in plastic limit of cohesive soils with glass cullet content.

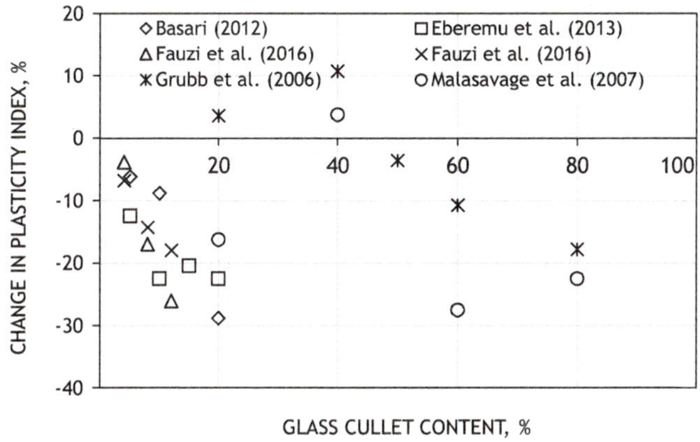

Figure 7.9 Change in plasticity index of cohesive soils with glass cullet content.

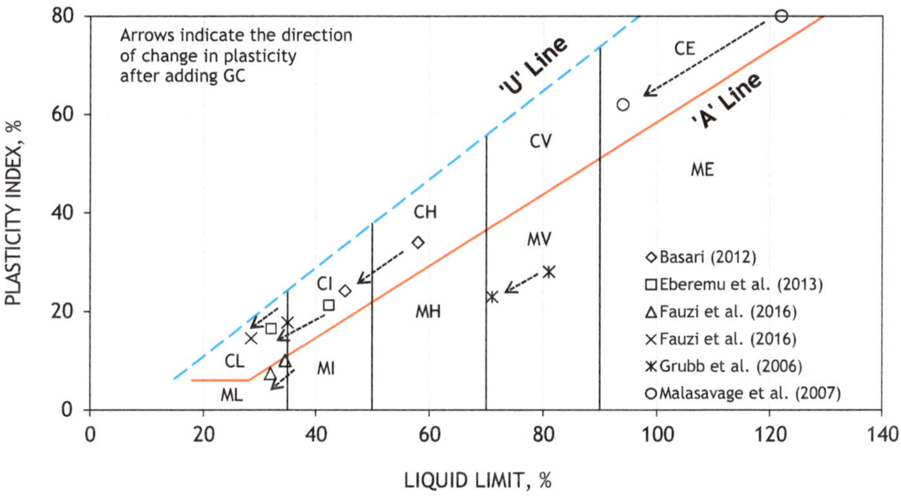

Figure 7.10 Plasticity of glass cullet (GC) on a Casagrande chart. *C*, clay; *M*, silty; *L*, low plasticity; *I*, intermediate plasticity; *H*, high plasticity; *V*, very high plasticity; *E*, extremely high plasticity. Organic materials localise below the 'A' line.

7.4 Mechanical Properties

7.4.1 Compaction

Compaction is a densification process of packing soil particles together to improve the engineering properties of the soil, particularly strength and bearing capacity. The degree of packing of soil particles is dependent on the soil type (such as clay, silty and sand), the range of particle sizes, the water content and the amount of compaction effort applied. The compaction test results are normally presented in terms of the dry density and moisture content relationship, which suggests the maximum amount of dry solid material that can be packed into a given volume (known as MDD) and the moisture content (known as OMC) at which this can be achieved for a given compaction effort.

The MDD and OMC of GC with a maximum size of 6 or 19 mm, obtained using both the standard and the modified compaction tests, are shown in Figure 7.11. The results show that the average MDD of GC using the modified compaction method is about 9.5% greater than that achieved using the standard compaction method. The OMC of GC is in the range of about 5%–13% for both the standard and the modified compaction tests.

The effects of GC on the MDD and OMC of either fine-grained or coarse-grained host material are shown in Figures 7.12 and 7.13, respectively. For coarse-grained (granular) soils, the addition of GC up to 100% content does not show any significant change in the MDD. The OMC of granular host soils remains almost unchanged up to 35%, beyond which it decreases as GC content increases. However, this reduction in OMC is considered to be small.

For fine-grained soils, the MDD increases as GC content increases, and this effect becomes more pronounced when the GC content is greater than 35%. The addition of GC can result in a significant reduction on the OMC of fine-grained soils; in some cases, this reduction can be greater than 50% when more than 40% GC is added (Figure 7.13).

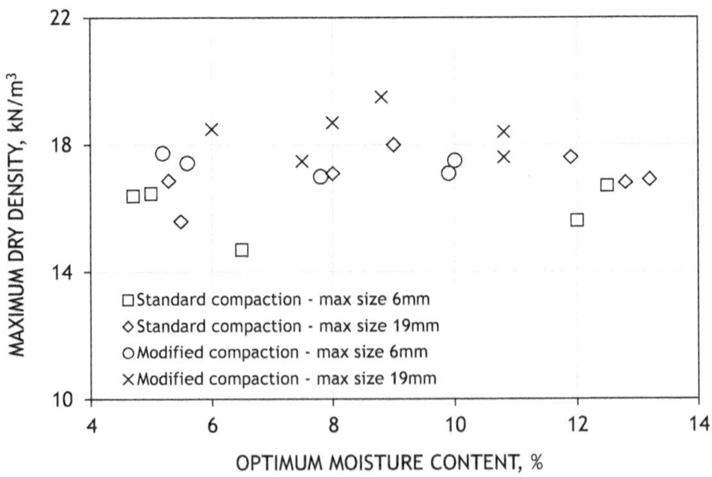

Figure 7.11 Maximum dry density and optimum moisture content of glass cullet at 6 and 19 mm maximum size.
Data taken from Dames and Moore (1993), Disfani et al. (2011a,b, 2012), Grubb et al. (2006a), PennDOT (2001a, 2001b), PennDOT (2006) and Wartman (2004a).

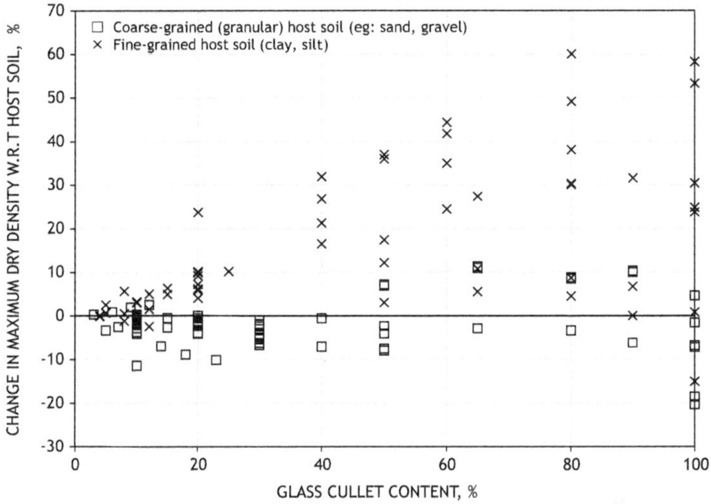

Figure 7.12 Effect of glass cullet on maximum dry density of both coarse- and fine-grained soils. Data taken from Ali and Arulrajah (2012), Ali et al. (2011a,b), Arulrajah et al. (2013, 2014), Basari (2012), Davidovic et al. (2012), Eberemu et al. (2013), Fauzi et al. (2016), Finkle and Ksaibati (2007), Grubb et al. (2006a), Malasavage et al. (2007), Nash et al. (1995b), Nuruzzaman and Hossain (2014), PennDOT (2001a), Vishnu and Chandrakaran (2016) and Wartman (2004a).

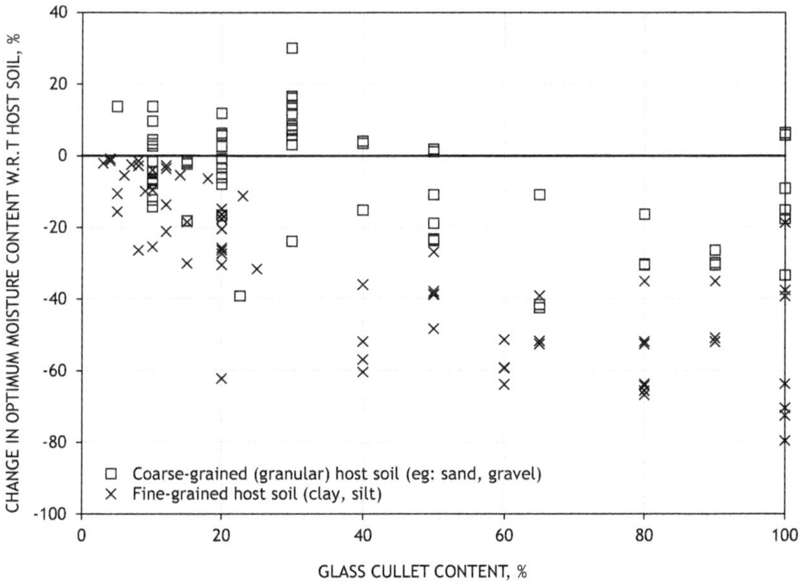

Figure 7.13 Effect of glass cullet on optimum moisture content of both coarse- and fine-grained soils.
Data taken from Ali and Arulrajah (2012), Ali et al. (2011a,b), Arulrajah et al. (2013, 2014), Basari (2012), Davidovic et al. (2012), Eberemu et al. (2013), Fauzi et al. (2016), Finkle and Ksaibati (2007), Grubb et al. (2006a), Malasavage et al. (2007), Nash et al. (1995b), Nuruzzaman and Hossain (2014) PennDOT (2001a), Vishnu and Chandrakaran (2016) and Wartman (2004a).

7.4.2 California Bearing Ratio

The CBR test was developed in the 1920s to assess the strength of road pavement construction materials, namely, subgrade, subbase and base course. It is a simple test that can be conducted on samples prepared in the laboratory and the field. CBR values have been used to determine the thickness of pavement layers universally in most road design methods.

Investigations into the determination of the CBR of GC can be divided into two groups: (i) GC only and (ii) a mixture of GC and soils (both fine and coarse grained). The relationship between CBR and MDD for GC specimens of maximum size 3–19 mm, prepared using both standard and modified compaction methods, is shown in Figure 7.14. The minimum CBR requirements for capping, subbase and base course layers given by the US Army (1994a) are also shown in the figure for comparison.

It can be seen that specimens prepared using the modified compaction tend to have a higher CBR compared with those prepared using the standard compaction technique. This is to be expected, as the compaction effort used in modified compaction is about 4.5 times that of standard compaction, resulting in a greater MDD and thus higher CBR values. Overall, it appears from Figure 7.14 that the CBR of GC can range from as low as 10% (at MDD of 15.5 kN/m^3) to as high as 75% (at MDD of 19.5 kN/m^3). Given that the MDD of GC normally fluctuates within 16–18 kN/m^3 (Figure 7.11), the CBR of GC is likely to be in the range of 20%–40%. Thus the material is suitable for use in capping and subbase layer construction.

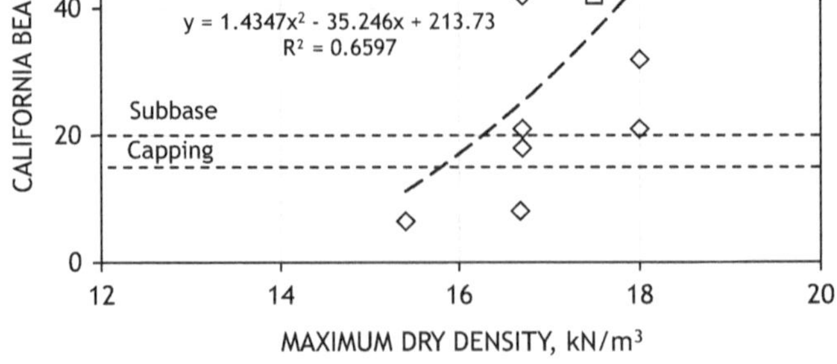

Figure 7.14 Relationship of California bearing ratio and maximum dry density of glass cullet.
Data taken from Disfani et al. (2011a,b, 2012), SCE Recycling (2013) and So et al. (2015).

Table 7.3 Strain to failure of materials comprising glass cullet

Reference	Material	Density (kN/m^3)	Confining Pressure (kPa)	Strain to Failure (%)
Dames and Moore (1993)	Crushed rock	95% MDD	5	2.0
			10	2.0
			20	2.7
	GC 15% and crushed rock 85% Maximum size 6.3 mm	95% MDD	5	2.8
			10	4.5
			20	5.0
	GC 15% and crushed rock 85% Maximum size 19 mm	95% MDD	5	3.0
			10	2.5
			20	4.5
	GC 50% and crushed rock 50% Maximum size 19 mm	95% MDD	5	2.5
			10	5.0
			20	5.5
Ali and Arulrajah (2012)	GC 10% and recycled crushed concrete 90%	MDD	50	1.6
			100	2.0
			200	2.5

GC, glass cullet; *MDD*, maximum dry density.

7.4.3 Stress–Strain Behaviour

In general, the stress–strain behaviour of soils may be grouped into two categories, namely, (i) dense sand and overconsolidated clays and (ii) loose sand and normally consolidated clays. Materials in the former category tend to dilate, whilst those in the latter tend to compress as the soils start to shear during loading to peak stress.

The limited data on strain to failure for GC (PennDOT, 2001a) suggest that a relatively loosely compacted GC (at 90% MDD) required strains of 11%–18% to reach peak strength. This behaviour is close to that of soils comprising loose and normally consolidated material, which can deform up to 20% or more to reach peak strength. The effect of blending up to 50% GC with crushed rock and crushed recycled concrete aggregate on strain to failure has been assessed by Dames and Moore (1993) and Ali and Arulrajah (2012), respectively. The results from these investigations, as given in Table 7.3, suggest that the strain to failure of well-compacted specimens containing GC is generally less than about 5%. This might indicate that the stress–strain behaviour of GC is not too dissimilar to that of granular materials.

7.4.4 Shear Strength

The shear strength of a soil is the maximum stress a soil can resist without shear failure. It has two components, cohesion and friction (internal friction angle), and both can be measured in terms of (i) total stress conditions, where no volume change is permitted, or (ii) effective stress conditions, where volume change can occur through the drainage of water within the test soil.

Shear Strength of Glass Cullet

Since GC is a granular material, its shear strength is measured in terms of effective angle of friction (ϕ′), which can be determined using either a direct shear test in accordance with ASTM D3080 (2011), BS 1377-7 (1990) or BS 1377-8 (1990) or a triaxial test in accordance with ASTM D2166 (2016), ASTM D2850 (2015), BS 1377-7 (1990) and BS 1377-8 (1990).

The angle of friction of a granular soil can be affected by particle shape and dry density, void ratio and confining pressure. For example, the ϕ′ of angular sand may be as low as 30 degrees in a loose state, but greater than 45 degrees in a dense state. The ϕ′ of rounded sand is at least 10% lower than that of angular sand.

The shear strength of GC has been investigated by a number of researchers under different test conditions. This has resulted in a large range of ϕ′ values, which can be divided into two groups based on the maximum aggregate size of the GC used, that is, (i) up to 6.3 mm, shown in Figure 7.15, and (ii) between 9 and 20 mm, shown in Figure 7.16. In the former group, the reported ϕ′ values of GC prepared at densities ranging from 12.8 to 18.4 kN/m^3 are spread over a wide range of 34–68 degrees, but mostly within the range of 40–48 degrees. In the latter group, the ϕ′ values of GC vary from 44 to 62 degrees, showing a near-normal distribution with a mean of about 50 degrees, and nearly 70% of the results are in the range from 46 to 54 degrees.

The void ratio is the ratio of volume of voids to volume of solids. Typically, well-graded sand and gravel may have a void ratio that varies from about 0.75 for poorly compacted soil with low dry density to 0.25 for well-compacted soil with high dry density. The information available on the relationship between angle of friction and

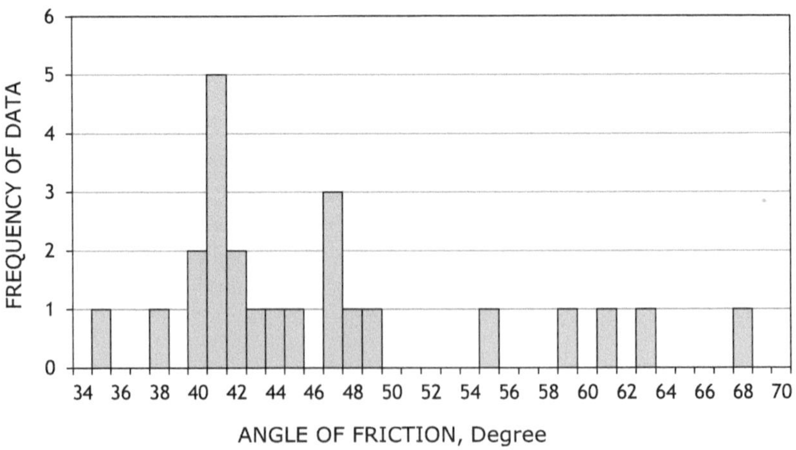

Figure 7.15 Angle of friction for glass cullet with a maximum size of 6.3 mm.
Data taken from Dames and Moore (1993), Disfani et al. (2011a,b, 2012), SCE Recycling (2013), So et al. (2015) and Wartman et al. (2004a).

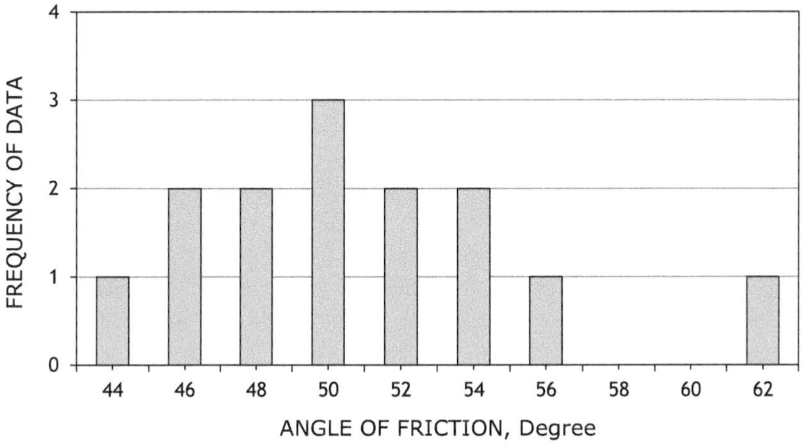

Figure 7.16 Angle of friction for glass cullet with a maximum size of 9–20 mm. Data taken from Disfani et al. (2011a), PennDOT (2001a), Piratheepan et al. (2013) and So et al. (2015).

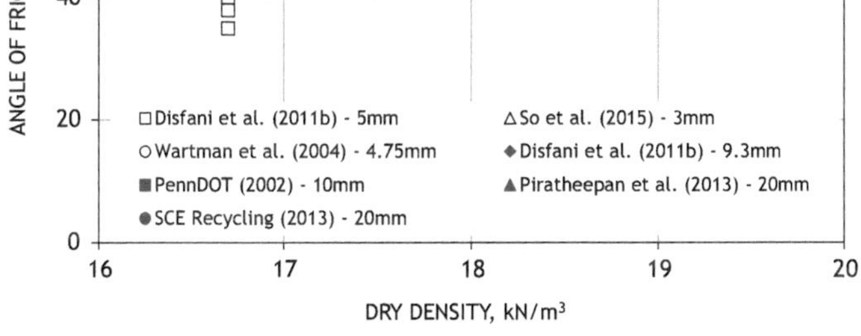

Figure 7.17 Relationship between angle of friction and dry density of glass cullet.

void ratio for GC is limited. Notwithstanding this, the relationship may be established between angle of friction and dry density (Figure 7.17), since the density of a soil is related to its voids volume. It appears that there is a general increasing trend in angle of friction of GC when its density increases from about 16.75 to 18.3 kN/m^3.

It was also found that the angle of friction of GC tended to decrease as the confining pressure increased from 30 to 300 kPa (Disfani et al., 2011a, 2011b, 2012; Wartman et al., 2004a). This observation is somewhat similar to that reported for cohesionless soils (Lamb and Whitman, 1979).

Shear Strength of Soils Containing Glass Cullet

GC has been used to modify the properties of other materials such as gravelly sand and crushed rock (Dames and Moore, 1993), kaolin, sandy silt and silty sand (PennDOT, 2001b), dredged materials (Grubb et al., 2006b), natural sand (Mavroulidou and Ahmed, 2011), sand (Basari, 2012) and laterite (Eberemu et al., 2013). The effects of GC on both cohesion and angle of friction have been expressed in relative terms with respect to the host material (the reference), as shown in Figures 7.18 and 7.19, respectively. In general, the addition of GC to these soils results in a reduction in cohesion. However, if the host material has low plasticity, such as quarry fines, as to be expected, GC has little effect on its cohesion (Figure 7.18). Depending on the angle of friction of the host material and the quantity of GC added, Figure 7.19 shows that the use of GC (i) increases the angle of friction of kaolin, laterite, silty sand and sandy silt, (ii) decreases the angle of friction of quarry fines and (iii) has no significant effect on the angle of friction of sand or a mixture of sand and gravel.

The information on the shear strength of GC tested in unconsolidated undrained condition has been limited. One study, however, undertaken by Grubb at al. (2006a,b), suggests that the use of 20%–80% GC in dredged materials resulted in 80% reduction in cohesion and 70% increase in angle of friction.

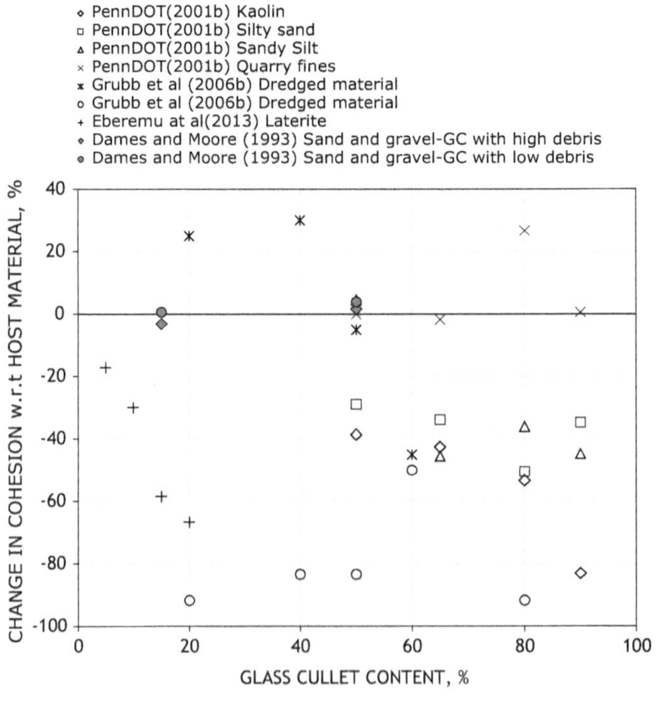

Figure 7.18 Effect of glass cullet on cohesion of host material.

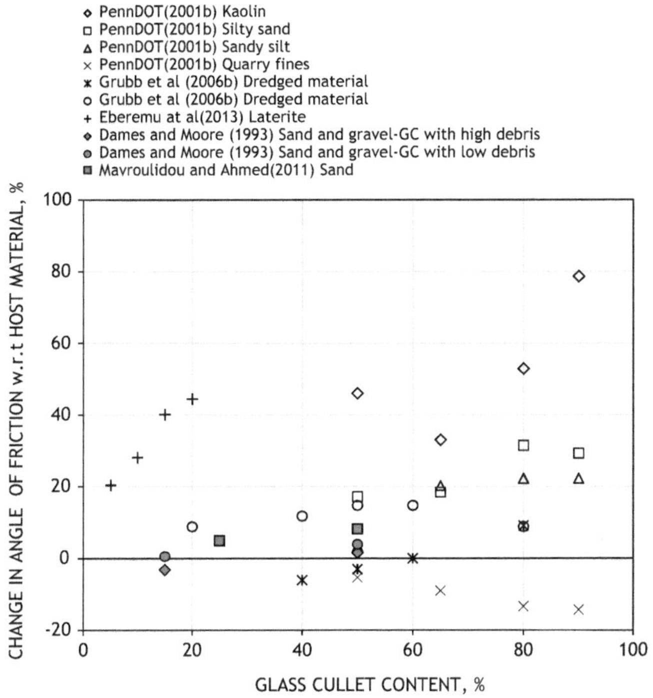

Figure 7.19 Effect of glass cullet on angle of friction of host material.

7.4.5 Soil Modulus

The often-used term 'modulus' may be defined in many ways, such as tangent, secant, chord, shear, cyclic, bulk, load or unload modulus, all of which can be determined from different parts of the stress–strain curve of a soil. Under a monotonic loading, perhaps the most pertinent measure is secant modulus, which is the slope of a straight line joining the origin to a point between 0.33 and 0.50 peak stress as recommended by Lamb and Whitman (1979). Modulus of soil is used for estimating the initial settlement of a soil when subjected to a load. Typically, a firm, becoming stiff, clay can have a modulus value that ranges from about 20 to >100 MPa; for loose sand and dense sand, the corresponding ranges are 10–30 and 30–70 MPa.

Figure 7.20 shows the relationship between confining pressure and secant modulus, at one-third of the peak strength, of GC samples for different sizes: <4.76, <9.7, <20 (as received) and 2.4–20.0 mm. It can be seen that the secant modulus of GC increases, at a decreasing rate, as the confining pressure increases up to 480 kPa. A similar trend is also observed in the relationship between confining pressure and initial tangent modulus for recycled concrete aggregate mix containing 10% GC (Ali and Arulrajah, 2012) and crushed-rock mixes containing 50% GC (Dames and Moore, 1993).

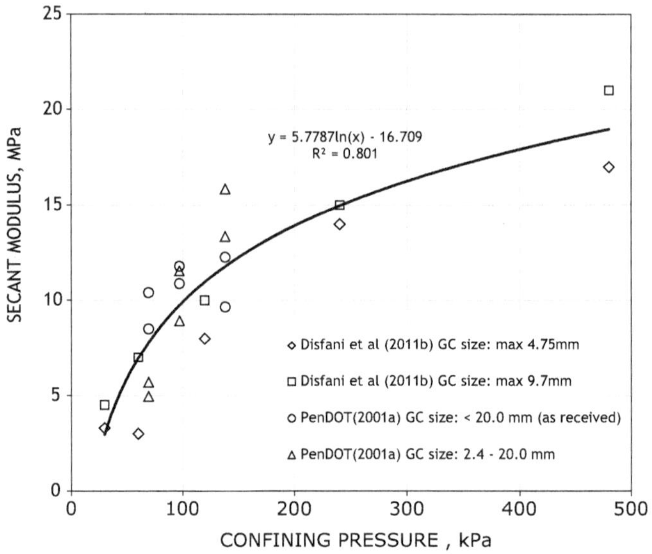

Figure 7.20 Relationship between confining pressure and secant modulus (at 0.33 × peak stress) of glass cullet for different sizes.

7.4.6 Resilient Modulus

The resilient modulus (M_R) is a measure of the elastic deformation of a soil under repeated or cyclic loading. In the limited research undertaken, the effects of adding 15% and 50% GC, with a maximum size of 6.35 and 19 mm, on the M_R of crushed rock has been evaluated using a modified AASHTO T294 (1994) test procedure (Dames and Moore, 1993). As the CBR test is easier and cheaper compared with the M_R test, the M_R values of crushed-rock mixes containing GC are plotted against the corresponding CBR values in Figure 7.21, together with the commonly used relationship between the two established by Powell et al. (1984).

The use of GC with maximum sizes of 6.35 and 19 mm, for a content up to 50%, led to a reduction in the M_R of crushed rock, with an average M_R value about 15% lower than that of the reference with 100% crushed rock (Figure 7.21). Notwithstanding this, it has been claimed that the M_R value of mixes containing up to 50% GC was still considered to be appropriate for use in a typical road pavement design (Dames and Moore, 1993).

In addition, the results shown in Figure 7.21 also suggest that the CBR and M_R relationship developed by Powell et al. (1984) does not fit well for 100% crushed-rock mixes and crushed-rock mixes containing GC. Perhaps further work in this area is required.

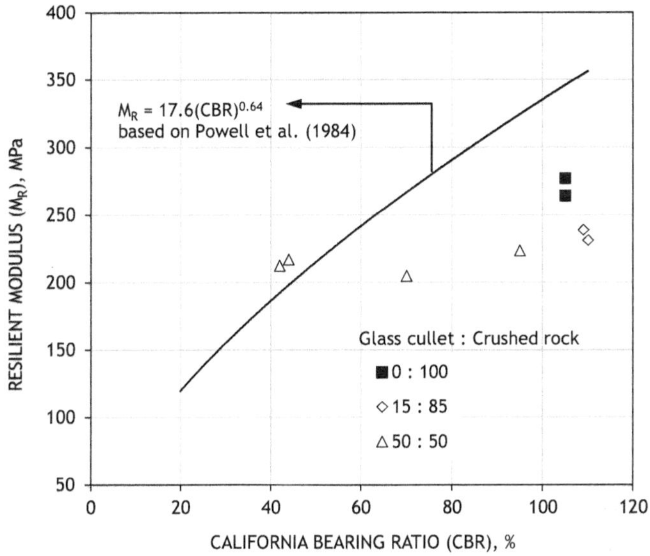

Figure 7.21 Relationship of California bearing ratio and resilient modulus of crushed-rock mixes containing glass cullet.
Based on Dames and Moore (1993).

7.4.7 Permeability

Soil is a three-phase material comprising solids (soil particles), air and water, the last occupying the void space between the particles. The ease with which water can flow through a soil is described as its permeability. The permeability of a soil is normally expressed as the coefficient of permeability (or hydraulic conductivity), which is defined as the quantity of flow through a unit area of soil under a unit pressure gradient.

Figure 7.22 shows the permeability of GC and soils containing up to 80% GC, together with the relevant types of permeability test and the typical permeability values and classifications. The permeability of GC, determined using the constant head test in accordance with ASTM D2434-68 (2006), has been found to range from 1.0×10^{-3} to 1.0×10^{-7} m/s, having an average value of 1.4×10^{-3} m/s. Compared with other types of soil, the permeability of GC is similar to that of coarse sand and fine sand/coarse silt, though it tends to behave like the former (Figure 7.22). In general, the permeability of GC is somewhere between high and medium, which suggests that it would provide good drainage properties and get rid of the potential buildup of excessive pore water pressure when used as a fill material.

When adding GC at up to 60% content to other coarse-grained and fine-grained soils, although it appears that the permeability of the resulting soils might be a few orders of magnitude away from that of the reference soils (without GC), the permeability behaviour and classification of the resulting soils remain essentially unchanged (Figure 7.22).

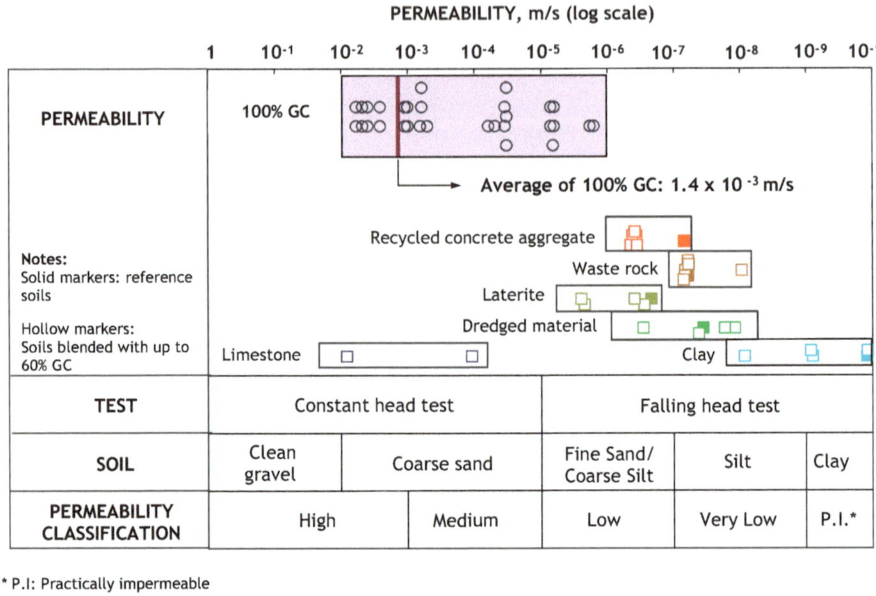

Figure 7.22 Permeability of glass cullet (GC) and soils containing GC.
Data taken from Ali and Arulrajah (2012), Ali et al. (2011a), Arulrajah et al. (2014), Bo and Arulrajah (2011), CWC, 1996b, Disfani et al. (2011a), Eberemu et al. (2013), Grubb et al. (2006b), Kunishima et al. (2000), Malasavage et al. (2007), Mikami et al. (2009), Nash et al. (1995b), PennDOT (2002), PennDOT (2006), Reddy (1999), So et al. (2015) and Wartman et al. (2004a).

7.4.8 Thermal Conductivity

The thermal conductivity of a material is defined as the amount of heat that can be transmitted per unit temperature gradient through unit thickness of a material, measured perpendicular to the source of heat. Conductivity can be measured using the steady-state method (ASTM C518, 2015) or transient-state method (ASTM Standard D5334, 2008), depending on the type of application. Apart from the inherent characteristics of a material, both the dry density and the degree of saturation are perhaps the most important factors affecting thermal conductivity, since the former relates to the packing of particles, which affects heat transfer through conduction, and the latter relates to heat transfer through convection. Thus, the thermal conductivity of a material increases as both its dry density and the degree of saturation increase.

The thermal conductivity of intact glass is about 1 W/mK. In the case of GC, its heat conductivity relies on particle-to-particle contact and any water present between the particles. Table 7.4 presents the thermal conductivity of GC at different moisture contents and that of other granular materials and geothermal grout. The results suggest

Table 7.4 Thermal conductivity of glass cullet (GC), natural granular materials and geothermal grout

Material	Aggregate Size	Thermal Conductivity, W/mK	Reference
GC (moisture 10.7%)	6.4 mm	0.315	Dames and Moore (1993)
GC (moisture 6.5%)	6.4 mm	0.26	
GC mixed with sand and gravel 50:50 (moisture 7.4%)	6.4 mm	0.463	
Sand and gravel	n.a.	0.638	
Sand—dry	Fine to coarse	0.15–0.27	Hamdhan and Clarke (2010)
Sand—saturated	Fine to coarse	2.75–3.72	
Sand and gravel	n.a.	0.77[a]	British Geological Survey (2011)
Saturated sand	n.a.	2.50[a]	
Silt	n.a.	1.67[a]	
Geothermal grout	n.a.	0.68–1.69	cetco.com
Geothermal grout—sand/cement	n.a.	1.5–2.3	Allan (1997)
Geothermal grout—sand/cement/ blast furnace slag or fly ash	n.a.	2.2–2.4	

[a]Typical values.

that the thermal conductivity of compacted GC with an approximate 6%–10% moisture content is about a quarter to a third of that of intact glass. The thermal conductivity of GC is marginally greater than that of dry sand but significantly smaller than that of saturated sand. It is reasonable to surmise that the thermal conductivity of GC is essentially similar to that of natural sand; thus the material may be potentially suitable for use as a replacement for sand in thermal grout. However, further exploration of this area would be needed.

7.5 Durability

Soils comprising various proportions of clays, silts, sands and gravels are used as foundation materials for most construction (buildings, roads, etc.) and at times are used to form structures such as embankments, levees and dams. These constructions are required to remain structurally sound over their design life, which is largely dependent on the durability of the soils. In this section, the durability of GC is assessed in terms of its performance under repeated loading and frost susceptibility.

7.5.1 Repeated Loading

Soils in the upper layers of road pavements, adjacent to machine foundations, or in embankments of roads and railways are likely to be subjected to dynamic loading. Most materials exhibit a critical threshold stress level (i.e., ratio of applied stress/strength at which settlement increases significantly).

There is limited information on the critical threshold value of GC under dynamic loading. However, it is important to note that the particle size distribution of GC before and after compaction showed little degradation (Ali and Arulrajah, 2012; Dames and Moore, 1993; PennDOT, 2001a; Wartman et al., 2004a). Thus, assuming that GC behaves in a manner similar to that of a granular soil, the working strength of GC under repeated loading is likely to be less than about 50% of the strength used under static loading. This is a rough estimation and it is strongly recommended that cyclic load tests be conducted to ascertain the true value of threshold stress for GC.

7.5.2 Frost Susceptibility

The frost susceptibility of soils can be tested in accordance with BS812-124 (2009) or ASTM D5918 (2013). It should be noted that neither of these tests gives an exact measure of heave owing to the differences in freezing regimes adopted under laboratory and natural climatic conditions. Therefore the results of these tests should be treated merely as an indication of the degree of frost susceptibility and loss of strength upon thawing.

Having said that, in general, a greater rate of frost heave suggests a greater susceptibility to frost action. Highly to very highly frost-susceptible soils have an average velocity of frost heave between 4 and 40 mm/day for materials with 30%–80% fraction less than 0.02 mm (US Army Corps of Engineers, 1984b). These soils generally comprise fine sand, silt and clayey silt with permeability in the approximate range of 1×10^{-7}–1×10^{-5} m/s. High-plasticity clays (with low permeability) and free-draining soils, such as sands and gravels, are less susceptible to frost action. The simplest guidance provided by Lichtberger (2005) suggests that soils with a percentage passing a 0.063-mm sieve greater than 15% are not likely to be frost susceptible. However, if the coefficient of uniformity ($C_u = D_{60}/D_{10}$) is greater than 15 and the percentage passing a 0.063-mm sieve is between 5% and 15%, the soils are considered to have an intermediate sensitivity to frost attack.

Results available on the frost susceptibility of GC are limited. However, given that GC is found to have (i) an average C_u value of 6.5 and usually not in the range of fine sand to clayey silt (Section 7.3.1), (ii) permeability mass loss not greater than 7.5% in the soundness test using sodium sulphate (Section 7.3.8) and (iii) permeability behaviour similar to that of coarse sand (Section 7.4.7), it is safe to assume that GC is not likely to be susceptible to freeze–thaw action.

7.6 Environmental Impact

As it is derived from glass recovered from municipal waste, GC is likely to have both organic and nonorganic contaminants arising from the remnants of materials stored in glass containers, as well as both ferrous and nonferrous metals used in packaging. These are sometimes referred to as debris and their presence in GC can be as high as 15% (CWC, 1996b). As noted by Chesner et al. (1998), clean and uncontaminated GC is likely to present consistent engineering properties. It is thus important to limit the amount of debris for practical purposes. For most geotechnical applications, the maximum debris level in GC has been capped at 5% (Chesner et al., 1998).

The environmental assessment of GC has been undertaken by several researchers in Australia (Disfani et al., 2012; Imteaz et al., 2012) and the United States (Wartman, 2004a; Cosentino et al., 1995) to examine its total and leachate concentrations of metal and nonmetal constituents and organic compounds. The findings emerging from the studies are given in Table 7.5.

Table 7.5 Environmental assessment of glass cullet (GC)

Reference	Findings
(a) Metals	
Disfani et al. (2012)	Tests on the total and leachate concentrations of arsenic, barium, beryllium, cadmium, chromium, copper, lead, mercury, nickel, selenium, silver and zinc in GC samples showed that the material was not hazardous.
Wartman (2004a)	Traces of barium, chromium, lead and mercury present in GC samples might be due to contamination or ingredients used in glassmaking, but the concentrations were below the US EPA drinking water standard and hazardous waste limits.
Imteaz et al. (2012)	The concentrations of lead, copper and zinc in GC samples were below the threshold limits specified in EPA Victoria, but the iron concentration in an acid-washed sample was above the limit.
(b) Nonmetals	
Disfani et al. (2012)	The contents of benzene, monocyclic aromatic hydrocarbons, polycyclic aromatic hydrocarbons, benzo(a)pyrene and PAHs (total) and their leached concentrations in acidic and alkaline solutions were below the EPA Victoria threshold values.
(c) Organic Compounds	
Cosentino et al. (1995)	The biological oxygen demand, total Kjeldahl nitrogen and total phosphorus concentrations of GC might be high due to organic contamination.

In general, the results suggest that the metal contents in GC and the released concentrations from GC were below the threshold specified by the Environmental Protection Agency (EPA) in Victoria, Australia, and the United States. Traces of barium, chromium, lead and mercury have been detected in GC samples; most probably this was due to contamination, though this might also attributed to the ingredients used in glassmaking (Wartman, 2004a). In only one case, the iron concentration in an acid-washed GC sample was found to exceed the threshold limit, but the source of the iron was unclear (Imteaz et al., 2012).

Nonmetallic compounds, such as benzene, monocyclic aromatic hydrocarbons, polycyclic aromatic hydrocarbons (PAHs), benzo(a)pyrene and PAHs (total), which are used in the manufacture of plastics, wood preservatives, insecticides, dyes, detergents and various other chemicals, might be present in GC. The total contents of these nonmetallic compounds and their leached concentrations in acidic and alkaline solutions were less than 0.1 mg/kg and 0.001 mg/L, respectively, and below the EPA Victoria threshold values (Disfani et al., 2012). The concentrations of organic compounds in GC, measured in the forms of biological oxygen demand, total Kjeldahl nitrogen and total phosphorus, could be high if it is contaminated (Cosentino et al., 1995).

7.7 Case Studies

Case studies involving the use of GC in various geotechnical applications are given in Table 7.6. The majority of the projects were undertaken in the United States in the 2000s and 2010s, and the suitability of GC for use as artificial beach/dune fill material (Makowski and Rusenko, 2007; Makowski et al., 2007, 2011, 2013) or in conjunction with dredged materials for the construction of embankments (Grubb et al., 2006a,b, 2008a,b) was investigated thoroughly. In the former case, it was found that the geotechnical properties of GC were compatible with beach

Table 7.6 Case studies on the utilisation of glass cullet (GC)

Application	Reference	Location	Work Undertaken and Findings
Trench backfill	DECC (2007a, 2007b)	Australia	Two field trials with GC as 25%, 50% and 100% natural sand replacement were performed. The material was easy to handle and compact. There were no problems with odour, skin contact or dust.
	Northwest EcoBuilding Guild (2014b)	USA	GC was used as a 100% replacement for sand and gravel. The material complied with all the standard requirements and was proved to be cost effective.

Table 7.6 Continued

Application	Reference	Location	Work Undertaken and Findings
Structural fill	Northwest EcoBuilding Guild (2014a)	USA	GC was used for the construction of the base of a sidewalk; the material presented no problems in terms of technical performance and safety.
Embankment	Grubb et al. (2006a,b, 2008a,b)	USA	Three trial embankments made with blends of dredged materials and GC at 20%, 50% and 80% were constructed. The use of GC increased the dry density and workability of dredged material. The cone penetrometer test (CPT) values of the blends were between 1 and 2 MPa. After 1 year, the CPT values were increased by two- to threefold. The nuclear density gauge was suitable for use to predict the field moisture content of the blends.
Bedding and jointing materials	Paving Expert (2017)	UK	GC of maximum size 8 and 1.6 mm was used as bedding and jointing materials, respectively, in a trial for paving blocks and concrete paving slabs. Overall trial showed that potential for such application was promising but further work on the grading of GC and health and safety issues was required.
Artificial beach/dune fill	Makowski and Rusenko (2007), Makowski et al. (2007, 2011)	USA	The geotechnical properties of GC were compatible with beach sand. The biological and abiotic analyses suggested that GC is safe and inert.
	Makowski et al. (2013)	USA	The use of GC as dune fill did not affect the growth of dune vegetation.
Soil improvement	Onitsuka et al. (2001)	Japan	Foamed waste glass was used as lightweight fill materials in slope restoration, subgrade or subbase construction and lime-bound materials.
	Moqsud and Hayashi (2007)	Japan	Blend of foamed waste glass and sand was used to improve the geoenvironmental conditions of a contaminated tidal flat.

sand and GC was biologically safe for marine applications. In the latter case, the dry density and workability of dredged materials were improved with the inclusion of GC, and the cone tip resistance of the resulting mixes was increased by two- to threefold after one year.

In addition, GC has been successfully used as a replacement for sand and fine gravel in small-scale trials in trench backfill applications in Australia (DECC, 2007a, 2007b) and as bedding and jointing materials in the United Kingdom (Paving Expert, 2017). Of added interest, foamed waste glass has been used as lightweight fill materials in a number of projects in Japan (Onitsuka et al., 2001; Moqsud and Hayashi, 2007).

Although not running in field trials, GC has shown some promising laboratory results for use in a granular form as a sand blanket layer in railway track foundations (Ghataora, 2001) and a beach nourishment material (Edge et al., 2002), and in a ground form as a cementitious material in mine fill applications (Archibald et al., 1995, 1999; Peyronnard and Benzaazoua, 2012).

7.8 Guidance on Using Glass Cullet

This section provides a general guide on using GC as a granular material in various potential geotechnical applications. The geotechnical characteristics and important engineering properties of GC, as discussed in the previous sections, are summarised in Table 7.7. The data shown cover the range of values found in the literature, together with the average and the size of the data. In brief, the geotechnical characteristics and engineering properties of GC are, by and large, comparable to those of natural sand.

Table 7.7 Characterisation and geotechnical properties of glass cullet

Property	Value Range [Average] (Size of Data)
Particle shape	Angular to rounded
Percentage passing 0.075 mm	Up to 5 (28)
Particle size D_{10} (mm)	0.11–2.10 (28)
D_{15} (mm)	0.19–4.80 (28)
D_{30} (mm)	0.14–6.91 (28)
D_{50} (mm)	0.20–12.0 (28)
D_{60} (mm)	0.22–13.1 (28)
D_{85} (mm)	2.20–30 (28)
D_{max} (mm)	3.50–40[a] (28)
Coefficient of uniformity, C_u	2.59–10.24 (28)
Coefficient of curvature, C_c	0.64–3.9 (28)
Specific gravity	2.14–2.52[b] [2.48] (28)

Table 7.7 Continued

Property	**Value Range [Average] (Size of Data)**
Standard compaction	
MDD (kN/m^3)	14.7–18.0 [16.5] (26)
OMC (%)	4.7–13.6 [9.9] (24)
Modified compaction	
MDD (kN/m^3)	17.0–19.6 [17.9] (34)
OMC (%)	5.2–11.2 [9.1] (34)
Permeability (m/s)	1.0×10^{-3} to 1.0×10^{-6} [4.2×10^{-4}] (40)
Shear strength: cohesion (kPa)	0
Max size 6.3 mm, angle of friction, φ (°)	35–68 [47] (24)
Max size 20 mm, angle of friction, φ (°)	44–62 [50] (14)
Coefficient of volume compressibility (m^2/MN)	No data available
Coefficient of consolidation, C_v (m^2/year)	σ_3 = 400 kPa; C_v = 457.6 (1) σ_3 = 800 kPa; C_v = 2825.6 (1)
Secant modulus at 33% of peak load (MPa)	
For confining pressure (kPa) of 30–100	3–10, increase with pressure (12)
100 to 250	10–15, increase with pressure (7)
250 to 500	18.5 (2)
Resilient modulus (MPa)	277 (1)

MDD, maximum dry density; *OMC*, optimum moisture content.
[a]Results based on bulk of the data. Maximum size as small as 0.4 mm (Basari, 2012) and greater than 100 mm (Mikami et al., 2009) has been recorded.
[b]Lower and higher values of specific gravity may be recorded if the glass cullet is contaminated.

A flow chart starting from the initial consideration of the possible application of GC, to a detailed assessment against requirements and to the final decision on material selection, is suggested in Figure 7.23. It is worth noting that if GC has to be modified, for example, through cleaning to remove debris or crushing to adjust particle size distribution, it is important to repeatedly examine the end product to ensure its compliance with the required standards, as the quality of the material may vary with time.

In a granular form similar to the size of natural fine aggregate, and provided that it is clean and has low debris content, GC is an ideal candidate for use in various geotechnical applications, such as (i) non-load-bearing and load-bearing fills for the construction of embankments and retaining walls and (ii) bedding and trench backfill for pipes and drainage medium. The technical performance requirements of these applications that need to be considered when GC is used are listed in Table 7.8. Among these, the particle size distribution of GC should be given utmost attention as it is a common factor affecting the compaction, strength, settlement and permeability of soils.

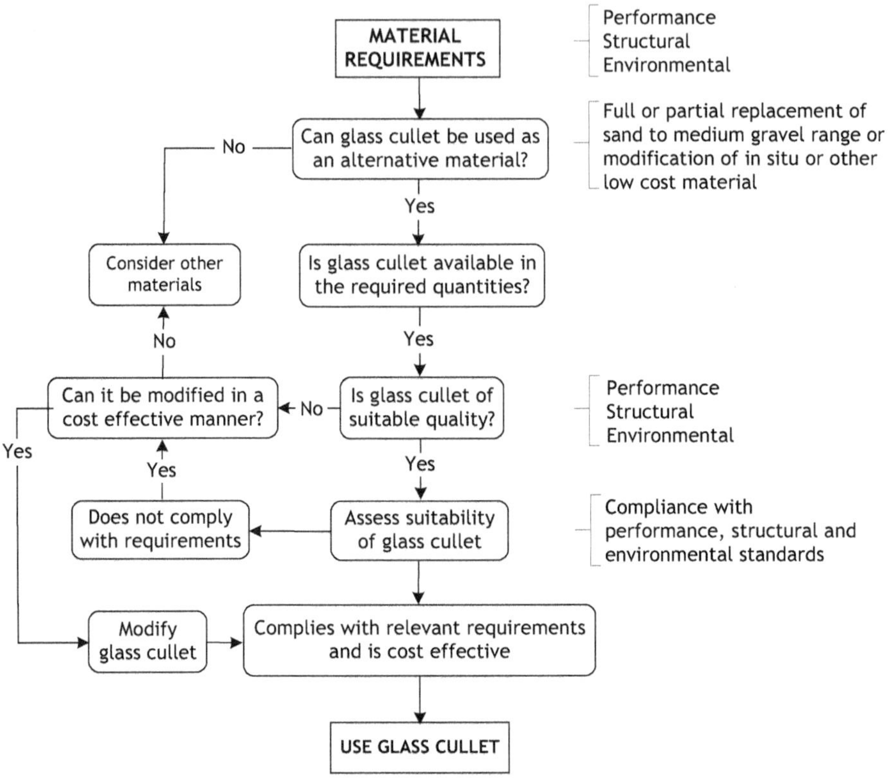

Figure 7.23 Flow chart for assessing suitability of glass cullet for any particular application.

7.9 Conclusions

This chapter assesses the performance of GC in geotechnical fields, including an initial characterisation of its geotechnical properties, followed by its mechanical and durability properties, as well as the associated environmental impact, case studies and guidance on using the material in the relevant field. The specific conclusions are provided below.

GC can be produced in a range of particle sizes, generally with the maximum particles not exceeding about 20mm. It should be noted that the shape of larger-sized GC is likely to be flaky and elongated, which can give rise to aggregate packing issues. Most of the studies relevant to the geotechnical field show that GC is normally used as a fine sand to medium gravel, with a maximum size of about 15mm and percentage passing a 0.06-mm sieve of less than 5%. The particle size distribution of the material tends to be uniformly or well graded, and its coefficient of uniformity is in the range from 3 to ~10, giving a mean value of 6.7. The change in particle size distribution of GC resulting from particle breakage due to compaction energy is insignificant, but this effect needs to be considered if the structural performance or the permeability of soil is critical.

Table 7.8 Technical performance requirements for various geotechnical applications of glass cullet

Application	Engineering Problem	Characterisation					Engineering Properties							Durability
		PSD	MDD	OMC	Gs	w	c	φ	m_v	C_v	Es	M_R	K	Frost-Sus
Non-load-bearing fill	Stability	x				x								
Load-bearing fill	Static load	x	x	x		x	x	x			x			
	Dynamic loading	x	x	x		x					x			
	Settlement	x	x	x	x	x	x	x						
Embankment	Slope stability	x	x	x		x	x	x						
	Bearing capacity	x	x	x	x	x			x	x				
Pipe bedding	Settlement	x	x	x										
Trench backfill	Bearing capability	x	x	x			x	x						
	Settlement								x	x				
Retaining wall backfill	Permeability	x											x	x
Drainage	Permeability	x	x	x									x	x

'x' indicates that GC needs to be tested for the requirement. *c*, cohesion; C_v, coefficient of consolidation; *Es*, secant modulus; *Frost-Sus*, frost susceptibility; *GC*, glass cullet; *Gs*, specific gravity; *K*, permeability; *MDD*, maximum dry density; M_R, resilient modulus; m_v, coefficient of volume compressibility; *OMC*, optimum moisture content; *PSD*, particle size distribution; *w*, moisture content; φ, friction angle.

The average specific gravity of GC is about 2.49, which is about 6% lower than that of natural sand and gravel. Glass is an impermeable material, and the water absorption of GC is close to zero. The soundness value of GC is less than 7.5%, suggesting that the material is resistant to freeze–thaw attack. The organic content in GC is affected by the quality of the cleaning process; on average, its content is about 1.3%.

The MDD and OMC of GC are $17 \pm 2\,kN/m^3$ and $9\% \pm 4\%$, respectively, regardless of its maximum size. These values are not too dissimilar to those of soil of similar aggregate size and compaction energy used. When GC is added to a coarse-grained (granular) soil, there is little change in either the MDD or the OMC of the mixture. However, if the host soil is fine grained, the use of GC can result in an increase in MDD and a reduction in OMC.

The friction angle of GC ranges from 34 to over 60 degrees, though the reported data have been mostly within the range of 40–50 degrees. The friction angle of GC can be affected by its voids content and confining pressure. When GC is added to fine-grained soils, it results in a reduction in the PI and the friction angle of the soil and a reduction in its cohesion. This effect is, however, insignificant in coarse-grained soils.

At 0.33 of peak strength, the secant modulus of GC decreases, at a decreasing rate, as confining pressure increases. The use of GC as a replacement for crushed rock at up to 50% reduces its M_R, though the value is still within an acceptable range. The CBR of GC is in the range of 20%–40%.

The permeability of GC is in the range from 1.0×10^{-3} to $1.0 \times 10^{-7}\,m/s$, having an average value of $1.4 \times 10^{-3}\,m/s$, which is comparable to that of coarse sand and fine sand/coarse silt. The thermal conductivity of GC is marginally greater than that of dry sand but significantly smaller than that of saturated sand. Although the available information on the durability of GC is limited, the material is assumed to have no critical issue under repeated loading and frost attack.

The results of total and leachate concentrations of metal and nonmetal constituents and organic compounds suggest that the material is safe and inert, but its debris or contaminant content should be kept to the minimum. In a granular form, GC has been used as a trench backfill, structural fill, artificial beach/dune fill and bedding and jointing materials in different field trials of various scales in Australia, the United Kingdom and the United States.

References

AASHTO M145-91, 2012. Standard Specification for Classification of Soils and Soil-Aggregate Mixtures for Highway Construction Purposes. American Association of State Highway and Transportation Officials, Washington, DC, USA.

AASHTO T294, 1994. Standard Method of Test for Resilient Modulus of Unbound Granular Base. Subbase Materials and Subgrade Soils. American Association of State Highway and Transportation Officials, Washington, DC, USA.

Ali M M Y and Arulrajah A, 2012. Potential use of recycled crushed concrete-recycled crushed glass blends in pavement subbase applications. In: GeoCongress 2012, Oakland, CA, USA, pp. 3662–3671.

Ali M M Y, Newman A, Arulrajah A and Disfani M M, 2011a. Application of recycled glass - crushed rock blends in road pavements. Australian Geomechanics 46 (1), 113–122.

Ali M M Y, Arulrajah A, Disfani M M and Piratheepan J, 2011b. Suitability of using recycled glass - crushed rock blends for pavement subbase applications. In: Proc. Geo-Frontiers 2011: Advances in Geotechnical Engineering, Dallas, Texas, pp. 1325–1334.

Allan M L, 1997. Thermal Conductivity of Cementitious Grouts for Geothermal Heat Pumps Report No: BNL-65129. Brookhaven National Laboratory, Upton, NY, USA. 58 pp.

Amlashi S M H, Vaillancourt M and Carter A, 2015. Current state of the art practice of use of glass in pavement structures. In: The 2015 Conference of the Transportation Association of Canada Charlottetown on Innovation in Pavement Materials and Surfacing Technology, Canada 19 pp.

Archibald J F, deGagne D O, Lausch P and DeSouza E M, 1995. Ground waste glass as a pozzolanic consolidation agent for mine backfill. Rock Mechanics 88 (995), 80–86.

Archibald J F, Chew J L and Lausch P, 1999. Use of ground waste glass and normal Portland cement for improving slurry and paste backfill support performance. Journal of the Canadian Institute of Mining, Metallurgy and Petroleum (CIM Bulletin) 92 (1030), 74–80.

Arnold G, Werkmeister S and Alabaster D, 2008. The effect of adding recycled glass on the performance of basecourse aggregate. NZ Transport Agency, Wellington, New Zealand. NZ Transport Agency Research Report 351, 40 pp.

ARRB, 2010. Specification for Recycled Crushed Glass as an Engineering Material. ARRB Group Limited, Australia. 39 pp.

Arulrajah A, Ali M M Y, Disfani M M, Piratheepan J and Bo M W, 2013. Geotechnical performance of recycled glass-waste rock blends in footpath bases. Journal of Materials in Civil Engineering 25 (5), 653–661.

Arulrajah A, Ali M M Y, Disfani M M and Horpibulsuk S, 2014. Recycled-glass blends in pavement base/subbase applications: laboratory and field evaluation. Journal of Materials in Civil Engineering 26 (7) 12 pp.

ASTM C88, 2013. Standard Test Method for Soundness of Aggregates by Use of Sodium Sulfate or Magnesium Sulfate. ASTM International, West Conshohocken, PA, USA.

ASTM C518, 2015. Standard Test Method for Steady-State Thermal Transmission Properties by Means of the Heat Flow Meter Apparatus, Insulation of Construction Materials. ASTM International, West Conshohocken, PA, USA.

ASTM D2166, 2016. Standard Test Method for Unconfined Compressive Strength of Cohesive Soil. ASTM International, West Conshohocken, PA, USA.

ASTM D2434–68, 2006. Standard Test Method for Permeability of Granular Soils (Constant Head). ASTM International, West Conshohocken, PA, USA.

ASTM D2850, 2015. Standard Test Method for Unconsolidated-Undrained Triaxial Compression Test on Cohesive Soils. ASTM International, West Conshohocken, PA, USA.

ASTM D3080, 2011. Standard Test Method for Direct Shear Test of Soils Under Consolidated Drained Conditions. ASTM International, West Conshohocken, PA, USA.

ASTM D4972, 2001. Standard Test Method for pH of Soils. ASTM International, West Conshohocken, PA, USA.

ASTM D5334, 2008. Standard Test Method for Determination of Thermal Conductivity of Soil and Soft Rock by Thermal Needle Probe Procedure. ASTM International, West Conshohocken, PA, USA.

ASTM D5918, 2013. Standard Test Methods for Frost Heave and Thaw Weakening Susceptibility of Soils. ASTM International, West Conshohocken, PA, USA.

Basari E, 2012. Effect of granular waste glass on soil properties. In: International Conference on New Developments in Soil Mechanics and Geotechnical Engineering, Nicosia, North Cyprus, pp. 479–484.

Bo M W and Arulrajah A, 2011. Use of recycled waste in highway engineering. In: Proceedings of the 3rd International Conference on Geotechnical Engineering for Disaster Mitigation and Rehabilitation, Semarang, Indonesia, pp. 263–270.

British Geological Survey, 2011. Temperature and Thermal Properties BGS Report No. GR_999999/1. British Geological Survey, Nottingham, UK. 7 pp.

BS 1377-3, 1990. Methods of Test for Soils for Civil Engineering Purposes. Chemical and electro-chemical tests. British Standard Institution, London, UK.

BS 1377-7, 1990. Methods of Test for Soils for Civil Engineering Purposes — Part 7: Shear Strength Tests (Total Stress). British Standard Institution, London, UK.

BS 1377-8, 1990. Methods of Test for Soils for Civil Engineering Purposes —Part 8: Shear Strength Tests (Effective Stress). British Standard Institution, London, UK.

BS 812-105.1, 1989. Method for Determination of Particle Shape; Flakiness Index. British Standard Institution, London, UK.

BS EN ISO 14688–1, 2002. Geotechnical Investigation and Testing — Identification and Classification of Soil — Part 1: Identification and Description. British Standards Institute, London, UK.

BS EN 13242:2002+A1, 2007. Aggregates for Unbound and Hydraulically Bound Materials for Use in Civil Engineering Work and Road Construction. British Standard Institute, London, UK.

BS 812-124, 2009. Testing Aggregates. Method for determination of frost heave. British Standards Institution, London, UK.

BS EN 1997-2, 2007. Eurocode 7. Geotechnical Design. Ground Investigation and Testing. British Standards Institute, London, UK.

CETCO, 2013. Geothermal Grout. CETCO. Technical data, 4 pp.

Chesner W H, Collins R J and MacKay M H, 1998. User Guidelines for Waste and By-product Materials in Pavement Construction. Federal Highways Administration, VA, USA. Publication No. FHWA-RD-97–148.

Cosentino P J, Kalajian E H, Heck III H H H and Sheikh C S, 1995. Developing Specification for Waste Glass and Waste-to-energy Bottom Ash as Highway Fill Materials, Volume 2 of 2 (Waste Glass) Report No. FL/DOT/RMC/06650–7754. Florida Department of Transportation, Tallahassee, FL, USA. 152 pp.

CWC, 1996a. Best Practices in Glass Recycling - The Behaviour of Glass Aggregate Under Structural Loads. Clean Washington Center, Seattle, USA. BP-GL4-01-05, 2 pp.

CWC, 1996b. Best Practices in Glass Recycling – Typical Geotechnical Parameters of Glass Aggregate. Clean Washington Center, Seattle, WA, USA. BP-GL4-01-02, 2 pp.

Dames and Moore, 1993. Glass Feedstock Evaluation Project – Engineering Suitability Evaluation. Clean Washington Center, Seattle, WA, USA. BP-GL-93–95, 110 pp.

Davidovic N, Bonic Z and Prolovic V, 2012. Waste glass as additive to clayey material in subgrade and embankment of road pavement. Architecture and Civil Engineering 10 (2), 215–222.

DECC, 2007a. Trial of Recycled Glass as Pipe Embedment Material – Part 1. Department of Environment and Climate Change NSW, Southern Sydney, NSW, Australia. 11 pp.

DECC, 2007b. Trial of Recycled Glass as Pipe Embedment Material – Part 2. Department of Environment and Climate Change NSW, Southern Sydney, NSW, Australia. 13 pp.

Disfani M M, Arulrajah A, Ali M M Y and Bo M W, 2011a. Fine recycled glass: a sustainable alternative to natural aggregates. International Journal of Geotechnical Engineering 5, 255–266.

Disfani M M, Arulrajah A, Bo M W and Hanhour R, 2011b. Recycled crushed glass in road work applications. Waste Management 31, 2341–2351.

Disfani M M, Arulrajah A, Bo M W and Sivakugan N, 2012. Environmental risks of using recycled crushed glass in road applications. Journal of Cleaner Production 20, 170–179.

Eberemu A O, Edeh J E and Gbolokun A O, 2013. The geotechnical properties of Lateritic soil treated with crushed glass cullet. Advanced Materials Research 824, 21–28.

Edge B L, Cruz-Castro O and Magoon O T, 2002. Recycled glass for beach nourishment. In: Proceedings of the 28th International Conference on Coastal Engineering, Cardiff, UK, pp. 3630–3641.

Fauzi A, Djauhari Z and Fauzi U J, 2016. Soil engineering properties improvement by utilization of cut waste plastic and crushed waste glass as additive. International Journal of Engineering and Technology 8 (1), 15–18.

Finkle I and Ksaibati K, 2007. Recycled Glass Utilization in Highway Construction. University of Wyoming, Wyoming, USA. 25 pp.

Ghataora G S, 2001. Utilisation of Glass Cullet as Replacement of Blanketing Sand in Railway Track Foundations Internal Report. University of Birmingham, Birmingham, UK. 122 pp.

Grubb D G, Davis A F, Sands S C, Carnivale III M, Wartmann J and Gallagher P M, 2006a. Field evaluation of crushed glass-dredged material blends. Journal of Geotechnical and Geoenvironmental Engineering 132 (5), 577–590.

Grubb D G, Gallagher P M, Wartman J, Liu Y and Carnivale III M, 2006b. Laboratory evaluation of crushed glass–dredged material blends. Journal of Geotechnical and Geoenvironmental Engineering 132 (5), 562–576.

Grubb D G, Wartman J, Malasavage N E and Mibroda J G, 2007. Turning mud into suitable fill: amending OH, ML-MH and CH soils with curbside-collected crushed glass (CG). In: Geo-Denver 2007 on New Peaks in Geotechnics, Denver, CO, USA 14 pp.

Grubb D G, Cadden A W and Miller D D, 2008a. Crushed glass-dredged material (CG-DM) blends: role of organic matter content and DM variability on field compaction. Journal of Geotechnical and Geoenvironmental Engineering 134 (11), 1665–1675.

Grubb D G, Wartman J and Malasavage N E, 2008b. Aging of crushed glass-dredged material blend embankments. Journal of Geotechnical and Geoenvironmental Engineering 134 (11), 1676–1684.

Hagerty M M, Hite D R, Ullrich C R and Hagerty D J, 1993. One-dimensional high-pressure compression of granular media. Journal of Geotechnical Engineering 119 (1), 1–18.

Hamdhan I N and Clarke B G, 2010. Determination of thermal conductivity of coarse and fine sand soils. In: Proceedings World Geothermal Congress 2010, Bali, Indonesia, pp. 1–7.

Henry K S and Morin S H, 1997. Frost susceptibility of crushed glass used as construction aggregate. Journal of Cold Regions Engineering 11 (4), 326–333.

Highways Agency, 2016. Series 600 – earthworks. Manual of contract documents for highway works. Specification for Highway Works, vol. 1. Highway Agency, London, UK.

Hoffman G, 2002. Use Guidelines for Embankments Incorporating Crushed Glass. PennDOT, Harrisburg, PA, US. Provisional Special Provision, 6 pp.

Huang P T, Patel M, Santagata M S and Bobet A, 2009. Classification of Organic Soils Report No. FHWA/IN/JTRP-2008/2. Indiana Department of Transportation, Indianapolis, IN, USA. 195 pp.

Imteaz M A, Ali M M Y and Arulrajah A, 2012. Possible environmental impacts of recycled glass used as a pavement base material. Waste Management & Research 30 (9), 917–921.

INDOT, 2012. Standard Specifications. Indiana Department of Transportation, Indianapolis, IN, USA.

Kang J, Hong S and Kim Y, 2009. A study of frost heave susceptibility of soil mixed with waste glass. In: JSSI & JSSE Joint Conference, Sapporo, Japan 1 pp.

Kunishima T, Aria K and Yaginuma H, 2000. Utilisation of glass cullet for sand-cushion layer of ILB pavement. In: JIPEA World Congress 2000, pp. 159–166.

Lamb T W and Whitman R V, 1979. Soil Mechanics. John Wiley & Sons, New York, USA.

Lee L T, 2007. Recycled Glass and Dredged Materials Report No. ERDC TN-DOESR-T8. U.S. Army Engineer Research and Development Center, Vicksburg, USA. 12 pp.

Li Y, Huang R, Chan L and Chen J, 2013. Effects of particle shape on shear strength of clay-gravel mixture. KSCE Journal of Civil Engineering 17 (4), 712–717.

Lichtberger B, 2005. Track Compendium: Formation, Permanent Way, Maintenance, Economics. Eurailpress Tetzlaff-Hestra GmbH & Co. KG, Hamburg, Germany.

Makowski C and Rusenko K, 2007. Recycled glass cullet as an alternative beach fill material: results of biological and chemical analysis. Journal of Coastal Research 23 (3), 545–552.

Makowski C, Thomson G, Foye P and Higgins S, 2007. Broward county beach demonstration project: from beers to beaches. In: Kraus N C and Rosati J D (Eds.), Proceedings of Sixth International Symposium in Coastal Engineering and Science of Coastal Sediment Process, New Orleans, LA, USA, pp. 1–13.

Makowski C, Finkl C W and Rusenko K, 2011. Using recycled glass cullet for coastal protection: a review of geotechnical, biological and abiotic analyses. Journal of Coastal Research 64, 1362–1366 (Special Issue).

Makowski C, Finkl C W and Rusenko K, 2013. Suitability of recycled glass cullet as artificial dune fill along coastal environments. Journal of Coastal Research 29 (4), 772–782.

Malasavage N E, Gallaghar P M, Grubb D G, Wartman J and Carnivale III M, 2007. Modifying a plastic clay with crushed glass: implications for constructed fills. Japanese Geotechnical Society, Soils and Foundations 47 (6), 1017–1027.

Mavroulidou M and Ahmed M U, 2011. Geotechnical properties of glass cullet. In: Proceedings of the 12th International Conference on Environmental Science and Technology, Rhodes, Greece, pp. 678–685.

Mikami T, Koseki J, Sato T and Yajima C, 2009. Use of granular material made from recycled glass as countermeasure against earthquake-induced uplift of underground pipes. Bulletin of ERS 42, 87–95.

Moqsud M A and Hayashi S, 2007. Effectiveness of recycled waste glass as a contaminated tidal mud improvement material. Daffodil International University Journal of Science and Technology 2 (2), 53–55.

Nash P T, Jayawickrama P, Tock R W, Senadheera S, Viswanathan K and Woolverton B, 1995a. Use of Glass Cullet in Roadway Construction, Phase 1 – Literature Review and Identification of Sources and Suppliers Report No. 0-1331-1. Texas Department of Transport, Austin, TX, USA. 54 pp.

Nash P T, Jayawickrama P, Tock R W, Senadheera S, Viswanathan K and Woolverton B, 1995b. Use of Glass Cullet in Roadway Construction: Laboratory Testing and Specification Development Report No. 0-1331-2F. Texas Department of Transport, Austin, TX, USA. 44 pp.

Northwest EcoBuilding Guild, 2014a. Crushed Glass Structural Fill at West Bay Business Park. Available from: http://www.ecobuilding.org/.

Northwest Ecobuilding Guild, 2014b. Crushed Glass Fill at Little Rock Housing Project. Available from: http://www.ecobuilding.org/.

Nuruzzaman M and Hossain M A, 2014. Effect of soda lime glass dust on the properties of clayey soil. Global Journal of Researches in Engineering (E) Civil and Structural Engineering 14 (5), 17–22.

Onitsuka K, Shen J, Hara Y and Sato M, 2001. Utilization of foaming waste glass as construction materials. In: Dhir R K, Limbachiya M C and Dyer T D (Eds.), Proceedings of the International Symposium on Recycling and Reuse of Glass Cullet, Dundee, Scotland, UK. Thomas Telford, London, UK, pp. 195–207.

Ooi P S K, Li M M W, Sagario L Q and Song Y, 2008. Shear strength characteristics of recycled glass. Transportation Research Record: Journal of the Transportation Research Board 2059, 52–62.

Paving Expert, 2017. Tameside Trial Alternative Sand, News and Views – News on the Street. Available from: http://www.pavingexpert.com/news219.htm.

PennDOT, 2001a. Laboratory Evaluation of Select Engineering-Related Properties of Glass Cullet. Department of Transportation, Harrisburg, PA, USA. 21 pp.

PennDOT, 2001b. Engineering Related Properties of Glass Blended With Various Soils. Pennsylvania Department of Transport, Harrisburg, Pennsylvania, USA. 32 pp.

PennDOT, 2002. Provisional Special Provision Embankments Incorporating Crushed Glass Report No. 400-02-02. Pennsylvania Department of Transport, Harrisburg, Pennsylvania, USA. 6 pp.

PennDOT, 2006. Crushed Glass Fact Sheet. Pennsylvania Department of Transport, Harrisburg, Pennsylvania, USA. 2 pp.

Peyronnard O and Benzaazoua M, 2012. Alternative by-product based binders for cemented mine backfill: recipes optimisation using Taguchi method. Minerals Engineering 29, 28–38.

Piratheepan J, Arulrajah A and Disfani M M, 2013. Large-scale direct shear testing of recycled construction and demolition materials. Advances in Civil Engineering Materials 2 (1), 25–36.

Powell W D, Potter J F, Mayhew H C and Nunn M E, 1984. The Structural Design of Bituminous Roads Report 1132. Transport and Road Research Laboratory, Berkshire, UK. 62 pp.

Prowell B D, Zhang J and Brown E R, 2005. Aggregate Properties and the Performance of Superpave-Designed Hot Mix Asphalt NCHRP Report 539. Transportation Research Board, Washington, DC, USA. 91 pp.

Reddy K, 1999. Use of glass cullet as backfill material for retaining structures. In: International Conference on Solid Waste Technology and Management, Philadelphia, PA, USA, pp. 1–8.

SCDOT, 2011. Supplemental Technical Specification for - Hot-Mix Asphalt Material Properties. South Carolina Department of Transportation, Columbia, SC, USA.

SCE Recycling, 2013. EnviroGlass (Recycled Glass Sand). Available from: http://sce-recycleit.com/wp-content/uploads/2013/11/EnviroGlass.pdf.

Shin C J and Sonntag V, 1994. Using Recovered Glass as Construction Aggregate Feedstock. Transportation Research Record 1437. , pp. 8–18.

So S T C, Lee R M K, Hui T and Shiu Y K, 2015. Study of using recycled glass cullet as an engineering fill in reclamation and earthworks in Hong Kong. Japanese Geotechnical Society Special Publication 2 (54), 1874–1879.

Soil and Environmental Engineers Inc. and Re-sourcing Associates Inc., 1998. A Tool Kit for the Use of Post-consumer Glass as a Construction Aggregate Report No. GL-97–5. Clean Washington Centre, Seattle, Washington, USA. 92 pp.

TRB, 2013. Recycled materials and byproducts in highway applications. Manufacturing and Construction Byproducts - A Synthesis of Highway Practice, vol. 8. Transportation Research Board, Washington, USA.

TxDOT, 1999. Use of Glass Cullet in Roadway Construction, Year of the Recycled Roadway Materials, February: Glass Cullet. Texas Department of Transportation, Austin, TX, USA. 29 pp. 101.

US Army Corps of Engineers, 1984a. Engineering and Design Airfield Flexible Pavement Mobilization Construction. Engineer Manual no. 1110-3-141. Department of Army, Washington, DC, USA. 102 pp.

US Army Corps of Engineers, 1984b. Pavement Criteria for Seasonal Frost Conditions Cost. Engineer manual no. 1110-3-138. Department of the Army, Washington, DC, USA. 98 pp.

Vishnu T C and Chandrakaran S, 2016. Evaluation of strength characteristics of crushed glass-soil mixes. Journal of Geotechnical Studies 1 (1), 1–15.

Wartman J, Grubb D G and Nasim A S M, 2004a. Select engineering characteristics of crushed glass. Journal of Materials in Civil Engineering 16 (6), 526–539.

Wartman J, Grubb D G and Strenk P, 2004b. Engineering properties of crushed glass-soil blends. In: Geotrans 2004, Los Angeles, CA, USA, pp. 732–739.

Use of Glass Cullet in Road Pavement Applications

8

Main Headings

- Unbound applications
- Hydraulically bound applications
- Bituminous bound applications
- Environmental impact
- Case studies

Synopsis

This chapter assesses the appropriateness and performance of glass cullet (GC) as an aggregate in road pavement, dealing with its use in unbound, hydraulically bound and bituminous bound applications. The California bearing ratio of GC has been found to be comparable to that of natural aggregate. In hydraulically bound applications, GC fine aggregate is shown to be compatible with other waste materials for use in subbase and base, such as steel slag, ladle furnace slag and foundry sand. The performance of GC in concrete rigid pavement was generally satisfactory. It is shown that the Marshall method can be adopted when designing bituminous mixtures made with GC in coarse, fine and filler aggregate size fractions. The use of GC at up to 10% does not adversely affect the stiffness modulus, rutting resistance or moisture damage resistance of bituminous mixtures. Antistripping agent is effective in improving the performance of mixtures containing GC. The material does not have a negative environmental impact. Case studies with GC in road pavement applications are also included.

Keywords: Glass cullet, Unbound applications, Hydraulically bound applications, Bituminous bound applications, Environmental impact, Case studies.

Sustainable Construction Materials: Glass Cullet. https://doi.org/10.1016/B978-0-08-100984-0.00008-4

8.1 Introduction

In general, road pavements are made of a series of layers of granular materials, each having a unique structural function. The construction begins with the subgrade, which acts as a stable foundation to the entire pavement. Next is the capping layer, which is included where subgrade is weak. This is followed by the subbase and base course (or base as known in the United States), to distribute the load transmitted from the upper surfacing layers, namely, the binder course and wearing course. The wearing course provides a smooth but adequate skid resistance for good vehicle riding quality, whilst the binder course helps to distribute the load applied onto the wearing course.

In road pavement construction, the use of aggregate is high, as it constitutes the main component in forming each layer of the pavement, except for the subgrade, which normally consists of local soil or nearby excavated materials. Indeed, of the total 2.6 billion tonnes of aggregate production in Europe in 2013, 20% was used in road-related work (European Aggregates Association, 2016). Although recycled and secondary aggregates (RSAs), such as recycled concrete aggregate, derived from construction and demolition waste, and blast furnace slag, a by-product of iron and steel making, are permitted in most standards (e.g., BS EN 13043, 2002; BS EN 13242:2002+A1, 2007), much more is needed to get RSA materials adopted in the standard specifications and accordingly processed for their use in road pavement works. As the broadly based technical requirements of an aggregate for use in the various layers of road pavement can vary greatly from subgrade to surface layers, this should work in favour of increasing the use of RSA materials in road pavement construction. Such a development should help to promote sustainability within the construction sector. This chapter deals with the potential use of glass cullet (GC) derived from soda lime glass as a replacement for natural aggregate, in the form of unbound, hydraulically bound and bituminous bound materials, in road pavement construction.

8.2 General Information

An overview of the effect of GC aggregate on the technical performance of unbound, hydraulically bound and bituminous bound mixes is summarised in Table 8.1. This includes studies from established organisations such as the Transportation Research Board, National Center for Asphalt Technology, National Virginia Transportation Research Council and Wisconsin Transportation Information Center, in the United States, as well as work that has been produced by the two groups of individual researchers from Japan and the United Kingdom. The overall impression is that the use of GC aggregate, at about 10%–20% as a replacement for natural aggregate, is not likely to adversely affect the technical performance of unbound, hydraulically bound and bituminous bound mixes. The literature also tends to suggest that, other than the wearing course, GC aggregate is suitable for use in the construction of all pavement layers, as well as in pavements with low traffic conditions.

Table 8.1 Overview of the use of glass cullet (GC) in unbound, hydraulically and bituminous bound applications

Reference	Main Findings
(a) Individual Researchers	
Huang et al. (2007), UK	GC aggregate in bituminous mixtures of lower content and smaller size generally resulted in satisfactory performance.
Jamshidi et al. (2016), Japan	GC aggregate can be used in hydraulically and bituminous bound mixes, but some technical issues remain to be addressed. The material can be used in unbound applications for low traffic conditions.
(b) Established Organisations in the United States	
Halstead (1993) for Virginia Transportation Research Council	GC aggregate can be used in unbound and bituminous bound applications.
Kandhal (1992) for National Center for Asphalt Technology	The use of GC aggregate in bituminous mix could give rise to stripping potential.
Transportation Research Board (2013)	GC aggregate has been used in road pavement applications in the USA for many years, but there are some technical issues remaining to be addressed.
Wisconsin Transportation Information Center (1999)	Use in hydraulically and bituminous bound mixes does not exhibit improvement in performance.

Some of the main points emerging from this overview are given below:

(a) In Unbound Mixes

Owing to the breakage of GC aggregate under load, it has been recommended that the material should be mixed with other high-quality materials for use in base and subbase layers, and the compatibility of the mixtures should also be assessed. However, GC aggregate can be used safely under low load-bearing conditions such as pedestrian pavements and walkways (Jamshidi et al., 2016).

(b) In Hydraulically Bound Mixes

When GC aggregate is used in hydraulically bound mixes, two main issues are commonly reported, i.e., reduction in consistence (workability) and risk of alkali–silica reaction (ASR) expansion (Jamshidi et al., 2016; Transportation Research Board, 2013). The former issue is due to the angularity of the material, which results in more aggregate interlocking. A higher amount of high-range water-reducing admixture is needed to maintain an adequate consistence of the mix. The latter issue is a detrimental reaction between the silica of GC aggregate and the alkaline solution of concrete, which results in an expansive gel in the presence of moisture. However, fly ash and ground granulated blast furnace slag (GGBS) have been proved to be effective in mitigating ASR expansion induced by GC aggregate in concrete.

(c) In Bituminous Mixes

The main concern of the use of GC aggregate in bituminous bound mixes is perhaps the susceptibility to moisture damage in the form of stripping, which involves the loss of adhesion between the GC aggregate and the asphalt (Halstead, 1993; Kandhal, 1992; Transportation Research Board, 2013; Wisconsin Transportation Information Center, 1999). It has been suggested that the inclusion of hydrated lime in bituminous mixes containing GC aggregate can retain its stripping resistance (Huang et al., 2007; Jamshidi et al., 2016).

8.3 Unbound Applications

Except for the uppermost layer of a road pavement, aggregate can be used in an unbound form without bitumen or Portland cement (as a binder) in any layer underneath, i.e., subgrade, capping layer, subbase and base course. The geotechnical properties and durability of GC aggregate, which have been previously discussed in Chapter 7, can be useful in studying the effect of GC aggregate as an unbound material in road pavements. As for this section, the analysis focuses on the California bearing ratio (CBR) value, which is one of the parameters used in road pavement design.

The CBR test is an empirical measure, used to determine indirectly the strength and stiffness of subgrade, subbase and base materials. In principle, the test compares the load required to cause a penetration into the test soil with that of the standard material of well-graded crushed stone, and the results are expressed in terms of percentage. The higher the CBR value, the higher the strength of a soil. For further details on the CBR tests, the reader is advised to refer to AASHTO T193 (2013), ASTM D1883 (2016) and BS 1377-9 (1990).

The CBR values of GC aggregate in two different forms, i.e., sand size fractions and a mixture of gravel and sand size fractions, are shown in Figure 8.1, together with those of different types of natural soils (O'Flaherty, 2002). It can be seen that, for a given compaction effort, the CBR value of GC gravel and sand mix is higher than that of GC-sand. In addition, the former material is more sensitive to the compaction effort, as it shows greater improvement in CBR value than the latter material when the compaction is changed from standard effort to modified effort. It appears that the CBR value of the sample with GC gravel and sand mix is not affected by the possible presence of flaky or elongated coarse aggregate, which can give rise to poor compaction and subsequently low load-bearing capability.

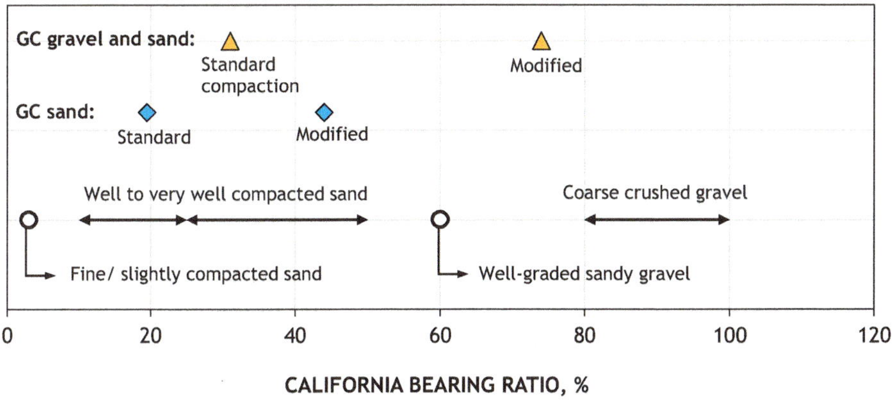

Figure 8.1 California bearing ratio values of glass cullet (GC) and typical sand and gravel. Data taken from Arulrajah et al. (2013, 2014) and Disfani et al. (2011).

Table 8.2 Capping thickness for subgrade with various California bearing ratios (CBR) (Highway Agency, 1995)

CBR	Capping Thickness, mm
≤2	600
2–5	350
5–15	150[a]
>15	No capping

[a]No capping for flexible and flexible composite, but subbase should increase to 225 mm.

Compared with other natural soils, the CBR values of GC-sand with standard and modified compaction at 20% and 44%, respectively, are within those of well to very well compacted sand. For GC gravel and sand mix, its CBR with standard compaction (30%) is within that of very well compacted sand, whilst its CBR with modified compaction (74%) is close to that of coarse crushed gravel.

Overall, the CBR of GC aggregate is comparable to that of natural soil of similar size fractions, indicating that GC aggregate is a viable material for use in road pavement construction. For example, in the *Design Manual for Roads and Bridges* in the United Kingdom (Highway Agency, 1995), the CBR of subgrade is used to determine the suitable thickness of the capping layer. A capping layer is not required for rigid and rigid composite pavements when the CBR of the subgrade is greater than 15%. However, for a subgrade with CBR less than 15%, the lower the CBR value, the larger the thickness of the capping layer (Table 8.2). As shown in Figure 8.1, the CBR of GC aggregate is generally greater than 15%, therefore its use as subgrade can eliminate the need for a capping layer, leading to a reduction in the total thickness of the pavement. On the other hand, if the subgrade consists of weak native soil, GC aggregate can be used as the capping material, which is required to have a minimum CBR of 15%, as specified by the Highway Agency (1995).

8.4 Hydraulically Bound Applications

8.4.1 Hydraulically Bound Mixtures

In the constructions of capping layer, subbase and base course, hydraulic binders such as Portland cement, fly ash, GGBS, lime and gypsum are sometimes added to the aggregate mixtures to improve their mechanical properties in terms of load-bearing capacity, stiffness and stability. The resulting materials, known as hydraulically bound mixtures, also offer other benefits, including reducing pavement thickness, enhancing long-term performance and saving construction cost.

The research on the use of GC aggregate in hydraulically bound mixtures, though limited in scope, has shown that its use in conjunction with other waste materials, namely steel slag, ladle furnace slag, foundry sand and fly ash, in Portland cement-bound mixtures is potentially feasible as subbase and base course materials (Pasetto and Baldo, 2016). In this study, six aggregate mixtures were produced with 4% Portland cement and different proportions of the waste materials, in which GC aggregate having a maximum aggregate size of 3 mm was used at 10% and 20%. The mechanical properties of these mixtures were not too dissimilar, and on average, the unconfined compressive strength, indirect tensile strength and dynamic elastic modulus at 7 days were about 7 MPa, 0.7 MPa and 11.5 GPa, respectively. A slight increase in the values was reported when measured at 28 and 90 days. Overall, although further work may still be required, the preliminary data seem to suggest that GC aggregate as well as other waste materials can be used as a replacement for natural materials in hydraulically bound mixtures.

8.4.2 Concrete Slab

Apart from hydraulically bound mixtures, hydraulic binders, mainly Portland cement or its combination with GGBS and fly ash, are also used in the construction of rigid pavements, where the top layer is a concrete slab. The concrete slab is either plain or reinforced, as well as being made with or without transverse joints. Typically, the concrete mixes used in rigid pavements have the following special technical requirements:

- Low consistence which is sufficient to allow proper casting.
- Fast setting and hardening times, as well as high early strength development.
- Good skid resistance under all weather conditions.

Table 8.3 summarises the results of fresh and hardened properties of pavement concrete containing GC fine aggregate from laboratory investigations undertaken by Dumitru et al. (2010) and Polley et al. (1998), in which the mixes were intended for use in concrete slabs of rigid pavement in field trials. The relevant studies on rigid pavement concrete are limited, since the material is part of a larger

Table 8.3 Effect of glass cullet fine aggregate on the properties of concrete used in rigid pavement

	Main Observations	
Property	**Dumitru et al. (2010)**	**Polley et al. (1998)**
GC fine aggregate	Container glass, with a PSD between 0.3 and 3 mm; used up to 60% content	Container glass, with maximum aggregate size less than 1.5 mm; used up to 25% content
Fresh properties	Slump was unaffected, but bleeding was slightly increased and setting times were prolonged	Higher water demand for a given consistence.
Mechanical properties	No significant change in strength, but better shrinkage resistance was observed	The variation in compressive strength ranged from 25% strength loss to 5% strength gain
Durability	Resistance to chloride remained unchanged, but resistance to abrasion was reduced; ASR expansion could be mitigated with the use of fly ash	Acceptable resistance to freeze–thaw attack; ASR expansion could be mitigated with the use of fly ash

ASR, alkali–silica reaction; *GC*, glass cullet; *PSD*, particle size distribution.

family of concrete, and readers are advised to refer to Chapter 4, which deals with the effects of the use of GC fine aggregate on the performance of concrete. In general, these two studies have shown that the use of GC fine aggregate as a partial replacement for natural fine aggregate in rigid pavement concrete results in satisfactory performance, but some areas require further attention, as discussed below.

Although prepared to sand size fractions, the particle size distribution (PSD) of the GC fine aggregate used in these studies did not cover the full range of fine aggregate sizes normally used in practice, i.e., between 0.063 and 4 mm. The material used was either (i) lacking in fines, in the study undertaken by Dumitru et al. (2010), where the PSD was between 0.3 and 3 mm, or (ii) lacking in coarser size fractions, in the study of Polley et al. (1998), where the maximum aggregate size was less than 1.5 mm. It would appear that, in these two studies, GC fine aggregate was used as a natural fine aggregate replacement without retaining the PSD of the natural fine aggregate, which in this case affected the aggregate packing of concrete, with consequent results:

- Fresh concrete showed adverse effects in the form of reduced slump and harshness, leading to an increase in the water demand of the mix, as observed by Polley et al. (1998). In another study undertaken by Dumitru et al. (2010), the bleeding and air content of mixes were found to slightly increase. In addition, the initial and final setting times were both prolonged by about 2 h, though the material was reported not to contain detectable sugar or organic impurities.

- Compressive, tensile and flexural strength of concrete was not affected up to 60% GC-sand replacement of natural sand (Dumitru et al., 2010). However, in the study of Polley et al. (1998), the inclusion of GC was found to reduce compressive strength because of the increase in the water content in the mix for a given cement content. The presence of contaminants in unwashed GC fine aggregate was also considered to be a factor in this. The strength reduction was attributed partially to the weak bond between the cement paste and the GC fine aggregate. Owing to the relatively high stiffness of the material, concrete containing GC fine aggregate showed higher resistance to shrinkage deformation than the reference natural fine aggregate concrete (Dumitru et al., 2010).
- The deterioration processes normally experienced in pavement concrete, such as chloride ingress, freeze–thaw attack and abrasion, in general appeared not to be affected by the use of GC fine aggregate (Dumitru et al., 2010; Polley et al., 1998). However, at 60% sand content, abrasion resistance was adversely affected. Both studies showed that the use of fly ash at 20%–40% content as a cement component is effective in mitigating the ASR expansion resulting from the use of fine GC aggregate.

8.5 Bituminous Bound Applications

8.5.1 Marshall Mix Design

The Marshall mix design is perhaps the most commonly used empirical method, others being the Hveem method and the Gyratory Testing Machine method, in the laboratory for optimising the bitumen content of hot-mix bituminous mixtures and in the field for controlling the quality of hot-mix bituminous mixtures. The method was first developed by Bruce Marshall of the Mississippi State Highway Department in the United States in the 1930s (Transportation Research Board, 2011), and since then the method has been developed further and is now commonly used worldwide and adopted by many standards that are recognised internationally, such as ASTM D6927 (2015) and BS EN 12697-34 (2012).

In the Marshall test, briefly, a series of hot-mix bituminous mixtures with different bitumen contents is prepared by compacting them in cylindrical mounds using a standard drop hammer. The optimum bitumen content of the mixtures is selected based on the specific requirements of the Marshall design parameters, namely, stability, flow, air voids content, voids in the mineral aggregate (VMA) and voids filled with bitumen (VFB).

Using the Marshall method, hot-mix bituminous mixtures containing up to 20% GC aggregate in various sizes as (i) coarse aggregate with maximum size less than 12.7 mm, (ii) fine aggregate with maximum size less than 4.75 mm and (iii) filler with maximum particle size less than 75 μm have been undertaken by a number of researchers. Figures 8.2–8.6 show the test results of the Marshall design parameters of GC aggregate mixtures with the same optimum bitumen content as the corresponding reference natural aggregate mixture, which is about 5.5% on average. The compaction level used was not clearly indicated in some studies; however, where available it was noted that the specimens had been compacted with 50 and 75 blows to simulate medium and heavy traffic conditions, respectively.

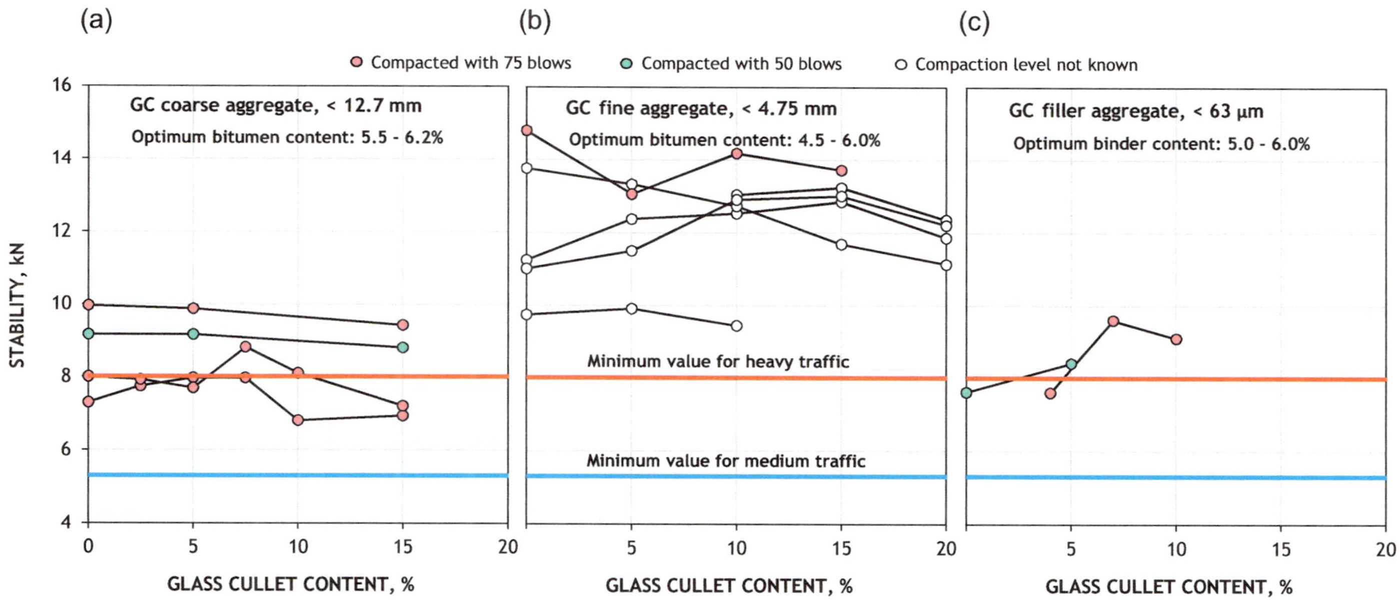

Figure 8.2 Marshall stability of mixes containing glass cullet (GC) with maximum aggregate size (a) less than 12.7 mm as coarse aggregate, (b) less than 4.75 mm as fine aggregate and (c) less than 63 μm as filler.
Data taken from Androjic and Dimter (2016), Arabani (2011), Arabani and Kamboozia (2013), Arabani et al. (2012), Behbahani et al. (2016), Ghasemi and Marandi (2013), Hughes (1990), Issa (2016), Jony et al. (2011), Raad (1992) and Wu et al. (2005).

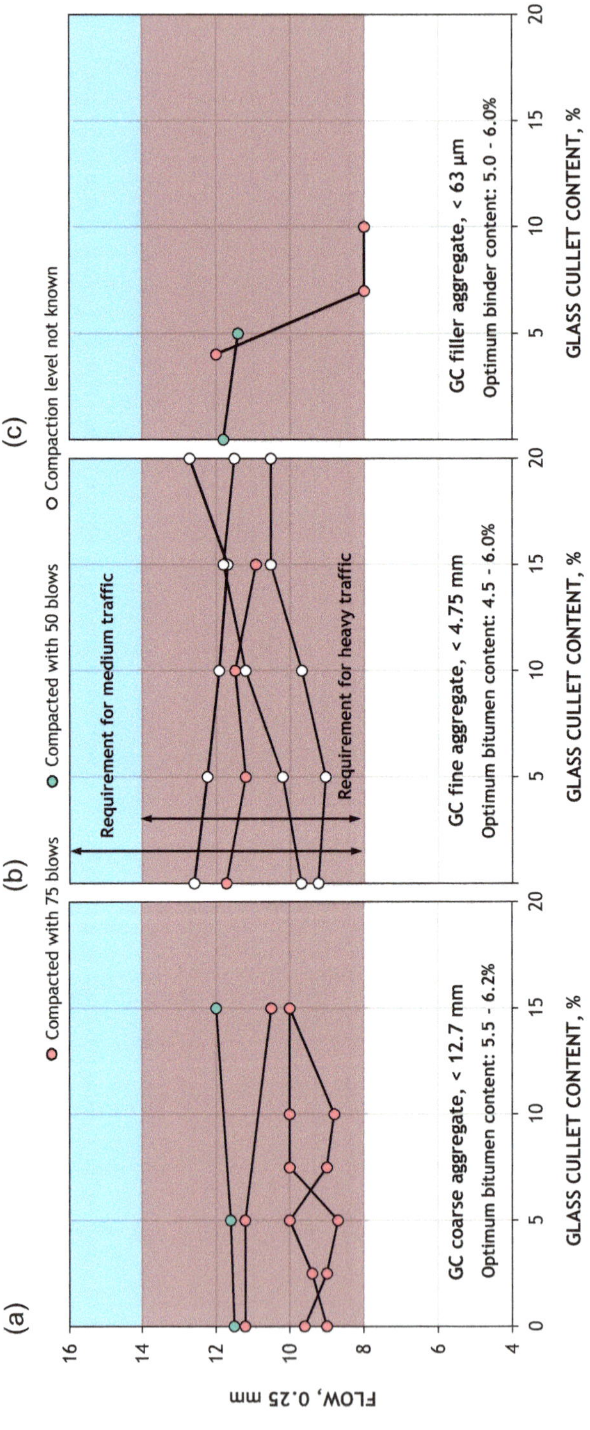

Figure 8.3 Marshall flow of mixes containing glass cullet (GC) with maximum aggregate size (a) less than 12.7 mm as coarse aggregate, (b) less than 4.75 mm as fine aggregate and (c) less than 63 µm as filler.
Data taken from Androjic and Dimter (2016), Arabani (2011), Arabani and Kamboozia (2013), Arabani et al. (2012), Behbahani et al. (2016), Ghasemi and Marandi (2013), Hughes (1990), Issa (2016), Jony et al. (2011) and Raad (1992).

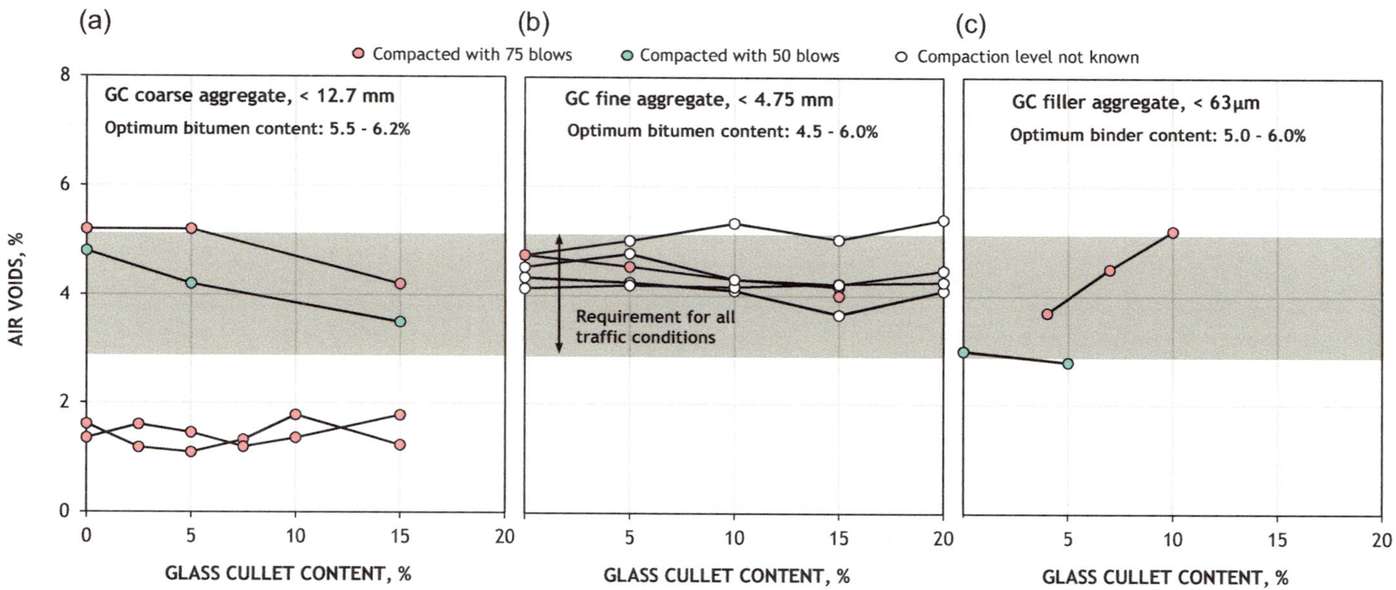

Figure 8.4 Air voids of mixes containing glass cullet (GC) with maximum aggregate size (a) less than 12.7 mm as coarse aggregate, (b) less than 4.75 mm as fine aggregate and (c) less than 63 μm as filler.
Data taken from Androkjic and Dimter (2016), Anochie-Boareng and George (2016), Arabani (2011), Arabani et al. (2012), Behbahani et al. (2016), Ghasemi and Marandi (2013), Hughes (1990), Issa (2016), Jony et al. (2011), Raad (1992) and Wu et al. (2005).

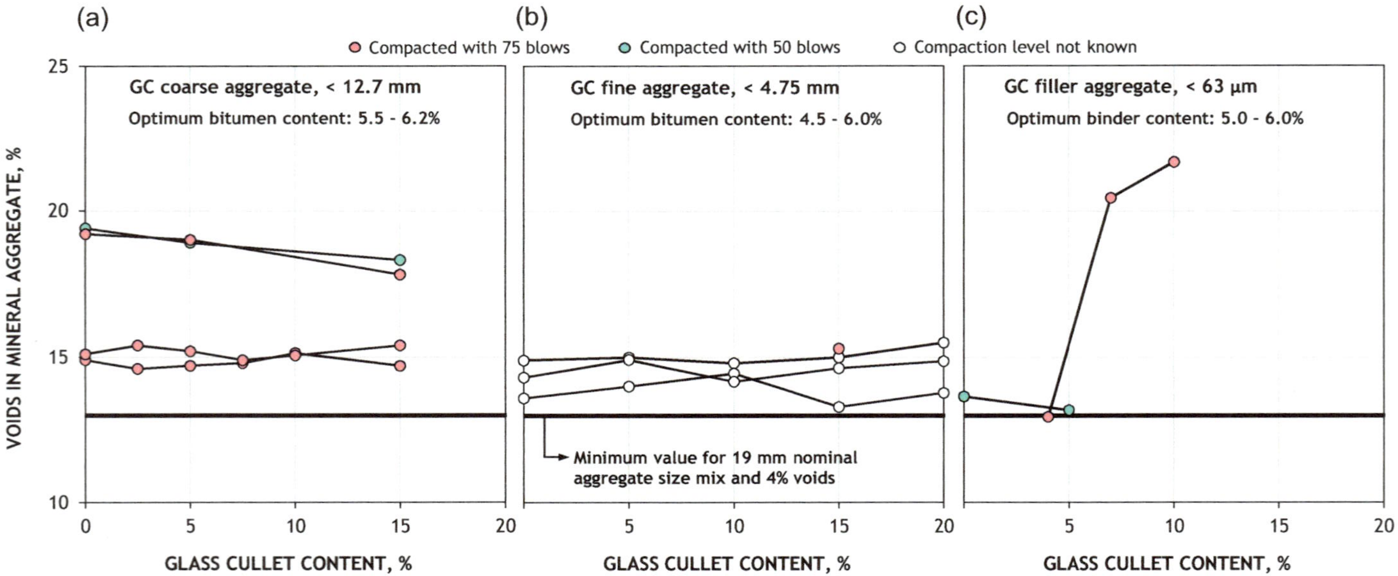

Figure 8.5 Voids in the mineral aggregate of mixes containing glass cullet (GC) with maximum aggregate size (a) less than 12.7 mm as coarse aggregate, (b) less than 4.75 mm as fine aggregate and (c) less than 63 µm as filler.
Data taken from Androjic and Dimter (2016), Anochie-Boareng and George (2016), Arabani (2011), Arabani et al. (2012), Behbahani et al. (2016), Ghasemi and Marandi (2013), Hughes (1990), Jony et al. (2011) and Raad (1992).

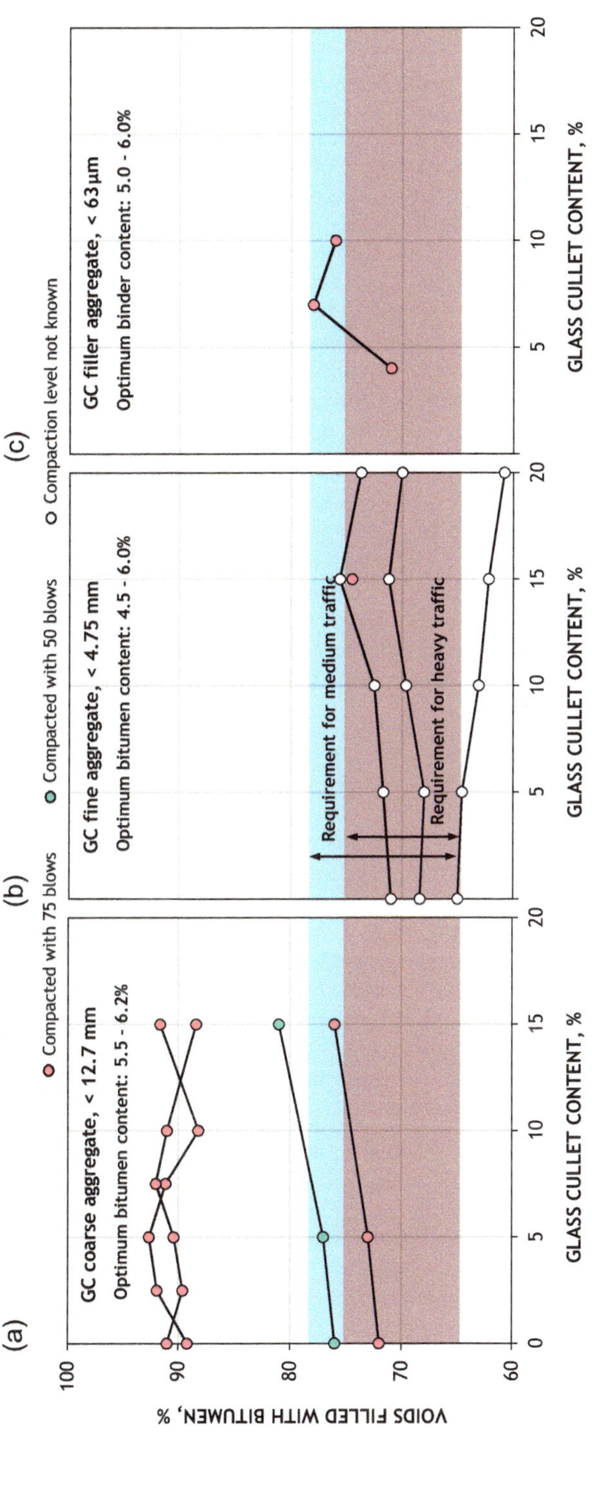

Figure 8.6 Voids filled with bitumen of mixes containing glass cullet (GC) with maximum aggregate size (a) less than 12.7 mm as coarse aggregate, (b) less than 4.75 mm as fine aggregate and (c) less than 63 μm as filler.
Data taken from Androjic and Dimter (2016), Anochie-Boareng and George (2016), Arabani (2011), Arabani et al. (2012), Behbahani et al. (2016), Ghasemi and Marandi (2013), Hughes (1990), Jony et al. (2011) and Raad (1992).

To evaluate the performance of the hot-mix bituminous mixtures containing GC aggregate, the requirements of the Marshall design parameters under medium and heavy traffic conditions specified by the Asphalt Institute (1997) are shown in Figures 8.2–8.6 for comparison. The following main points can be observed from these figures:

1. **Marshall Stability:** This is a measure to evaluate the strength of hot-mix bituminous mixtures and is a function of interaggregate friction and the viscosity of bitumen. The results shown in Figure 8.2(a and b) suggest that the use of GC as coarse aggregate and fine aggregate, up to 20%, does not significantly affect the Marshall stability of the mixture. Therefore, satisfying the minimum Marshall stability depends on the initial value of the reference mixtures made without GC aggregate. The Marshall stability values of all the mixtures made with GC as coarse aggregate and fine aggregate met the specified requirements, apart from two series of mixtures (see Figure 8.2(a)), in which the initial value was close to or below the minimum value of 8 kN for heavy traffic conditions. On the other hand, when GC was used as a filler aggregate up to 10%, the material tended to increase the Marshall stability of the mixture.

2. **Marshall Flow:** The Marshall flow is a measure of elastic plus plastic deformation of a mixture during the stability test. High flow values imply that the bitumen content of a mixture is excessive. In general, the Marshall flow values of hot-mix bitumen mixtures remain essentially unchanged when GC coarse aggregate and fine aggregate are used, and all are within the specification limits for medium and heavy traffic conditions (Figure 8.3(a and b)). A similar observation is also made for the mixtures containing GC filler, though in one case the flow value decreased significantly with the use of GC filler at 7% and 10% (Figure 8.3(c)).

3. **Air Voids:** All hot-mix bitumen mixtures contain a small amount of air voids, which are trapped within the bitumen and between aggregates during compaction. The air voids content of hot-mix bitumen mixtures is normally designed to 4% by volume, with an allowable range of 3%–5%. The air voids content of a mixture is affected by its bitumen content, PSD and aggregate packing as well as the quality of compaction. Figure 8.4 shows that, except for the two series of mixtures made with GC coarse aggregate, the air voids contents of mixtures made with GC coarse aggregate, fine aggregate and filler aggregate fluctuated within the limits of 3%–5%, or slightly exceeded the upper limit. As the PSD of the total aggregate was not kept constant when GC aggregate was added to the mixtures in most studies, such variations in the air voids contents are to be expected. However, the low air voids content of mixtures containing GC coarse aggregate, as well as of the corresponding reference natural aggregate mixture (Figure 8.4(a)), might indicate that their bitumen content was excessive.

4. **Voids in the Mineral Aggregate:** The total voids between aggregate particles is known as VMA, which is the sum of air voids content and effective bitumen content (the portion that is not absorbed by aggregates). This parameter indicates the space available for bitumen to coat each aggregate with an adequate bitumen film thickness. The minimum VMA specified by the Asphalt Institute (1997) varies depending on both the nominal maximum aggregate size and the designed air voids of mixtures. The minimum VMA requirement for mixtures with 19-mm aggregate size and 4% air voids is selected as an example in this case, as shown in Figure 8.5. In general, the VMA values of all the mixtures made with GC aggregate are above this minimum value. However, for its use as filler aggregate at more than 5%, the increase in the VMA of GC filler aggregate mixtures seems to be too high (Figure 8.5(c)) and could be uneconomical, as this demands higher bitumen content.

5. **Voids Filled with Bitumen:** The space present between aggregate particles and filled with bitumen is referred to VFB and is a function of VMA, air content and the effective binder content. This VFB value is used to describe the richness of the bitumen in a mixture. Except for the case in which the initial VFB of the reference natural aggregate mixture is high, the results in Figure 8.6 suggest that the VFB of mixtures containing GC coarse, fine and filler aggregates, up to 5%, is within the specified requirements for medium and heavy traffic conditions. It should be mentioned that owing to the impermeable characteristics of GC material, which is less likely to absorb bitumen, in practice, the use of GC aggregate can lead to an increase in both the effective bitumen content for a given bitumen and the VFB of the mixture.

Overall, GC can be used in the form of coarse, fine and filler aggregate, as a replacement for natural aggregate in hot-mix bituminous mixtures designed using the Marshall method. The Marshall design parameters are mostly unaffected when the material is used at 5% as coarse aggregate, 20% as fine aggregate and 5% as filler aggregate. However, it should be mentioned that the use of GC in coarse size fractions may prove problematic, as the material tends to be flaky and elongated, which can adversely affect the Marshall properties (Androjic and Dimter, 2016; Jasim et al., 2014). In addition, the PSD of the combined aggregate mix needs to remain unchanged when GC aggregate is used in order not to significantly affect the aggregate particle packing as well as the void-related properties.

8.5.2 Stiffness Modulus

The stiffness modulus measures the elastic response of a granular material under an applied load. In the field of road pavements, this property can be determined by several methods, amongst which the indirect tensile stiffness modulus test, which is a nondestructive method, has been commonly used to test hot-mix bituminous mixtures containing GC fine aggregate. In such studies, the replacement of natural fine aggregate by GC fine aggregate was commonly kept up to 20% content, except in one case in which the GC content was 100% (Airey et al., 2004). Some of these mixtures contained antistripping agent (Airey et al., 2004; Arabani, 2011; Behbahani et al., 2014; Shafabakhsh and Sajed, 2014). Table 8.4 summarises the parameters and results of the tests that were conducted at temperatures ranging from 5°C to 55°C.

In general, the results suggest that, for a given test temperature, the use of GC fine aggregate, up to about 10% and without the addition of antistripping agent, does not adversely affect the stiffness modulus of the mixtures. In some cases, although the air voids content of the GC fine aggregate mixture is about 0.5%–1.0% lower than that of the reference mixture, the expected higher stiffness modulus due to denser aggregate packing does not seem to emerge. However, when an antistripping agent was added to these mixtures, the stiffness modulus tended to improve, to where the modulus value could be similar to or higher than that of the reference mixtures made without antistripping agent.

In addition, although the stiffness modulus of both mixtures made with and without GC fine aggregate decreases as temperature increases, the response to the change in temperature of both mixtures is not the same. There is an indication that the reduction in stiffness modulus of mixtures made with GC fine aggregate is less than that of

Table 8.4 Effect of glass cullet fine aggregate on stiffness modulus of hot-mix bituminous mixtures

Reference	GC, %	Test Temp., °C	Antistripping Agent[a], %	Air Voids	Effect on Stiffness Modulus
AFGC (2011)	0–5	–	–	–	No significant effect
Airey et al. (2004)	0, 100	20	0, 0.3	Lower than reference	No significant effect
Arabani (2011)	0–20	5, 25, 40	3–5	Comparable to/lower than reference	Increase in value at higher GC content
Behbahani et al. (2015)	0, 10	5, 25, 40	0–4.5	–	Value increases as antistripping is increased
Hughes (1990)[b]	0–15	22.2	–	Lower than reference	No significant effect
Sivasubramaniam (2011)	0–5	15, 25, 55	–	Comparable to/lower than reference	No significant effect/increase in value
Shafabakhsh and Sajed (2014)	0–20	5, 25, 40	Used, but content unknown	–	Increase in value at higher GC content

GC, glass cullet; *reference*, reference mixtures made without GC; *Temp*, temperature.
[a]Added to mixtures made with fine GC aggregate, except for Shafabakhsh and Sajed (2014), in which antistripping agent was added to reference mixture.
[b]GC is used as coarse aggregate.

mixtures made without GC fine aggregate, but such effect is not consistent across all the studies. Further work in this area is therefore needed to understand the temperature effect on the stiffness modulus of hot-mix bituminous mixtures containing GC fine aggregate to varying proportions.

8.5.3 Rutting Resistance

Rutting, or permanent deformation, is one of the primary failure mechanisms in bituminous bound road pavements. It is an accumulation of small unrecovered strains, which are caused by repeated traffic load through consolidation, lateral movement of the upper portion of the pavement layer or the combination of both (Kandhal and Cooley, 2003). The effect of GC fine aggregate on the rutting resistance of hot-mix bituminous mixtures

Table 8.5 Rutting resistance of hot-mix bituminous mixtures made with glass cullet (GC) fine aggregate

Reference	GC, %	Test Method	Main Observation
(a) Deformation Strain			
Airey et al. (2004)	0, 100	Confined repeated-load axial test at 40°C, DD 185 (1994)	Final strain of GC mixture made with and without antistripping agent[a] is higher than that of the reference
Behbahani et al. (2015)	0, 10	Static creep test, FHWA (2001)	Final strain of GC mixtures is higher than that of the reference, but it reduces as the antistripping agent[b] content increases
(b) Rut Depth and Rut Rate			
Fitzsimons and Gibney (2004)	0–20	Wheel tracking test at 60°C, PrEN 12697-22 (1999)	Rut depth and rut rate of GC mixtures with up to 10% GC below the limit
Lanchance-Tremblay et al. (2014, 2016)	0–20	Wheel tracking test at 60°C, LC26-410 (2001)	Rut depth of GC mixtures exceeded the limit
(c) Dynamic Stability			
Wu et al. (2005)	0–20	Dynamic stability test at 60°C, JTG E20-T0719 (1993)	Dynamic stability is essentially unchanged at 5% GC inclusion

[a]0.3% antistripping agent.
[b]0%–4.5% antistripping agent.

has been evaluated in the form of deformation strain, rut depth and rut rate and dynamic stability under a static or dynamic load. A description of the work undertaken and the main observations of the results are summarised in Table 8.5. The emerging findings are as follows:

- The deformation strain of GC fine aggregate mixture made without antistripping agent is higher than that of reference mixture (Airey et al., 2004; Behbahani et al., 2014). The addition of an antistripping agent to GC fine aggregate mixtures has been shown to improve its rutting resistance, but its effectiveness depends on the amount used. Based on the available studies, it appears that the use of antistripping agent, in an amount of at least 2.5%, in GC fine aggregate mixtures could result in higher rutting resistance than the reference mixture.

- In the wheel tracking test, the results of Fitzsimons and Gibney (2004) show that the rut rate and rut depth of mixtures containing up to 10% GC fine aggregate correspondingly meet the maximum values of 5 mm/h and 7 mm specified in the highway specifications in the United Kingdom. On the other hand, the work undertaken by Lanchance-Tremblay et al. (2016) shows that the rut depth of mixtures containing 10% and 20% GC fine aggregate exceeds the limit of 10% specified in the highway specifications in Canada.

- The dynamic modulus of a mixture is essentially unchanged at 5% GC fine aggregate inclusion, beyond which the modulus value decreases as GC fine aggregate increases, up to 20% (Wu et al., 2005).

Collectively, the results tend to suggest that the rutting resistance of hot-mix bituminous mixture is not adversely affected with the use of GC fine aggregate up to about 10%, beyond which the rutting resistance decreases as the GC fine aggregate content increases. The addition of antistripping agent, for more than 2.5%, can be effective to enhance the rutting resistance of mixtures containing GC fine aggregate.

8.5.4 Moisture Damage Resistance

Moisture damage resistance of hot-mix bituminous mixtures is associated with two interrelated mechanisms:

(i) The first is the loss of cohesion, which involves the softening and weakening of the binding strength of bitumen. As this is essentially concerned with the binder's performance, little information about it can be found in the literature studying the use of glass in bituminous mixtures, and therefore for further information on this the readers are advised to refer to Hicks (1991).

(ii) The second mechanism is the loss of adhesion in the form of stripping due to the physical separation of bitumen and aggregate. Hot-mix bituminous mixtures containing GC aggregate are prone to stripping because the material is smooth and of an impermeable nature. This causes the bitumen to adhere less firmly to the GC particles in the mix. The high affinity of GC for water because of it high silica content (Hicks, 1991) is the other factor in the poor performance of the material in hot-mix bituminous mixtures.

The effect of GC aggregate on the moisture damage resistance of hot-mix bituminous mixtures made with or without antistripping agent has been investigated by many researchers, using several test methods. Most of these studies involved immersion mechanical tests, which measure the change in a mechanical property of a mixture after subjecting it to water immersion at a high temperature within the range of 40°C–60°C. Both indirect tensile strength and Marshall stability tests have commonly been adopted in such studies. A detailed description of the work undertaken in this area and the main findings are presented in Table 8.6. The information has been separated based on GC aggregate maximum size, with coarse aggregate of maximum size of 12.7 mm and fine aggregate of maximum size of 4.75 mm.

It would appear from Table 8.6 that all the investigations conducted before the year 2000 have been on the use of GC coarse aggregate in hot-mix bituminous mixtures, and after the year 2000, the maximum size of the GC aggregates used was changed to 4.75 mm. This is perhaps a useful change as GC aggregate at coarse size fractions would be prone to being flaky and elongated, which can be expected to give rise to breakage of the material and poor aggregate compaction (Malisch et al., 1975; Larsen, 1989).

Table 8.6 Moisture damage resistance of hot-mix bituminous mixtures made with glass cullet (GC) coarse and fine aggregates

Reference	GC, %	Antistripping Agent	Test Method	Findings
(a) GC Coarse Aggregate With Maximum Size of 12.7 mm				
Hughes (1990)	0–15	1.0% hydrated lime	Indirect tensile strength test	Little or no effect on moisture resistance
Malisch et al. (1970)	100	Proprietary antistripping agent; slow setting cationic emulsion	Static stripping test	No stripping when antistripping agent is added
Malisch et al. (1975)	100	1%–4% commercial antistripping agents; 0.9%–5.4% hydrated lime; 6.5% limestone dust	Compressive strength test at 60°C	Hydrated lime is the most effective antistripping agent in improving the moisture resistance
Maupin (1998)	0–20	Chemical additive; hydrated lime	Indirect tensile strength test at 60°C	Both antistripping agents increase the moisture resistance
Nash et al. (1995)	5	0.5% commercial antistripping agents; 0.5% hydrated lime	Asphalt stripping test	Commercial antistripping agent shows better performance
Raad (1992)	15	0.25% commercial antistripping agent	Coating and stripping test at 25°C; Marshall stability test at 60°C	No sign of stripping or loss of stability
(b) GC Fine Aggregate With Maximum Size of 4.75 mm				
Airey et al. (2004)	100	0% and 0.3% antistripping agent	Stiffness modulus test at 60°C	Moisture resistance of GC mixtures is less than that of reference
Anochie-Boateng and George (2016)	0, 15	3% hydrated lime	Modified Lottman indirect tension test at 60°C	Moisture resistance is unchanged
Behbahani et al. (2014)	0, 10	0%–4.5% Zycosoil	Modified Lottman indirect tension test at 60°C	Moisture resistance is increased with the use of antistripping agent

Continued

Table 8.6 Continued

Reference	GC, %	Antistripping Agent	Test Method	Findings
Fitzsimons and Gibney (2004)	0–20	–	Stiffness test at a cycle of 60°C, 5°C and 20°C	Moisture resistance is slightly reduced
Lanchance-Tremblay et al. (2016)	0, 10	–	Marshall stability test at 60°C	Low moisture resistance but meets the minimum local requirement
Su and Chen (2002)	0–15	0%–3% hydrated lime	Stability test	Moisture resistance increases as lime increases
Wu et al. (2005)	0–20	–	Stability test, freeze–thaw splitting test, water-immersed rutting test at 45°C	Low moisture resistance but mixtures with up to 10% GC meet the local requirement
	10	0.4% coconut oil ethanolamine, 2% hydrated lime	Stability test, freeze–thaw splitting test	Both antistripping agents increase the moisture resistance

For both coarse and fine GC aggregates, most of the studies investigating the stripping effect of the GC aggregate have been limited to the use of the material up to 20%, with a few cases in which the GC aggregates fully replace natural aggregate (Airey et al., 2004; Malisch et al., 1970, 1975). Hydrated lime has been the most commonly used antistripping agent in these investigations, followed by the synthetic type of antistripping agents. Other materials such as limestone dust, nanotechnology Zycosoil, slow setting cationic emulsion and coconut oil ethanolamine have also been tested as antistripping agents.

Overall, the information summarised in Table 8.6 suggests that, though its moisture damage resistance is reduced (Lanchance-Tremblay et al., 2015; Wu et al., 2005) without the addition of antistripping agent, it is possible to use GC fine aggregate up to 10% in hot-mix bituminous mixtures and still comply with the local standard requirements. When antistripping agent is used, the moisture damage resistance of mixtures made with GC coarse and fine aggregates is generally improved. Depending on the type and amount of antistripping agent applied, the moisture damage resistance of mixtures could be similar to or higher than that of mixtures made with natural aggregate (Anochie-Boateng and George, 2016; Behbahani et al., 2015; Hughes, 1990; Maupin, 1998; Su and Chen, 2002).

It is not possible to compare the effectiveness of antistripping agents in improving the moisture damage resistance of mixtures made with GC, as other than hydrated lime, the type of antistripping agent used is different in each individual study. Notwithstanding this, hydrated lime appears to be effective in increasing the moisture damage resistance of bituminous mixtures made with GC aggregates in most of the studies examined (Anochie-Boateng and George, 2016; Hughes, 1990; Malisch et al., 1975; Maupin, 1998; Su and Chen, 2002). Although the effect of hydrated lime can vary with type of bitumen used, based on the available data, the use of hydrated lime as an antistripping agent, at least at 1%, is most likely to result in GC aggregate bituminous mixtures with satisfactory performance in comparison with natural aggregate bituminous mixtures.

8.6 Environmental Impact

GC can be derived from different sources of waste glass, amongst which the soda lime type, such as container glass and flat glass, is the most abundant and is the obvious choice for developing for use as an aggregate in the construction industry. Although soda lime glass is a safe material, to be sure, for health and safety reasons it is still required to be tested for its environmental impact to ensure its use does not bring any harm to nature and/or human life.

The environmental assessment of the use of soda lime GC as a fine aggregate in the unbound base layer of a pavement has been undertaken by Arulrajah et al. (2016) and Imteaz et al. (2012). The chemical composition of GC has been determined for a range of toxic elements, namely, arsenic, barium, cadmium, chromium, copper, mercury, molybdenum, nickel, lead, selenium, silver and zinc. The results showed that the toxic elements present in soda lime GC were well below the regulatory limits, with many elements being less than 5 mg/kg. The material can be classified as 'Category C' waste, which is the least hazardous type based on the waste classification specified by the Environment Protection Agency Victoria in Australia (EPA Victoria, 2007). Moreover, the leaching characteristics of soda lime GC were determined using both acidic and alkaline solutions in accordance with Australian standard leaching procedure (AS 4439.3, 1997) and it has been found that the amounts of the heavy metals released from GC in both buffer solutions were below the threshold of hazardous waste specified by EPA Victoria (2007).

Although the research on the environmental impact of soda lime GC aggregate in hydraulically and bituminous bound applications has been limited, given the benign nature of the material, it is very likely that similar positive results would be obtained for the material in such applications. Thus, it could be argued that no further testing of the material for environmental impact is required, but to be sure, environmental protection authorities the world over are most likely to demand further proof that the material is benign and perfectly safe to use.

8.7 Case Studies

Most of the case studies undertaken during the period of 1970–2016 that deal with the use of GC aggregate in both coarse and fine size fractions in road pavements, as an unbound, hydraulically bound and bituminous bound material, are summarised in Table 8.7. Most of these field applications have been undertaken in the United States, mainly in the unbound and bituminous forms, followed by Australia. There have also been a few case studies using GC aggregate in road pavement applications in Canada, the United Kingdom and Taiwan. The material was probably first used in practice in bituminous bound application as early as the 1970s and 1980s (Malisch et al., 1975; Larsen, 1989); during the 1990s there was an increase in the interest in its use as an unbound aggregate. On the other hand, the use of GC aggregate in hydraulically

Table 8.7 Case studies using glass cullet (GC) in various road pavement applications

		GC		
Reference	**Country**	**Size, mm**	**Content, %**	**Application**
(a) Unbound Applications				
Arulrajah et al. (2013)	Australia	<4.75	15, 30	Base for footpath
HDR Engineering (1997)	USA	(Coarse)	Mostly 10	Base and subbase for road
Krivit (1999)	USA	<12.7	Up to 100	Base and subgrade for car park and road
Reindl (2003)	USA	n.a.	n.a.	Base for sidewalk
Senadheera et al. (2005)	USA	n.a.	0–15	Base for road
TxDOT (1999)	USA	<15	10, 20	Base for road
Wisconsin Transportation Information Center (1999)	USA	n.a.	n.a.	Base for road
(b) Hydraulically Bound Applications				
Polley et al. (1998)	USA	<4	10, 20	Concrete slab
Dumitru et al. (2013), Song et al. (2016)	Australia	0.3–3	45	Concrete slab
Wood et al. (2011)	Australia	(Fine)	n.a.	Concrete slab

Continued

Table 8.7 Continued

		GC		
Reference	**Country**	**Size, mm**	**Content, %**	**Application**
(c) Bituminous Bound Applications				
City of Armadale (2010)	Australia	<2.36	n.a.	Car park
Emersleban and Meyer (2012, 2014)	USA	(Fine)	100	Roads of light-to-heavy traffic conditions
Malisch et al. (1975) and Larsen (1989)	Canada, USA	<9.5 to <25	20–70	Roads of light-to-heavy traffic conditions
Maupin (1997)	USA	n.a.	15	Residential roads
Nicholls and Lay (2002)	UK	<20	30	Binder course and base course
Su and Chen (2002)	Taiwan	<4.75	10	Surface course
Wood et al. (2011)	Australia	n.a.	n.a.	Road

bound applications began to emerge only in the 2000s. A more detailed discussion of these case studies involving the use of GC aggregate in road pavement applications is provided Section 10.3.4; however, for completeness of this chapter, some of the main points are given below:

- In unbound applications, GC aggregate can be used at up to 30% as a base material in the construction of pavement. Although information regarding the aggregate size used is generally lacking, it suggests that GC has tended to be used in coarse size fractions in the case studies reported in Table 8.7. No adverse effect of the use of GC aggregate in unbound form has been reported.
- In hydraulically bound applications, GC has been used as a replacement for natural fine aggregate in the construction of concrete slabs in rigid pavements. The response is generally positive, with satisfactory long-term performance of concrete recorded in terms of compressive strength, water absorption and resistance to chloride ingress, freeze–thaw attack and abrasion. No sign of ASR was found in concrete slab containing GC fine aggregate (Dumitru et al., 2013; Song et al., 2016); however, the time duration for which the data have been reported is 5 years, which is too short for the ASR symptoms to become visible in every situation.
- In bituminous bound applications, earlier case studies suggest that GC was normally used as a coarse aggregate, with maximum size up to 25 mm, and it tended to result in many technical problems such as loss of adhesion, lower skid resistance and breakage of the material (Malisch et al., 1975; Larsen, 1989). However, in recent years, the material has been used as a fine aggregate with maximum size less than 4 mm, with no adverse effect on the technical performance of bituminous mixtures being reported (City of Armadale, 2010; Emersleban and Meyer, 2012, 2014; Su and Chen, 2002).

8.8 Conclusions

This chapter examines the potential for the use of GC as a replacement for natural aggregate, in unbound, hydraulically bound and bituminous bound forms, in the construction of road pavements. In general, the use of GC aggregate in these applications is technically feasible but requires some modification to the mix design to prevent any undue risk.

The CBR of GC is essentially comparable to that of natural gravel and sand. At a standard compaction effort, the CBR value of GC in a sand form and a mix of gravel and sand form is 20% and 30%, respectively, in comparison to the 10%–50% CBR value for natural sand. The corresponding values at modified compaction effort for GC are 44% and 74%. The results suggest that GC can be used in subgrade as a replacement for in situ soil, potentially allowing for a reduction in the thickness of pavements, as well as in the capping layer as a granular material.

Although limited data are available, GC fine aggregate has been compatible with steel slag, ladle furnace slag, foundry sand and fly ash, in producing Portland cement bound mixtures for use in the construction of subbase and base course. The resulting mixtures were reported to give satisfactory unconfined compressive strength, indirect tensile strength and dynamic elastic modulus. In concrete slab applications, the use of GC fine aggregate, because of particle packing, has been reported to affect the consistence, compressive strength, and abrasion resistance of concrete, but nonetheless the overall performance remained acceptable.

In bituminous bound applications, GC can be used in the form of coarse, fine and filler aggregate, as a replacement for natural aggregate, in hot-mix bituminous mixtures designed using the Marshall method. GC fine aggregate can reduce the stiffness modulus, rutting resistance and moisture damage resistance of hot-mix bituminous mixtures, though its inclusion at up to 10% generally gives an acceptable performance. The addition of antistripping agent has been shown to enhance the engineering performance of mixtures made with GC aggregate. In the case of moisture damage resistance, the use of hydrated lime at at least 1.0% is generally found to be effective to make GC aggregate bituminous mixtures less susceptible to stripping.

The environmental assessment of GC aggregate has shown that the material is safe to use, with the amounts of toxic elements present in GC, as well as their released concentrations, being below the regulatory limits for hazardous waste specified by EPA Victoria (2007). As an aggregate component, GC has been used in road pavement constructions mainly in the United States and Australia. The use of GC, in both fine and coarse size fractions, as unbound and hydraulically bound material has been shown to be satisfactory; it is well suited to fine aggregate bituminous bound applications.

References

AASHTO T193, 2013. Standard Method of Test for the California Bearing Ratio. American Association of State Highway and Transportation Officials, Washington, DC, USA.

AFGC, 2011. Performance of Asphalt Mixtures Containing Recovered Crushed Glass. Australian Food and Grocery Council, Fact Sheet. 2 pp.

Airey G D, Collop A C and Thom N H, 2004. Mechanical performance of asphalt mixtures incorporating slag and glass secondary aggregates. In: Proceedings of the 8th Conference on Asphalt Pavement for South Africa (CAPSA'04), Sun City, South Africa 13 pp.

Androjic I and Dimter S, 2016. Properties of hot mix asphalt with substituted waste glass. Materials and Structures 49, 249–259.

Anochie-Boateng J and George T, 2016. Use of waste crushed glass for the production of hot-mix asphalt. In: Fourth International Conference on Sustainable Construction Materials and Technologies, Las Vegas, USA 11 pp.

Arabani M, 2011. Effect of glass cullet on the improvement of the dynamic behaviour of asphalt concrete. Construction and Building Materials 25, 1181–1185.

Arabani M and Kamboozia N, 2013. The linear visco-elastic behaviour of glasphalt mixture under dynamic loading conditions. Construction and Building Materials 41, 594–601.

Arabani M, Mirabdolazimi S M and Ferdowsi B, 2012. Modelling the fatigue behaviours of glasohalt mixtures. Scentia Iranica A 19 (3), 341–345.

Arulrajah A, Ali M M Y, Disfani M M, Piratheepan J and Bo M W, 2013. Geotechnical performance of recycled glass-waste rock blends in footpath bases. Journal of Materials in Civil Engineering 25 (5), 653–661.

Arulrajah A, Ali M M Y, Disfani M M and Horpibulsuk S, 2014. Recycled-glass blends in pavement base/subbase applications: laboratory and field evaluation. Journal of Materials in Civil Engineering 26 (7), 12.

Arulrajah A, Imteaz M, Horpibulsuk S, Du Y and Shen J, 2016. Recycled concrete aggregate/municipal glass blends as a low-carbon resource material for footpaths. Road Materials and Pavement Design 1–14.

AS 4439.3, 1997. Australian Standard.Wastes, Sediments and Contaminated Soils. Part 3: PREPARATION of Leachates – Bottle Leaching Procedure. Standards Australia, Homebush, NSW, Australia.

Asphalt Institute, 1997. Mix Design Methods for Asphalt Concrete and Other Hot-Mix Types MS-02. Asphalt Institute, Lexington, KY, USA. 141 pp.

ASTM D1883, 2016. Standard Test Method for California Bearing Ratio (CBR) of Laboratory-compacted Soils. ASTM International, West Conshohocken, PA, USA.

ASTM D6927, 2015. Standard Test Method for Marshall Stability and Flow of Asphalt Mixtures. ASTM International, West Conshohocken, PA, USA.

Behbahani H, Ziari H, Kamboozia N, Khaki A M and Mirabdolazimi S M, 2015. Evaluation of performance and moisture sensitivity of glasphlat mixtures modified with nanotechnology zycosoil as an anti-stripping additive. Construction and Building Materials 78, 60–68.

Behbahani H, Ziari H and Kamboozia N, 2016. Evaluation of the visco-elasto-plastic behavior of glasphalt mixtures through generalized and classic Burger's models modification. Construction and Building Materials 118, 36–42.

BS EN 1377–1379, 1990. Methods for Test for Soils for Civil Engineering Purposes. In-situ tests. BSI, London, UK.

BS EN 12697–34, 2012. Bituminous Mixtures. Test methods for hot mix asphalt. BSI, London, UK.

BS EN 13043, 2002. Aggregates for Bituminous Mixtures and Surface Treatments for Roads, Airfields and Other Trafficked Areas. BSI, London, UK.

BS EN 13242:2002+A1, 2007. Aggregates for Unbound and Hydraulically Bound Materials for Use in Civil Engineering Work and Road Construction. BSI, London, UK.

City of Armadale, 2010. Information Bulletin. City of Armadale, Armadale, Western Australia, Issue no. 05/2010 Available from: https://www.armadale.wa.gov.au.

DD 185, 1994. Method for Assessment of Resistance to Permanent Deformation of Bitumen Aggregate Mixtures Subject to Unconfined Uniaxial Loading. BSI, London, UK.

Disfani M M, Arulrajah A, Bo M W and Hankour R, 2011. Recycled crushed glass in road work applications. Waste Management 31 (11), 2341–2351.

Dumitru I, Song T, Caprar V, Brooks P and Moss J, 2010. Incorporation of recycled glass for durable concrete. In: Second International Conference on Sustainable Construction Materials and Technologies, Ancona, Italy 9 pp.

Dumitru I, Song T, Bornstein B, Brooks P and Moss J, 2013. Field trials using recycled glass as natural sand replacement and powdered glass as cementitious materials replacement in concrete pavement. In: Third International Conference on Sustainable Construction Materials and Technologies, Kyoto, Japan 9 pp.

Emersleben A and Meyer N, 2012. The use of recycled glass for the construction pavements. In: GeoCongress 2012, Oakland, CA, USA, pp. 1642–1649.

Emersleben A and Meyer M, 2014. Sustainable pavement construction by the use of recycled glass. International Journal of Geotechnical Engineering 8 (4), 436–440.

EPA Victoria, 2007. Classification of Wastes. Environmental Protection Agency of Victoria, Melbourne, Australia. Information Bulletin, Publication No. 448.

European Aggregates Association, 2016. Annual Review 2015–2016. European Aggregates Association, Brussels, Belgium. 32 pp.

FHWA, 2001. Test Method for Determining the Creep Compliance, Resilient Modulus and Strength of Asphalt Material Using the Indirect Tensile Test Device. FHWA Protocol P07. Version 1.1, Washington, DC, USA.

Fitzsimons L and Gibney L, 2004. Recycled glass in hot rolled asphalt wearing course mix. In: Limbachiya M C and Roberts J J (Eds.), Proceedings of the International Conference on Sustainable Waste Management and Recycling: Glass Waste, London, UK. Thomas Telford, London, UK, pp. 187–197.

Ghasemi M and Marandi S M, 2013. Laboratory studies of the effect of recycled glass powder additive on the properties of polymer modified asphalt binders. International Journal of Engineering 26 (10), 1183–1190.

Halstead W J, 1993. Use of Waste Glass in Highway Construction Report No. VTRC 93–TAR2. Virginia Transportation Research Council, Charlottesville, VI, USA. 23 pp.

HDR Engineering Inc., 1997. Glass Cullet Utilization in Civil Engineering Applications Technical Report. Nebraska State Recycling Association, Omaha, NE, USA. 30 pp.

Hicks R G, 1991. Moisture Damage in Asphalt Concrete. National Cooperative Highway Research Program Synthesis 175. Transportation Research Board, Washington, DC, USA. 90 pp.

Highway Agency, 1995. HA 44/91: design and preparation of contract documents. In: Design Manual for Roads and Bridges (DMRB), vol. 4. Highway Agency, London, UK. Section 1, Part 1, 96 pp.

Huang Y, Bird R N and Heidrich O, 2007. A review of the use of recycled solid waste materials in asphalt pavements. Resources, Conservation and Recycling 52, 58–73.

Hughes C S, 1990. Final Report – Feasibility of Using Recycled Glass in Asphalt Report No. VTRC 90–R3. Virginia Transportation Research Council, Charlottesville, Virginia, USA.

Imteaz M A, Younus Ali M M and Arulrajah A, 2012. Possible environmental impacts of recycled glass used as a pavement base material. Waste Management & Research 30 (9), 917–921.

Issa Y, 2016. Effect of adding crushed glass to asphalt mix. Archives of Civil Engineering LXII (2), 35–44.

Jamshidi A, Kurumisawa K, Nawa T and Igarashi T, 2016. Performance of pavements incorporating waste glass: the current state of the art. Renewable and Sustainable Energy Reviews 64, 211–236.

Jasim A, 2014. Using waste glass as secondary aggregates in asphalt mixtures. International Journal of Advanced Research 2 (1), 41–46.

Jony H H, Al-Rubaie F and Jahad I, 2011. The effect of using glass powder filler on hot asphalt concrete mixtures properties. Engineering and Technology Journal 29 (1), 44–57.

JTG E20-T0719, 1993. Standard Test Method of Bitumen and Bituminous Mixture for Highway Engineering. China Communications Press, Beijing, China.

Kandhal O S, 1992. Waste Materials in Hot Mix Asphalt – an Overview NCAT Report No. 92-6. National Center for Asphalt Technology, AL, USA. 16 pp.

Kandhal P S and Cooley Jr. L A, 2003. Accelerated Laboratory Rutting Tests: Evaluation of the Asphalt Pavement Analyzer National Cooperative Highway Research Program Report 508. Transportation Research Board, Washington DC, USA. 116 pp.

Krivit D, 1999. Ramsey County Demonstrations of Recycled Glass as an Aggregate Supplement Final report. Dan Krivit and Associates, St. Paul, MN, USA. 23 pp.

Lanchance-Tremblay E, Vaillancourt M and Perraton D, 2014. Evaluation of the performance of an asphalt mix with crushed glass. In: 59th Annual Conference of the Canadian Technical Asphalt Association, Winnipeg, Manitoba 15 pp.

Lanchance-Tremblay E, Vaillancourt M and Perraton D, 2016. Evaluation of the impact of recycled glass on asphalt mixture performances. Road Materials and Pavement Design 17 (3), 600–618.

Larsen D A, 1989. Feasibility of Utilising Waste Glass in Pavements Report No. 343-21-89-6. Connecticut Department of Transportation, Wethersfield, CT, USA. 27 pp.

LC 26-410, 2001. Résistance à la deformation des enrobés à l'essai d'orniérage. Ministère des Transports du Québec, Quebec, Canada, (Equivalent, EN 12697–22).

Malisch W R, Day D E and Wixson B G, 1970. Use of Domestic Waste Glass as Aggregate in Bituminous Concrete. Highways Research Board, Washington, DC, USA, pp. 1–10. Highway Research Record 307.

Malisch W R, Day D E and Wixson B G, 1975. Use of Domestic Waste Glass for Urban Paving Summary Report EPA-670/2-75-053. National Environmental Research Center, Office of Research and Development, US Environmental Protection Agency, Cincinnati, OH. 46 pp.

Maupin Jr. G W, 1997. Final Report: Glasphalt Test Sections in Virginia Report No. FHWA/A/VTRC 98–R6. Virginia Transportation Research Council, Charlottesville, VA, USA. 10 pp.

Maupin Jr. G W, 1998. Final Report: Effect of Glass Concentration on Stripping of Glasphalt Report No. VTRC 98–R30. Virginia Transportation Research Council, Charlottesville, VA, USA. 11 pp.

Nash P, Jayawickrama P, Tock R W, Senadheera S, Viswanathan K and Woolverton B, 1995. Use of Glass Cullet in Roadway Construction: Laboratory Testing and Specification Development Report No.: 0-1331-2F. Texas Department of Transportation, Austin, TX, USA. 44 pp.

Nicholls J C and Lay J, 2002. Glassphalt. The use of recycled glass in asphalt by RMC Aggregates (UK) Ltd. Aggregate Information Service. Digest No. 72, 4 pp.

O'Flaherty C A (Ed.), 2002. Highways. Butterworth-Heinemann, Woburn, MA, USA. 553 pp.

Pasetto M and Baldo N, 2016. Recycling of waste aggregate in cement bound mixtures for road pavement bases and sub-bases. Construction and Building Materials 108, 112–118.

Polley C, Cramer S and de la Cruz R, 1998. Potential for using waste glass in Portland cement concrete. Journal of Materials in Civil Engineering 10 (4), 210–219.

PrEN 12697-22, 1999. Bituminous Mixtures – Test Methods for Hot Mix Asphalt – Part 22: Wheel Tracking. Draft European Standard.

Raad L, 1992. Potential Utilization of Recycled Waste Glass in Alaskan Pavements Report No. INE/TRC/QRP-92.01. Alaska Cooperative Transportation and Public Facilities Research Program. 23 pp.

Reindl J, 2003. Reuse/recycling of Glass Cullet for Non-container Uses. Dane County Department of Public Works, Madison, WI, USA. 136 pp.

Senadheera S, Rana A and Nash P T, 2005. Characterization of the Behaviour of Granular Material Containing Glass Cullet. Available from: http://www.vegvesen.no.

Shafabakhsh G H and Sajed Y, 2014. Investigation of dynamic behaviour of hot mix asphalt containing waste materials; case study: glass cullet. Case Studies in Construction Materials 1, 96–103.

Sivasubramaniam S, 2011. Performance of Asphalt Mixtures Containing Crushed Glass Fines Report No.: 21/2009/BA967-Rev1. GHD Geotechnics, Artarmon, NSW, Australia.

Song T, Dumitru I and Bornstein B, 2016. Long term durability properties of concretes with glass sand and glass powder. In: Fourth International Conference on Sustainable Construction Materials and Technologies, Las Vegas, NV, USA 9 pp.

Su N and Chen J, 2002. Engineering properties of asphalt concrete made with recycled glass. Resources, Conservation and Recycling 35, 259–274.

Transportation Research Board, 2011. A Manual for Design of Hot Mix Asphalt with Commentary. Transportation Research Board, Washington, DC, USA. National Cooperative Highway Research Program Synthesis 673, 272 pp.

Transportation Research Board, 2013. Recycled Materials and Byproducts in Highway Applications. Manufacturing and Construction Byproducts, vol. 8. Transportation Research Board, Washington, DC, USA. NCHRP Synthesis 435, 99 pp.

TxDOT, 1999. Use of Glass Cullet in Roadway Construction, Year of the Recycled Roadway Materials, February: Glass Cullet. Texas Department of Transportation, Austin, TX, USA. 29 pp.

Wisconsin Transportation Information Center, 1999. Using Recovered Materials in Highway Construction. Wisconsin Transportation Bulletin No. 20. WI, USA, 8 pp.

Wood M, Jeffreys C and Richardon P, 2011. Waverley Council: Recycled Glass in Roads. Department of Sustainability, Environment, Water, Population and Communities, National Waste Policy, Australia. 2 pp.

Wu S, Yang W and Xue Y, 2005. Preparation and properties of glass-asphalt concrete. In: Seventh International Conference on the Bearing Capacity of Roads, Railways and Airfields, Trondheim, Norway 11 pp.

Use of Glass Cullet in Ceramics and Other Applications

9

Main Headings

- Glass cullet in ceramics
- Glass cullet as aesthetic finishing material
- Glass cullet as filtration medium
- Glass cullet as epoxy composite
- Glass cullet as blast abrasive
- Glass cullet as paint filler
- Glass cullet in elastomeric roof coating
- Other applications

Synopsis

This chapter deals with the use of glass cullet as a new material in alternative markets, other than the previously covered concrete, geotechnical and road pavement applications. The results of several experimental investigations have been compiled and analysed on the use of waste glass in the manufacture of ceramic-based products, namely, glass-ceramics, foam glass-ceramics, ceramic bricks and tiles, porcelain products and glazing material. Other applications of waste glass have also been identified and succinctly addressed in this chapter, such as the production of aesthetic finishing materials, filtration medium, epoxy resins, glass fibres, blast abrasive, paint filler, elastomeric roof coatings, amongst others.

Keywords: Glass cullet, Waste glass, Sustainability, Ceramics, Aesthetic finishing material, Filtration medium, Epoxy, Glass fibres, Blast abrasive, Paint filler, Roof coating.

Sustainable Construction Materials: Glass Cullet. https://doi.org/10.1016/B978-0-08-100984-0.00009-6

9.1 Introduction

To increase the potential uses of glass cullet (GC), alternative markets, other than concrete, geotechnical and road pavement applications, need to be developed and exploited. From the literature review, several other applications using GC as a raw material have been identified, of which the use of GC in ceramic-based applications has been given emphasis in experimental investigations and with very promising results. The scope of this chapter is to highlight the several other secondary and alternative applications in existence for GC. Although these are extremely interesting applications, a succinct appraisal of the successful use of waste glass will be made throughout this chapter, in line with the approach adopted in the previous chapters.

9.2 Glass Cullet Use in Ceramics

9.2.1 Glass-Ceramics

Glass-ceramics are polycrystalline materials, based on inorganic silicates embedded in a glass matrix and arranged by the controlled crystallisation of appropriate glass compositions. This crystallisation is achieved by subjecting the parent glass to carefully regulated heat treatments at specific temperatures, with the aim of optimising the nucleation rate and growth rate of crystals, for specific periods of time. Depending on the production procedure, desired application and raw materials, the resulting glass-ceramic can be a high-strength material, with zero porosity and high temperature stability, among several other positive features (Chinnam et al., 2013).

Since the manufacture of glass-ceramics requires energy-intensive heat treatments, the production process is generally expensive and with significant environmental impacts. For this reason, various routes for the production of glass-ceramics that allow incorporating waste glass have been devised over time, with the purpose of reducing the costs of production (Rawlings et al., 2006):

- **Conventional (two-stage) method**: Glass is devitrified in a two-stage heat treatment to induce a high nucleation rate and a high crystal growth rate. After having obtained a high nucleation rate, prompted by an initial low-temperature heat treatment, the glass forms high-density nuclei throughout its interior. In the second stage, the glass is subjected to a higher-temperature heat treatment to promote the nuclei's growth at a reasonable rate.
- **Modified conventional (single-stage) method**: Where there is a significant similarity of the curves for nucleation rate and growth rate, then it is possible to subject glass to a single-stage heat treatment. This rate curve overlap is possible by optimising the glass composition.

- **Petrurgic method**: Some glass-ceramics are produced by subjecting the original glass to very slow and controlled cooling from its molten state, without the need of an intermediate temperature. With this method, the nucleation and crystal growth can both occur during the cooling stage.
- **Powder method**: In this process, the glass needs to be ground to a very fine state, which is then shaped by cold compaction followed by a high-temperature treatment to sinter the compacted material. This method is used only if there are obvious advantages, since a high sintering temperature is required and the final product's performance is similar to that obtained by other methods.

Table 9.1 presents the details of some of the most relevant studies found on the use of GC derived from different types of waste glass in the production of glass-ceramics.

Bulk Density

The results presented in Table 9.2, summarising the findings of several studies undertaken to determine the effects of incorporating GC and other waste materials on the bulk or apparent density of glass-ceramics, suggest that an increase in the GC content normally results in a greater densification of glass-ceramics. Of the few exceptions to this are the results of Hanpongpun et al. (2007), indicating that the bulk density of glass-ceramics decreases with increasing GC, probably due to the corresponding reduction in zinc waste content, which has higher specific gravity.

A similar phenomenon may have also occurred in the study of Bernardo et al. (2012), in which, despite a slight reduction in density, the incorporation of GC induced greater crystallisation with subsequent decrease in porosity and enhanced mechanical performance of the product.

For the same sintering temperature, as the sintering time increases, the density of the glass-ceramics containing waste glass also tends to increase (Lu et al., 2016a; Ponsot et al., 2014). The results of Lu et al. (2016a) suggest that exposing the material to a rapid heating rate contributes to greater densification and mechanical performance of glass-ceramics. However, it causes a drop in the degree of crystallinity due to the reduction in surface nucleating sites. Furthermore, as the sintering temperature increases, the bulk density of the ceramic tends to increase with it, whereas the opposite can be observed in terms of the glass-ceramic's porosity (Fan and Li, 2013, 2014; Zhang et al., 2011; Zhang and Liu, 2013; Furlani et al., 2010).

Titanium dioxide (TiO_2) is normally used as a nucleation agent in glass-ceramics. Yang et al. (2015) incorporated increasing amounts of TiO_2 in mixes with 50% soda lime–silicate glass and 50% stainless steel glass. Although the addition of TiO_2 resulted in large bulk crystals, with increasing TiO_2 content, short rod-like crystals were also observed, which demonstrated improved phase separation and nucleation. Thus, it was concluded that TiO_2 can be considered as an effective nucleation agent as it increased both the quantity and the uniformity of the crystal nuclei.

Table 9.1 Use of various types of glass cullet in glass-ceramics

	Glass Cullet		
Reference	**Type**	**Content, %**	**Other Materials**
Bernardo et al. (2012)	Soda lime–silicate, borosilicate	25, 50	Fly ash, mixed boron waste, electric arc furnace slag from Fe–Ni production
Fan and Li (2013, 2014)	LCD	70	30% calcium fluoride sludge, 30% basic oxygen furnace slag, addition of 5%–20% MgO, 10% and 20% Al_2O_3
Furlani et al. (2010)	Energy-saving lamps	10–40	Steel slag
Furlani et al. (2011)	Container glass	24–36	36%–54% incinerated paper sludge, 10%–40% red/yellow clay or kaolin
Hanpongpun et al. (2007)	Container and flat glass	40–80	Zinc hydrometallurgy waste
Ljatifi et al. (2015)	Container glass	30	56% electrofurnace slag, 7% dust and 7% converter slag from ferronickel smelting plant
Lu et al. (2016a)	Soda lime–silicate	70	30% fly ash and addition of 5%–25% magnesia
Ponsot et al. (2014)	CRT	64	36% nepheline glass
Yang et al. (2015)	Soda lime–silicate	50	50% stainless steel slag and addition of 3%–9% TiO_2
Zhang and Liu (2013)	Flat glass	100	Addition of 12%–25% kaolin as a crystallisation promoter
Zhang et al. (2011)	Flat glass	80–90	Kaolin, talc, $BaCO_3$, ZnO

CRT, cathode ray tube; *LCD*, liquid crystal display.

Porosity and Water Absorption

From an overall perspective, the results in Table 9.3, which presents a summary of the results of several studies on the influence of GC on the permeability of glass-ceramics, it can be concluded that the incorporation of GC lowers the porosity and water absorption of the product. The results of Bernardo et al. (2012) suggest that the incorporation of borosilicate GC may result in lower porosity than when using soda lime–silicate GC. In the study of Hanpongpun et al. (2007), in which GC of different colours was used (i.e., clear and brown), suggesting slightly different chemical compositions between the two, it was observed that this is unlikely to affect the porosity and water absorption properties of glass-ceramics. Nevertheless,

Table 9.2 Potential variations in the density of glass-ceramics made with glass cullet and other waste materials

Reference	Results
Bernardo et al. (2012)	The apparent density of glass-ceramics made with borosilicate was higher than when soda lime–silicate GC was used. The apparent density decreased with increasing GC content.
Fan and Li (2013, 2014)	The addition of 5%–10% MgO enhanced the sinterability of LCD GC, which resulted in a higher degree of densification of the glass-ceramic. However, further increase in MgO content resulted in lower apparent densities. The addition of Al_2O_3 hindered the viscous sintering, which led to a decrease in apparent density.
Furlani et al. (2010)	The bulk density increased with increasing GC content. At 1040°C, the density of specimens containing 10% and 40% GC was 2.38 and 2.76 g/cm^3, respectively.
Furlani et al. (2011)	The apparent density of glass-ceramics made with GC from containers increased and decreased with increasing kaolin and yellow clay content, respectively. No significant changes were noted when red clay was introduced.
Hanpongpun et al. (2007)	The bulk density decreased with increasing GC content. The incorporation of amber GC caused slightly lower density values.
Ljatifi et al. (2015)	The use of 30% GC from container bottles allowed production of glass-ceramics with a maximum density of 3.16 g/cm^3.
Lu et al. (2016a)	The glass-ceramics reached a maximum density of 1.96 g/cm^3 when 70% GC was used with 5% MgO and it decreased thereon.
Ponsot et al. (2014)	Apart from the increased density when CRT was used, it also decreased the optimum sintering temperature.
Zhang and Liu (2013)	The density of GC ceramics increased with increasing kaolin-based crystallisation promoter content.
Zhang et al. (2011)	The density of glass-ceramics peaked for a crystallisation promotor content of 12%–15% and GC content of 85%–88%, at 950°C.

CRT, cathode ray tube; *GC*, glass cullet; *LCD*, liquid crystal display.

it should be noted that not only do the compositions of all glass-ceramics change between research studies, but the GC types and other waste materials are different as well, making it difficult to reach specific conclusions concerning the effects of GC type and content.

In some studies, in which magnesium oxide (MgO) was used as a modifying agent, the porosity of glass-ceramics decreased when a content of 5%–10% MgO was used, but started to increase afterwards (Fan and Li, 2013, 2014; Lu et al., 2016a). It was also observed that the addition of magnesium oxide in conjunction with aluminium oxide resulted in even greater porosity (Fan and Li, 2014).

Table 9.3 Water absorption and porosity of glass-ceramics made with glass cullet and other waste materials

Reference	Results
Bernardo et al. (2012)	The porosity decreased with increasing GC content. Borosilicate GC-based glass-ceramics exhibited lower porosity than when soda lime–silicate GC was used.
Fan and Li (2013, 2014)	The water absorption and open porosity were lowest when 5% MgO content was used and decreased even further with increasing sintering temperature.
Furlani et al. (2010)	An increasing GC content decreased water absorption, the magnitude of which was greater for increasing temperatures. The use of 30%–40% GC as steel slag replacement resulted in glass-ceramics with water absorption values below 1% at 1100°C.
Furlani et al. (2011)	An increasing GC content resulted in higher water absorption in comparison with mixes with greater red/yellow clay or kaolin content, but was offset by increasing sintering temperatures.
Hanpongpun et al. (2007)	The water absorption and apparent porosity remained unchanged regardless of the GC content and its colour.
Lu et al. (2016a)	Higher MgO contents increased porosity, in which the pores changed from a spherical shape to interconnected irregularity.
Ponsot et al. (2014)	Although the addition of GC from CRT resulted in higher density, it also caused slightly higher porosity.
Zhang and Liu (2013)	Despite the increased density with increasing kaolin-based crystallisation promoter, the water absorption also increased.

CRT, cathode ray tube; *GC*, glass cullet.

In contrast to the results of the other studies described in Table 9.1, those obtained by Furlani et al. (2011) showed that the water absorption of glass-ceramics increased notably with the inclusion of GC. This is perhaps because GC was used in conjunction with incinerated paper sludge to replace both red/yellow clay and kaolin (but with a greater emphasis on the paper sludge content) and resulted in higher porosity. However, this difference in porosity was offset as the firing temperature increased.

Shrinkage

The shrinkage and water absorption of glass-ceramics are physical parameters that are used to draw the sintering curves, thereby leading to the optimisation of firing cycles and thus the production of materials with the required properties. The shrinkage of glass-ceramics made with mercury-free neon lamp GC, incinerated paper sludge and red/yellow quartzitic clay or high-grade kaolin was evaluated by Furlani et al. (2011). The results of shrinkage as a function of the sintering temperature showed that all compositions displayed a flat trend up to 1080°C, beyond which the shrinkage started to increase up to 1140°C, and its magnitude was greater for specimens containing a greater amount of GC. Glass-ceramics containing 24% and 28% GC mostly displayed shrinkage values of less than 10% regardless of the sintering temperature. The results of Rozenstrauha et al. (2013), however, showed that the addition of 20% GC resulted in a shrinkage value of 18.1%, whereas that of glass-ceramics containing 10% GC was 19.6%.

Compressive and Flexural Strengths

The compressive and flexural strengths of glass ceramics containing GC mixed with other waste materials have been assessed in several studies, the results of which are given in Table 9.4. Owing to the differences in mix design, chemical composition of the materials, firing temperature and other parameters used in each study, the data are discussed according to the relative results obtained within each study.

Table 9.4 Compressive and flexural strengths of glass-ceramics made with glass cullet and other waste materials

Reference	Results
Bernardo et al. (2012)	The flexural strength increased from 34 to 65 MPa when the soda lime–silicate GC content increased from 25% to 50%. It increased from 78 to 96 MPa for the same contents of borosilicate GC.
Fan and Li (2013, 2014)	The flexural strength was highest when 5%–10% MgO content was used and increased even further at higher sintering temperature.
Furlani et al. (2010)	The flexural strength increased with increasing GC content. At 1040°C, the flexural strength values of specimens containing 10% and 40% GC were 18 and 62 MPa, respectively.
Furlani et al. (2011)	The integration of 20%–30% red/yellow clay or kaolin peaked the flexural strength (for 32%–28% GC content), which started to decrease thereafter.
Hanpongpun et al. (2007)	Increasing the GC content resulted in lower flexural strengths. Brown GC caused slightly lower flexural strength values than clear GC.
Lu et al. (2016a)	For the same GC content, the flexural strength peaked at 10% MgO content and increased further as the sintering time and rate increased.
Ponsot et al. (2014)	Compared to other nepheline glass-ceramics, when 64% CRT was used, the flexural strength decreased.
Yang et al. (2015)	The addition of 7% TiO_2 caused a significant increase in flexural strength (i.e., from around 50 MPa to almost 150 MPa).
Zhang and Liu (2013)	The compressive and flexural strengths peaked when 15% kaolin-based crystallisation promoter was used alongside the GC (247 and 119 MPa, respectively), but decreased afterwards.
Zhang et al. (2011)	At a sintering temperature of 950°C, the compressive and flexural strengths peaked at a crystallisation promoter content of 12%–15%, which also corresponds to the maximum density values.

CRT, cathode ray tube; *GC*, glass cullet.

Concerning the effects of incorporating different GC types and contents, the results of some investigations (Bernardo et al., 2012; Furlani et al. (2010); Rozenstrauha et al., 2013) showed that the flexural strength of glass-ceramics may increase with increasing GC content and more so if the particle size of the GC decreases (Lu et al., 2014). However, the opposite has been observed in other studies (Furlani et al., 2011; Hanpongpun et al., 2007). This inconsistency may be due to the other types of waste materials also used, the variation in which makes a comparison viable only on a case-to-case basis. Nevertheless, owing to the enhanced densification prompted by the incorporation of GC, as shown previously, the use of GC is likely to increase the flexural strength of glass-ceramics (Fan and Li, 2013, 2014; Zhang et al., 2011; Zhang and Liu, 2013).

Nevertheless, Bernardo et al. (2012) observed that the flexural strength of glass-ceramics made with borosilicate GC was consistently higher than that of specimens made with soda lime GC. The colour of soda lime GC may have some influence on flexural strength due to the slight variations in chemical composition, which may affect the crystallisation of the material (Hanpongpun et al., 2007).

Regarding the influence of incorporating clay or kaolin alongside GC from container glass, Furlani et al. (2011) observed that glass-ceramics made with increasing amounts of red/yellow quartzitic clay or high-grade kaolin result in an increase in flexural strength. Zhang and Liu (2013) and Zhang et al. (2011) observed similar results after incorporating 12%–15% kaolin, which acted as a crystallisation promoter.

The addition of 5%–10% MgO in glass ceramics made with 70% GC from crushed liquid crystal display (LCD) and 30% calcium fluoride sludge (Fan and Li, 2013) yielded the highest flexural strength, but it decreased for higher MgO content. The same authors, in a later study (Fan and Li, 2014), noticed further increases in the flexural strength of glass-ceramics when 30% LCD and 70% slag with a MgO content of 20% were used. Lu et al. (2016a) also observed peak flexural strength values after incorporating 5%–10% MgO. The enhanced mechanical performance prompted by the incorporation of MgO is closely related to the porosity, crystallinity and crystal shape anisotropy of the sintered glass-ceramics (Lu et al., 2016a). However, it was found that the combined use of MgO with Al_2O_3 decreases the flexural strength of glass-ceramics as the latter component hinders the viscous sintering (Fan and Li, 2014).

Other Properties

Vickers hardness is frequently measured to determine the resistance to deformation, densification and fracture of ceramics. Generally, it has been reported that the incorporation of GC results in an increase in Vickers hardness (Zhang and Liu, 2013; Zhang et al., 2011; Furlani et al., 2010) and may increase further with the addition of clay and kaolin (Furlani et al., 2011).

When assessing the leachability of glass-ceramics made with waste glass from crushed energy-saving lamps, Furlani et al. (2010) observed that the leached heavy metals concentrations of the resultant products were very low and decreased with increasing GC content. The authors explained that the low leachability is mainly due to the material's microstructure, which consisted of a polycrystalline framework surrounded by a vitreous phase.

9.2.2 Foam Glass-Ceramics

Foam glass-ceramics possess a wide variety of interesting properties, including reasonable compressive strength despite their extremely low density, chemical resistance and incombustibility, among other features (Binner, 2005; Scarinci et al., 2005). These materials are known to have diverse applications, such as catalyst support for gas or liquid metal filtration, mineral casting, electromagnetic radiation absorber, thermal and acoustic insulation, membranes in combustion technology and lightweight load-bearing structures (Chinnam et al., 2013; Chakartnarodom and Ineure, 2014; Petrella et al., 2010; Benhaoua et al., 2015; Sommariva and Weinberger, 2015; Kazmina et al., 2015).

The production of foam glass-ceramics is similar to that normally applied for dense glass-ceramics, but the former requires the incorporation of a foaming agent (mostly carbon-based substances), which is initially mixed together with the glass prior to any heat treatment. To potentiate and maintain the effervescence effect, the reaction or decomposition of the foaming agent should occur close to the softening temperature of the glass (Scarinci et al., 2005; Rawlings et al., 2006). The performance of the final foam glass-ceramic products mainly depends on the foaming agents, the initial size of the glass particles and the heat treatment schedule (Scarinci et al., 2005).

Even though GC has been considered as the most suitable starting material for the process (Colombo et al., 2003), this material is also known to partially crystallise during the heating process, which interferes with the foaming process, giving undesirable effects. Although this crystallisation enhances the mechanical properties of the final product, it also limits the expansion prompted by the foaming agent, resulting in a material with a lower than expected porosity (Chinnam et al., 2013). Greater crystallisation has been observed when GC is mixed with the foaming agent and other glasses that are easier to crystallise (Tulyaganov et al., 2006; Bernardo et al., 2009; Fernandes et al., 2009).

Table 9.5 presents the mix constituents used in producing foam glass-ceramics containing GC in experimental studies. GC, which was obtained from different types of materials, contributed to the total amount or a great portion of glassy material. Silica-rich materials, such as fly ash, ground granulated blast furnace slag (GGBS) and clay, were used to make up the total glass material. Several types of foaming agents were used, e.g., calcium carbonate ($CaCO_3$), silicon carbide (SiC), aluminium nitride (AlN) and manganese dioxide (MnO_2), giving varying degrees of success.

Density

Table 9.6 presents the findings from research on the density of foam glass-ceramics made with different types and amounts of GC. The use of GC generally results in foam glass-ceramics with significantly low density. Although the effectiveness of different types of GC on the density of foam glass-ceramics is uncertain, since few comparative studies have been found, GC from crushed cathode ray tubes (CRTs) appears to result in lower density compared with soda lime–silicate GC (Benzerga et al., 2015). However, the density variation from incorporating GC of different types becomes negligible in comparison

Table 9.5 Type and content of glass cullet and other materials used in foam glass-ceramics

Reference	Glass	Content, %[a]	Other Materials
Apkar'yan et al. (2015)	Lamps, containers and window glass	73–91	9%–27% ceramic filler; 3% sawdust as organic additive; 5% coke as foaming agent
Ayadi et al. (2011)	Soda lime–silicate	100	1% $CaCO_3$ as foaming agent
Benzerga et al. (2015)	CRT and soda lime–silicate glass	0–100	Carbon (C), silicon carbide (SiC) and aluminium nitride (AlN) as foaming agents
Chakartnarodom and Ineure (2014)	Soda lime–silicate glass	40–80	20%–60% fly ash/rice husk ash; $CaCO_3$ as foaming agent
Ding et al. (2015)	E-glass	45–65	35%–55% GGBS; $CaCO_3$ as foaming agent
Fernandes et al. (2009)	Flat glass	50–90	10%–50% fly ash; $CaCO_3$ and $CaMg(CO_3)_2$ as foaming agents
Francis et al. (2013)	–	70	30% GGBS; 1%–10% SiC as foaming agent
Gong et al. (2016)	Container glass	100	3%–20% dried porcine bones as foaming agent
Konig et al. (2015)	CRT	100	0.5%–1.5% activated charcoal and 1.8%–9% MnO_2 as foaming agents
Llaudis et al. (2009)	Flat glass	100	Si_3N_4 as foaming agent and MnO_2 as foaming promoter
Marangoni et al. (2014b)	Soda lime–silicate	40–60	40%–60% basalt scoria
Petersen et al. (2014)	CRT	100	2%–22% soda ash (Na_2CO_3) as foaming agent
Petersen et al. (2015)	CRT	100	7% MnO_2 as foaming agent
She et al. (2014)	LCD and window glass	70–80 and 20–30	1%–3% SiC
Taurino et al. (2014)	Borosilicate glass from washing machines	100	SiC
Wang et al. (2016)	Soda lime–silicate	100	0.5%–5% KNO_3 as foaming promoter; 0.5% SiC as foaming agent
Zilli et al. (2015)	Soda lime–silicate	70–95	5%–30% sewage sludge from ceramic tile factory; 6%–12% graphite as foaming agent

CRT, cathode ray tube; *GC*, glass cullet; *GGBS*, ground granulated blast furnace slag; *LCD*, liquid crystal display.
[a]Amounts correspond to the total amount of glass material and exclude foaming agents/promoters.

Table 9.6 Density of foam glass-ceramics made with glass cullet and other waste materials

Reference	Density
Apkar'yan et al. (2015)	The density of foam glass-ceramics varied between 0.20 and 0.29 g/cm^3 and reached minimum density for a clay content of 9%.
Ayadi et al. (2011)	The apparent density decreased from 1.05 to 0.25 g/cm^3 when the sintering temperature increased from 750°C to 900°C.
Benzerga et al. (2015)	The apparent density decreased with increasing CRT content to a minimum of 0.34 g/cm^3 when 70% CRT and 30% SLS were used. The use of AlN allowed the lowest density values (0.35 g/cm^3), followed by SiC (0.49 g/cm^3) and C (0.85 g/cm^3).
Chakartnarodom and Ineure (2014)	The density increased with increasing GC content and increased further if mixed with fly ash rather than rice husk ash.
Ding et al. (2015)	The bulk density decreased with increasing GC content.
Fernandes et al. (2009)	The apparent density was lowest for 1%–2% $CaCO_3$ or $CaMg(CO_3)_2$ (0.36–0.41 g/cm^3) and for temperature around 850°C.
Francis et al. (2013)	A combination of sintering temperature of 1000°C and 4%–6% SiC content allowed a minimum density of 0.76 g/cm^3.
Konig et al. (2015)	The apparent density decreased with decreasing GC particle size and deceased further with increasing amount of MnO_2. A minimum apparent density of 0.14 g/cm^3 was obtained after combining GC with average particle size below 15 μm, 1% charcoal and 5.4% MnO_2.
Llaudis et al. (2009)	For a fixed Si_3N_4 content of 3.5%, as the amount of MnO_2 increased, the bulk density of the foam glass-ceramics decreased (as low as 0.2 g/cm^3).
Marangoni et al. (2014)	The apparent and bulk densities of foam glass-ceramics decreased with increasing GC content.
Petersen et al. (2014)	A minimum density of 0·28 g/cm^3 was obtained when the glass powder was foamed with 6–14 wt.% Na_2CO_3.
Taurino et al. (2014)	The apparent density decreased with increasing sintering temperature. Exposure to 800°C resulted in 0.88 g/cm^3, whilst 850°C–900°C resulted in densities around 0.50 g/cm^3.
Wang et al. (2016)	The addition of KNO_3 as foaming promoter greatly improved the foaming process and reduced the bulk density. Samples sintered at 950°C showed a bulk density around 0.15 g/cm^3.

CRT, cathode ray tube; *GC*, glass cullet; *SLS*, soda lime–silicate.

with the effect of the foaming agent. Benzerga et al. (2015) observed that the use of AlN and carbon-based foaming agents, for the same content, resulted in foam glass-ceramics with density values of 0.35 and 0.85 g/cm^3, respectively. However, little difference was noticed in the density of foam glass-ceramics made with either $CaCO_3$ or $CaMg(CO_3)_2$ as a foaming agent (Fernandes et al., 2009).

Owing to the emissions of carbon dioxide, carbon monoxide and sulphur oxide from the sintering process of foam glass-ceramics containing carbon-based foaming agents, the use of silicon nitride (Si_3N_4) (Llaudis et al., 2009), manganese oxide (MnO_2) (Llaudis et al., 2009; Konig et al., 2015; Petersen et al., 2015) and potassium nitrate (KNO_3) (Wang et al., 2016) as foaming agents and/or promoters has begun to attract research attention. The combined use of Si_3N_4 and MnO_2 (Llaudis et al., 2009) may lead to an exponential decrease in the bulk density of foam glass-ceramics. This is due to the oxidation of Si_3N_4, caused by MnO_2, and consequently the formation of N_2 gas. When either of those two foaming agents is used alone, the density of the end products shows a linear reduction (Llaudis et al., 2009). In the study undertaken by Konig et al. (2015), MnO_2 was used in conjunction with a very small amount of charcoal. The results showed that the moderate foaming effect of the charcoal was greatly improved by the addition of MnO_2, giving foam glass-ceramics with an apparent density as low as 0.131 g/cm^3. It has been suggested that the release of O_2 during the decomposition of MnO_2 starts in the same sintering temperature range for GC from crushed CRTs (Petersen et al., 2015). The growth of foam glass-ceramics is thus caused by increasing O_2 pressure, and when the gas pressure exceeds the forces of surface tension and viscosity, the pore wall stretches and the pore expands.

Instead of using a foaming agent, Marangoni et al. (2014b) produced self-foaming glass-ceramics by using basalt scoria containing high iron oxide content, which reduces with increasing temperature, consequently releasing O_2. Despite reasonably low bulk and apparent density being achieved, the results suggest that the foaming mechanisms in self-foaming glass-ceramics are not as effective compared with those induced by a foaming agent.

The density of foam glass-ceramics was found to decrease substantially with decreasing average particle size of GC (Ketov, 2001). Foam glass-ceramics made with 400-μm-sized GC particles exhibited a density around 0.53 g/cm^3, which decreased further to around 0.13 g/cm^3 when GC with an average particle size of 10 μm was used. Similar observations made by Brusatin et al. (2004) showed that, when using soda lime–silicate-based GC and SiC as foaming agent in producing foam glass-ceramics, the pore diameter decreased and homogeneity increased as the initial glass powder size decreased from 300 to 38 μm.

Porosity and Water Absorption

The porosity and water absorption of foam glass-ceramics, which are inversely proportional to their density, appear to slightly decrease with increasing GC content (Table 9.7). However, the type and amount of the foaming agent used have a greater influence on the degree of foaming of the glass-ceramics and, consequently, on their porosity and water absorption, compared with GC. In the work undertaken by Francis et al. (2013), for a given firing temperature, the water absorption of foam glass-ceramics made with 70% GC with 30% GGBS was found to increase with SiC

Table 9.7 Porosity and water absorption of foam glass-ceramics made with glass cullet and other waste materials

Reference	Porosity and Water Absorption
Chakartnarodom and Ineure (2014)	The water absorption was lowest when 60% GC from crushed windows was combined with 40% fly ash.
Ding et al. (2015)	The water absorption increased with increasing GC content.
Francis et al. (2013)	Foamed glass-ceramics containing 4%–6% SiC content showed a maximum water absorption of 75% after being exposed to a sintering temperature of 1000°C.
Konig et al. (2015)	A maximum open porosity of around 90% was obtained after sintering CRT GC to a temperature of 840°C, containing 1% charcoal and 5% MnO_2 as foaming agents.
Llaudis et al. (2009)	For a fixed Si_3N_4 content of 3.5%, as the amount of MnO_2 increased, the water absorption of the foam glass-ceramics increased up to 23.2%.
Marangoni et al. (2014b)	The total porosity increased with increasing GC content and was directly proportional to the bulk and apparent density.
Petersen et al. (2015)	The water absorption increased with increasing hold time and exposure to higher temperatures. The closed porosity, which varied between 6% and 97%, decreased with increasing treatment time and temperature.
Taurino et al. (2014)	The total porosity, which followed an inversely proportional relationship to density, increased with sintering temperature.
Zilli et al. (2015)	Increasing exposure time to increasing sintering temperatures resulted in increasing porosity. However, for a temperature of 900°C, an exposure of 15 min led to the same porosity levels as when samples were exposed for 120 min.

CRT, cathode ray tube; *GC*, glass cullet.

as a foaming agent for a content up to around 5%, beyond which it decreased. It was explained that the reduction in water absorption caused by higher SiC content was due to the presence of a higher amount of SiO_2, which resulted in the formation of a glassy phase and incomplete oxidation of SiC. In another study (Llaudis et al., 2009), for the same 3.5% content of Si_3N_4, as the MnO_2/Si_3N_4 molar ratio increased from 0 to 6, the water absorption of the foam glass-ceramics increased from less than 0.1% to 23.2%.

In the study of Petersen et al. (2015), the addition of 7% MnO_2 to GC from crushed CRTs enabled the production of foam glass-ceramics with porosity levels in the range of 79%–92% with different degrees of closed porosity (6%–97%), which decreased with increasing treatment temperature and time.

Compressive and Flexural Strengths

Table 9.8 presents descriptions of the findings that emerged from research on the compressive and flexural strength of foam glass-ceramics containing GC. It is clear that the influence of GC on these properties depends on the nature of the additional glassy material. The results of experimental studies in which soda lime–silicate-based GC was used (Apkar'yan et al., 2015; Chakartnarodom and Ineure, 2014; Gong et al., 2016; Marangoni et al., 2014b) showed that compressive strength decreased with increasing GC content. The opposite was observed with increasing red clay (Apkar'yan et al., 2015) or basalt scoria content (Marangoni et al., 2014b), which was a result of greater densification. It was observed that the strength properties may also improve in the increasing presence of GC when combined with fly ash (Chakartnarodom and Ineure, 2014) or porcine bone ash (Gong et al., 2016). The mechanical performance of foam glass-ceramics may also improve by reducing the average particle size of the GC, which results in a decrease in pore diameter and a more homogeneous distribution of pores (Brusatin et al., 2004).

At a given foaming agent content, the strength properties of foam glass-ceramics are related to the degree of foaming, and low-strength values correspond to the optimum foaming and thus a material with minimum density and maximum porosity. Francis et al. (2013) observed that, for a sintering temperature of 900°C,

Table 9.8 Compressive and flexural strengths of foam glass-ceramics made with glass cullet and other waste materials

Reference	Compressive and Flexural Strength
Apkar'yan et al. (2015)	The use of GC from crushed lamps, containers and window glass with red clay gave compressive strength values in the range of 0.82–2.5 MPa, which increased with increasing clay content.
Chakartnarodom and Ineure (2014)	The compressive strength increased with increasing GC content and increased further when mixed with fly ash compared with rice husk ash.
Gong et al. (2016)	The flexural strength decreased with increasing porcine bone content (3% gave 29.7 MPa and 20% gave 16.7 MPa).
Llaudis et al. (2009)	The compressive strength, which was directly proportional to the bulk density, decreased from 44 to 0.5 MPa when the MnO_2/Si_3N_4 molar ratio increased from 0 to 6, for a fixed 3.5% Si_3N_4 content.
Marangoni et al. (2014b)	Increasing GC content from 40% to 60% led to a decrease in compressive strength from 15 to 2 MPa, which corresponded to decreases in density.
Zilli et al. (2015)	The compressive strength decreased with increasing sintering temperatures. Samples with 90% GC and 10% sewage sludge and 10% graphite showed compressive strength between 12.5 and 6.4 MPa after increasing the sintering temperature from 800°C to 950°C.

GC, glass cullet.

the compressive strength of foam glass-ceramics increased with increasing SiC content, but the opposite was perceived at 1000°C. The effect of SiC content on the strength properties of foam glass-ceramics made with GC from crushed CRTs has been assessed (Guo et al., 2010), and it was found that minimum compressive and flexural strengths of 11 and 5 MPa, respectively, were achieved at 5% SiC content.

Thermal Conductivity

The effect of GC on the thermal conductivity of foam glass-ceramics has been tested and the main findings emerging from the studies are described in Table 9.9. It has been found that the use of GC can result in foam glass-ceramics having significantly low thermal conductivity values, some of which are comparable to commercially available materials that are routinely used for insulation of buildings (e.g., expanded polystyrene, polyurethane foam, wood fibre and cork boards). However, foam glass-ceramics have the distinct advantage of being noncombustible and exhibiting excellent dimensional stability for a wide range of temperatures (Foamglas, 2016). In a field trial undertaken using foam glass-ceramics as insulating material in a detached house, it was shown that the thermal loss of a foam-glass-ceramics-insulated house can be reduced by around 60% compared with that of an uninsulated house (Ayadi et al., 2011).

It has been previously observed that an increase in sintering temperatures increases the total porosity but will decrease the closed porosity of foam glass-ceramics (Petersen et al., 2015). However, since the degree of closed porosity has no influence on the overall thermal conductivity of the material, it does not pose a problem in terms of insulation.

Table 9.9 Thermal conductivity of foam glass-ceramics made with glass cullet and other waste materials

Reference	Thermal Conductivity
Apkar'yan et al. (2015)	The thermal conductivity of glass-ceramics, which was directly proportional to its density, varied between 67 and 87 mW/(m°K) for density values of 0.20 and 0.28 g/cm^3.
Konig et al. (2015)	The thermal conductivity of foam samples varied between 42 and 50 mW/(m°K) for a density range between 0.130 and 0.175 g/cm^3.
Petersen et al. (2015)	The thermal conductivity of specimens with density values between 0.25 and 0.40 g/cm^3 varied between 53 and 66 mW/(m°K).
Zilli et al. (2015)	Increasing sintering temperatures resulted in lower thermal conductivity. Samples made with 90% GC and 10% sewage sludge resulted in a thermal conductivity of 101 mW/(m°K) after being exposed to a temperature of 950°C for 30 min.

GC, glass cullet.

In the work of Konig et al. (2015), which produced foam glass-ceramics with very low thermal conductivity using crushed CRTs, it was found that the optimum concentrations of activated charcoal and MnO_2 for maximum degree of foaming were between 0.5%–1% and 5.4%–7.2%, respectively. Within these concentration ranges, a small variation in density and open porosity was observed. The authors also stated that the foaming process was very sensitive to the sintering temperature and foaming time, the optima of which were 800°C and 5 min, respectively.

9.2.3 Bricks

Physical Properties of Unfired Clay Bodies

The physical properties of fresh unfired clay bodies, which are discussed in this section, are important because several physical changes take place before the firing process and these have significant influence on the quality and properties of the end product. The information available on the fresh properties of bricks manufactured with GC was found reported in a single study (Dondi et al., 2009). Clay bricks were made with carbonate-rich and fine-grained clay and two different types of GC, from end-of-life PC monitors and TV sets (glass from the screen and CRTs), at 2% and 5% levels. The results of the physical properties of the unfired clay bodies are given in Table 9.10, each of which is discussed below.

(i) Working Moisture

This property corresponds to the amount of water required to achieve the desired consistence during the brick forming process and influences the drying shrinkage of bricks at the later drying process. The results in Table 9.10 show that the working moisture content, assessed in accordance with ASTM C324 (1992), remained almost unchanged with the incorporation of 2% and 5% screen glass and CRT.

Table 9.10 Effects of different glass cullet types on the physical properties of unfired clay bricks

		Screen Glass		CRT	
Property	**Control**	**2%**	**5%**	**2%**	**5%**
Working moisture, %	33.3	33.7	33.4	33.9	32.3
Pfefferkorn index, %	43.9	44.8	42.2	50.6	46.8
Shrinkage weight loss, %	35	37	32	37	32
Drying sensitivity index	0.37	0.44	0.46	0.37	0.43
Hygroscopicity, wt.%	1.57	1.52	1.72	1.61	1.5
Dry flexural strength, N/mm^2	6.4	5.5	4.3	6.4	4.7

CRT, cathode ray tube.
Based on Dondi et al. (2009).

(ii) Pfefferkorn Plasticity Index

Fresh clay bodies may exhibit different degrees of plasticity, which vary according to the subsequent forming process. The plasticity of clay materials, which is influenced by the particle size distribution, fineness and moisture content, among other factors, may be measured and characterised by stress–strain relationships, the Atterberg method, the Pfefferkorn test and other methods. The Pfefferkorn plasticity index test results, as given in Table 9.10, do not show significant changes when GC from crushed screens was incorporated into the clay bodies, but a slight increase was observed when GC from crushed CRTs was used, which corresponds to higher shear resistance (Dondi et al., 2009).

(iii) Weight Loss with Shrinkage

The results in Table 9.10 suggest that the incorporation of 2% and 5% GC from crushed screens and CRTs into clay bodies has limited influence on weight loss with shrinkage.

(iv) Drying Behaviour

Prior to firing the bricks at high temperatures, excess moisture must be removed completely to avoid cracking, owing to the pressure induced by trapped water. The drying behaviour of clay bodies can be assessed by means of the Adamel apparatus, which is graphically represented by the Bigot curve, drying sensitivity index and drying shrinkage. The drying sensitivity index, which determines the tendency of drying of clay bodies or the easiness of drying, is calculated as the weight loss against the drying shrinkage strain. It is preferable to have clay bodies with a low drying sensitivity index, as they can be safely dried and are unlikely to crack. The results in Table 9.10 suggest that the incorporation of GC may increase the drying sensitivity index of unfired clay bricks.

(v) Dry Flexural Strength

During the production of bricks, dried clay bodies are inevitably subjected to forces during handling and from the weight of other clay bodies placed above them when set in the kiln, and thus they must exhibit considerable strength to withstand those loads. The dry flexural strength of clay bodies is influenced by the particle size of clay materials, mineral composition, porosity, drying temperature and rate. In the study of Dondi et al. (2009), the dry flexural strength of clay bodies was measured in accordance with ASTM C674 (1994). It was found that the inclusion of 2% and 5% GC from crushed screens and CRTs caused a decrease in the dry flexural strength of unfired clay bodies (Table 9.10).

(vi) Hygroscopicity

Hygroscopicity is the ability of a material to absorb moisture from the air and is of importance in brick manufacturing as, after the drying process, clay bodies may absorb the air's moisture, though it is insufficient to turn the bodies to plastic. It can be seen from Table 9.10 that the use of 2% and 5% GC from crushed screens and CRT GC has little influence on the hygroscopicity of clay bodies.

Deformation of Clay Bodies Before and After Firing

Bricks usually undergo two types of shrinkage during the production cycle, namely, drying shrinkage and firing shrinkage. Drying shrinkage is caused by evaporation during the drying process of unfired bricks. Firing shrinkage is caused by the fusion and chemical reactions of the mixture during the firing process. Examples of factors contributing to shrinkage include the type of clay, amount of water added and firing temperature. Shrinkage affects the designed brick size and makes the clay bodies prone to cracking, leading to poorer performance of the resulting brick. The effects of using GC, at up to 45% content, on the drying shrinkage (Dondi et al., 2009; Loryuenyong et al., 2009), firing shrinkage (Dondi et al., 2009; Lin, 2007a; Loryuenyong et al., 2009; Rahman et al., 2015; Phonphuak et al., 2016) and under-load firing shrinkage (Smith, 2004a,b; Harrison, 2005) of bricks have been analysed.

(i) Drying Shrinkage

The results shown in Figure 9.1, which presents the relative drying shrinkage of unfired clay bodies, suggest that drying shrinkage decreases with increasing GC content. At higher GC contents, the drying shrinkage appears to be more stabilised compared with that of the control samples without GC (Loryuenyong et al., 2009). However, further research in this area is required to ascertain this. Furthermore, in the study of Dondi et al. (2009), the results suggest that GC from both crushed CRTs and screens exerts similar effects on the drying shrinkage of unfired bodies.

(ii) Firing Shrinkage

Figure 9.2 presents the shrinkage values of bricks containing GC fired at different temperatures from various studies (Dondi et al., 2009; Loryuenyong et al., 2009; Lin, 2007a; Phonphuak et al., 2016; Rahman et al., 2015). Most of the results suggest that increasing incorporation of GC may lead to slightly greater shrinkage

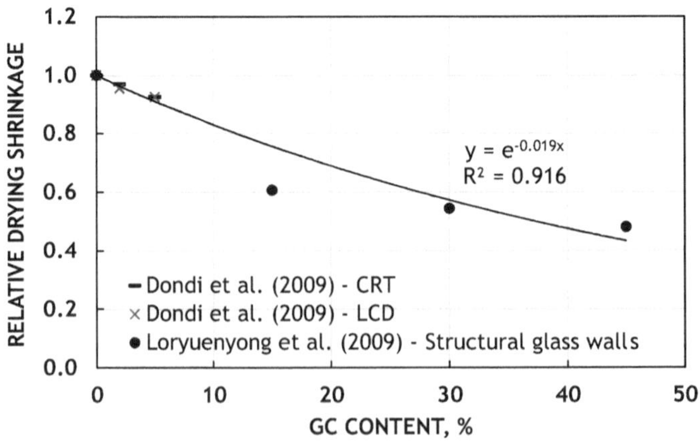

Figure 9.1 Relative drying shrinkage of unfired clay bricks made with increasing GC content. *CRT*, cathode ray tube; *GC*, glass cullet; *LCD*, liquid crystal display.

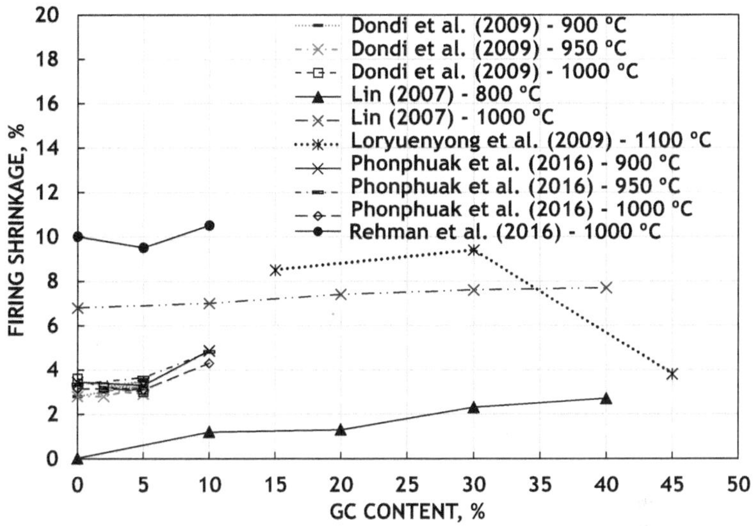

Figure 9.2 Effect of GC content on the firing shrinkage of bricks. *GC*, glass cullet.

of the brick after firing. An increase of almost 3% in shrinkage was observed in bricks containing 40% GC compared with that of control specimens fired at 800°C (Lin, 2007a). Nevertheless, the shrinkage of bricks containing up to 10% GC is comparable to that of conventional bricks (Dondi et al., 2009; Lin, 2007a; Phonphuak et al., 2016; Rahman et al., 2015). Even though the firing shrinkage increases significantly with increasing firing temperatures, the absolute increase in shrinkage with increasing GC content appears to be the same regardless of the temperature. Furthermore, according to the CNS 382 R2002 (2007) criteria *apud* Lin (2007a), the firing shrinkage limit for good-quality bricks is below 8%, and this is met by almost all GC bricks.

(iii) Firing Shrinkage Under Load

In some studies (Smith, 2004a,b; Harrison, 2005), the authors evaluated the deformation of brick specimens containing GC as a fluxing agent under a load during firing. Powdered glass was added as a fluxing agent during the firing process to seal smaller capillaries, thereby resulting in lower water absorption (Smith, 2004b).

The testing consisted of loading briquettes, which were cut from dried unfired bricks, into a kiln, to simulate the typical loading of a brick at the bottom of a 16-brick-high setting pattern on a kiln car. According to the authors, the purpose of the test was to assess the sintering and densification process in the form of a linear change in height of the stack of briquettes. The under-load firing shrinkage curves of bricks made with GC and various types of clays, such as Etruria marl (EM) (Smith, 2004a,b), carboniferous shale (CS) (Smith, 2004b) and Quaternary glacial lacustrine clay (QGLC)

(Smith, 2004b), have been reported. The results of bricks containing 5% and 10% ground mixed-coloured container GC showed expansions similar to those of the controls when heated up to 800°C. The contraction of clay bodies containing 5% and 10% GC also demonstrated a pattern similar to that of the controls at temperatures between 850°C and 1200°C. It was explained that, as the incorporation of 10% GC could result in equivalent linear shrinkage at a lower firing temperature (compared with the normal top firing temperature), it is feasible to produce end products made with GC with desirable physical properties using less energy.

Similar findings were reported in the study by Harrison (2005), in which brick samples were prepared from buff clay (fire clay) and red clay (EM) with the addition of 5% and 10% GC with particle size less than 150 μm. The results showed that the incorporation of GC as a fluxing agent achieved the desired shrinkage more quickly at lower firing temperatures. This indicates the potential savings of both firing time and firing temperature with the use of GC, thereby increasing the production rate and reducing production costs.

Physical Properties of Fired Clay Bodies

(i) Porosity and Water Absorption

Bricks normally contain some degree of porosity, which is influenced by the properties of the constituent clay materials, their particle size, the degree of compaction and the heat treatment procedure. Although porosity in bricks is an undesirable yet unavoidable characteristic, which increases water absorption and reduces mechanical performance, high-porosity bricks are manufactured for thermal insulation purposes.

Table 9.11 summarises the results of the apparent porosity of clay bricks containing up to 45% GC and fired at various temperatures ranging from 900°C to 1200°C. In the study of Dondi et al. (2009), in which 2% and 5% GC from crushed screens and CRTs were used, no significant changes were observed in the porosity regardless of the firing temperature. In contrast, Loryuenyong et al. (2009) evaluated bricks made with GC (up to 45% content from structural glass walls) and noticed a decrease in porosity, which can also be seen in Figure 9.3 in terms of water absorption. However, unexpectedly high porosity and water absorption were observed at 45% GC content, which was due to an increased number of open pores as a result of an outflow of glass particles from the brick's surface (Loryuenyong et al., 2009).

The influence of GC on the water absorption of bricks, expressed in relative terms, with respect to those made without GC, is shown in Figure 9.3. Taken together, the studies cover several parameters, such as the type of GC used, its content (in this case, results were only for the range of 0%–45%) and the firing temperature (from 800 to 1200°C). In general, the results show that the water absorption of bricks decreases with increasing GC content. However, the water absorption values of bricks made with lower GC levels tended to be very similar to those of the control bricks. The tests carried out by Dondi et al. (2009), in accordance with ASTM C373 (1994), showed that the water absorption values of clay bricks containing 2% and 5% GC from crushed

Table 9.11 Effects of glass cullet on the apparent porosity of clay bricks at various firing temperatures

	Brick		Apparent Porosity, %				
			Firing Temperature, °C				
Reference	**Type**	**GC, %**	**900**	**950**	**1000**	**1100**	**1200**
Dondi et al. (2009)	Control	0	30.4	31.0	30.1	–	–
	Screen glass	2	31.5	31.4	29.8	–	–
		5	31.3	30.6	29.6	–	–
	CRT	2	31.2	31.9	30.5	–	–
		5	29.7	30.6	29.2	–	–
Loryuenyong et al. (2009)	Structural glass walls	15	–	–	–	4.5	–
		30	–	–	–	4.0	–
		45	–	–	–	24.0	–

CRT, cathode ray tube; *GC*, glass cullet.

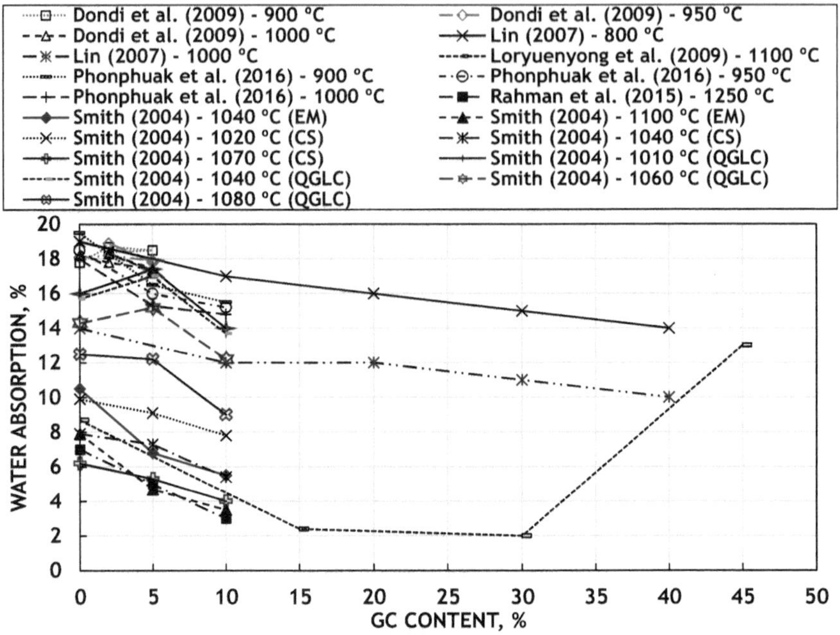

Figure 9.3 Water absorption of bricks containing increasing GC content. *CS*, carboniferous shale; *EM*, Etruria marl; *GC*, glass cullet; *QGLC*, Quaternary glacial lacustrine clay.

screens and CRTs were similar to those of the reference bricks regardless of the firing temperature. In some cases, the water absorption of the bricks remained the same or slightly increased when 5% GC was incorporated, but the value decreased at 10% GC content (Smith, 2004a).

The influence of incorporating higher GC content, such as up to 40% (Lin, 2007a) and 45% (Loryuenyong et al., 2009), has been explored. Both studies exhibited similar reduction in water absorption for GC content up to 30%, but, thereafter, different results were observed. Whilst the values reported by Lin (2007a) continued to decrease, those determined by Loryuenyong et al. (2009) (GC obtained from crushed structural glass walls) were shown to increase by more than two times those of the control sample. It was explained that the increased water absorption of those clay bricks was a result of GC flowing out from the brick specimens' surface, which left open pores. Notwithstanding this, further research is needed to ascertain the optimum GC level without adversely affecting the performance of bricks.

According to Lin (2007a), as per the requirements of Chinese National Standard CNS 1127-R3042 (1999), the water absorption of first-, second- and third-class bricks must be less than 15%, between 15% and 19% and above 23%, respectively. This might suggest that it is possible to produce higher-class clay bricks containing 30%–40% GC, fired at a lower firing temperature. It has been shown that bricks containing up to 40% GC and fired at 800°C showed the same water absorption as control bricks made without GC fired at 1000°C (Lin, 2007a).

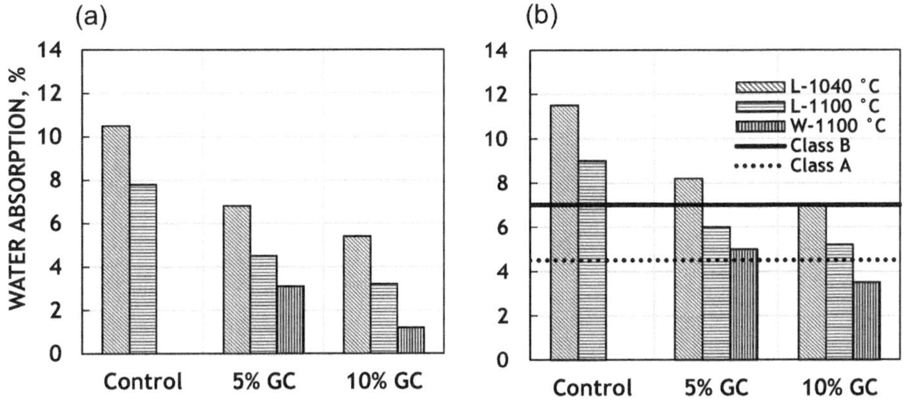

Figure 9.4 Effect of GC on water absorption of clay bricks made with Etruria marl at different firing temperatures. (a) 24-h cold water immersion test. (b) 5-h boiling water immersion test. *GC*, glass cullet; *L*, laboratory firing; *W*, works firing.

According to BS-EN-771-1 (2011), the water absorption of class A engineering bricks and DPC 1 (damp proof course) and class B engineering bricks and DPC 2, conforming with EN-772-7 (1998), must be below 4.5% and 7.0%, respectively.

Smith (2004b) produced bricks made with three different types of clay (EM, CS, QGLC) containing 0%, 5% and 10% mixed-colour container GC. The water absorption of these bricks, obtained from a 24-h cold water soaking, as shown in Figure 9.3, decreased with increasing mixed-colour container GC content and even further with increasing firing temperature. It was explained that the reduction of water absorption indicated a high degree of fluxing of the GC, resulting in a reduction of porosity. In Figures 9.4–9.6, a comparison is made between the water absorption values of bricks (Smith, 2004a,b) made with EM, CS and QGLC, respectively, determined using 24-h cold water and 5-h boiling water immersion tests. In contrast to the bricks made with QGLC, the water absorption values of bricks made with EM and CS clay, containing 5% and 10% GC, fired at the right firing temperature met the BS-EN-771-1 (2011) requirements, when measured according to EN-772-7 (1998) (boiling water immersion method). Therefore, it appears that the use of GC as a fluxing agent may help to achieve the same water absorption requirements at a firing temperature lower than that of clay bricks made without GC.

The effects of incorporating 5% and 10% GC on the water absorption of clay bricks made with fire clay (buff clay) and EM (red clay), at a firing temperature of 1060°C–1140°C, have been explored (Harrison, 2005). The GC used was of the soda lime–silicate glass type and had a maximum particle size of 150 μm. The water absorption results determined using the cold water immersion method as per BS-EN-771-1 (2011) are shown in Figure 9.7. Similar to the previous findings (Figures 9.4–9.6), the incorporation of GC decreased the water absorption of clay bricks made with fire clay and EM clay, the effect of which increased with the firing temperature.

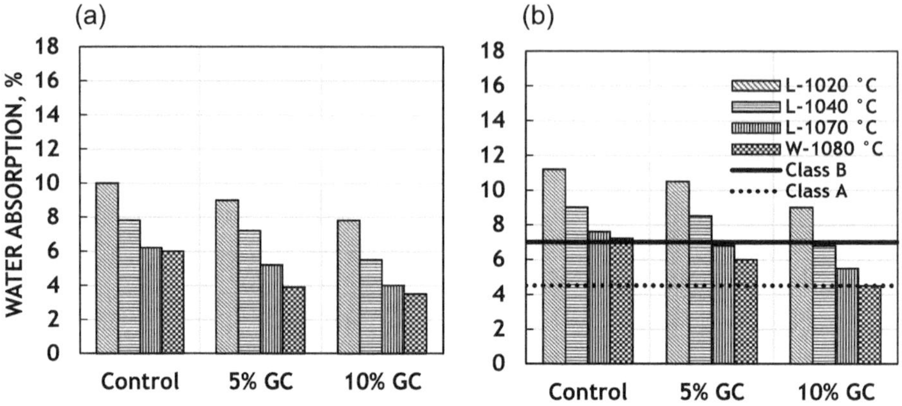

Figure 9.5 Effect of GC on the water absorption of clay bricks made with carboniferous shale at different firing temperatures. (a) 24-h cold water immersion test. (b) 5-h boiling water immersion test. *GC*, glass cullet; *L*, laboratory firing; *W*, works firing.

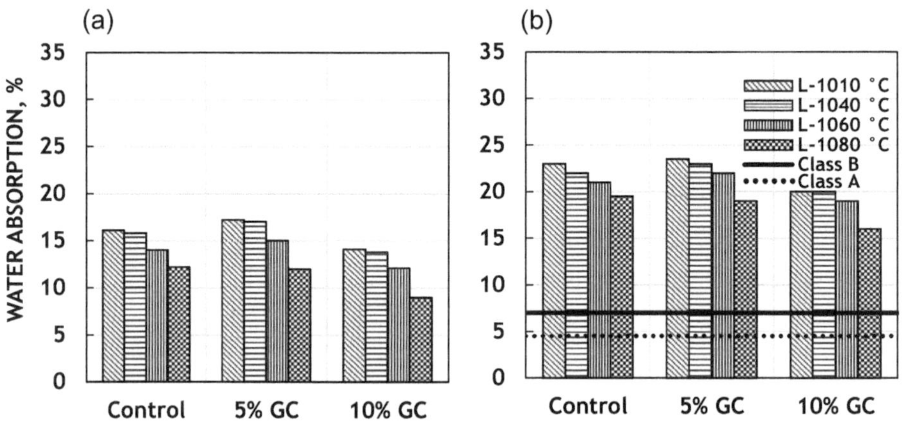

Figure 9.6 Effect of GC on the water absorption of clay bricks made with Quaternary glacial lacustrine clay at different firing temperatures. (a) 24-h cold water immersion test. (b) 5-h boiling water immersion test. *GC*, glass cullet; *L*, laboratory firing.

(ii) Density

In addition to the specific gravity of the constituent materials, the bulk density of clay bodies, which are porous, is influenced by their total porosity. Bricks with high bulk density and low porosity are expected to have reasonably good mechanical and durability-related performance.

Figure 9.8 presents the effects of GC from various sources on the bulk density of clay bricks, fired at different temperatures. The results suggest that the use of GC increases the bulk density of clay bricks, especially at higher GC content. This effect is more prominent with increasing firing temperatures (Lin, 2007a; Loryuenyong et al., 2009; Phonphuak et al., 2016).

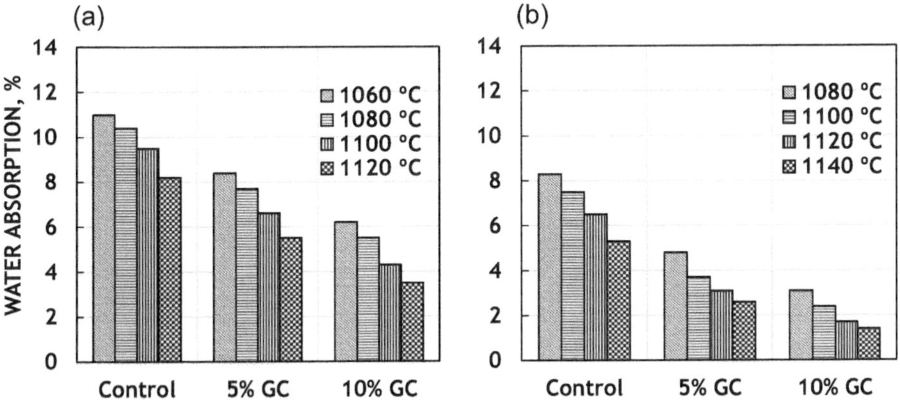

Figure 9.7 Water absorption of clay bricks made with GC fired at different temperatures. (a) Fire clay (buff clay). (b) Etruria marl. *GC*, glass cullet.

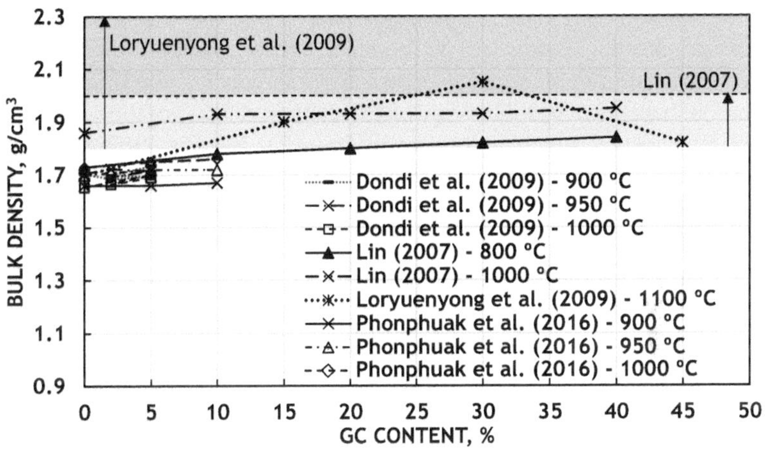

Figure 9.8 Bulk density of bricks made with increasing GC content. *GC*, glass cullet.

For comparison, the typical bulk density of bricks, as suggested by Lin (2007a), is in the range of 1.8–2.0 g/cm^3, and by Loryuenyong et al. (2009), in the range of 1.8–2.3 g/cm^3, both shaded in Figure 9.8. This shows that, to achieve the typical bulk density range, clay bricks may be produced by either increasing the firing temperature or using higher GC content. Since the former consumes greater energy usage and thus generates higher CO_2 emissions, the use of GC is a preferred option as it offers a more sustainable route.

It was found that the relative increase in bulk density is independent of the type of GC used, whether crushed LCD (Lin, 2007a), CRT (Dondi et al., 2009) or soda lime–silicate glass (Phonphuak et al., 2016). However, the results of Loryuenyong et al. (2009) showed that, when GC from crushed structural walls was used, the bulk density of clay bricks increased with increasing GC content up to 30%, beyond which it decreased but was within the typical bulk density range. In the same study, Loryuenyong et al. (2009) also showed that GC was more effective in increasing the bulk density of clay bricks than rice husks.

Mechanical Properties of Fired Clay Bodies

(i) Compressive Strength

According to BS-EN-771-1 (2011), although there are no strength classes for clay bricks, the mean compressive strength of the product must be declared by the manufacturer and classified as category I, in which the probability of failing to achieve the declared strength does not exceed 5%, or category II, in which the level of confidence placed on category I does not apply. There are, however, strength classes for engineering bricks, referenced in the UK *National Annex* to BS-EN-771-1 (2011). These classes were based on the now withdrawn standard BS-3921 (1985), which specified minimum compressive strength values of 50 and 70 MPa for class B and A engineering bricks, respectively, but increased those values to 75 and 125 MPa, respectively.

In view of the potential mechanical performance enhancement with increasing use of waste glass and with the subsequent opportunity of reducing the firing temperature, several authors have analysed the effect of GC content on the compressive strength of clay bricks, the results of which can be seen in Figure 9.9. It covers different types of GC, incorporation levels up to 45%, firing temperatures in the range of 800°C–1140°C and various clay materials. Overall, the results tend to suggest that an increase in GC content leads to an increasing compressive strength. Furthermore, based on the BS-EN-771-1 (2011) requirements for class A and B engineering bricks, it is clear that, compared with most control specimens, which did not belong to any of those classes, the incorporation of a small amount of GC is able to gain sufficient strength for structural purposes. The incorporation of GC may allow the production of clay bricks with compressive strength values similar to those of control bricks, but at a lower firing temperature.

Clay bricks with up to 40% GC (Lin, 2007a) showed no major change in compressive strength at a firing temperature of 800°C, but the mechanical performance increased at higher firing temperatures of 900°C and 1000°C. Lin (2007a) also observed that the compressive strength of all clay bricks containing up to 40% GC met the minimum strength requirements of first-class and second-class bricks, i.e., 15 and 10 MPa, respectively, according to Chinese National Standard CNS 382-R2002 (2007).

Only a few cases in which higher incorporation levels led to a strength decrease were observed. Loryuenyong et al. (2009), after having fired ball clay bricks to a temperature of 1100°C with GC from crushed structural glass walls, observed a decline in performance when more than 15% GC (up to 45%) was used. The authors explained this as a result of GC oozing out from the bricks' surface. Nevertheless, according to ASTM C902 (2015), the resulting bricks met the minimum required compressive strength of a paving brick subjected to light traffic, which is 17.2–20.7 MPa (2500–3000 psi) in a less severe environment.

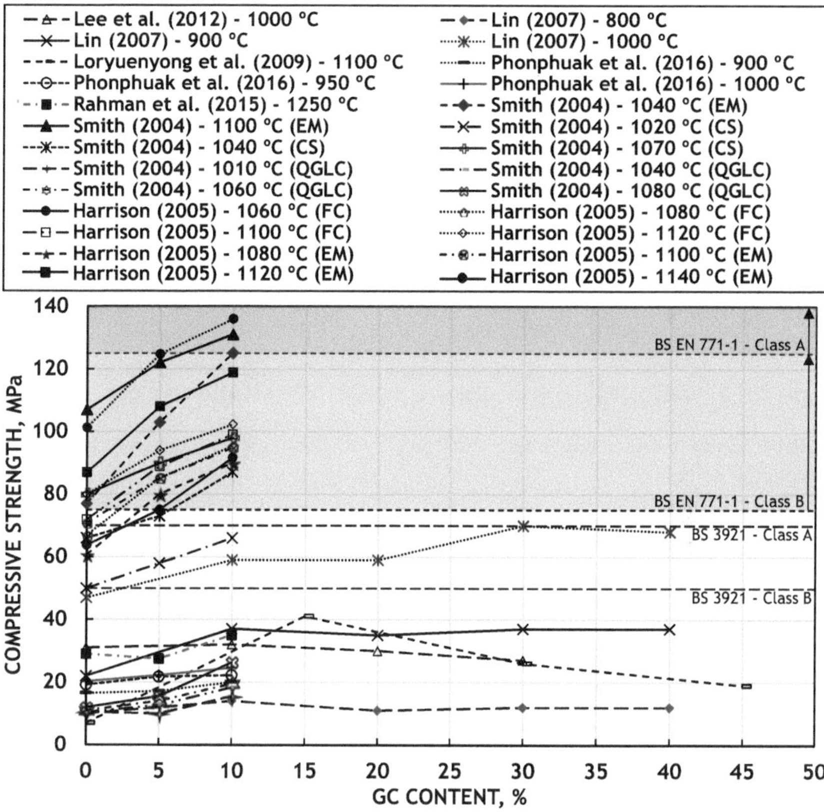

Figure 9.9 Effect of GC on the compressive strength of clay bricks. *CS*, carboniferous shale; *EM*, Etruria marl; *FC*, fire clay; *GC*, glass cullet; *QGLC*, Quaternary glacial lacustrine clay.

Figure 9.10 presents the compressive strength of clay bricks containing 0%, 5% and 10% mixed-colour container GC (soda lime–silicate glass), made with three different types of clay (EM, CS and QGLC), fired at different temperatures (Smith, 2004b). Regardless of the firing temperature and type of clay used, all bricks exhibited an increase in compressive strength with higher incorporation of GC. Figure 9.11, which presents the same analysis for the results reported by Harrison (2005), shows the same trend. Bricks made with EM showed the highest compressive strength values compared with CS and QGLC (Harrison, 2005; Smith, 2004b). After incorporation of 10% GC, at a relatively low firing temperature of 1040°C, the compressive strength of the resultant bricks increased by over 60%, meeting the strength requirement for class A engineering bricks (Smith, 2004b). Bricks made with CS clay and 10% GC fired at a temperature of 1040°C were class B engineering bricks, whereas the control made without GC did not belong to any class. Similar findings were also reported by Harrison (2005) for bricks made with fire clay (buff clay) and EM (red clay).

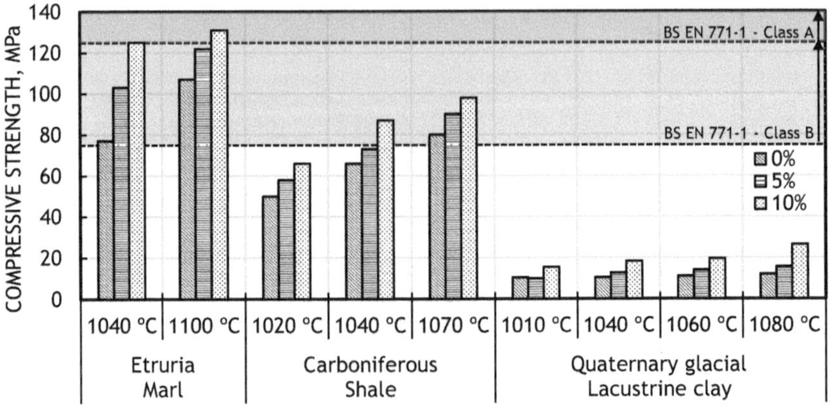

Figure 9.10 Effect of increasing temperature on clay bricks made with different types of clay and glass cullet contents.
Adapted from Smith (2004b).

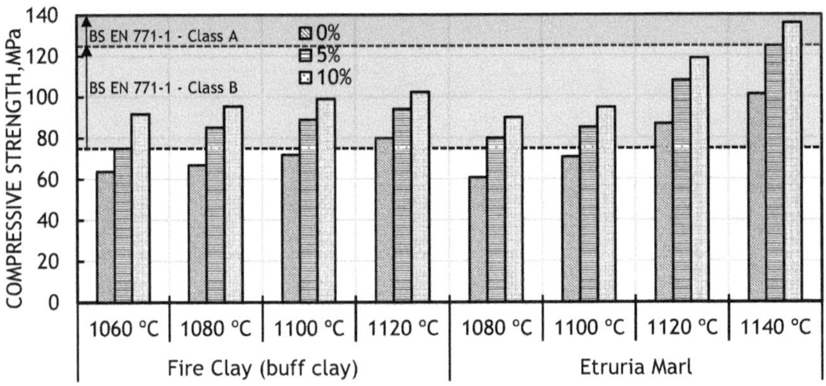

Figure 9.11 Effect of increasing temperature on clay bricks made with different types of clay and glass cullet contents.
Adapted from Harrison (2005).

(ii) Flexural Strength

The results on the flexural strength of clay bricks containing GC from crushed CRTs and screen panels (Dondi et al., 2009) and structural glass walls (Loryuenyong et al., 2009), manufactured at different firing temperatures, are presented in Table 9.12. The results obtained by Dondi et al. (2009) showed that the use of 2% and 5% GC from crushed CRTs and screen panels did not present any significant change in the flexural strength of clay bricks, fired at temperatures of 800–1000°C.

The flexural strength of clay bricks, evaluated by Loryuenyong et al. (2009), fired at 1100°C and made with GC from structural glass walls that were crushed to a particle size of less than 1 mm, increased when 15% GC was used, but thereafter it decreased up to 45% GC content.

Table 9.12 Flexural strength of clay bricks made with glass cullet

Reference	GC Type	GC, %	Clay Type	Temp., °C	Results
Dondi et al. (2009)	CRT, screen panels	2, 5	Carbonate-rich and fine-grained clay	800, 900 and 1000	No significant change in flexural strength with GC inclusion
Loryuenyong et al. (2009)	Structural glass walls	0–45	Ball clay	1100	Strength increased at 15% GC and decreased thereafter up to 45%

CRT, cathode ray tube; *GC*, glass cullet.

9.2.4 Tiles

Tiles can be manufactured from a wide range of materials, such as ceramic, natural stone, metal and glass, and are commonly used for covering and cladding of roofs, floors and walls. Depending on the application, tiles require specific properties, for example, roof tiles must have good freeze–thaw resistance, floor tiles should possess good abrasion resistance and wall tiles should be good at waterproofing. Given the compatibility of glass in the manufacturing process of tiles, several experimental studies have been undertaken on the use of GC in the production of tiles (Akristos, 2005; Anderson, 2008; Zhu et al., 2003; Coventry et al., 1999; CWC, 1999; Dondi et al., 2009; Hwang et al., 2006; Lin, 2007b; Matteucci et al., 2002; Raimondo et al., 2007; Youssef et al., 1998).

Physical Properties of Unfired Tiles

Only one study was found to investigate the effect of incorporating GC from crushed PC display screens and CRTs from TV sets on the physical properties of unfired roof tiles (green bodies) made with low-carbonate and coarse-grained clay (Dondi et al., 2009). The results on the physical properties of tiles made with GC, presented in Table 9.13, show trends similar to those observed in bricks made with GC given in Table 9.10.

The working moisture for extrusion of tiles, determined in accordance with ASTM C324 (1992), showed no significant change when 2% and 5% GC from crushed screens or CRTs were used (Dondi et al., 2009). A similar trend was also reported in tiles made with 5% and 10% soda lime–silicate glass from crushed containers and float glass (Matteucci et al., 2002) and screen glass or CRT (Raimondo et al., 2007) as partial replacement for sodium feldspar.

The results of the Pfefferkorn plasticity index test (Table 9.13), which was carried out in accordance with Van der Velden (1979), show that the index decreased with increasing content of crushed screen or CRT GC at 2% and 5% levels. The authors observed that, for a given water content, the shear resistance of extruded bodies containing GC was higher than that of the control specimens.

Table 9.13 Effect of glass cullet on the physical properties of unfired roof tiles

		Screen Glass		CRT	
Physical Properties	**Control**	**2%**	**5%**	**2%**	**5%**
Working moisture, wt.%	23.5	23.1	23.6	24.9	24.8
Pfefferkorn index, wt.%	40.6	32.4	31.3	32.6	30.3
Drying shrinkage, cm/m	6.1	6.1	6.5	6.8	6.6
Weight loss with shrinkage, %	48	50	44	50	53
Drying sensitivity index	0.33	0.34	0.40	0.50	0.50
Hygroscopicity, wt.%	1.94	1.88	1.64	2.01	1.57
Dry flexural strength, N/mm^2	8.9	9.1	7.0	6.8	7.0

CRT, cathode ray tube.
Values sourced from Dondi et al. (2009).

As presented in Table 9.13, the results of the drying shrinkage of unfired tiles suggest that the inclusion of both types of GC causes a slight increase in drying shrinkage. This is probably due to the greater moisture loss of GC specimens, which was reflected by greater weight loss. In addition, the drying sensitivity index also suggests that CG specimens are more sensitive to the drying process.

The results of Table 9.13 show that the incorporation of GC from crushed PC display screen panels and CRTs from TV sets tends to slightly reduce the hygroscopicity of unfired bodies. It should be noted that high hygroscopicity could result in considerable expansion due to reabsorption of moisture.

As roofing tiles are thin, it is important for unfired bodies to have sufficient dry flexural strength to avoid damage throughout the firing process. The results obtained by Dondi et al. (2009) suggest that the incorporation of a small amount of GC can result in unfired bodies with slightly lower flexural strength than that of the reference, and thus additional care needs to be taken when handling dried bodies containing GC. However, the opposite was observed in the results of Matteucci et al. (2002).

Physical Properties of Fired Tiles

The porosity of tiles, which is directly related to moisture absorption, must be kept to a minimum to prevent the growth of unwanted organisms and thus deterioration of both performance and aesthetic value. The effect of increasing GC inclusion on the porosity, water absorption and bulk density of tiles, fired at temperature ranging from 900 to 1220°C, has been studied (Dondi et al., 2009; Hwang et al., 2006; Lin, 2007a; Matteucci et al., 2002; Youssef et al., 1998; Raimondo et al., 2007; Kim et al., 2016; Lin et al., 2015). A description of the work undertaken involving the

Table 9.14 Experimental campaigns on the use of glass cullet in glass-ceramics

Reference	GC Type	GC, %	Other Materials	Tile	Firing Temp., °C	Sintering Time, h
Dondi et al. (2009)	TV CRT and PC display screen panel	0–5	Low-carbonate and coarse-grained clay	Roofing tiles	900–1000	4
Hwang et al. (2006)	Mixed-colour container	0–50	Bentonite clay	Ceramic tiles	800–1100	2
Lin (2007b)	TFT-LCD	0–50	White clay	Ceramic tiles	900–1200	6
Matteucci et al. (2002)	Container and float	0–10	Ball clay	Porcelain stoneware tiles	1120–1200	1
Raimondo et al. (2007)	CRT and TV/PC screen display	0–10	Ball clay	Porcelain stoneware tiles	1180–1220	1
Youssef et al. (1998)	Soda lime–silicate	0–50	Mixes of Bahria, Aswan and Sinai clay	Wall and flooring tiles	950–1100	1, 2, 3
Kim et al. (2016)	LCD	0–40	Clay, quartz, feldspar	Ceramic tiles	1000–1150	1
Lin et al. (2015)	Solar panel	0–40	White clay	Ceramic tiles	800–1100	2

CRT, cathode ray tube; *GC*, glass cullet; *TFT-LCD*, thin-film transistor–liquid crystal display.

use of GC in tiles is presented in Table 9.14, covering the materials used, the firing temperature and time, and the types of tiles manufactured. It appears that the effects of GC incorporation can be separated into two main groups, based on the content of GC used, i.e., up to 10% by weight (Dondi et al., 2009; Matteucci et al., 2002; Raimondo et al., 2007) and up to 50% by weight (Lin, 2007a; Youssef et al., 1998; Kim et al., 2016; Lin et al., 2015).

Figure 9.12 presents the water absorption of ceramic tiles made with GC content up to 10%. The results suggest that the inclusion of low levels of GC has marginal influence on the water absorption, though, in some cases, it results in a reduction. However, when part of the sodium feldspar was replaced with GC from screen glass, the resulting ceramic tiles exhibited an increase in water absorption (Raimondo et al., 2007). As this finding contradicted the majority of studies, further testing would have to be carried out to ascertain whether the specific chemical composition of the GC used has an influence on the porosity of ceramic tiles.

The water absorption and porosity of ceramic tiles decreased with increasing firing temperature up to 1220°C, regardless of the GC content for up to 10% GC. However, Hwang et al. (2006) observed a significant increase in water absorption and porosity of ceramic tiles fired at 1100°C, when GC was used up to 10%, but the values decreased at higher GC contents. The authors explained that this could be due to an expanding behaviour as a result of the formation of gas bubbles in the ceramic tiles.

In contrast to what is observed in Figure 9.12 for ceramic tiles containing small amounts of GC, Figure 9.13 clearly shows that the incorporation of greater amounts of GC leads to a reduction in water absorption and porosity, the effect of which becomes more significant with increasing firing temperature.

Youssef et al. (1998) also evaluated the effect of sintering time on the water absorption and porosity of ceramic tiles at a given firing temperature. The purpose of maintaining the materials at a constant firing temperature is to allow ample time for various reactions to take place during the vitrification process. It

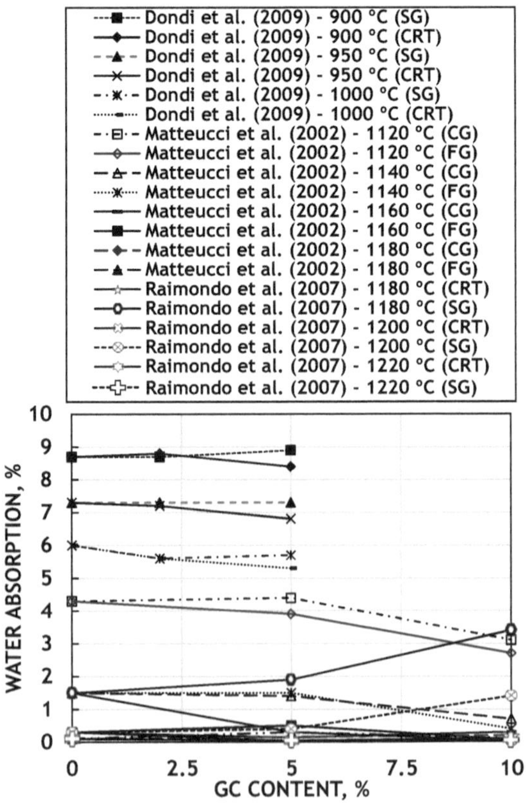

Figure 9.12 Water absorption of ceramic tiles made with GC. *CG*, container glass; *CRT*, cathode ray tube; *FG*, float glass; *GC*, glass cullet; *SG*, screen glass.

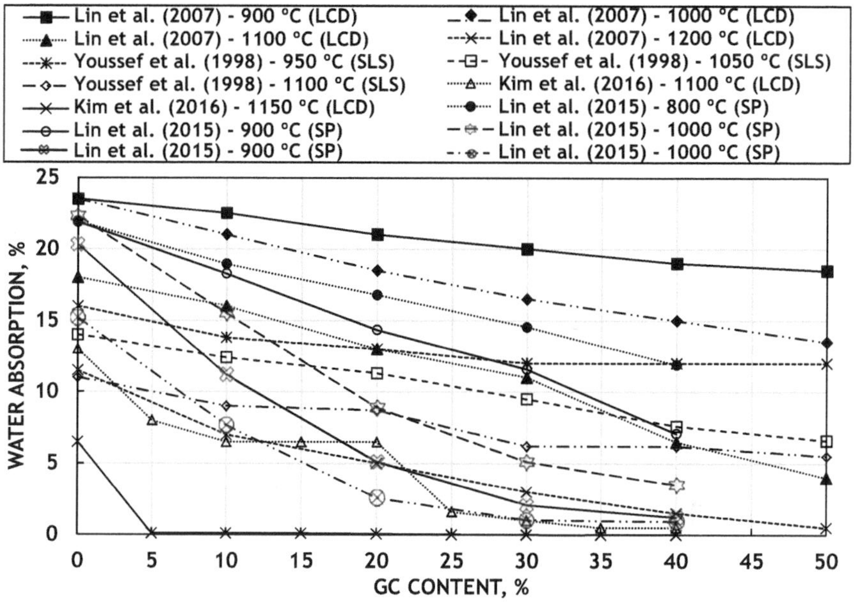

Figure 9.13 Water absorption of ceramic tiles made with GC. *GC*, glass cullet; *LCD*, liquid crystal display; *SLS*, soda lime–silicate; *SP*, solar panel.

was found that, at a given temperature, an increase in sintering time decreased both the apparent porosity and the water absorption. This decrease was greater for control tiles in comparison with that of tiles containing 40% and 50% soda lime–silicate GC.

Contradictory findings have been observed with regard to the effect of incorporating up to 10% GC on the bulk density of ceramic tiles, as shown in Figure 9.14. The majority of the results obtained by Dondi et al. (2009) and Matteucci et al. (2002) show that the bulk density of ceramic tiles remains unchanged or increased slightly when GC is used; however, those of Raimondo et al. (2007) suggest the opposite.

However, when GC is used at an even greater content, Figure 9.15 shows that ceramic tiles containing up to 50% GC exhibit a slightly higher bulk density than reference ceramics tiles. This can be associated with the reduction in water absorption and porosity. However, Kim et al. (2016) observed a decrease in the bulk density of ceramic tiles when up to 40% GC from crushed LCD was incorporated. This is probably due to the difference in the densities of the materials, as LCD glass, which had a density of 2.45 g/cm^3, was used to replace feldspar, which had a density varying between 2.55 and 2.76 g/cm^3.

The Vickers hardness of GC-containing ceramic tiles obtained by Lin (2007b) and Lin et al. (2015) is shown in Figure 9.16. In general, the incorporation of GC leads to an increase in the hardness of the resulting material, regardless of the type of GC used. However, in only one case, the Vickers hardness of ceramic

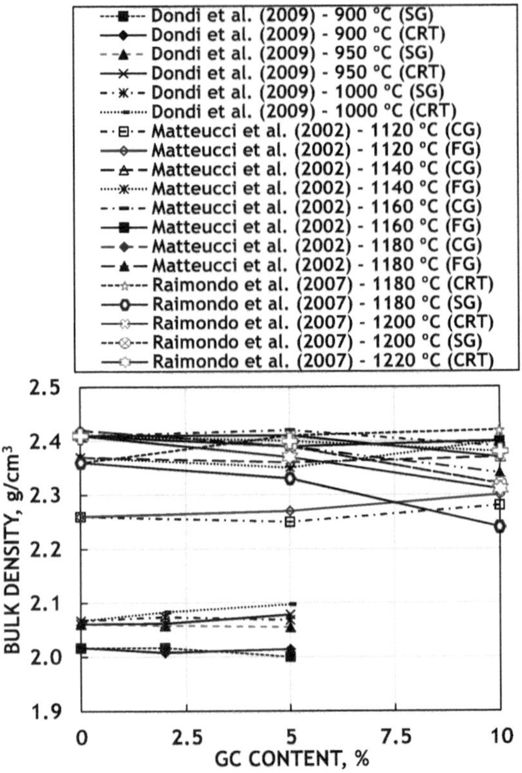

Figure 9.14 Bulk density of ceramic tiles made with GC. CG, container glass; *CRT*, cathode ray tube; *FG*, float glass; *GC*, glass cullet; *SG*, screen glass.

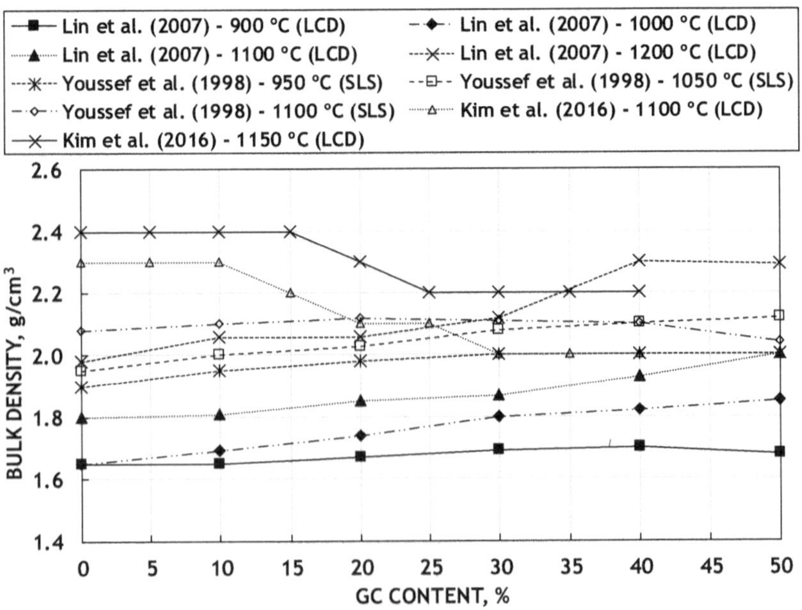

Figure 9.15 Bulk density of ceramic tiles made with GC. *GC*, glass cullet; *LCD*, liquid crystal display; *SLS*, soda lime–silicate.

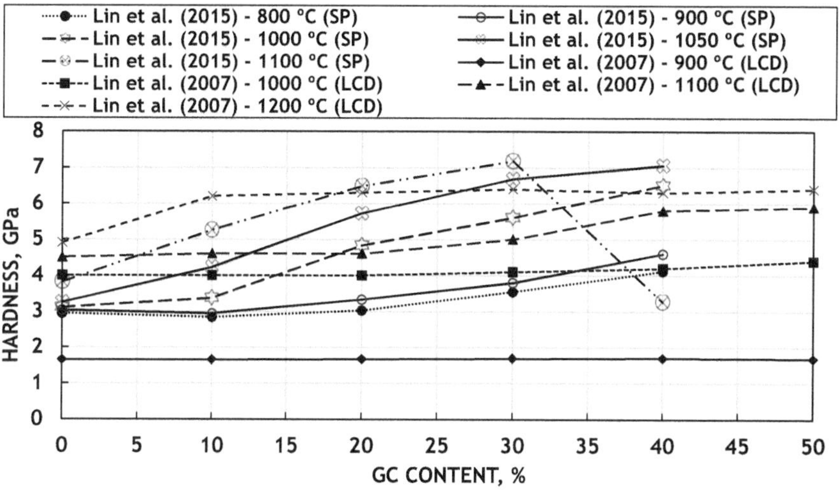

Figure 9.16 Vickers hardness of ceramic tiles made with GC. *GC*, glass cullet; LCD, liquid crystal display; *SP*, solar panel.

tiles was found to decrease with the incorporation of 40% GC from crushed solar panels (Lin et al., 2015). This might be treated as an experimental error since it did not correlate well with the corresponding findings on water absorption and porosity.

The results for the relative mass loss due to abrasion of GC-containing ceramic tiles, reported in two studies (Lin, 2007b; Lin et al., 2015), appear to be contradictory, as shown in Figure 9.17. The 2015 study suggests that the abrasion resistance of ceramic tiles increases with increasing GC (from crushed solar panels) content. The 2007 study showed that the incorporation of GC from crushed LCD led to a decrease in the resistance to abrasion for replacement levels between 10% and 40% at firing temperatures in the range of 1000–1100°C. However, the resistance to abrasion of ceramic tiles increased when either 50% GC content was used or firing was done at a higher sintering temperature of 1200°C.

The influence of using GC from crushed soda lime–silicate glass (Bourtsalas et al., 2015a,b; Faria and Holanda, 2016) and screen glass (Galan-Arboledas et al., 2016) on the properties of glass-ceramic tiles has been explored. The results showed that the use of GC as a replacement for fine incinerated bottom ash at up to 30% (Bourtsalas et al., 2015a,b), and clay at 5% (Faria and Holanda, 2016) and 10% (Galan-Arboledas et al., 2016), enhanced the densification process, subsequently leading to a decrease in porosity and water absorption.

Shrinkage is another factor that influences the quality of ceramic tiles. This factor is affected by firing temperature and, normally, an increase in the firing temperature leads to an increase in shrinkage. Figures 9.18 and 9.19 present the effect of incorporating GC on the firing shrinkage of ceramic tiles. At lower GC level

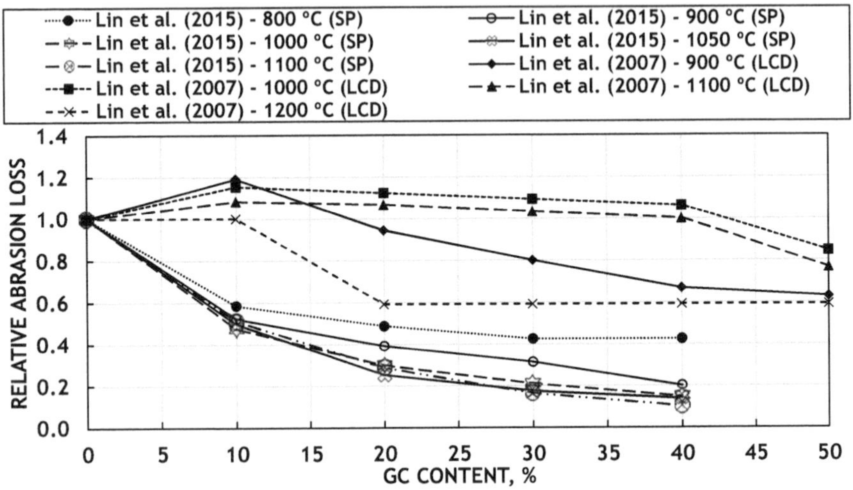

Figure 9.17 Relative abrasion mass loss of GC-containing ceramic tiles. *GC*, glass cullet; *LCD*, liquid crystal display; *SP*, solar panel.

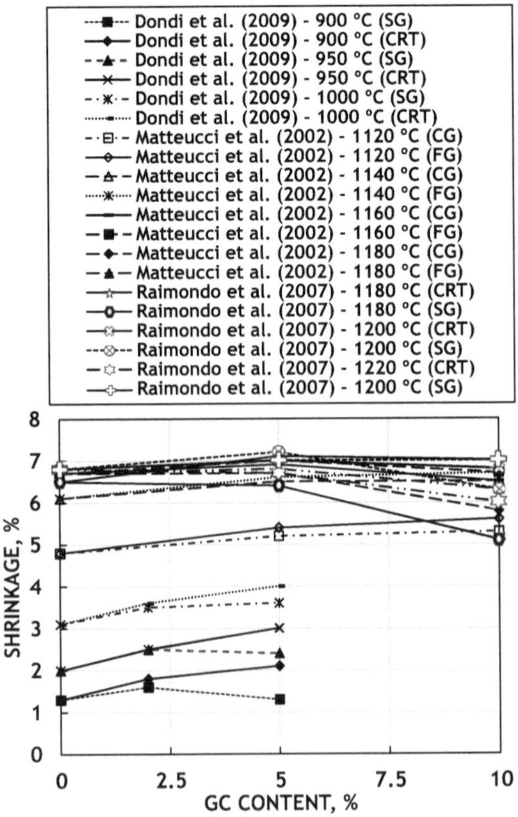

Figure 9.18 Shrinkage of ceramic tiles with GC. *CG*, container glass; *CRT*, cathode ray tube; *FG*, float glass; *GC*, glass cullet; *SG*, screen glass.

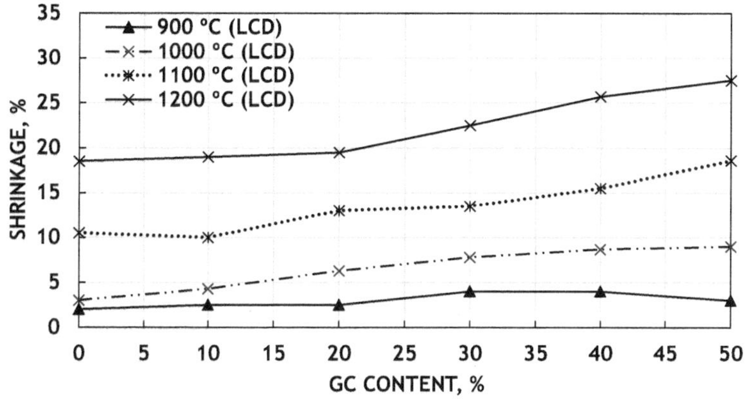

Figure 9.19 Shrinkage of ceramic tiles with GC. *GC*, glass cullet; *LCD*, liquid crystal display. Adapted from Lin (2007b).

(Figure 9.18), ceramic tiles generally show a slight increase in firing shrinkage, although there were some contradictory results (Raimondo et al., 2007). It is noted that replacing the raw materials with GC from crushed CRTs consistently resulted in a slightly higher shrinkage in tiles compared with replacement by glass from crushed PC display screen panels, but the explanation for this is not available (Dondi et al., 2009).

The increase in firing shrinkage was notably greater at higher GC levels and sintering temperatures. As shown in Figure 9.19, the firing shrinkage remained mostly unaffected by the incorporation of GC from crushed LCD screens at sintering temperature of 900°C. However, the firing shrinkage of tiles began to increase as sintering temperature increased from 900 to 1200°C (Lin, 2007b). Thus, the combination of these two factors must be closely controlled to minimise the shrinkage of tiles during the sintering process.

However, Hwang et al. (2006) showed an opposite trend, i.e., the increasing GC levels caused greater shrinkage at 800°C, but the value was unchanged or even lower for higher sintering temperatures.

Mechanical Performance of Fired Tiles

Tiles, such as those used in flooring, must possess adequate mechanical performance, as they are subjected to loading. However, the strength of tiles should not be considered as the dominant decision-making criterion for the mix design of all tiles, as the durability-related performance of tiles, such as those used in roofing and walls, is equally important.

Research on the compressive strength of wall and flooring tiles made with mixes of Bahria, Aswan and Sinai clay and soda lime–silicate GC at replacement levels up to 50%, manufactured at a range of firing temperatures from 950 to 1100°C at constant 3-h sintering time, was conducted by Youssef et al. (1998). The results suggest that the compressive strength of the resultant tiles increased with increasing GC content, and, at a given GC content, the improvement in strength is greater at higher firing temperatures (Figure 9.20). Ceramic tiles fired at 1100°C showed a peak compressive strength of almost 90 MPa at 30% GC inclusion, beyond which it decreased, but was still maintained at 50%–80% higher than that of the reference tiles.

In the experimental research of Anderson (2008), although no control tiles (without GC) were produced, the mean flexural strength of tiles made with GC was 1360 N, exceeding the minimum requirement of 600 N, given in BS-EN-1304 (2013).

The results obtained by Dondi et al. (2009) for the production of tiles containing relatively low replacement levels of 2% and 5% GC from crushed PC display screen panels and CRT from old TV sets showed that the flexural strength was marginally affected by the incorporation of GC.

Lin (2007b) observed that, when GC was used at up to 50%, the flexural strength of the resultant tiles fluctuated within a narrow range of around 2 MPa. However, at 1200°C, all GC-containing tiles exhibited similar or higher flexural strength values compared with the control specimens. A peak flexural strength was observed at the 30%–40% GC level, which corresponded to an increase of 30% relative to the control tiles.

Figure 9.21 presents the flexural strength of ceramic tiles containing up to 10% soda lime–silicate glass from containers and float glass as a partial replacement for sodium feldspar. Tiles sintered in a temperature range of 1120–1140°C showed similar flexural strength regardless of the GC content. However, at higher sintering temperatures (1160–1180°C), the flexural strength of GC-containing tiles is slightly lower than that of the reference. This trend was observed in both types of GC, which is only natural since they exhibit similar chemical compositions.

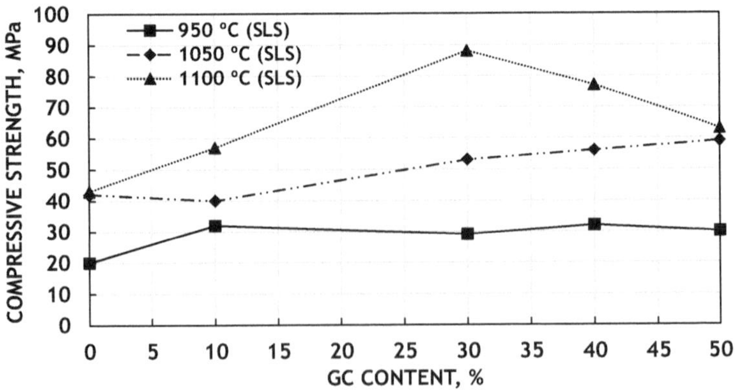

Figure 9.20 Compressive strength of GC-containing ceramic tiles. *GC*, glass cullet; *SLS*, soda lime–silicate.
Adapted from Youssef et al. (1998).

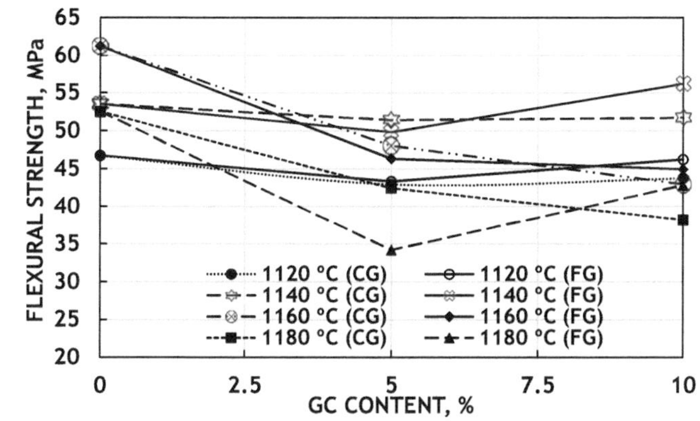

Figure 9.21 Flexural strength of ceramic tiles made with GC. *CG*, container glass; *FG*, float glass; *GC*, glass cullet.
Adapted from Matteucci et al. (2002).

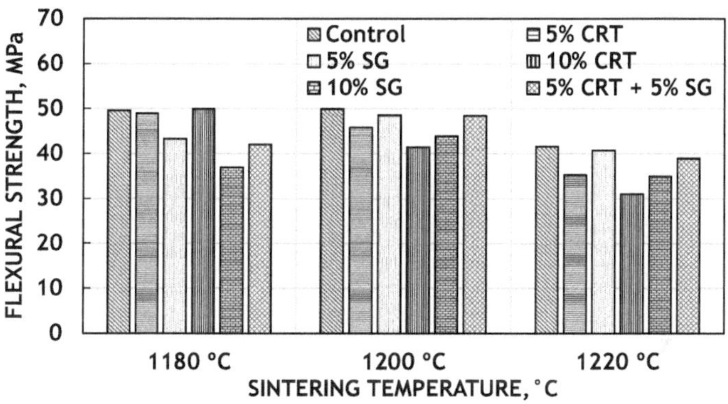

Figure 9.22 Flexural strength of ceramic tiles made with glass cullet. *CRT*, cathode ray tube; SG, screen glass.
Adapted from Raimondo et al. (2007).

Figure 9.22 presents the flexural strength of ceramic tiles made with 5% and 10% GC from crushed TV and PC screen panels and from CRTs. At a sintering temperature of 1180°C, the incorporation of CRT glass had little influence on the flexural strength of ceramic tiles, whilst screen glass caused a clear reduction. However, at 1200°C, the opposite was observed, but with a smaller difference in strength relative to the control. At a sintering temperature of 1220°C, the flexural strength of all tiles decreased and more so for CRT-containing specimens. The mixture of 5% CRT and 5% SG resulted in flexural strength values similar to those of the controls.

The flexural strength of glass-ceramic tiles containing GC from crushed soda lime–silicate glass (Cetin et al., 2015; Lu et al., 2016a,b; Marangoni et al., 2014a; Ponsot et al., 2015a) and borosilicate glass (Cetin et al., 2015; Marangoni et al., 2014a) was found to vary from 35 to almost 100 MPa.

Heavy Metal Leachability

The results for leaching behaviour of roof tiles containing 5% GC from crushed PC display screen panels and CRTs from old TV sets, assessed using the toxic characteristic leaching procedure, showed that the released concentrations of lead, barium and strontium in the leachates were considerably low.

Significant immobilisation of lead was also observed by Raimondo et al. (2007). Ceramic tiles containing up to 10% GC from crushed screen glass showed similar concentrations of released lead in the leachate compared with control tiles. Tiles containing 10% CRT, however, showed slightly higher concentration of released lead (0.37 and 0.71 mg/kg for 5% and 10% CRT, respectively) as CRTs have a high lead content.

In addition, the introduction of soda lime–silicate glass in glass-ceramic tiles containing incinerated bottom ashes (Bourtsalas et al., 2015b) or municipal solid waste ashes (Ponsot et al., 2015a) resulted in reductions in the leached heavy metals concentrations. This is probably due to the ability to encapsulate heavy metals resulting from the enhanced liquid-phase sintering of the other glass material when GC is used.

9.2.5 Porcelain

Porcelain is a ceramic material, which is made by heating kaolin and other constituent materials (i.e., clays, feldspar or flint, and silica) in a kiln to very high temperatures, varying between 1200°C and 1400°C. The strength, toughness and porosity of porcelain are affected by the degree of vitrification and the formation of mullite within the clay body. Owing to a combination of low porosity, normally observed in glass, and the ability to retain its shape when heated, normally observed in clay, porcelain exhibits reasonably good mechanical and durability-related performance (Shashidhar and Reed, 1990).

Several experimental studies have been conducted on the production of porcelain made with different types of glass, such as CRTs (Andreola et al., 2008) and LCDs (Kim et al., 2015), but mostly soda lime–silicate glass from containers or flat glass (Rambaldi et al., 2007; Wannagon et al., 2012; Bragança and Bergmann, 2005; Carus and Bragança, 2013; Tarvornpanich et al., 2005; Jackson et al., 2004; Tucci, 2004; Yürüyen and Toplan, 2009; Luz and Ribeiro, 2007), as a partial replacement for feldspar (fluxing agent) at up to 35% (summarised in Table 9.15). Fluxing agent is added to promote the vitrification process of clay bodies at lower firing temperatures. There has been great interest in using waste glass as a fluxing agent, owing to the similarities in chemical composition between GC and feldspar.

Andreola et al. (2008) evaluated the effect of incorporating up to 35% GC from crushed CRTs as a replacement for Na–feldspar in the production of porcelain tiles. The results showed that the incorporation of the low-viscosity CRT glass offered an opportunity to reduce the amount of feldspar since it had a positive effect on the quartz dissolution and the formation of the liquid phase. GC from crushed CRTs, in small

Table 9.15 Use of glass cullet in the production of porcelain

	Glass				
Reference	**Type**	**Content, %**	**Particle Size, μm**	**Optimum Content, %**	**Main Physical Tests**
Andreola et al. (2008)	CRT	0–35	<63	5	Porosity, bulk density, water absorption, shrinkage, Young's and shear moduli, Poisson ratio
Bragança and Bergmann (2005)	Soda lime–silicate	25	<45	25	Bulk density, water absorption, shrinkage, fracture toughness and flexural strength
Carus and Bragança (2013)	Soda lime–silicate	10	<32	10	Water absorption, bulk density, shrinkage, flexural strength, slip casting characteristics
Kim et al. (2015)	LCD	0–25	<50	25	Water absorption, bulk density, shrinkage, XRD
Luz and Ribeiro (2007)	Soda lime–silicate	0–20	<425	5	Porosity, bulk density, water absorption, flexural strength, abrasion resistance
Rambaldi et al. (2007)	Soda lime–silicate	6	<100	6	Bulk density, water absorption, shrinkage, pyroplastic deformation
Tarvornpanich et al. (2005)	Soda lime–silicate	0–25	<186	6.25	Bulk density, water, absorption, shrinkage, densification behaviour
Tucci (2004)	Soda lime–silicate	0–20	<45	10	Porosity, water absorption, shrinkage, Young's modulus, flexural strength
Wannagon et al. (2012)	Soda lime–silicate	0–20	<150	20	Porosity, bulk density, water absorption, shrinkage, flexural strength
Yürüyen and Toplan (2009)	Soda lime–silicate	10–25	–	10	Bulk density, sintering kinetic

CRT, cathode ray tube; *LCD*, liquid crystal display; *XRD*, X-ray diffraction.

amounts (up to 5%, by weight), can be used as a replacement for conventional fluxing agents, as it improves the densification process (lower final open and total porosity and higher apparent density were observed) and mechanical performance (slightly improved in Young's and shear moduli). However, a replacement level between 5% and 35% can result in negative effects, as the glass would react with the other raw materials, modifying their microstructure, causing the formation of a new crystalline phase (barian orthoclase).

Bragança and Bergmann (2005), who studied the influence of replacing feldspar (25% by weight) with soda lime–silicate GC on the properties of porcelain bodies found that the GC acted as a very effective flux and produced a similar crystalline structure, although a slight decline in mechanical performance was observed.

In another study (Carus and Bragança, 2013), the effect of incorporating soda lime–silicate glass as a replacement for feldspar on the properties of porcelain made of a mixture of calcined bone (50%), kaolin (25%) and feldspar (25%) was evaluated. Compared with feldspar, the use of waste glass as a fluxing agent may require lower firing temperatures. The mechanical strength and thermal expansion of porcelain remained unchanged when GC was used as fluxing agent, and the final appearance had a superior whiteness, as the glass did not contain ferrous oxides.

Kim et al. (2015) produced porcelain bodies with waste glass from crushed LCD used as a replacement for feldspar. At a sintering temperature of 1100°C, the increasing incorporation of GC resulted in a decrease in water absorption to a point that it became comparable to that of the control porcelain, which was fired at a temperature of 1200°C. A similar trend was observed in terms of the thermal expansion coefficient. It should also be noted that the increasing incorporation of LCD led to an increase in the glass phase, but a reduction in the mullite and quartz phases.

Luz and Ribeiro (2007) evaluated the effect of the partial replacement of feldspar and quartz with soda lime–silicate glass powder, from the cutting and washing process of glass pieces. Again, the results suggested that glass powder was an effective fluxing agent. Its incorporation accelerated the densification process and resulted in an end product with lower open porosity and water absorption, but higher shrinkage values and high closed porosity. Replacing a small amount of feldspar with 5% glass powder showed good mechanical performance and Weibull's modulus.

Rambaldi et al. (2007) replaced part of the potassium and sodium feldspathic sands with soda lime–silicate glass powder in the production of whiteware bodies. The results showed that the incorporation of glass powder up to 6%, by weight, could produce porcelain bodies with properties, mineralogical composition and firing behaviour similar to those of traditional porcelain.

Tarvornpanich et al. (2005) produced porcelain bodies containing 0%–25%, by weight, of nepheline syenite, a rock that consists largely of nepheline and alkali feldspar, which was replaced by soda lime–silicate GC. The results showed that the batch containing 6.25% GC, fired at 1100°C, had the best physical properties, giving near 0% water absorption and bulk density of 2.40 g/cm^3. The performance of glass-containing porcelains was comparable to that of commercial porcelains fired at 1200°C, suggesting that the use of GC in the production of porcelain bodies had the benefits of reducing material cost and lowering firing temperatures. In another study (Jackson et al., 2004), in which soda lime–silicate glass was also used as a replacement for nepheline syenite, it was found that GC could act as an extremely strong flux, as only around 9% GC was required to achieve performance similar to that of porcelain made with 15% mineral flux. However, using GC as the sole flux generated an unacceptably short firing range, whereas a mixture of GC and mineral flux allowed a greater window in terms of attaining the maximum density and minimum water absorption.

Tucci (2004) analysed the effect of replacing up to 20% of sodium feldspar with soda lime–silicate glass on the properties of porcelain. The authors observed that the optimum replacement level was 10% GC, at which a considerable decrease in firing temperature and improved mechanical performance were observed. Similar findings were made by Wannagon et al. (2012) in the production of porcelain bodies containing glaze sludge and by Yürüyen and Toplan (2009) in the production of porcelain containing 25% fly ash and 50% kaolin in addition to the potassium feldspar and waste glass.

9.2.6 Glaze

A glaze is a finishing coating layer melted and fused permanently with the surface of clay through a crystalline structure during firing. Aluminium is an important constituent in the glaze mix, as it acts as a stabiliser, allowing the glaze to attach firmly to the ceramic substrate. Engobe is an interlayer that sometimes is applied first on a substrate under the glaze layer in most ceramic tiles, to increase the substrate's opacity and attenuate physical–chemical differences, increase the compatibility between the substrate and the glaze and reduce defects on the ceramic material's surface (Dal Bo et al., 2014). Depending on the glaze content and the desired finishing, the surface finish of glaze can be glossy or matte, and it can be transparent, opaque, white or coloured. Different types of glazes and engobes can be produced by varying the proportions of the constituents, such as alumina and silica, e.g., increasing the alumina and decreasing the silica produces a matte glaze.

Andreola et al. (2005, 2007b) carried out an experimental investigation on the feasibility of incorporating CRT glass into ceramic glaze formulations, as a substitute for 'ceramic frits,' in the production of pigmented, silk-screened and flame-hardened glazes, which are used for floor tile coatings. The aesthetic and chemical characterisations of the ceramic tiles glazed with CRT glass were similar to those of the control specimens. A life-cycle assessment on the use of CRT glass in a ceramic glaze formulation that was carried out and compared with that of commercial frit (Andreola et al., 2007a) suggested that CRT reduces the overall environmental impact of the glazing process.

In the study of Hreglich et al. (2001), crushed glass from CRTs, up to 35% content by weight, was used in the production of a matte ceramic glaze. The results showed that the chemical composition of the resulting glaze was quite similar to that of conventional glazes commonly used for wall and floor tiles. In addition, after the thermal treatment, the surface of the tiles exhibited a high-quality, warm and silk-like appearance, similar to that of commercial products.

Dal Bo et al. (2014) also evaluated the influence of replacing the ceramic frit with recycled glass in the production of ceramic engobes. A response surface methodology was used to model the engobe properties according to the content of raw materials in the mix design. The results showed that the linear shrinkage of ceramic engobes increased with increasing recycled glass content. It was also found that a high content of feldspar and glass in the engobe mix design led to a reduction in water absorption owing to the higher sintering degree of the raw materials.

Caki et al. (2013) used waste glass as partial replacement for sodium and potassium feldspar in the production of ceramic glazes. An increase in GC content led to an increase in the thermal expansion coefficient and a slight change in colour of the final glaze. However, the use of waste glass without colourant additions resulted in cracked surfaces. It has been suggested (Ponsot et al., 2015b) that the colour, as well as the shrinkage and viscous flow of the glaze, can be adjusted with the addition of zircon and recycled borosilicate glass.

Siikamäki (2004) carried out a comparative analysis between the performance of glazes of earthenware containing 86%–96% CRT and a commercial glaze (containing lead oxide as flux). Three different types of CRT, obtained from colour TV panel glass, colour PC panel and a mixture of the two, were tested in the study. Two series of glazes, i.e., clear and coloured, were prepared and chemical durability and glossy surface quality were measured in the former, whilst colouring effect was measured in the latter. Although the CRT glazes exhibited good alkaline resistance, the control glaze showed a glossier surface than the CRT glazes, as it contained lead oxide (from flux), which is an essential component to produce glossy surfaces. Furthermore, the evaluation of the colouring properties of the coloured glazes, which were made into seven different colour families (light red, brownish red, brown, turquoise, lilac, bluish velvet and black), showed that the performance of the three CRT glazes were similar. However, it was also pointed out that the brightness of the colour hues in the CRT glazes was decreased owing to the absence of lead oxide.

Siikamäki and Hupa (2001) replaced feldspar with CRT glass at three different contents (4.5%, 9% and 14.5%, by weight) as glaze raw material for tableware. The firing behaviour and colouring properties of glazes containing three different contents of colour PC panel were also evaluated. The results showed that the replacement of feldspar with coloured CRT glass could produce glazes that were fully comparable to the reference samples. In addition, the results of the leaching tests carried out on the glaze containing 14.5% CRT glass showed that the released concentrations of lead, barium and cadmium were well below the maximum allowable limits. Although not regulated, the leachability of barium from CRT glaze should be tested, owing to the potential presence of barium oxide.

The experimental results of glazes produced with E-fibreglass indicated that the thermal expansion coefficients of those glazes were lower than that of the substrate, which gives a convenient mass–glaze concordance (Capatina et al., 2012). Also, scanning electron microscope images of the support–glaze interface indicated a well-defined interface and the presence of an intermediate phase between the glaze and the substrate.

Dima et al. (2012) evaluated the possibility of using waste glass as a substitute for 'ceramic frits' in the manufacture of glazes for various ceramic products. The results showed clear increases in the thermal expansion coefficients of glazes in the temperature range of 20°C–300°C. For this reason, the authors pointed out that the incorporation of GC into the production of glazes must be made with great care to ensure a good compatibility between glaze and substrate. A linear increase in the thermal expansion coefficient of glaze with increasing waste glass content was also observed by Nandi et al. (2015). However, the results also showed that the increasing incorporation of glass decreased the glass transition temperature, coupling temperature and softening point of the obtained frits.

9.3 Glass Cullet as Aesthetic Finishing Material

Owing to the appealing colours of GC, mainly from waste container glass, the material has a great potential for use as a decorating feature in construction materials, such as tiles, bricks and panels (Coventry et al., 1999; Hewitt, 1997; Wright, 2003; Holcroft and Pudner, 2007).

Depending on the desired finishing, GC can be crushed into different particle sizes and used as inert fillers, with a combination of cement/polymers and additives, in the production of tiles, bricks and panels. The resulting products have low production cost and high value in the recycling market. Owing to their great versatility, durability and hard-wearing and slip-resistant properties, cementitious or polymeric panels containing GC can be used in a wide variety of applications, such as furniture (e.g., work tops), flooring and interior and external claddings (Hewitt, 1997; Coventry et al., 1999; Holcroft and Pudner, 2007). Some of these applications may be made in situ and still have the same aesthetic value, yet lower production costs, compared with other normally used decorative materials, such as marble and terrazzo (Coventry et al., 1999).

9.4 Glass Cullet as Filtration Medium

Crushed and graded waste glass can be used to replace natural sand, anthracite and garnet filtration systems for drinking water, municipal and industrial wastewater, farming, swimming pools and fisheries (Wright, 2003). Although there have been several reports on the successful application of GC as filtration media in such systems (WRAP, 2006a,b; AquaEnviro, 2006; Horan and Lowe, 2007; AquaTest and Stuth, 1997; Aquatic Commercial Industries, 1998; Elliott, 2001a,b), its use as a drinking water filtration medium requires an individual approval (Wright, 2003).

Gray & Osborne, Inc. (1996), evaluated the use of GC as a filtration medium in slow sand filtration. A comparison was made between the performance of the glass-based medium and that of natural sand. The results indicated that GC can act as an effective filter medium for slow-rate filtration for raw water sources. Similar results were observed in a pilot trial (AquaEnviro, 2006; Horan and Lowe, 2007), in which a comparative evaluation was performed on a natural sand filtration system, a commercial glass-based system and a GC-based system made of three different particle sizes (coarse, medium and fine). The results showed that the performance of the GC-based and commercial glass-based systems was superior compared with that of a natural sand filtration system, in terms of the total amount of suspended solids and the amount of treated influent. As far as the GC size is concerned, fine GC could produce high-quality effluent, which was also observed by others (WRAP, 2006b; Elliott, 2001a,b), but it became blinded very rapidly. Coarse GC, on the other hand, could process the influent three times faster, but with a reduction in the quality of the final effluent. It has been suggested that the use of recycled GC could reduce the amount of medium required by 10% (AquaEnviro, 2006; Horan and Lowe, 2007). In another field test (Aquatic Commercial Industries, 1998), in which GC was used as a filtration medium in an athletic club's swimming pool, the amount of filtration medium could be reduced by about 20%. When properly processed and graded, the use of GC as a filtration medium improved the clarity of the water and increased the efficiency of backwash.

Filtration is essential in micro-irrigation systems to avoid clogging. By evaluating the performance of four granular filtration media, namely silica sand, crushed recycled glass, surface-modified glass and glass microspheres, Bove et al. (2015) observed that the pressure losses for the silica sand, crushed glass and modified glass were 39%, 27% and 10% lower, respectively, compared with that of microspheres. The authors explained that the different pressure drops between the media were due to the shape of the GC particles (sharper and more angular than silica sand particles) as a result of the manufacturing process.

Though only limited information was available, Holcroft and Pudner (2007) also recognised the potential of using GC as a granular filtration medium in the treatment of drinking water and municipal and industrial wastewater. The results of trials and small-scale commercial operations in several areas in the United Kingdom showed the potential use of recycled GC as a natural sand replacement in an active filtration medium for water treatment plants. Indeed, several water companies expressed interest in switching to this material, provided that the cost of using GC matched that of natural sand.

In another study (AquaTest and Stuth, 1997), in which the use of GC was evaluated for the biological treatment of residential wastewater, it was observed that GC can be used as a viable alternative in standard sand filter design, provided that the material is properly graded.

In the construction of wells, the use of artificial sand packs may be needed when the surrounding geological conditions are poor, for example, inadequate particle size distribution of soils, which can influence the flow rate of the aquifer. The primary purpose of an artificial sand pack in this case is to filter out the fine particles from the formation adjacent to the well (Wright, 2003). To provide good drainage, it has been suggested that high levels of cleanliness of GC are required prior to its use as a filter pack in a water well.

9.5 Glass Cullet as Epoxy Composite

Epoxy resins are known for the ability to adhere to most substrates and also possess excellent mechanical performance, high chemical resistance and heat insulation. The materials have a wide range of applications, including structural adhesives, coating of metals, electrical components, high-tension electrical insulators and fibre-reinforced plastic material.

In some cases, aggregates are mixed with epoxy to increase the volume, lower the cost of the composite, improve the aesthetic finishing and increase the surface's resistance to wear (CWC, 1996a). In view of the inert, hard and sterile properties of glass, crushed and graded GC has been used as an aggregate in a number of epoxy-related applications, such as standard grit substitute in friction floors, in countertops and in flooring (e.g., terrazzo) (Wright, 2003; CWC, 1996a).

9.6 Glass Cullet as Glass Fibres

Glass fibres, which are mainly silica-based materials, exhibit useful properties, such as transparency, hardness, strength, flexibility, stiffness and resistance to chemical attack, among others. These materials may be used in the manufacture of printed circuit boards, structural composites and many other special-purpose products (Wallenberger et al., 2001).

There have been some studies on the possibility of incorporating GC into the glass fibre manufacturing process. Coventry et al. (1999) suggested that recycled waste glass may be easily incorporated in the manufacture of glass fibres, which can then be used in thermal and acoustic insulation or for reinforcing cement-, gypsum- or resin-based products, without any loss of performance.

In the work undertaken by Hreglich et al. (2001), 10% GC from crushed CRT panel glass was incorporated in the manufacture of glass fibre. A slight adjustment to the mixture containing the CRT was made to maintain the same viscosity curve and workability as those of the reference glass fibre. After the mixture was heated for 8 h at 1450°C, the results showed that the quality of glass fibres made with 10% CRT was comparable to that of the reference in terms of the viscosity and spinning temperature. Nevertheless, it was also suggested that further research has to be carried out concerning the human health implication of the resulting glass fibres, owing to the relatively high barium and strontium content in CRTs.

Lancellotti et al. (2001) investigated the mechanical performance of glass fibres made with 100% GC, 100% incinerator ash and a 50/50 combination of the two materials. Glass fibres made with 50% GC and 50% incinerator ash were sized with aminopropyl-triethoxysilane using a double roller applicator to protect the glass filaments. All mixes were drawn into fibres at a winding drum rate of 3200 rpm and at a temperature of 1275°C. The results showed that the ultimate tensile strength, elastic modulus and elongation at rupture of GC-containing fibres were slightly lower than those of typical commercial fibres. Because of this, the authors inferred that further work is needed to improve the performance of glass fibres made with waste glass before their introduction into commercial applications.

9.7 Glass Cullet as Blast Abrasive

Abrasive blasting is normally applied in the automotive industry, shipyards and other industries that involve surface preparation and coating, where abrasive materials are propelled at high speed and pressure onto a substrate. The commonly used abrasive materials include smelter slags (e.g., copper slags and nickel slags), metallic abrasives (e.g., aluminium shot and steel grit) or mineral abrasives (e.g., sand and garnet).

AERCO (1998) prepared a report for the Clean Washington Center (CWC) studying the use of GC from a mixture of post-consumer container glass and post-industrial flat glass as material for blast abrasive. Three different sizes of GC were evaluated, namely, No. 16 (1.18 mm), No. 30 (600 μm) and No. 40 (425 μm), and the performance of GC was compared with that of copper and nickel slag abrasives for several parameters, such as particle size, nozzle blast pressure, nozzle size, substrate and coating thickness. In each case, the cleaning rate, consumption rate, surface profile, productivity and cost were identified. The results showed that the use of GC as a blast abrasive resulted in comparable or slightly better performance compared with slag-type abrasives. In addition, GC abrasives passed the California Air Resources Board testing and certification for listing in the Federal Qualified Product List. In addition, a number of inherent advantages of GC abrasives have been identified (AERCO, 1998):

- Many coatings can be cleaned exceptionally well due to the angular shape of GC.
- The abrasives have lower dust generation, thereby reducing post-blast cleanup costs.
- GC abrasive may be reused more times than some other lower cost abrasives. Based on an evaluation of the grain size distribution, Hupka et al. (1999) also observed that, after each blast-cleaning cycle, GC can be reused more often than copper slag.
- Unlike silica sand, GC contains negligible amounts of crystalline silica, reducing the risk of health hazard from silicosis (i.e., occupational lung disease).
- Unlike slag-type abrasives, GC does not contain high amounts of heavy metals and thus has less environmental impact.

Compared with copper slag, the use of GC as a blast abrasive does not cause any sparks upon impact, which reduces fire occurrences (Shipbuilders and Shiprepairers Association, 2006). Another identified positive feature of using GC as a blast abrasive (CWC, 1996b) is that the effectiveness of GC is independent of the colour of the GC, suggesting potential cost and effort savings for sorting of post-consumer container glass.

9.8 Glass Cullet as Paint Filler

The role of filler in paint is to reduce the cost of raw materials used in paint production, by increasing the volume of the paint, and, in some formulations, it is used to improve coating stability, rheology and other properties of paints.

Wright (2003) discussed the use of GC as a paint filler, based on the information obtained from the CWC. It was stated that recycled waste glass can be used as a competitive alternative to calcium carbonate and titanium dioxide, which are commonly used fillers in the paint industry. However, since glass is less opaque than those conventional fillers, to obtain the optical properties of paint, GC needs to be ground to particle sizes of 0.2–0.4 mm.

9.9 Glass Cullet in Elastomeric Roof Coating

A report of the CWC, prepared by Bailey and Associates (1998), presents information on the potential use of crushed GC as a feasible substitute for naturally occurring crystalline silica in the manufacture of elastomeric roof coatings. As the dust generated from natural crystalline silica is hazardous and carcinogenic, and GC, which is amorphous in nature, is not in this respect, the latter can be considered as a less hazardous alternative for the production of these construction materials.

A basic laboratory formulation was developed to evaluate the physical properties, viscosity and water resistance of roof coating made with different types of post-consumer container glass and post-industrial glass, of varying colours and particle sizes. The ingredients used in the manufacture of the roof coating included binder, cosolvent, dispersing agents, thickener, pigments (titanium dioxide, fire retardant and recycled glass), defoamer, preservative and antifungal agent. After the pilot trial, it was decided that clear glass of size less than No. 150 mesh would be used for further research.

During the laboratory test, four series of elastomeric roof coating mixes were prepared: two samples made with clear container GC, using two different types of binders; another two consisted of existing commercial elastomeric roof coating products, serving as control samples. Four plywood panels were painted with three coating layers of each mix and placed on a roof for eight months of weather exposure. The panels were monitored for dirt collection, discolouring, cracking, blistering and peeling.

The results showed that the samples made with GC did not show significant deterioration compared with the control commercial samples. Results of tests carried out on the elongation, reflection and gloss of these panels also showed that the performance of GC-containing roof coatings was comparable to that of the control panels.

In addition, one of the GC roof coating mixes was selected and used to coat an approximately 950-square-foot roof with three layers of coating and compared with a commercial product under the same exposure conditions. Although no quantitative results were made available, it was stated that the long-term performance in terms of discolouration, cracking, blistering, dirt accumulation and fungal growth, which was inspected quarterly, was satisfactory, and waste glass is suitable for use in elastomeric roof coatings.

9.10 Other Applications

Several other applications for the use of GC have been identified and are listed here:

- Magnetic glass (Romero-Perez et al., 2001)
- Adsorbent (Grutzeck and Marks, 1999; Pant, 2009; Petrella et al., 2010; Wei et al., 2011)
- Asphalt roof shingles (Kiletico et al., 2015)
- Calcium silicate cation exchanger (Coleman et al., 2014)
- Vitrified clay pipes (El-Shimy et al., 2014)
- Electromagnetic radiation absorber (Kazmina et al., 2015)
- Sea turtle nesting substrate (Makowski and Rusenko, 2007)
- Heat-resistant enamel coating (Tur and Terenshenko, 2001)
- Catalyst (Alfaro et al., 2011; Lee et al., 2011)
- Tableware (Hreglich et al., 2001)

9.11 Conclusions

This chapter deals with the incorporation of waste glass in ceramic applications including bricks, tiles, porcelain and glaze, as well as other alternative markets such as finishing material, filtration medium, epoxy composite, glass fibres, abrasive, paint filler and elastomeric roof coating. The specific conclusions are provided next.

The use of waste glass for the manufacture of glass-ceramics, which is an extremely resilient material that can be used in a wide range of applications, is more energy efficient compared with the use of natural raw materials. Despite its enhanced mechanical performance, there has been limited research on the development of functional glass-ceramics made of waste residues targeted for use in structural applications. Owing to the chemical composition variability of different types of GC and other glassy materials used in the production of glass-ceramics, the optimum combination ratio in the mix should be determined to yield maximum densification and minimum water absorption and porosity levels. This also applies to the firing temperature and sintering time, as an increasing amount of GC was found to have beneficial effects on the degree of crystallisation and thus the mechanical performance and durability of the end products.

Waste glass is also a valuable resource in the production of foam glass-ceramics that is lightweight and has good insulating properties and high resistance to chemical attack. Foam glass-ceramics can be used in a variety applications, including insulation of walls, ceilings, roofs and chimneys. As far as the waste glass properties are concerned, when waste glass is used in the production of foam glass-ceramics, the effectiveness of the foaming process can be affected by the fineness and type of waste glass used and also its reaction with other glass materials. However, its effectiveness also depends on the foaming agent and sintering temperatures and times.

It was found that the incorporation of up to 45% GC content caused a decrease in the drying shrinkage of ceramic bricks prior to firing. However, a reverse trend was observed after exposing the GC-containing bricks to high sintering temperatures, after which they exhibited greater firing shrinkage values, with enhanced densification and low levels of porosity of the microstructure. The same improvement was observed in the mechanical properties, in that an increasing amount of ground waste glass contributed to higher compressive and flexural strength values at lower temperatures compared with control bricks without glass. This presents a very interesting opportunity to increase the productivity and reduce the environmental burden of brick manufacturing, by lowering the sintering period and temperature, whilst adding value to the use of waste glass.

The use of GC in the production of ceramic tiles, which is in many ways similar to that of ceramic bricks, has also shown to have significantly beneficial outcomes. Incorporating amounts of up to 50% by weight of GC in ceramic tile production has shown increased densification and lower water absorption and porosity, often with greater resistance to abrasion, despite greater shrinkage after firing. However, due to the contradicting results on mechanical performance (i.e., some observed higher compressive strength with increasing GC content up to 50%, whilst others, for lower replacement levels, observed decreasing flexural strength), further research is needed. It was also observed that the sintering process of ceramic tiles made with GC containing high amounts of heavy metals is an extremely effective way of immobilising them into a valuable product.

The results of various investigations on the use of GC in porcelain production have shown that the material is a feasible and valuable substitute for feldspar as a fluxing agent. The incorporation of ground waste glass is capable of producing porcelain bodies with performance similar to that of corresponding control bodies fired at higher temperatures, thereby demonstrating its high value in reducing the firing temperature and thus lowering both production costs and the environmental impacts of the porcelain industry.

Existing literature on the use of ground waste glass in the production of engobes and glazes of ceramic products has also demonstrated the feasibility of GC inclusion in this application. Apart from exhibiting chemical and mineralogical compositions similar to those of conventionally used glazes, those containing GC also exhibited high-quality finish. Furthermore, the results of a life-cycle assessment also showed that the inclusion of CRT glass in a ceramic glaze formulation would reduce the overall environmental impact of the glazing process.

Although less emphasis has been given to the use of GC in other applications, the results in the literature have shown the possible and novel applications of GC in diverse fields such as aesthetic finishing material, filtration media, epoxy composites, glass fibres, blast abrasives, paint fillers, elastomeric roof coating and other industries.

References

AERCO, 1998. Testing and Certification of Industrial Abrasives Manufactured from Recycled Glass. Clean Washington Center, Seattle, USA.

Akristos, 2005. Full Scale Operational Trials Involving the Use of Recycled Glass in the Heavy Clay Industry (Bricks and Tiles). Waste & Resources Action Programme, Oxon, UK.

Alfaro S O, Rodríguez-González V, Zaldívar-Cadena A A and Lee S W, 2011. Sonochemical deposition of silver–TiO_2 nanocomposites onto foamed waste-glass: evaluation of Eosin Y decomposition under sunlight irradiation. Catalysis Today 166 (1), 166–171.

Anderson M, 2008. Commercial Viability of Masonry Products from CRT Panel Glass. Waste & Resources Action Programme, Oxon, UK.

Andreola F, Barbieri L, Corradi A, Ferrari A M, Lancellotti I and Neri P, 2007a. Recycling of EOL CRT glass into ceramic glaze formulations and its environmental impact by LCA approach. International Journal of Life Cycle Assessment 12 (6), 448–454.

Andreola F, Barbieri L, Corradi A and Lancellotti I, 2005. Cathode ray tube glass recycling: an example of clean technology. Waste Management & Research 23 (4), 314–321.

Andreola F, Barbieri L, Corradi A and Lancellotti I, 2007b. CRT glass state of the art – a case study: recycling in ceramic glazes. Journal of the European Ceramic Society 27 (2–3), 1623–1629.

Andreola F, Barbieri L, Karamanova E, Lancellotti I and Pelino M, 2008. Recycling of CRT panel glass as fluxing agent in the porcelain stoneware tile production. Ceramics International 34 (5), 1289–1295.

Apkar'yan A S, Gubaidulina T A and Kaminskaya O V, 2015. Foam-glass ceramic based filtering material for removing iron and manganese from drinking water. Glass and Ceramics 71 (11–12), 413–417.

AquaEnviro, 2006. Full Scale Trials of Recycled Glass as Tertiary Filter Media for Wastewater Treatment. Waste & Resources Action Programme, Oxon, UK.

AquaTest and Stuth, 1997. Crushed Glass as a Filter Medium for the Onsite Treatment of Wastewater. Clean Washington Center, Seattle, USA.

Aquatic Commercial Industries, 1998. Evaluation of Recycled Crushed Glass Sand Media for High-rate Sand Filtration. Clean Washington Center, Seattle, USA.

ASTM C324, 1992. Standard Test Method for Free Moisture in Ceramic Whiteware Clays. American Society for Testing and Materials, West Conshohocken, Pennsylvania, USA.

ASTM C373, 1994. Standard Test Method for Water Absorption, Bulk Density, Apparent Porosity, and Apparent Specific Gravity of Fired Whiteware Products, Ceramic Tiles, and Glass Tiles. American Society for Testing and Materials, West Conshohocken, Pennsylvania, USA.

ASTM C674, 1994. Standard Test Methods for Flexural Properties of Ceramic Whiteware Materials. American Society for Testing and Materials, West Conshohocken, Pennsylvania, USA.

ASTM C902, 2015. Standard Specification for Pedestrian and Light Traffic Paving Brick. American Society for Testing and Materials, West Conshohocken, Pennsylvania, USA.

Ayadi A, Stiti N, Boumchedda K, Rennai H and Lerari Y, 2011. Elaboration and characterization of porous granules based on waste glass. Powder Technology 208 (2), 423–426.

Bailey & Associates, 1998. Recycling of Crushed Glass into Coating Products. Clean Washington Center, Seattle. USA.

Benhaoua F, Ayadi A, Stiti N, Lamri Y and Ratni H, 2015. Elaboration and characterization of a cellular glass based cullet loaded with carbon fibers. Materials Letters 160, 278–281.

Benzerga R, Laur V, Lebullenger R, Le Gendre L, Genty S, Sharaiha A and Queffelec P, 2015. Waste-glass recycling: a step toward microwave applications. Materials Research Bulletin 67, 261–265.

Bernardo E, Pontikes Y and Angelopoulos G N, 2012. Optimisation of low temperature sinter crystallisation of waste derived glass. Advances in Applied Ceramics 111 (8), 472–479.

Bernardo E, Scarinci G, Bertuzzi P, Ercole P and Ramon L, 2009. Recycling of waste glasses into partially crystallized glass foams. Journal of Porous Materials 17 (3), 359–365.

Binner J, 2005. Ceramics foams. In: Cellular Ceramics: Structure, Manufacturing, Properties and Applications. Wiley-VCH Verlag GmbH & Co. KGaA, Weinheim, Germany.

Bourtsalas A, Vandeperre L, Grimes S, Themelis N, Koralewska R and Cheeseman C, 2015a. Properties of ceramics prepared using dry discharged waste to energy bottom ash dust. Waste Management & Research 33 (9), 794–804.

Bourtsalas A, Vandeperre L J, Grimes S M, Themelis N and Cheeseman C R, 2015b. Production of pyroxene ceramics from the fine fraction of incinerator bottom ash. Waste Management 45, 217–225.

Bove J, Arbat G, Duran-Ros M, Pujol T, Velayos J, Ramirez de Cartagena F and Puig-Bargues J, 2015. Pressure drop across sand and recycled glass media used in micro irrigation filters. Biosystems Engineering 137, 55–63.

Bragança S R and Bergmann C P, 2005. Waste glass in porcelain. Materials Research 8 (1), 39–44.

Brusatin G, Bernardo E and Scarinci G, 2004. Production of foam glass from glass waste. In: Sustainable Waste Management and Recycling: Glass Waste. Kingston University, London, UK, pp. 67–82.

BS-3921, 1985. Specification for Clay Bricks. British Standards Institution (BSI), UK.

BS-EN-771-1, 2011. Specification for Masonry Units. Clay Masonry Units. Comité Européen de Normalisation (CEN), Brussels, Belgium.

BS-EN-1304, 2013. Clay Roofing Tiles and Fittings. Product Definitions and Specifications. Comité Européen de Normalisation (CEN), Brussels, Belgium.

Caki M, Kaya S Y and Gunhan B, 2013. The use of glass waste in stoneware glazes. Ceramics Technical 37, 30–37.

Capatina C, Simonescu C M and Florea G, 2012. Glazes from wastes obtaining. Metalurgia International 17 (1), 16–21.

Carus L A and Bragança S R, 2013. Bone China formulated with waste glass. Advances in Applied Ceramics 112 (3), 169–175.

Cetin S, Marangoni M and Bernardo E, 2015. Lightweight glass-ceramic tiles from the sintering of mining tailings. Ceramics International 41 (4), 5294–5300.

Chakartnarodom P and Ineure P, 2014. Foam glass development using glass cullet and fly ash or rice husk ash as the raw materials. In: Traditional and Advanced Ceramics. Trans Tech Publications, pp. 73–78.

Chinnam R K, Francis A A, Will J, Bernardo E and Boccaccini A R, 2013. Review. Functional glasses and glass-ceramics derived from iron rich waste and combination of industrial residues. Journal of Non-crystalline Solids 365, 63–74.

Chinese National Standards (CNS), 1999. CNS 1127 Method of Test for General Types of Bricks for Building. CNS Catalog, R3042. Bureau of Standards, Metrology and Inspection. Ministry of Economic Affairs, Republic of China.

Chinese National Standards (CNS), 2007. CNS 382 Common Bricks. CNS Catalog, 2002. Bureau of Standards, Metrology and Inspection. Ministry of Economic Affairs, Republic of China.

Coleman N J, Li Q and Raza A, 2014. Synthesis, structure and performance of calcium silicate ion exchangers from recycled container glass. Physicochemical Problems of Mineral Processing 50 (1), 5–16.

Colombo P, Brusatin G, Bernardo E and Scarinci G, 2003. Inertization and reuse of waste materials by vitrification and fabrication of glass-based products. Current Opinion in Solid State & Materials Science 7 (3), 225–239.

Coventry S, Woolveridge C and Hillier S, 1999. The Reclaimed and Recycled Construction Materials Handbook. CIRIA, London, UK.

CWC, 1996a. Best Practices in Glass Recycling – Epoxy Applications for Recycled Glass. Clean Washington Center.

CWC, 1996b. Best Practices in Glass Recycling – Using Glass as a Blasting Abrasive. Clean Washington Center.

CWC, 1999. Testing of Recycled Glass and Inorganic Binder Paving Tiles. Clean Washington Center.

Dal Bo M, Bernardin A M and Hotza D, 2014. Formulation of ceramic engobes with recycled glass using mixture design. Journal of Cleaner Production 69, 243–249.

Dima V, Eftimie M, Volceanov A, Voicu G and Dinca R, 2012. Possibilities for recovery of some glass waste from cathode ray tubes. Revista Romana De Materiale-Romanian Journal of Materials 42 (1), 76–81.

Ding L, Ning W, Wang Q, Shi D and Luo L, 2015. Preparation and characterization of glass-ceramic foams from blast furnace slag and waste glass. Materials Letters 141, 327–329.

Dondi M, Guarini G, Raimondo M and Zanelli C, 2009. Recycling PC and TV waste glass in clay bricks and roof tiles. Waste Management 29 (6), 1945–1951.

El-Shimy Y N, Amin S K, El-Sherbiny S A and Abadir M F, 2014. The use of cullet in the manufacture of vitrified clay pipes. Construction and Building Materials 73, 452–457.

Elliott R W, 2001a. Evaluation of the use of crushed glass as filter medium: Part 1. Water Engineering & Management 148 (7), 13–18.

Elliott R W, 2001b. Evaluation of the use of crushed glass as filter medium: Part 2. Water Engineering & Management 148 (8), 17–20.

EN-772-7, 1998. Methods of Test for Masonry Units. Determination of Water Absorption of Clay Masonry Damp Proof Course Units by Boiling in Water. Comité Européen de Normalisation (CEN), Brussels, Belgium.

Fan C S and Li K C, 2013. Production of insulating glass ceramics from thin film transistor-liquid crystal display (TFT-LCD) waste glass and calcium fluoride sludge. Journal of Cleaner Production 57, 335–341.

Fan C S and Li K C, 2014. Glass-ceramics produced from thin-film transistor liquid-crystal display waste glass and blast oxygen furnace slag. Ceramics International 40 (5), 7117–7123.

Faria K C P and Holanda J N F, 2016. Thermal behavior of ceramic wall tile pastes bearing solid wastes. Journal of Thermal Analysis and Calorimetry 123 (2), 1119–1127.

Fernandes H R, Tulyaganov D U and Ferreira J M F, 2009. Preparation and characterization of foams from sheet glass and fly ash using carbonates as foaming agents. Ceramics International 35 (1), 229–235.

Foamglas, 2016. FOAMGLAS® ONE™ Insulation. Available at: www.industry.foamglas.com/en/industry/products/foamglas_one_insulation/physical_properties/.

Francis A A, Abdel Rahman M K and Daoud A, 2013. Processing, structures and compressive properties of porous glass-ceramic composites prepared from secondary by-product materials. Ceramics International 39 (6), 7089–7095.

Furlani E, Tonello G and Maschio S, 2010. Recycling of steel slag and glass cullet from energy saving lamps by fast firing production of ceramics. Waste Management 30 (8–9), 1714–1719.

Furlani E, Tonello G, Maschio S, Aneggi E, Minichelli D, Bruckner S and Lucchini E, 2011. Sintering and characterisation of ceramics containing paper sludge, glass cullet and different types of clayey materials. Ceramics International 37 (4), 1293–1299.

Galan-Arboledas R J, Cotes T, Martinez C and Bueno S, 2016. Influence of waste addition on the porosity of clay-based ceramic membranes. Desalination and Water Treatment 57 (6), 2633–2639.

Gong Y, Dongol R, Yatongchai C, Wren A W, Sundaram S K and Mellott N P, 2016. Recycling of waste amber glass and porcine bone into fast sintered and high strength glass foams. Journal of Cleaner Production 112, 4534–4539.

Gray & Osborne, 1996. Evaluation of Crushed Recycled Glass as a Filtration Medium in Slow Sand Filtration Recycling Technology Assistance Partnership (ReTAP), Report prepared for the Clean Washington Center (CWC).

Grutzeck M W and Marks J A, 1999. Synthesis of double-layer silicates from recycled glass cullet: a new type of chemical adsorbent. Environmental Science & Technology 33 (2), 312–317.

Guo H W, Gong Y X and Gao S Y, 2010. Preparation of high strength foam glass–ceramics from waste cathode ray tube. Materials Letters 64 (8), 997–999.

Hanpongpun W, Jiemsirilers S and Thavorniti P, 2007. Effects of clear and amber cullet on physical and mechanical properties of glass-ceramics containing zinc hydrometallurgy waste. Journal of Solid Mechanics and Materials Engineering 1 (11), 1305–1312.

Harrison T W, 2005. Full Scale Operational Trials Involving the Use of Recycled Glass Additions to Clay Bricks Fired Through a Tunnel Kiln. Waste & Resources Action Programme, Oxon, UK.

Hewitt A, 1997. Glass and cement – a marriage of convenience. Concrete Quarterly (183), 13 pp.

Holcroft C and Pudner M, 2007. Glass Market Development Assessments for the English Regions and the Nations. Waste & Resources Action Programme, Oxon, UK.

Horan N J and Lowe M, 2007. Full-scale trials of recycled glass as tertiary filter medium for wastewater treatment. Water Research 41 (1), 253–259.

Hreglich S, Falcone R and Vallotto M, 2001. The recycling of end of life panel glass from TV sets in glass fibres and ceramic productions. In: Proceedings of the International Conference on Recycling and Reuse of Glass Cullet, UK, pp. 123–134.

Hupka J, Rzechula J and Korytkowska M, 1999. Investigation of cullet as abrasive material. In: XXXVI International Symposium on Physicochemical Problems of Mineral Processing, Wroclaw, Poland, pp. 75–82.

Hwang J -Y, Huang X, Garkida A and Hein A, 2006. Waste colored glasses as sintering aid in ceramic tiles production. Journal of Minerals and Materials Characterization and Engineering 5 (2), 119–129.

Jackson P R, Hancock P and Cartlidge D, 2004. Recycled bottle glass as a component of ceramic sanitaryware. In: Proceedings of the International Conference on Sustainable Waste Management and Recycling: Glass Waste. Kingston University, London, UK, pp. 313–320.

Kazmina O, Suslyaev V, Dushkina M, Semukhin B and Iop, 2015. Characterization of porous glass-ceramic material as absorber of electromagnetic radiation. In: Proceedings of the International Scientific Conference on Radiation-Thermal Effects and Processes in Inorganic Materials, Tomsk, Russia. https://doi.org/10.1088/1757-899x/81/1/012036.

Ketov A A, 2001. An experience of reuse of a glass cullet for production of foam structure material. In: Proceedings of the International Conference on Recycling and Reuse of Glass Cullet, UK, pp. 85–91.

Kiletico M J, Hassan M M, Mohammad L N and Alvergue A J, 2015. New approach to recycle glass cullet in asphalt shingles to alleviate thermal loads and reduce heat island effects. Journal of Materials in Civil Engineering 27 (8), 9 pp.

Kim K, Kim K and Hwang J, 2015. LCD waste glass as a substitute for feldspar in the porcelain sanitary ware production. Ceramics International 41 (5), 7097–7102.

Kim K, Kim K and Hwang J, 2016. Characterization of ceramic tiles containing LCD waste glass. Ceramics International 42 (6), 7626–7631.

Konig J, Petersen R R and Yue Y, 2015. Fabrication of highly insulating foam glass made from CRT panel glass. Ceramics International 41 (8), 9793–9800.

Lancellotti I, Barbieri L, Corradi A, Brusatin G and Scarinci G, 2001. Glass cullet: scrap or new raw material? In: Proceedings of the International Conference of Recycling and Reuse of Glass Cullet, UK, pp. 93–102.

Lee S W, Obregón-Alfaro S and Rodríguez-González V, 2011. Photocatalytic coatings of silver–TiO_2 nanocomposites on foamed waste-glass prepared by sonochemical process. Journal of Photochemistry and Photobiology A: Chemistry 221 (1), 71–76.

Lee J, Cho S, Han B, Seo Y, Kim B and Heo S, 2012. Recycling of TV CRT panel glass by incorporating to cement and clay bricks as aggregates. In: 2012 International Conference on Ecology. Waste Recycling and Environment Advances in Biomedical Engineering, Beijing,China, pp. 257–267.

Lin K -L, 2007a. The effect of heating temperature of thin film transistor-liquid crystal display (TFT-LCD) optical waste glass as a partial substitute partial for clay in eco-brick. Journal of Cleaner Production 15 (18), 1755–1759.

Lin K -L, Lee T -C and Hwang C -L, 2015. Effects of sintering temperature on the characteristics of solar panel waste glass in the production of ceramic tiles. Journal of Material Cycles and Waste Management 17 (1), 194–200.

Lin K L, 2007b. Use of thin film transistor liquid crystal display (TFT-LCD) waste glass in the production of ceramic tiles. Journal of Hazard Materials 148 (1–2), 91–97.

Ljatifi E, Kamusheva A, Grozdanov A, Paunovic P and Karamanov A, 2015. Optimal thermal cycle for production of glass-ceramic based on wastes from ferronickel manufacture. Ceramics International 41 (9), 11379–11386.

Llaudis A S, Tari M J O, Ten F J G, Bernardo E and Colombo P, 2009. Foaming of flat glass cullet using Si_3N_4 and MnO_2 powders. Ceramics International 35 (5), 1953–1959.

Loryuenyong V, Panyachai T, Kaewsimork K and Siritai C, 2009. Effects of recycled glass substitution on the physical and mechanical properties of clay bricks. Waste Management 29 (10), 2717–2721.

Lu J, Cong X and Lu Z, 2016a. Influence of magnesia on sinter-crystallization, phase composition and flexural strength of sintered glass-ceramics from waste materials. Materials Chemistry and Physics 174, 143–149.

Lu J, Lu Z, Peng C, Li X and Jiang H, 2014. Influence of particle size on sinterability, crystallisation kinetics and flexural strength of wollastonite glass-ceramics from waste glass and fly ash. Materials Chemistry and Physics 148 (1–2), 449–456.

Lu Z, Lu J, Li X and Shao G, 2016b. Effect of MgO addition on sinterability, crystallization kinetics, and flexural strength of glass-ceramics from waste materials. Ceramics International 42 (2), 3452–3459.

Luz A P and Ribeiro S, 2007. Use of glass waste as a raw material in porcelain stoneware tile mixtures. Ceramics International 33 (5), 761–765.

Makowski C and Rusenko K, 2007. Recycled glass cullet as an alternative beach fill material: results of biological and chemical analyses. Journal of Coastal Research 233, 545–552.

Marangoni M, Ponsot I, Kuusik R and Bernardo E, 2014a. Strong and chemically inert sinter crystallised glass ceramics based on Estonian oil shale ash. Advances in Applied Ceramics 113 (2), 120–128.

Marangoni M, Secco M, Parisatto M, Artioli G, Bernardo E, Colombo P, Altlasi H, Binmajed M and Binhussain M, 2014b. Cellular glass-ceramics from a self foaming mixture of glass and basalt scoria. Journal of Non-crystalline Solids 403, 38–46.

Matteucci F, Dondi M and Guarini G, 2002. Effect of soda-lime glass on sintering and technological properties of porcelain stoneware tiles. Ceramics International 28 (8), 873–880.

Nandi V S, Raupp-Pereira F, Montedo O R K and Oliveira A P N, 2015. The use of ceramic sludge and recycled glass to obtain engobes for manufacturing ceramic tiles. Journal of Cleaner Production 86, 461–470.

Pant D, 2009. Waste glass as absorbent for thin layer chromatography (TLC). Waste Management 29 (7), 2040–2041.

Petersen R R, Konig J, Smedskjaer M M and Yue Y, 2014. Foaming of CRT panel glass powder using Na_2CO_3. Glass Technology: European Journal of Glass Science and Technology, Part A 55 (1), 1–6.

Petersen R R, Konig J and Yue Y, 2015. The mechanism of foaming and thermal conductivity of glasses foamed with MnO_2. Journal of Non-crystalline Solids 425, 74–82.

Petrella A, Petruzzelli V, Basile T, Petrella M, Boghetich G and Petruzzelli D, 2010. Recycled porous glass from municipal/industrial solid wastes sorting operations as a lead ion sorbent from wastewaters. Reactive and Functional Polymers 70 (4), 203–209.

Phonphuak N, Kanyakam S and Chindaprasirt P, 2016. Utilization of waste glass to enhance physical-mechanical properties of fired clay brick. Journal of Cleaner Production 112, 3057–3062.

Ponsot I, Bernardo E, Bontempi E, Depero L, Detsch R, Chinnam R K and Boccaccini A R, 2015a. Recycling of pre-stabilized municipal waste incinerator fly ash and soda-lime glass into sintered glass-ceramics. Journal of Cleaner Production 89, 224–230.

Ponsot I, Dal Mas G, Bernardo E, Dal Maschio R and Sglavo V M, 2014. Double strengthening by ion exchange of sintered nepheline glass-ceramics: a new simplified method. Journal of Ceramic Processing Research 15 (6), 411–417.

Ponsot I, Detsch R, Boccaccini A R and Bernardo E, 2015b. Waste derived glass ceramic composites prepared by low temperature sintering/sinter-crystallisation. Advances in Applied Ceramics 114, S17–S25.

Rahman A, Urabe T, Kishimoto N and Mizuhara S, 2015. Effects of waste glass additions on quality of textile sludge-based bricks. Environmental Technology 36 (19), 2443–2450.

Raimondo M, Zanelli C, Matteucci F, Guarini G, Dondi M and Labrincha J A, 2007. Effect of waste glass (TV/PC cathodic tube and screen) on technological properties and sintering behaviour of porcelain stoneware tiles. Ceramics International 33 (4), 615–623.

Rambaldi E, Carty W M, Tucci A and Esposito L, 2007. Using waste glass as a partial flux substitution and pyroplastic deformation of a porcelain stoneware tile body. Ceramics International 33 (5), 727–733.

Rawlings R D, Wu J P and Boccaccini A R, 2006. Glass-ceramics: their production from wastes – a review. Journal of Materials Science 41 (3), 733–761.

Romero-Perez M, Rincón J M, Oliver C J R G, D'Ovidio C and Esparza D, 2001. Magnetic properties of glasses with high iron oxide content. Materials Research Bulletin 36 (7–8), 1513–1520.

Rozenstrauha I, Lodins E, Krage L, Drille M, Balode M, Putna I, Filipenkov V, Chinnam R K and Boccaccini A R, 2013. Functional properties of glass–ceramic composites containing industrial inorganic waste and evaluation of their biological compatibility. Ceramics International 39 (7), 8007–8014.

Scarinci G, Brusatin G and Bernardo E, 2005. Glass foams. In: Cellular Ceramics: Structure, Manufacturing, Properties and Applications. Wiley-VCH Verlag GmbH & Co., KGaA, Weinheim, Germany.

Shashidhar N and Reed J S, 1990. Recycling fired porcelain. American Ceramic Society Bulletin 69 (5), 834–841.

She L, Guo Y, Ruan J, Qiao Q and Sun S, 2014. Study on the critical control of foam insulation materials produced by waste TFT-LCD glass. In: Proceedings of the 3rd Asian Pacific Conference on Mechanical Components and Control Engineering, Tianjin, China, pp. 74–78.

Siikamäki L A R, 2004. Glaze for low-fired ceramics from end-of-life cathode-ray-tube glass. In: Proceedings of the International Conference on Sustainable Waste Management and Recycling: Glass Waste. Kingston University, London, UK, pp. 281–290.

Siikamäki R and Hupa L, 2001. Utilisation of EOL CRT-glass as a glaze raw material. In: Proceedings of the International Conference of Recycling and Reuse of Glass Cullet, UK, pp. 135–145.

Smith A S, 2004a. Recycled glass as a brick fluxing agent. In: Proceedings of the International Conference on Sustainable Waste Management and Recycling: Glass Waste. Kingston University, London, UK, pp. 149–165.

Smith A S, 2004b. To Demonstrate Commercial Viability of Incorporating Ground Glass in Bricks With Reduced Emissions and Energy Savings. Waste & Resources Action Programme, Oxon, UK.

Sommariva L and Weinberger K, 2015. Energy and natural resources saving in the production of expanded glass granules. In: Proceedings of the 12th International Conference on Chemical & Process Engineering (ICHEAP12). Milan, Italy, pp. 2437–2442.

Shipbuilders and Shiprepairers Association (SSA), 2006. Full Scale Operational Trials Involving the Use of Recycled Glass as a Blasting Medium in the Marine Sector. Waste & Resources Action Programme, Oxon, UK.

Tarvornpanich T, Souza G P and Lee W E, 2005. Microstructural evolution on firing soda-lime-silica glass fluxed whitewares. Journal of the American Ceramic Society 88 (5), 1302–1308.

Taurino R, Lancellotti I, Barbieri L and Leonelli C, 2014. Glass-ceramic foams from borosilicate glass waste. International Journal of Applied Glass Science 5 (2), 136–145.

Tucci A, 2004. Use of soda-lime scrap-glass as a fluxing agent in a porcelain stoneware tile mix. Journal of the European Ceramic Society 24 (1), 83–92.

Tulyaganov D, Fernandes H, Agathopoulos S and Ferreira J, 2006. Preparation and characterization of high compressive strength foams from sheet glass. Journal of Porous Materials 13 (2), 133–139.

Tur N N and Terenshenko O P, 2001. To the development of heat resistant enamel coating based on physico-chemical properties of glasses. In: Proceedings of the International Conference of Recycling and Reuse of Glass Cullet, UK, pp. 109–112.

Van der Velden J H, 1979. Analysis of the Pfefferkorn test. Ziegelindustrie International 9, 532–542.

Wallenberger F T, Watson J C and Li H, 2001. Glass fibers. Composites 21, 27–34.

Wang X, Feng D, Zhang B, Li Z, Li C and Zhu Y, 2016. Effect of KNO_3 on the microstructure and physical properties of glass foam from solid waste glass and SiC powder. Materials Letters 169, 21–23.

Wannagon A, Sornlar W and Choeycharoen P, 2012. Crystalline phases and physical properties of modified stoneware body with glaze sludge. Ceramics International 38 (6), 4485–4494.

Wei Y L, Lin C Y, Ko K W and Wang H P, 2011. Preparation of low water-sorption lightweight aggregates from harbor sediment added with waste glass. Marine Pollution Bulletin 63 (5–12), 135–140.

WRAP, 2006a. Farm demonstrates recycled glass filtration system. Filtration & Separation 43 (4), 10 pp.

WRAP, 2006b. Recycled glass filters – an option for wastewater processing? Filtration & Separation 43 (2), 12 pp.

Wright L, 2003. Development of a Regional Waste Recovery/Processing Sector. Wellington City Council, Ministry for the Environment & Ministry of Economic Development, New Zealand.

Yang J, Zhang S, Liu B, Pan D, Wu C and Volinsky A A, 2015. Effect of TiO_2 on crystallization, microstructure and mechanical properties of glass-ceramics. Journal of Iron and Steel Research International 22 (12), 1113–1117.

Youssef N F, Abadi M F and Shater M A O, 1998. Utilization of soda glass (cullet) in the manufacture of wall and floor tiles. Journal of the European Ceramic Society 18 (12), 1721–1727.

Yürüyen S and Toplan H Ö, 2009. The sintering kinetics of porcelain bodies made from waste glass and fly ash. Ceramics International 35 (6), 2427–2433.

Zhang W and Liu H, 2013. A low cost route for fabrication of wollastonite glass–ceramics directly using soda-lime waste glass by reactive crystallization–sintering. Ceramics International 39 (2), 1943–1949.

Zhang W Y, Gao H and Xu Y, 2011. Sintering and reactive crystal growth of diopside–albite glass–ceramics from waste glass. Journal of the European Ceramic Society 31 (9), 1669–1675.

Zhu H Y, Byars E A and Meyer C, 2003. Use of waste glass for construction products: legislative and technical issues. In: Proceedings of the International Symposium of Recycling and Reuse of Waste Materials. University of Dundee, Scotland, UK, pp. 827–838.

Zilli M, Arcaro S, Cesconeto F R, de Oliveira Maia B G, Pereira F R and Novaes de Oliveira A P, 2015. Production and characterisation of ceramic foams from industrial solid waste. In: Proceedings of the 12th International Conference on Chemical & Process Engineering ICHEAP12, Milan, Italy, pp. 1783–1788.

Environmental Impact, Case Studies and Standards and Specifications

10

Main Headings

- Environmental impact
- Case studies
- Standards and specifications
- Cement clinker and cementitious material
- Concrete
- Geotechnical applications
- Road pavements

Synopsis

This chapter deals with the environmental impact, case studies and standards and specifications relating to the use of glass cullet (GC) as raw feed in Portland cement clinker production, in a ground form, as a pozzolanic material in combination with Portland cement in concrete application, and in a granular form, as an aggregate component in concrete, geotechnical and road pavement applications. The heavy metals released from GC (soda lime type) used in these applications have been found to be below the regulatory limits specified in the US Code of Federal Regulations. The material is alkaline, is noncorrosive and contains very low organic content if properly processed. The case studies in this respect, in general, have also produced positive results. The status of this material in these applications with respect to the relevant standards and specifications, mainly in the United Kingdom and the United States, are also discussed. On the other hand, cathode ray tube glass has been found to be a hazardous material because of its high lead content.

Keywords: Glass cullet, Environmental impact, Case studies, Standards and specifications, Cement clinker, Cementitious material, Concrete, Geotechnical applications, Road pavements.

Sustainable Construction Materials: Glass Cullet. https://doi.org/10.1016/B978-0-08-100984-0.00010-2

10.1 Introduction

The life cycle of construction materials, from the initial material extraction, processing, delivery, use in construction and maintenance to the final disposal, consumes large amounts of non-renewable resources and contributes to CO_2 emissions and the demand for landfills. Without a change in practices, the construction industry will continue to have a major impact on the environment. As the environmental regulations of the construction industry become more stringent, initiatives have been taken to improve its sustainability. Undoubtedly, one of the most commonly considered initiatives to achieve this includes the use of recycled and secondary materials as an alternative to natural materials. In some cases, their inclusion as construction materials can also enhance the performance of the resulting products. A classic example of this is fly ash, generated at coal-fired electric power stations, as a replacement for Portland cement in concrete. This not only reduces the energy consumption and CO_2 emissions of Portland cement production, but also improves the fresh and hardened properties of the concrete. Fly ash has now been accepted as one of the common cements in most standards, such as EN 197-1 (2011).

Other secondary materials, such as glass cullet (GC), which has a long history of use as an aggregate in construction (Stroup-Gardiner and Wattenberg-Komas, 2013) and technical information available in the literature supporting its use in a ground form as a cement component or in a granular form as an aggregate component in concrete, geotechnical and road pavement applications (Chapters 4–8), are yet to be adopted in the standards. However, the wise use of GC as construction material also requires its safe use. Thus, it is important to examine its environmental credentials, as well as those of the resultant products. In addition, the evidence presented as case studies, as well as testing and evaluating the material for its compliance with the national and international standards and specifications, would undoubtedly help in establishing a strong case for the adoption of GC use in the construction industry.

This chapter provides a comprehensive analysis of the environmental impact of, case studies on and standards and specifications on the use of GC in a ground form, as a cement component in concrete applications, and in a granular form, as an aggregate component in concrete, geotechnical and road pavement applications.

10.2 Environmental Impact

The environmental impact associated with the use of GC can be assessed in many ways, depending on the extent and nature of the environmental impact being studied. Most importantly, and perhaps as the first concern, the material itself should not pose a threat to the environment and human health by releasing harmful elements. In this section, the focus of the environmental impact addressed is in terms of the toxic element leaching of GC itself and when used in various construction applications: concrete, geotechnical, road pavement applications.

10.2.1 The Material: Glass Cullet

Glass is a versatile material with many industrial and domestic applications, and its intended properties can be formulated by modifying its chemical composition. Owing to this versatility, different types of glass are manufactured for their use in many industrial and domestic applications. As shown previously in Section 3.5.1, GC derived from soda lime glass, which represents the largest sector of glass industry for the production of containers and flat glass, does not normally contain high concentration of the toxic metals adopted by the US Code of Federal Regulations (2017) in classifying the hazardous waste.

On the other hand, electronic glass, such as cathode ray tubes (CRTs), which contain lead, and liquid crystal displays (LCDs), which contain lead and mercury, used in the production of televisions, laptops, computers, and other similar gadgets, is classified as hazardous waste by the Environmental Agency (2017) in the United Kingdom. However, the US EPA (2016) appears to classify only CRTs as hazardous and no information can be found relating to other glasses. On balance, and to avoid any risk, the use of CRT and LCD GC is likely to be prohibited within the construction materials.

Figure 10.1 shows heavy metal leached concentrations for the container, LCD and CRT glasses. The data have been obtained using different test methods, namely (i) toxicity characteristics leaching procedure (TCLP) covered in US EPA, 1992, (ii) synthetic precipitation leaching procedure, covered in US EPA, 1994, and (iii) one- or two-stage batch test covered by BS EN 12457 Parts 1–4, 2002. Of the three, TCLP, which is designed to simulate leaching under landfill conditions, is more commonly used. For comparison, the regulatory maximum levels for a series of heavy metal leached concentrations using TCLP, as given in the US Code of Federal Regulations (2017) document, are also clearly shown in Figure 10.1.

As can be seen from Figure 10.1, the heavy metals released from GC of the container type are below the specified limits. Other elements that are not listed in the figure, namely, aluminium, calcium, magnesium, potassium, silica and sodium, can be released to various degrees from container-type GC (Remarque et al., 2001; Mirzahosseini and Riding, 2014). However, these elements are not used in the US Code of Federal Regulations (2017) for the classification of the hazardous waste.

The heavy metal leached concentrations of GC derived from LCDs are also shown in Figure 10.1 and are within the regulatory levels. In several cases, leached lead and mercury concentrations are not even detectable (Lin et al., 2008; Wang et al., 2014), though all these results have been produced from one country and further confirmation may therefore be needed. For GC derived from CRTs, the results suggest that the leachability of heavy elements can be affected by the pH of the leachant, as the released concentrations of barium and lead are the highest in buffer solution at pH 13 compared with pH between 5 and 7.5. However, as the data amount to two test results reported from a single study, further studies are needed to establish the role of pH in the leachability of heavy metals from CRT glass.

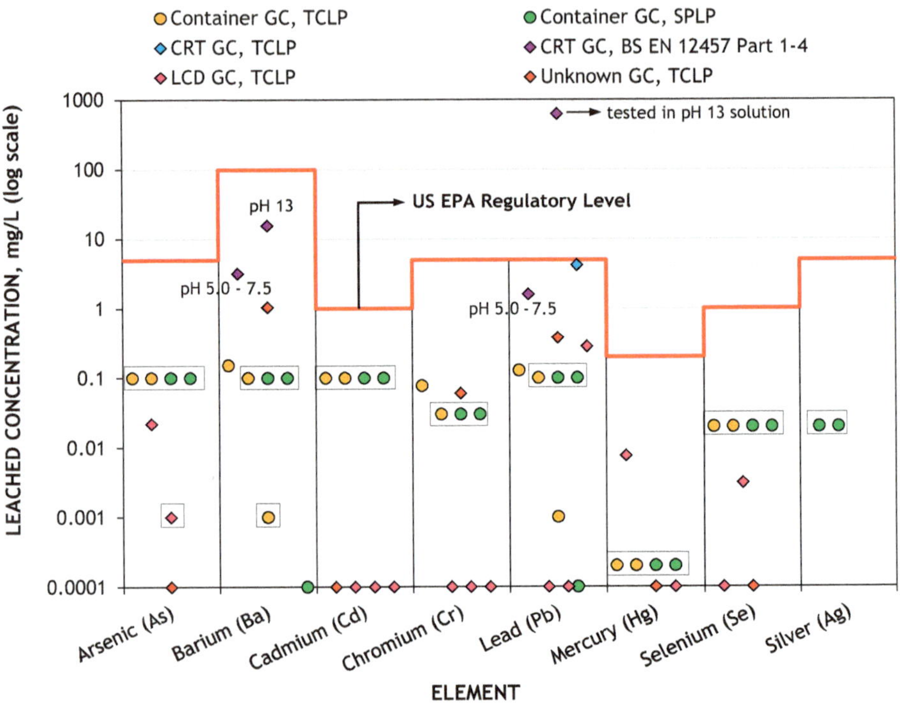

Figure 10.1 Leached element concentrations of various glass cullets using different test methods. Note: for the boxed data, the leached concentration is less than the value shown. Data at 0.0001 mg/L: undetectable level. *CRT*, cathode ray tube; *GC*, glass cullet; *LCD*, liquid crystal display; *SPLP*, synthetic precipitation leaching procedure; *TCLP*, toxic characteristic leaching procedure.
Data taken from Lin et al. (2008, 2009, 2016), Lin et al. (2017), Morrison (2004), PennDOT (2001), Romero et al. (2013), Wang and Hou (2011), Wang et al. (2014) and Wartman et al. (2004).

Overall, it can be assumed that, unless contaminated, GC derived from soda lime glass will not endanger the environment when used as construction material. On the other hand, glasses laden with significant amount of toxic elements, i.e., CRT and LCD, are not likely to be specified as a component of cement and aggregate in the manufacture of construction materials, except when subjected to a strict adherence to special terms and conditions. However, as the results presented in Figure 10.1 have been derived from a very limited number of studies, further work is needed to determine both the understanding and suitability of the tests used and the confidence one can place in such results.

10.2.2 Glass Cullet Use in Concrete Applications

The environmental impact of GC in concrete applications covers its use in cement clinker production, in finely ground form as a cement component and as a fine aggregate component. Most probably because of its high lead content, many of the leaching studies have been conducted on concrete containing GC derived from

CRTs. It would appear that, with the chemical composition of soda lime glass, it can be assumed to be benign in the concrete environment when used as a part of cement or aggregate, and therefore little has been published in the area of environmental impact.

As Raw Feed in Cement Clinker Production

The ability to encapsulate lead released from CRT funnel glass when used as part of the raw feed, at a low content, for cement clinker production has been reported in a study undertaken in Thailand (Lairaksa et al., 2013). Raw cement clinker mixtures containing 0%–0.5% CRTs were heated to 1480°C in an electric furnace for 90 min, and the resultant clinker was ground to a particle fineness less than 75 μm. The lead encapsulation was analysed using X-ray fluorescence spectroscopy, as well as microwave-assisted digestion atomic absorption spectroscopy (AAS), because of its higher sensitivity. The results obtained from this study are shown in Figure 10.2. Although there is a considerable difference in the two sets of values, none of the clinker mixtures achieved full encapsulation of lead. The release of lead from the clinker containing 0.1% CRT funnel glass was the lowest. The investigation was a preliminary study and further work is needed to ensure full encapsulation of lead in cement clinker and no lead contamination in the cement kiln emissions.

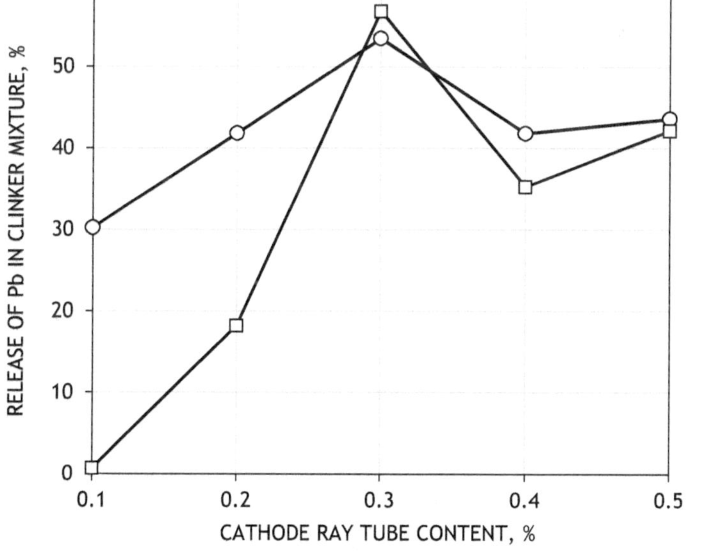

Figure 10.2 Release of Pb in cement clinker containing cathode ray tube funnel glass determined using different methods.
Based on Lairaksa et al. (2013).

As Cement Component

Using finely ground glass cullet (GGC) from CRTs as a cement component is shown to have released lead concentrations below the regulatory limit (Moncea et al., 2013). This CRT glass consisted of over 95% funnel GC and a small amount of panel GC and was ground to a Blaine fineness of around 300 m^2/kg. This ground CRT was used at about 40% in two separate mortar specimens containing Portland cement and a mix of Portland cement and 20% ground granulated blast furnace slag (GGBS). The lead release potential was studied using a monolith tank leaching test in accordance with the Dutch NEN 7345 (1995). After 64 days of monitoring, the cumulative values of lead released from the test mortars, given in Table 10.1, were determined by the AAS method. Although the lead leaching from the CRT mortar containing GGBS was almost double that from CRT mortar containing Portland cement, the results obtained with both the cements were still well below the limit of 100 mg/m^2 suggested by the Dutch Building Material Decree (DBMD, 1995).

Comparing the cases in which CRT funnel GC is used as raw feed for clinker production and separately in the ground form as a cement component, because the material ends up finally in a cementitious system, the latter application may offer more benefits. These would be due to the possibility of incorporating a relatively larger content of ground CRT into a cement mixture, as well as being free from the risk of lead contamination that can occur in the emissions from clinker production.

As Fine Aggregate Component

The leaching behaviour of concrete and mortar of 0.4–0.6 water/cement ratio, made with GC as a fine aggregate up to 100% replacement and cured for 7 to 28 days, has been investigated. The bulk of these studies were undertaken in Hong Kong, with a few in the United Kingdom, the United States and Thailand. As CRT funnel GC was the subject of the assessment in all these studies, the testing focused on the release of lead and barium, mostly in accordance with the TCLP (US EPA, 1992).

The leaching of barium from the test mixes was no more than 1 mg/L, which is well below the US Code of Federal Regulations (2017) hazardous limit of 100 mg/L. Figure 10.3 shows the lead leaching results, which for ease of comparison are separated based on (i) whether the GC samples were treated with acid, (ii) the GC type, whether it is CRT or container glass, and (iii) whether the test mixes contained fly ash or GGBS.

It is evident from Figure 10.3 that the released concentration of lead increases as CRT funnel GC content increases, regardless of whether the CRT glass has been treated. However, for untreated CRT funnel GC, the amount of leaching tends to exceed the US Code of Federal Regulations (2017) regulatory limit of 5 mg/L when the CRT content exceeds 20%. The inclusion of fly ash or GGBS as a replacement for Portland cement

Table 10.1 Cumulative lead released from mortar containing ground cathode ray tubes after 64 days

	Specimen 1	Specimen 2
Cement component[a], %	PC, 38.5; CRT, 28.2	PC, 24.7; GGBS, 13.9; CRT, 28.2
Water/cement ratio	0.5	0.5
Cumulative lead released, mg/m²	0.24	0.51

[a]*CRT*, cathode ray tube (funnel); *GGBS*, ground granulated blast furnace slag; *PC*, Portland cement.
Based on Moncea et al. (2013).

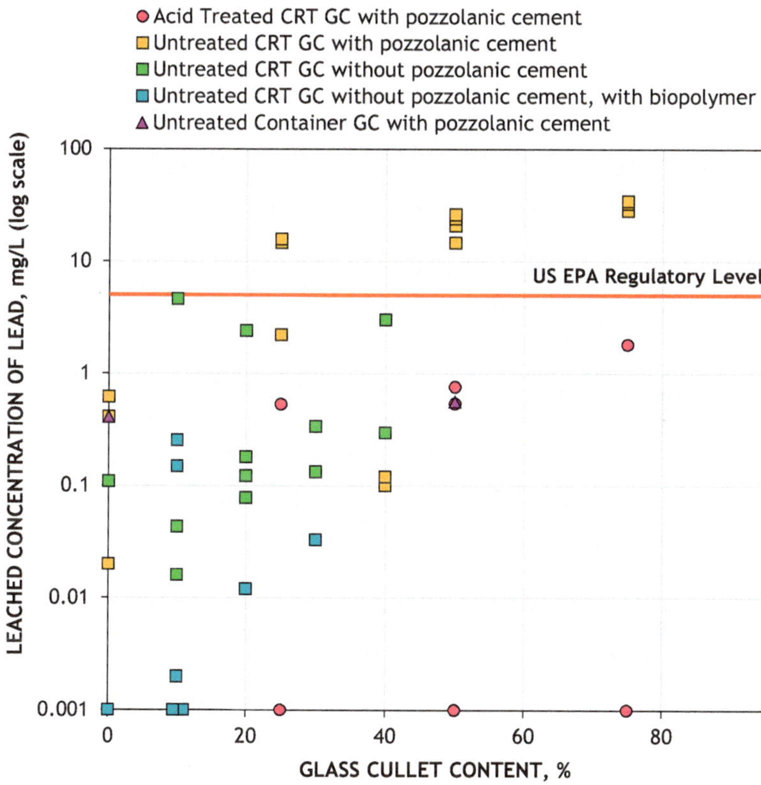

Figure 10.3 Lead leaching from concrete or mortar containing CRTs and container GC as a fine aggregate. *CRT*, cathode ray tube; *GC*, glass cullet.
Data taken from Kim et al. (2009), Ling and Poon (2011, 2012a,b, 2013), Maschio et al. (2013), Morrison (2004), Romero (2013), Romero et al. (2013), Sua-iam and Makul (2013) and Zhao et al. (2013).

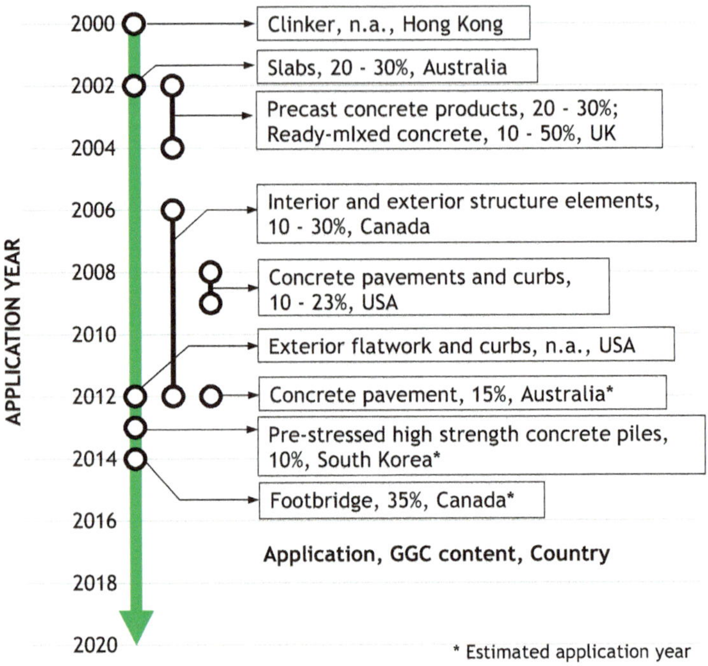

Figure 10.4 Use of ground glass cullet (GGC) in clinker production and as cement component for different concrete applications.
Data taken from Byars et al. (2003, 2004a,b), Chen et al. (2002), Dumitru et al. (2013), Jang et al. (2014), Nassar and Soroushian (2011), Omran and Tagnit-Hamou (2016), Peyvandi et al. (2013), Shayan and Xu (2006), Song et al. (2016), Soroushian (2012), Tagnit-Hamou and Bengougam (2012) and Tagnit-Hamou et al. (2015).

does not seem to help with reducing the leachability of lead from mixes containing CRT funnel GC. Notwithstanding this, some preliminary studies have suggested that the release of lead from CRT funnel GC can be controlled to below the limit (Figure 10.4), as described below:

- Treating crushed CRTs with 5% nitric acid for 3 h to extract lead from its surface (Ling and Poon, 2011, 2012b).
- Adding biopolymer solutions to the concrete and mortar, such as a mixture of xanthan and guar gum (Kim et al., 2009) and a mixture of boric acid and guar gum (Kim et al., 2009; Romero, 2013; Romero et al., 2013), which have the benefit of encapsulating the lead element.

10.2.3 Glass Cullet Use in Geotechnical Applications

When used as fill material and topsoil in geotechnical applications, GC will be in contact with soil, rainwater and groundwater. This might create conditions conducive to leaching of toxic metals from GC, which are hazardous to life, and migration of organic compounds (for example, sugar residue from contaminated container glass), which can cause bacterial growth in soil and water. Thus, it is important to investigate the environmental suitability of GC for use as a fine aggregate in geotechnical applications.

Table 10.2 Chemical and environmental testing on container glass cullet (GC)

Measurement	Main Observation	Notes
pH value	7.1–10.4	n.a.
Total contaminant contents and leachate concentrations	Mostly below the regulatory limits	Attention may be needed regarding chromium(VI), iron and lead
Organic material	Generally not a concern for uncontaminated container GC	Affected by processing quality

Data based on Cosentino et al. (1995); Dames and Moore (1993); DECC (2007); Disfani et al. (2011a,b, 2012); Imteaz et al. (2012); Soil and Environmental Engineers and Re-sourcing Associates (1998).

A series of chemical and environmental tests on container GC was conducted in the United States during the 1990s and in Australia in the late 2000s and early 2010s. In total, about 40–60 container GCs were examined, with samples sourced from kerbside collection being more than those from recycling plants. The information provided in the literature on the material as to whether it was in a ready-to-use form is not clear, as the test samples were reported to contain various amounts of non-glass materials such as paper, plastic and metal pieces.

The results (as summarised in Table 10.2) suggest that container GC is safe for the environment, but care must be taken for possible contamination due to the presence of non-glass materials. The main points to emerge from these tests are described in the following Sections (a)–(c). It should be mentioned that relevant tests on container GC have not been conducted in recent years, which is not too surprising because of the benign nature of the material.

(a) pH Value

The pH value of container GC is in the range of 7.1–10.4, indicating that the material can be of slight to modest alkaline nature. According to the US Code of Federal Regulations (2017), a solid waste is classified as hazardous if it is corrosive, with a pH value either ≤2 or ≥12.5. Thus, container GC is considered to be noncorrosive.

(b) Contaminant Constituents and Leachate Concentration

The analysis of contaminant constituents and leachates of container GC has been undertaken to assess its heavy metal and aromatic hydrocarbon contents and the leachability of harmful substances. As different tests and contaminant limits were used in Australia and the United States, for convenience the results for the two countries are discussed separately.

In Australia, the EPA Victoria (EPA Victoria, 2007) specifies thresholds of (i) contaminant constituents for waste used as a fill material and (ii) contaminant constituents and leachate concentrations in classifying hazardous waste (Disfani et al., 2012). It should also be noted that, for the leaching test, the material was tested in acidic (pH 5) and alkaline (pH 9.2) solutions in accordance with the Australia Standard

Leaching Protocol (ASLP) (AS 4439.3, 1997). Container GC has been examined against these requirements, and some of the results from Disfani et al. (2012) are given in Table 10.3. It can be seen that, except for chromium, the contaminant constituents of container GC and its leachate concentrations (in both acidic and alkaline solutions) are well below the thresholds set for fill materials and inert waste. However, in the case of chromium, as the measurements were in terms of total chromium, i.e., sum of chromium(III) and chromium(VI), and the specified requirement is in terms of chromium(VI), the exact amount of chromium(VI) present in this container GC could not be ascertained. Given that chromium(III) oxide is added to glass to give a green colour, it is very likely that the detected chromium was chromium(III).

Another series of tests conducted in Australia (Imteaz et al., 2012) show that, except for iron (leaching under acid solution), all the heavy metals, namely, copper, lead and zinc, released from container GC are below the regulatory levels in accordance with EPA Victoria (2007). However, the study did not confirm whether or not this was due to the presence of contaminants in the GC sample, for example, metal bottle caps.

Table 10.3 Contaminant constituents and leachate concentrations of container glass cullet

	Constituent, mg/kg of Dry Weight			Leached Concentration, mg/L		
Contaminant	**Result**	**Limit[a]**	**Limit[b]**	**In pH 5**	**In pH 9.2**	**Limit[b]**
Arsenic	<5	20	500	<0.01	<0.1	0.35
Cadmium	0.5	3	100	0.004	<0.02	0.1
Chromium	<5	1[c]	500	<0.01	<0.1	2.5
Copper	6	100	5000	0.12	<0.1	100
Lead	12	300	1500	0.19	<0.1	0.5
Mercury	<0.05	1	75	<0.001	<0.01	0.05
Nickel	<5	60	3000	<0.01	<0.1	1
Selenium	<5	10	50	<0.01	<0.1	0.5
Silver	<5	10	180	<0.01	<0.1	5
Zinc	34	200	35000	0.79	0.1	150
Cyanide	<5	50	2500	4	<0.05	<0.05
MAHs	<0.1	7	50	<0.001	<0.001	n.a.
Benzene	<0.1	1	4	<0.001	<0.001	0.05
PAHs	<0.1	20	50	<0.01	<0.01	n.a.
BaP	<0.1	1	5	<0.001	<0.001	0.0005

BaP, benzo[a]pyrene; *MAHs*, monocyclic aromatic hydrocarbons; *n.a*, not available; *PAHs*, polycyclic aromatic hydrocarbons.
[a]Limit for use as fill material.
[b]Limit for solid inert waste.
[c]For chromium(VI).
Based on Disfani et al. (2012).

The American research (Dames and Moore, 1993), on the other hand, made a number of observation, showing that:

(i) The metal concentrations detected in container GC, namely antimony, arsenic, beryllium, cadmium, chromium, cobalt, copper, lead, mercury, nickel, selenium, silver, thallium and zinc, were similar to or below those typically found in granite rock.

(ii) The results of leaching tests, conducted in accordance with the ASTM D4793 (1993) method, showed that the metals released from container GC are comparable to the naturally occurring metals in groundwater and surface water.

(iii) The lead content and leachable lead amount from container GC, except in one case in which the values for both were excessively high, were either undetectable or at low level, which is again similar to granite, with the leaching of lead being below 0.5 mg/L, which is substantially less than the limit of 5 mg/L specified by the US Code of Federal Regulations (2017).

(iv) The semivolatile organic compounds, such as phthalates and polycyclic aromatic hydrocarbons and others, which are attributed to non-glass materials, do not appear to present a concern.

Overall, although the hazardous waste classifications and leaching methods adopted in various countries can be different, the chemical constituents and leachate concentrations of container GC itself are such that they will comply with the regulatory limits. As other non-glass materials can be present in container GC due to improper sorting, it is recommended to have the material tested for its chemical composition leaching characteristics.

(c) Organic Material

Organic material is one of the pollutants in waste water. Analyses such as biological oxygen demand (BOD), chemical oxygen demand and total organic carbon have been conducted on the leachate from container GC to evaluate the organic material present. Different observations have been reported. For example, for the leachate of container GC tested by Dames and Moore (1993) and Imteaz et al. (2012), the organic material was either undetectable or at a safe level. On the other hand, the BOD of leachate studied by Cosentino et al. (1995) was greater than that of untreated domestic waste water. However, in a different study (DECC, 2007), the high BOD value was to be related to the presence of paper and food residues in the GC test samples.

This reinforces the point that to produce container GC of usable quality, the material needs to be properly processed, which includes sorting and removal of contaminants from a GC stock, as well as washing it to remove any food residues present.

10.2.4 Glass Cullet Use in Road Pavement Applications

The only study on the environmental impact of GC as a fine aggregate in road pavement applications has been undertaken in conjunction with coarse recycled concrete aggregate (RCA) and for its use as a footpath base layer (Arulrajah et al., 2016). The chemical composition and the leaching potentials were examined for a range of heavy metals in three sets of materials, namely GC alone, RCA alone and a mixture of 85% RCA and 15% GC. For the leaching test, the materials were subjected to two buffer solutions, acidic and alkaline, and the tests were performed in accordance with the ASLP (AS 4439.3, 1997). The heavy metals considered in these tests were arsenic, barium, cadmium, chromium, copper, mercury, molybdenum, nickel, lead, selenium, silver and zinc. The tests showed that heavy metals in all three waste materials were below the specified levels for 'Category C,' the least hazardous, in accordance with EPA Victoria (2007). For the leaching characteristics of these materials, the tests suggested that except for molybdenum, which was not tested, the amounts of the heavy metals released from GC, RCA and the mixture of the two were below the regulated limits specified by EPA Victoria (2007), and therefore all three materials can be categorised as nonhazardous waste.

In summary, despite the limited number studies reported for the environmental impact of using GC in the field of road pavements, based on the evidence of the data presented in Section 10.2.3, it can be concluded that container GC is a safe material for use in geotechnical applications and therefore it can be argued that a similar environmental impact would be the outcome of using GC in unbound applications in road pavements. Although it is not too difficult to deduce that GC will perform in a similar manner when used in hydraulically and bituminous bound applications, for precautionary reasons it would be advisable to confirm this by additional tests undertaken for these two applications.

10.3 Case Studies

Case studies are an important part of developing confidence in the acceptance of new materials for their use in practice, as they provide additional valuable information to support laboratory studies regarding the responses and adaptations relating to the use of the material in real situations. In this section, the case studies cover pilot-scale projects and field testing that were carried out worldwide where GC was used (i) in a ground form, as raw feed in clinker material and cement component, and (ii) as a fine aggregate in concrete, geotechnical and road pavement applications.

10.3.1 Glass Cullet Use as Clinker Raw Feed and Cement Component

Owing to its chemical composition and pozzolanic properties, finely ground GC is a viable material for use as raw feed in cement clinker production and as a replacement for Portland cement. Several case studies have been reported in the literature; however, they have mainly tended to be in the area of GC use as a cement component rather than as raw feed in the manufacture of Portland cement clinker, as can be seen from Figure 10.4. The first relevant work took place in 2000 in Hong Kong with two plant trials using about 120°tonnes of container GC as raw feed in clinker manufacturing (Chen et al., 2002). Since then, though several studies were undertaken, they tended to be concentrated in the period leading to 2015, with most of the work undertaken in Australia, Canada, Hong Kong, South Korea, the United Kingdom and the United States. Canada has the highest number of reported case studies. The GGC content in these studies was generally kept between 10% and 30%, except in one case in which GGC was used at 50%. The type of glass used was predominantly container glass, except for one study in South Korea in which LCD glass was used (Jang et al., 2014).

In general, the outcomes of the case studies have been encouraging, as the GGC mixes are reported to meet performance requirements. The main points to emerge from these case studies (Figure 10.4) are described next:

- As raw feed in clinker production (Chen et al., 2002), the inclusion of GC did not cause significant change in the pollutant gas levels. It is reported that the chemical properties and strength of the cement produced were similar to those grounds from normal clinker. The total alkali content of the cement, although it showed a slight increase, was comparable to the previous records of the plant.
- When GGC was used as a cement component, the early compressive strength and other mechanical properties up to 28 days of concrete mixes with equal water/cement ratios were lower than those of reference concrete mixes without GGC. However, at the later ages, the mechanical properties of GGC concrete were similar to or higher than those of reference concrete (Byars et al., 2004a; Nassar and Soroushian, 2011; Omran and Tagnit-Hamou, 2016). Improvement in water absorption and resistance to chloride ingress have been observed in concrete made with GGC (Shayan and Xu, 2006; Soroushian, 2012).
- The use of GGC in special concrete applications has been explored. In South Korea, 10% GGC was used in manufacturing steam-cured prestressed high-strength concrete piles (Jang et al., 2014). In Canada, a high-performance fibre-reinforced concrete containing 35% GGC was designed and used for the construction of a footbridge on a university campus (Tagnit-Hamou et al., 2015).
- In one case, the long-term performance of an in situ concrete made with GGC cement was monitored (Song et al., 2016). The concrete was made with 15% GGC and used to construct a section of a 100-m-long pavement in the south of Sydney, Australia, in 2012. A reference concrete made without GGC was cast for comparison. After five years in service, both pavements were still in a reasonable condition and showed similar friction resistance. Cored samples were taken to evaluate the in situ compressive strength, water absorption and chloride resistance. Although the results showed that the performance of GGC concrete was slightly lower than that of the reference concrete, no disruption in the GGC concrete pavement was observed.

10.3.2 Glass Cullet Use as Fine Aggregate in Concrete

The reported case studies using GC as a fine aggregate in concrete have been limited. This is perhaps due to the concern over the deleterious alkali–silica reaction (ASR) that may be induced in concrete with the use of GC. However, as discussed previously in Section 5.7.5, the potential risk of ASR in GC as a fine aggregate can be minimised with the use of pozzolanic materials such as fly ash and GGBS, as well as finely ground GC itself. Given this, it would be useful to consider developing case studies to investigate the ASR effect of fine aggregate in concrete using Portland–pozzolanic cement combinations, such as CEM II, CEM III and CEM IV cements in compliance with EN 197-1 (2011).

The descriptions and the main findings of the case studies carried out with GC fine aggregate in concrete, mainly in Australia and the United Kingdom, during the 2000s, are given in Table 10.4. The information available suggests that GC has been used at up to 100% as a replacement for natural fine aggregate in ready-mixed concrete, trial slabs, concrete pavements, precast products and mortar. Soda lime glass in the form of container glass was used in these applications and the material was normally prepared to an aggregate size less than 3 mm. In general, satisfactory performance in terms of consistence (workability), strength, water absorption and durability of concrete was achieved. Indeed, owing to its high modulus of elasticity and hardness,

Table 10.4 Case studies involving the use of glass cullet (GC) sand in concrete and mortar applications

	Application Details					
Reference	**Location**	**Year**	**Used in**	**GC Size**	**GC Content**	**Main Findings**
Bell (2006)	UK	2005	Concrete blocks	<6 mm	10%–20%	GC fine aggregate concrete blocks showed less shrinkage
Boniface (2006)	UK	2005–06	Mortar	150–300 μm	75%–100%	GC fine aggregate in mortar products generally resulted in satisfactory performance, although the consistence was compromised in a few cases
Byars et al. (2003, 2004a,b,c)	UK	2002–04	Precast products	<5 mm	15%–100%	Most of the products met the requirements and the initial ASR assessment showed no signs of deleterious expansion

Table 10.4 Continued

	Application Details					
Reference	**Location**	**Year**	**Used in**	**GC Size**	**GC Content**	**Main Findings**
Byars et al. (2004a,b)	UK	2002–04	Ready-mix concrete	1–3 mm	50%	The compressive strength of GC fine aggregate concrete was similar to that of natural sand concrete
Dumitru et al. (2013) and Song et al. (2016)	Australia	2012	Concrete pavement	0.3–3 mm	45%	The condition of GC fine aggregate concrete was good after 5 years of service; no signs of ASR
Polley et al. (1998)	USA	1994	Concrete pavement	0.61 mm (median)	10%–20%	Good durability in terms of abrasion and freeze–thaw resistance
Sagoe-Crentsil et al. (2001)	Australia	n.a.	Concrete slab	<2.46 mm	20%	The casting procedure was the same as normal; the fresh and hardened concrete properties remained almost unchanged
Shayan and Xu (2006)	Australia	2002	Concrete slab	0.15–2.36 mm	40%–75%	The hardened properties of concrete containing GC fine aggregate and were satisfactory

ASR, alkali–silica reaction.

the inclusion of GC fine aggregate in concrete has also been shown to improve the shrinkage resistance (Bell, 2006; Dumitru et al., 2013; Shayan and Xu, 2006) and abrasion resistance (Polley et al., 1998). However, when used in mortar applications, concern was expressed as the flow of some GC fine aggregate mixes was compromised (Boniface, 2006).

In addition, it has been reported that fresh GC fine aggregate concrete can be handled in the same way as natural aggregate concrete for placing, compaction, finishing and curing of concrete (Sagoe-Crentsil et al., 2001). This suggests that no change in concrete casting procedures is required when GC fine aggregate is used. This again works in favour of acceptance, within the current practice, of the use of GC fine aggregate in concrete. This would be good for promoting the use of recycled materials and sustainability in the broader sense.

Regarding the ASR concern, two separate studies have observed that concrete containing GC fine aggregate did not show ASR expansion in most cases. This was first reported from a pilot-scale project undertaken in the United Kingdom by Byars et al. (2004a). Concrete mixes made with GC fine aggregate (up to 100% content) and with or without GGC cement (up to 25%) were cast in large numbers of precast concrete products. The ASR expansion of these mixes was determined in accordance with BS 812-123 (1999). After 1 year of monitoring, no ASR-induced expansion was observed for all these mixes, except where high-alkali Portland cement (>1% Na_2O_{eq}) was used. The second case was reported by Song et al. (2016) for a field trial undertaken in Australia. In this trial, about one-third of a 100-m-long pavement was cast with a 0.45 water/cement ratio concrete, which contained 45% GC fine aggregate, but no pozzolanic material. After 5 years of service, the results obtained from petrographic examination of cored samples suggested that there was no sign of ASR in the concrete. Overall, although these ASR tests appear to be encouraging, to be certain, additional studies with a longer follow-up period are highly recommended to draw any firm conclusions about the risk of ASR with GC fine aggregate use in concrete under field conditions.

10.3.3 Glass Cullet Use as Aggregate in Geotechnical Applications

The case studies on the use of GC in geotechnical applications are summarised in Table 10.5, providing information on the size and content of GC used; the country involved, which is largely the United States, with a few from Australia and Japan; and the type of application. The United States seems to have a long history of using GC in geotechnical applications, possibly starting back in 1990. The material has been used mostly at 100%, either as sand or as a mixture of sand and gravel, with maximum aggregate size up to 19 mm. No adverse effects on the geotechnical performance have been reported when GC of gravel size was used, although the material was flaky or elongated. GC has been used in a wide range of applications (Table 10.5), which can be broadly categorised as follows:

- Construction backfill as general fill material
- Utility construction as pipe bedding and trench backfill
- Drainage as drainage aggregate
- Landfill construction as landfill cover

Table 10.5 Case studies using glass cullet (GC) in geotechnical applications

	GC			
Reference	**Size, mm**	**Content, %**	**Country**	**Used in**
DECC (2007)	Sand size[a]	25, 50, 100	Australia	Backfill
Grubb et al. (2006a, 2008a,b) and Lee (2007)	<9.5	20, 50, 80	USA	Embankment
HDR Engineering (1997)	n.a.	100	USA	Bedding, backfill
Makowski et al. (2007, 2011)	0.33–0.89	25, 50, 75	USA	Beach fill
Makowski et al. (2013)	0.40	100	USA	Artificial dune
Moqsud and Hayashi (2007)[b]	n.a.	15	Japan	Marine fill material
Northwest EcoBuilding Guild (2014a)	Sand and gravel size[a]	100	USA	Structural fill
Northwest EcoBuilding Guild (2014b)	<9.5	100	USA	Fill material
Onitsuka et al. (2001)[b]	n.a.	100	Japan	Lightweight backfill
PennDOT (2005)	<19	100	USA	Bedding, backfill
PennDOT (2006)	<9.5	100	USA	Fill material
Reindl (2003)	Sand and gravel size[a]	Up to 100	USA	Drainage aggregate, fill material, landfill cover, trench fill
Siddiki et al. (2004)	<12	100	USA	Backfill
Stevens (2007)	<9.5	100	USA	Backfill
TxDOT (1999)	n.a.	100	USA	Bedding

[a]GC was used either as a replacement for sand or in combination with sand, but its size was not known.
[b]Waste formed glass.

The general responses were positive for geotechnical applications with GC aggregate, although the details of the field assessments were normally not provided, except in one pilot project undertaken in Philadelphia, Pennsylvania, USA (Grubb et al., 2006a,b). Following on the successful laboratory evaluation (Grubb et al., 2006c), three trial embankments were constructed using GC–dredged material blends with GC content at 20%, 50% and 80%. The dimensions of each trial embankment were approximately 3.6 m high°×°3.6 m wide°×°15.2 m long with 3:1 ramps and 2:1 side slopes. The embankments were constructed in accordance with the procedures given in PennDOT (2004), but with some modifications. The in situ assessment, based on the cone penetration test, showed that the addition of GC improved the geotechnical performance of dredged material. Indeed, the blends of GC with dredged materials have been reported to offer more advantages than the conventional cement stabilised dredged materials in terms of technical performance as well as the construction strategy and cost.

10.3.4 Glass Cullet Use as Road Pavements

The case studies dealing with the use of GC aggregate in road pavements, in the form of unbound, hydraulically bound and bituminous bound applications, are summarised in Table 10.6. The record shows that the material was probably first tested for its potential use in the bituminous bound mixtures in a large number of field studies in the United States and Canada in the 1970s and 1980s and that this was followed by its

Table 10.6 Case studies using glass cullet aggregate in road pavement applications

		Glass Cullet		
Reference	**Country**	**Size, mm**	**Content, %**	**Main Findings**
(a) Unbound Applications				
Arulrajah et al. (2013)	Australia	<4.75	15, 30	Met the specified requirements
City of Armadale (2010)	Australia	n.a.	n.a.	n.a.
HDR Engineering (1997)	USA	n.a.	Mostly 10	Satisfactory performance
Krivit (1999)	USA	<12.7	Up to 100	Satisfactory performance
Reindl (2003)	USA	n.a.	n.a.	n.a.
Senadheera et al. (2005)	USA	n.a.	0–15	Stiffness improved
TxDOT (1999)	USA	<15	10, 20	Satisfactory performance
Wisconsin Transportation Information Center (1999)	USA	n.a.	n.a.	n.a.
(b) Hydraulically Bound Applications				
Polley et al. (1998)	USA	<4	10, 20	Good durability
Dumitru et al. (2013) and Song et al. (2016)	Australia	0.3–3	45	Satisfactory performance after 5 years of service
Wood et al. (2011)	Australia	n.a.	n.a.	n.a.

Table 10.6 Continued

Reference	Country	Glass Cullet Size, mm	Glass Cullet Content, %	Main Findings
(c) Bituminous Bound Applications				
Malish et al. (1975) and Larsen (1989)	Canada, USA	<9.5 to <25	20–70	Satisfactory performance except for ravelling
Maupin (1997)	USA	n.a.	15	Satisfactory performance except for ravelling
Nicholls and Lay (2002)	UK	<20	30	Met the specified requirements
Su and Chen (2002)	Taiwan	<4.75	10	Satisfactory performance
Wood et al. (2011)	Australia	n.a.	n.a.	Satisfactory performance

use in the 1990s, again mainly in the United States, in the unbound aggregate form for base and subbase construction applications. Comparatively, the field investigation for the use of GC in hydraulically bound applications has been less reported, with only a few recent cases found in Australia.

Overall, the GC content in the mixes, apart from a few early case studies in which the replacement of natural aggregate by GC aggregate exceeded 50%, in recent times has normally been less than 30%. The aggregate size of GC used, however, has tended to vary depending on the application. In unbound and bituminous bound forms, the material has normally been used as a mixture of sand and gravel, with maximum aggregate size up to 25 mm, whilst in hydraulically bound mixtures, it has been used as a fine aggregate, with maximum size of 4 mm.

Case Studies for Unbound Applications

Due to the brevity of the available information, it is estimated that the references listed in Table 10.6, in total, cover about 15 field records of using GC aggregate in the unbound form. Although the details of the field evaluations on the performance of GC aggregate in these applications were not normally provided, or were even absent, in the literature, many studies have briefly indicated that the results were generally satisfactory (HDR Engineering, 1997; Krivit, 1999; TxDOT, 1999). The general impression formed from the available information is that GC aggregate can be a viable material for use in unbound and bound applications, without compromising their technical performance. Indeed, some data have shown that the addition of GC aggregate results in improved performance of the mixes in terms of stiffness (Senadheera et al., 2005) and workability (Arulrajah et al., 2013).

It should, however, be mentioned that the improvement in the workability of unbound mixes in the construction of footpath bases (Arulrajah et al., 2013) was attained probably because of improved aggregate packing resulting from the mixture of GC aggregate with a maximum size of 4.75 mm and crushed rock with a maximum size of 25 mm. Additional in situ investigations with blends of 15% and 30% GC aggregate with crushed rock also showed that the mixes met the recommended requirements in terms of field compaction, grading and strength.

Case Studies for Hydraulically Bound Applications

Except for Wood et al. (2011), for which the field results were not available, the response of the case studies on the use of GC as a fine aggregate in hydraulically bound applications has generally been positive (Polley et al., 1998; Dumitru et al., 2013; Song et al., 2016). In a field trial undertaken by Polley et al. (1998) for a sidewalk pavement construction in the United States, the placement of hydraulically bound mixes containing 10% and 20% GC fine aggregate was reported to be of acceptable standard, although water was added during placement to ease the finishing of the concrete. The trial pavement containing GC aggregate showed good abrasion resistance and freeze–thaw resistance during three winters of exposure.

In another separate study of a concrete pavement made with and without 45% GC fine aggregate constructed in Australia, the field performance of the pavement was reported by Dumitru et al. (2013) after one and by Song et al. (2016) five years. The early records showed that the performance of the GC fine aggregate pavement was satisfactory, although visual inspection suggested that the GC fine aggregate mix was less resistant to wear. However, after 5 years of service, the skid resistance of the GC fine aggregate mix was reported to meet the minimum requirement of the local road authorities Song et al. (2016). Regarding long-term performance, road pavement made with hydraulically bound GC fine aggregate had a better performance than normal concrete mixes in terms of compressive strength, water absorption and resistance to chloride ingress.

Case Studies for Bituminous Bound Applications

In bituminous bound applications, the responses of the field studies with GC aggregates have been mixed, with the results on balance tending to be slightly unfavourable for most of the work reported in the three reports covering 50 case studies undertaken in the United States and Canada from the 1970s to the 1990s (Malish et al., 1975; Maupin, 1997; Larsen, 1989). The early records identified some common problems in the bituminous mixes, which were made with GC aggregates of maximum size up to 25 mm, such as:

(i) Loss of adhesion between GC aggregate and asphalt

(ii) Lower skid resistance

(iii) Breakage of GC coarse aggregate and subsequent ravelling (surface deterioration)

However, these problems have not been reported in the more recent studies undertaken in Australia, Taiwan and the United Kingdom (Wood et al., 2011; Nicholls and Lay, 2002; Su and Chen, 2002). Indeed, the conclusion reached was that the GC aggregates in bituminous mixes were of satisfactory performance. However, the main difference was that in these mixes, the maximum size of the GC aggregate was considerably smaller, close to that of fine aggregate, and this appears to be the main factor in producing performance comparable to that of the natural aggregate mixes, though in one case the maximum GC aggregate size was 20 mm. Thus, further case studies are needed to clarify the aggregate size affect.

Overall, although there are very limited case study data available for the bituminous mixes, and clearly more work is needed, aggregate sizes larger than the fine aggregate maximum size of 4 mm are unlikely to produce satisfactory bituminous mixes, because of the flaky nature of the material. Thus, it would be advisable to limit the maximum size of GC aggregate for use in bituminous mixes. However, that said, some further precautionary steps may also be considered, such as adding hydrated lime in the mix to improve the bond between GC aggregate and asphalt (see Chapter 8).

10.4 Standards and Specifications

When evaluating the potential of recycled and secondary materials for use in construction, it is important that they comply with the standards and specifications to ensure that the performance and quality of the end products are of an acceptable standard. This section discusses the major current standards and specifications, mostly those adopted in Europe and the United States, to gauge the suitability of using GC in concrete construction as a cement component, filler aggregate and fine aggregate and in geotechnical and road pavement applications as an aggregate component. Only soda lime glass, in the form of container glass and flat glass, is considered, as other types of glass are unlikely to be used as secondary materials in these applications.

10.4.1 Ground Glass Cullet Use as Cement Component

Although known to possess pozzolanic properties (see Chapter 4), GGC has not been recognised for use as a cement component, for example, as the common cements in European Standard EN 197-1 (2011) or as supplementary cementitious materials in American Standard ASTM C150 (2012) or, for that matter, in any other national standard. However, sufficient research has been undertaken for the material to be adopted in the manufacturing of cement in two ways:

(i) At 5%, as a minor additional constituent in compliance with EN 197-1 (2011) or as inorganic processing addition in compliance with ASTM C150 (2012).

(ii) At up to 35% as a major constituent, such as in CEM II cements in accordance with EN 197-1 (2011), with GGC content determined by the nature of the application, or in Portland–pozzolana cement in accordance with ASTM C595 (2013), where the GGC content may be increased up to 40%.

In supporting the adoption of GGC as a pozzolanic cement, the ASTM C1709 (2011) five-stage development program, which includes the necessary laboratory tests and field trials, can be used. The main actions required at each of the five stages as specified in the standard are as follows:

Stage I—Characterisation of the material

Stage II—Determination of suitable fineness

Stage III—Testing to ASTM C618 (2015) for fly ash, ASTM C989 (2016) for GGBS or Line up with Testing above. ASTM C1240 (2015) for silica fume

Stage IV—Concrete performance tests

Stage V—Field trials and long-term performance durability

There is a large volume of test data available in the literature to suggest that GGC can comply with requirements of stages I to V, in terms of the material properties, cement characteristics and concrete performance, as well as the field trials, though to a lesser extent. This information has previously been analysed and presented in Chapters 3 and 4, which deal with the material characteristics and performance of concrete containing GGC, respectively, as well as in Sections 10.2.1 and 10.3.1, which deal with leaching of heavy metals and the case studies on the use of GGC as a cement component, respectively.

To gauge the suitability of GGC as a pozzolanic material, the characteristics of GGC and its effects on the performance of concrete, in relation to the tests and properties required at each stage, are summarised in Table 10.7. For ease of reference, the inventory of where the relevant information can be found in this book is presented in Table 10.8.

Table 10.7 Evaluation of the suitability of ground glass cullet for use in concrete in accordance with ASTM C1709 (2011)

Stage	Evaluation
I	**Characterisation of the material** • The major chemical constituents of GGC of the soda lime type are SiO_2, Na_2O and CaO. • Other chemical constituents, which are important in cement chemistry, such as Al_2O_3, Fe_2O_3, MgO and K_2O, are also present in GGC but in small quantities. • PbO, which affects the setting time of cement, may be present due to contamination.
II	**Determination of suitable fineness** • The material needs to be crushed and ground to a minimum fineness of $300\,m^2/kg$ to exhibit pozzolanic properties, but a figure of $350\,m^2/kg$ may initially prove more attractive.

Table 10.7 Continued

Stage	Evaluation
III	**Testing to ASTM** C618 (2015) **for fly ash, ASTM** C989 (2016) **for GGBS or ASTM** C1240 (2015) **for silica fume** • The chemical and physical properties of GGC meet the requirements specified for fly ash. • The additional tests on GGC for its chlorides, soluble alkalis and leachable heavy metals have shown that those contents are not of concern.
IV	**Determination of concrete performance** • The material has been tested in concrete mixtures with cement content mainly varying from 200 to 500 kg/m^3 for their fresh and hardened properties. • Its use as a Portland cement replacement has been up to 70%, but normally in the range of 10%–30%. • No issues have been reported for its use in concrete containing water-reducing admixtures, air-entraining agents and other chemical admixtures. • In general, the use of GGC does not adversely affect the fresh properties of concrete, though it can increase the setting time and decrease the temperature of the mix. • The effect of GGC on the hardened properties of concrete is, by and large, similar to that of fly ash. Improvements in strength, permeation and resistance to both chloride ingress and sulphate attack have been reported. The material is also effective in reducing the ASR-induced expansion in concrete due to reactive aggregate.
V	**Conducting field trials and long-term field performance and durability evaluations** • Detailed information on the field trials and the long-term field performance of concrete made with GGC, though limited, has been encouraging. • Only one field study has suggested that the use of GGC with Portland cement did not compromise the long-term performance of concrete after 5 years of service, though in this case GGC content was 15%.

ASR, alkali–silica reaction; *GGBS*, ground granulated blast furnace slag; *GGC*, ground glass cullet.

It appears that there are a few required tests and properties of GGC, mainly in Stage III, that have not been commonly tested or made available in the literature. This information, though needed for completeness of the evaluation programme, is not likely to affect the development of GGC for use as a pozzolanic cement. However, the field trials using GGC cement have been limited, and this needs to be addressed. Overall, based on the current state of knowledge, it can be concluded that GGC is a safe pozzolanic material, which can be used as a cement component in concrete construction.

In addition, two specifications dealing with the use of GGC as a pozzolanic in concrete, mortar or grout have been proposed by two established organisations in the United Kingdom: (i) Concrete Technology Unit (CTU) as prepared by Dhir et al. (2003a) and (ii) Waste and Resources Action Programme (WRAP) as prepared by Byars et al. (2004b).

Table 10.8 Tests and properties required in ASTM C1709 (2011) and the corresponding available data in this book

Stage	Test and Properties	Available Data
I	Chemical analysis	Section 3.6.1
II	Suitable fineness	Sections 3.7.5 and 4.3.4
III	**Chemical requirements**	
	• Sum of SiO_2, Al_2O_3 and Fe_2O_3	Section 3.6.2
	• SO_3	Section 3.6.3
	• Moisture content	Not available
	• Loss on ignition	Section 3.6.5
	Physical requirements	
	• Fineness	Section 3.7.5
	• Water requirement	Not available
	• Soundness	Section 4.3.3
	• Strength activity index	Section 4.3.4
	Uniformity requirements	Not available
	Supplementary optional physical requirements	Not available
	Chlorides	Section 3.6.1
	Free calcium oxide	Not available
	Soluble alkalis	Section 3.6.4
	Leachable heavy metals	Section 10.2.1
	Air void stability	Not available
IV	**Fresh properties**	
	• Slump	Section 4.4.1
	• Bleeding	Section 4.4.2
	• Fresh density	Section 4.4.3
	• Air content	Section 4.4.4
	• Setting time	Section 4.4.5
	• Temperature	Section 4.4.6
	Hardened properties	
	• Compressive strength	Section 4.5.1
	• Flexural strength	Section 4.5.2
	• Modulus of elasticity	Section 4.5.3
	• Drying shrinkage	Section 4.5.6
	• Air void system parameters	Not available

Table 10.8 Continued

Stage	Test and Properties	Available Data
	• Sulphate resistance	Section 4.7.4
	• Alkali–silica reaction	Section 4.7.5
	Others (for specialised needs)	
	• Heat of hydration	Section 4.4.6
	• Permeation	Sections 4.6.1 to 4.6.4
	• Chloride ingress resistance	Section 4.7.1
	• Freeze–thaw resistance	Section 4.7.6
	• Scaling resistance	Not available
V	Field trials and long-term performance and durability	Section 10.3.1

These proposed specifications have suggested limits for the chemical and physical properties of GGC cement, as well as the associated test methods, as summarised in Table 10.9. In general, the requirements are quite similar to those for fly ash specified in BS EN 450-1 (2012), but with some differences and few new additional tests. A comparison of the requirements for these two materials is given below:

- **Loss of ignition (LOI)**: An LOI limit of no more than 1.5% is proposed for GGC cement. (This is considerably less than those for Portland cement in the form of CEM I cement and category S fly ash, with the lowest LOI limit for category A (BS EN 450-1, 2012) at 5.0%. Thus, the proposed limit for LOI is much more stringent than those for both CEM I cement and category S fly ash.)
- **Chloride and sulphate contents**: The requirements for chloride and sulphate contents are no more than 0.10% and 3%, respectively.
  (These are similar to those for CEM I cement and fly ash.)
- **Alkali content**: Because both sodium and potassium have a role in ASR, and because these elements can be found in GGC of the soda lime type, the alkali content of GGC cement has been proposed not to exceed 2%.
  (This limit appears to be more stringent than that for fly ash, which is 5% maximum.)
- **Lead content**: The proposed maximum limit is 0.1%.
  (Fly ash is not required to be tested for lead content. However, the test might be needed for GGC because of the potential contamination with lead-bearing material.)
- **Silica content**: The proposed minimum SiO_2 content for GGC is 68%.
  (Perhaps the minimum sum of SiO_2, Al_2O_3 and Fe_2O_3 of 70% specified for fly ash shall be used, instead of SiO_2 content alone.)
- **Moisture content**: The test for moisture content is not needed for fly ash. For GGC, the moisture content has been proposed not to exceed a 1% limit.

Table 10.9 Proposed specifications for use of ground glass cullet in concrete, mortar and grout

	WRAP by Byars et al. (2004b)		CTU by Dhir et al. (2003a)	
Property	**Limit**	**Test Method**	**Limit**	**Test Method**
(a) Chemical				
Loss on ignition, %	≤1.5	BS EN 196-2 (1995)	n.a.	n.a.
Chloride content, %	≤0.10	BS EN 196-21 (1992)	≤0.10	BS EN 196-21 (1992)
Sulphate content (SO_3), %	≤3	n.a.	n.a.	n.a.
Alkali content (Na_2O_{eq}), %	n.a.	n.a.	≤2.0	BS EN 196-21 (1992) Clause N.A.5.1[a]
Lead content (PbO), %	≤0.1	n.a.	n.a.	n.a.
Total silica (SiO_2), %	≥68	BS EN 196-2 (1995)	n.a.	n.a.
(b) Physical				
Moisture content, %	n.a.	BS 3892-1 (1997) Annex C	≤1.0	BS 6699 (1992) Appendix E
Fineness (Blaine method), m^2/kg	≥300	BS EN 196-6 (1992)	>200	BS EN 196-6 (1992)
Initial setting, min	<120 more than reference cement[b]	BS EN 196-3 (1995)	≤30 more than CEM I 42.5[b]	BS EN 196-3 (1995)
Soundness, mm	≤10[b]	BS EN 196-3 (1995)	≤10[b]	BS EN 196-3 (1995)
Strength Activity Index, %				
At 28 days	≥75[c]	BS EN 196-1 (1995)	n.a.	n.a.
At 90 days	≥85[c]		n.a.	n.a.

CTU, Concrete Technology Unit; *GGC*, ground glass cullet; *n.a.*, not available; *WRAP*, Waste and Resources Action Programme.
[a]Tentative method.
[b]At 30% GGC replacement.
[c]At 25% GGC replacement.

- **Fineness**: Between the two proposed values for GGC, the minimum requirement of 300 m^2/kg Blaine fineness is preferable to ensure the material is sufficiently fine to exhibit pozzolanic properties.
  (No comparison can be made with fly ash, as its fineness requirement is given in the form of the mass proportion in percentage retained on a 0.045-mm mesh sieve.)

- **Setting time**: The maximum limit for 30% GGC content on initial setting time is not to exceed 30 min longer than CEM I 42.5 or 120 min longer than reference cement. (This again cannot be compared with the setting time requirement for fly ash, as the content of the material used in the test is different, i.e., 25% as specified for fly ash.)
- **Strength activity index**: This is to be a minimum of 75% and 85% at 28 and 90 days with respect to the strength of the reference cement. (These are similar to those for fly ash.)

10.4.2 Glass Cullet Use as Fine Aggregate and Filler Aggregate in Concrete

Although manufactured aggregates are allowed for use in concrete in both the European standard EN 12620 (2008) and the American standard ASTM C33 (2016), GC aggregate does not quite fit in with the provided definitions of a manufactured aggregate. In the European standard, manufactured aggregate is an aggregate of mineral origin derived from an industrial process involving thermal and other modifications, whilst in the American standard, the material is obtained from crushing rock, gravel, iron blast furnace slag or hydraulic-cement concrete.

Notwithstanding this, owing to its near-zero porosity and relatively high hardness characteristics, if properly designed, GC aggregate can be used as a suitable substitution for natural aggregate, which can result in many benefits to the engineering properties of concrete. However, due to the flaky nature of the material at coarser sizes, GC is more suitable for use either as a (i) fine aggregate with aggregate size no larger than 4 mm (EN 12620, 2008) or 4.75 mm (ASTM C33, 2016) or (ii) filler aggregate with aggregate size predominantly less than 0.063 mm (EN 12620, 2008). In the absence of specifications for GC fine aggregate and filler aggregate, it could be argued that their compliance with the geometrical, physical and chemical requirements of the existing standards for concrete aggregates, such as EN 12620 (2008) and ASTM C33 (2016), should be sufficient for the material to be used in concrete.

In fact, during the 2000s, some well-established organisations, in the United Kingdom and Australia, proposed specifications specific to GC aggregate for use mainly as a fine aggregate, but some included the use of GC as a coarse aggregate and filler aggregates as well. The involved organisations were the CTU, UK (Dhir et al., 2003b), WRAP, UK (Byars et al., 2004a; Collins et al., 2004), and Commonwealth Scientific and Industrial Research Organisation, Australia (Sagoe-Crentsil et al., 2001). Three major requirements for GC aggregate as given in these proposed specifications, which are seen to be particularly related to the properties of the material, are described below.

(a) Aggregate Size

The material has been proposed for use mainly as a fine aggregate, with maximum aggregate size less than 4 mm (Dhir et al., 2003b) or 2.46 mm (Sagoe-Crentsil et al., 2001). Its use as a filler aggregate, with aggregate size passing 0.063 mm sieve, has also been proposed (Dhir et al., 2003b). The specifications proposed by Byars et al. (2004a) and Collins et al. (2004) included the use of GC aggregate as a coarse aggregate as well, but the allowable maximum aggregate size was not indicated. However, it shall be mentioned that, because the material can be flaky or elongated at larger sizes, its use as a coarse aggregate in concrete is considered to be inappropriate.

(b) Foreign Material

As the material is prepared from waste glass, foreign materials such as papers, plastics, metal pieces and other non-glass contaminants may be present in GC aggregate. The content of foreign material in GC aggregate has been limited to 1% (Byars et al., 2004a) or 2% (Sagoe-Crentsil et al., 2001) maximum by mass.

(c) Constituents That Alter the Rate of Setting and Hardening of Concrete

As stated in EN 12620 (2008), the presence of organic materials or other substances in aggregate can have an effect on stiffening time and compressive strength of concrete. The content of organic materials has been limited so as not to result in:

- Increase in the stiffening time of mortar test specimens by more than 120 min.
- Decrease in the compressive strength of mortar test by more than 20%.

The same requirements have been proposed for the use of GC aggregate (Dhir et al., 2003b; Collins et al., 2004), but with a more stringent requirement of a maximum 10% reduction in compressive strength (Byars et al., 2004a).

For GC aggregate, the most concerning organic substance is sugar, which normally comes from food residue. It has been recommended that GC should be cleaned prior to its use in concrete, and possibly with a high-temperature wash to expedite the removal of sugar (Sagoe-Crentsil et al., 2001).

Apart from these proposed specifications for ensuring the quality of GC aggregate, Dhir et al. (2003b) and Sagoe-Crentsil et al. (2001) have also prepared a guidance relating to the use of GC aggregate in concrete. It includes several topics, such as handling of material, mix design, concrete casting and specification clauses. As far as mix design is concerned, in the case of GC fine aggregate, it has been proposed that the material can be used as a part or full replacement for natural fine aggregate (Dhir et al., 2003b), although Sagoe-Crentsil et al. (2001) limited its inclusion to a maximum content of 20%. In the case of GC filler aggregate, the material can be used for no more than 20% of total fine aggregate content (Dhir et al., 2003b).

10.4.3 Glass Cullet Use as Aggregate in Geotechnical and Road Pavement Applications

With the exception of the material specification adopted by the American Association of State Highway and Transportation Officials (AASHTO M 318-02, 2015), the use of GC as one of the aggregate sources for geotechnical applications, as well as in road pavement applications in unbound, hydraulically bound and

Table 10.10 Physical properties of and deleterious substances in glass cullet aggregate used as base material (AASHTO M 318-02, 2015)

Physical Property/Deleterious Substance	Requirement/Limit
Glass type	Container glass
Maximum aggregate size	9.5 mm
Content of flat and elongated aggregate	<1%
Content of other types of glass	<5%
Content of deleterious substances (e.g., food residue)	<1%
Content of soil-like materials	<2%

bituminous bound forms, has not been generally accepted in the major standards, such as the American ASTM standards and the European CEN standards. This specification deals with GC aggregate and its use in the base course of road pavement. The physical properties and deleterious substance limits on GC aggregate used as a granular base course material are summarised in Table 10.10. This shows that GC can be used as one of the aggregate sources for road pavement applications, provided that the material is clean and has been prepared in fine aggregate size fractions. The AASHTO specification also recommends mixing of GC aggregate with other soil aggregates, but only up to 20% content. However, a higher GC aggregate content may be used where the performance of the mixture has been validated with the California bearing ratio test, resilient modulus test and field evaluation.

On the other hand, a large number of countries, 21 to be precise, have national standard specifications relevant to the use of GC aggregate in geotechnical and road pavement applications, including three states in Australia (New South Wales, Queensland and Victoria), Canada (Ontario), New Zealand, the United Kingdom and 16 states in the United States (Alaska, California, Connecticut, Florida, Idaho, Maryland, Massachusetts, Minnesota, New Hampshire, New Jersey, New York, Oregon, Pennsylvania, South Carolina, Virginia and Washington). All these standard specifications are up to date, with most of the latest versions being issued during 2015–17.

The information available in these national standards has been sorted in terms of application, glass type, maximum GC aggregate content, maximum aggregate size and maximum debris content (contaminant), in Table 10.11 for Australia and New Zealand, Table 10.12 for Canada and the United Kingdom and Table 10.13 for the United States. It should be mentioned that in some applications there could be several types of aggregate grading designs but GC aggregate is permitted only in some certain specific types. In addition, the definition of debris varies across the countries/states, and its limit may also vary depending on the type of contaminant and the GC aggregate's intended use. For further information on these matters the readers are advised to refer to the standard specifications mentioned. However, the main points to emerge from these tables are described below.

Table 10.11 Specifications for using glass cullet in geotechnical applications in Australia and New Zealand

Country	References	Applications	GC Properties					Additional References
			Allowable Type	Prohibited Type	Max. GC Content, %	Max. Size, mm	Max. Debris[a], %	
Australia								
New South Wales	RMS (2014)	Unbound and bound base and subbase	Container glass	CRT, lighting, lab. glass, ceramics	10	<5	2	RMS (2011) and NSW EPA (2014)
	RMS (2013a, 2013b, 2013c, 2013d)	Heavy-duty, light-duty dense graded asphalt; crumb rubber asphalt; stone mastic asphalt			2.5			
Queensland	TMR (2016a)	Unbound base and subbase	Container glass, window or building glass	Vehicle, CRT, lab. glass	5	<9.5	≤3	TMR (2016b)
Victoria	VicRoads (2017)	Asphalt intermediate and base course	Container glass	CRT, lab. glass, lighting	–	<4.75	2	VicRoads (2011)
New Zealand								
New Zealand	Transit New Zealand (2006a)	Base, subgrade	Container, window, plain ceramics	CRT, lighting, vehicle glass	5	<9.5	5	Transit NZ (2006b)
	NZ Transport Agency (2014)	Dense graded and stone mastic asphalt[b]	–	–	–	(use as coarse agg.)	–	–

CRT, cathode ray tube (funnel); *GC*, glass cullet.
[a]Debris can be plastic, metal pieces, papers, wood or other contaminants (definition of debris varies across states/countries); the limit of debris content varies depending on debris type, the value given here is the overall maximum.
[b]Needs approval from Transit New Zealand's Engineering Policy Manager.

Table 10.12 Specifications for using glass cullet (GC) in geotechnical applications in Canada and the United Kingdom

Country	References	Applications	GC Properties: Allowable Type	GC Properties: Prohibited Type	GC Properties: Max. GC Content, %	GC Properties: Max. Size, mm	GC Properties: Max. Debris[a], %	Additional References
Ontario, Canada	MTO (2013a)	Base, subbase and backfill	Post-consumer glass/ceramics	–	15	<13.2	1	–
	MTO (2013b)					–		
United Kingdom[b]	Department of Transport (2004)	Pipe bedding material, filter material	–	–	–	(use as fine agg.)	–	Department of Transport (2009)
		Embankment, fill material, capping	–	–	25	–	–	Department of Transport (2016a)
		Unbound mixtures	–	–	25	–	–	Department of Transport (2016b)
		Hydraulically bound aggregate for subbase and base[c]	–	–	40	–	–	Department of Transport (2016c)
		Bituminous bound layer	–	–	–	–	–	Department of Transport (2008)
		Concrete material[c]	–	–	–	–	–	Department of Transport (2016d)

[a]Debris can be plastic, metal pieces, papers, wood or other contaminants (definition of debris varies across states/countries).
[b]Recycled materials that comply with BS EN 13043 (2002) are permitted to be used.
[c]Use of GC aggregate may be permitted by the overseeing organisation.

Table 10.13 Specifications for using glass cullet in geotechnical applications in the United States

State	References	Applications	GC Properties					Additional References
			Allowable Type	Prohibited Type	Max. GC Content, %	Max. Size, mm	Max. Debris[a], %	
United States								
Alaska	Alaska DOT & PF (2015)	Embankment, base, surface course, subbase	Container, window, plain ceramics	Vehicle glass, lighting, CRT, porcelain, lab. glass	10	<9.5	2	–
California	Caltrans (2015a)	Subbase, base	–	–	–	–	–	Caltrans (2015b)
Connecticut	ConnDOT (2016)	Embankment	–	–	25	<25.0	–	–
		Bituminous concrete	Container	–	5	<9.5	1	
Florida	FDOT (2017)	Superpave asphalt concrete	–	–	15	<9.5	–	–
Idaho	IDT (2012)	Embankment	Container	–	15	–	–	–
		Base	Container	–	15	<19.0 or <7.9	–	
Maryland	SHA (2008)	Hot-mix asphalt	–	–	–	<19.5	1	–
		Bioretention soil mix			–	(Use as coarse sand)	–	

State	References	Applications	GC Properties					Additional References
			Allowable Type	Prohibited Type	Max. GC Content, %	Max. Size, mm	Max. Debris[a], %	
Massachusetts	MassDOT (2015)	Borrow materials, subbase, base, binder course, bituminous material	Container	–	10	<9.5	5	–
Minnesota	MnDOT (2016)	Base course	–	–	10	<19.0	–	Grading and Base Unit (2016)
New Hampshire	NHDOT (2016)	Base course	Container	–	–	<9.5	5	AASHTO M318 (2015)
New Jersey	NJDOT (2007)	Hot-mix asphalt base and intermediate course	Container	–	10	<9.5	3	–
New York	NYCDOT (2017)	Embankment, backfill	–	–	–	<9.5	5	
		Subbase course	–	–	30			
Oregon	ODOT (2015)	Nonstructural fill	–	–	100	<12.7	10	–
		Backfill, drainage blanket material	–	–	–	<12.7	10	–

Continued

Table 10.13 Continued

			GC Properties					
State	**References**	**Applications**	**Allowable Type**	**Prohibited Type**	**Max. GC Content, %**	**Max. Size, mm**	**Max. Debris[a], %**	**Additional References**
Pennsylvania	PennDOT (2016)	Unbound material	–	–	10	–	–	–
South Carolina	SCDOT (2007)	Embankment			25	<12.7	1	
		Hot-mix asphalt intermediate and base course	–	–	15	<9.5	–	–
Virginia	VDOT (2016)	Drainage, embankment, backfill	–	–	90 (100 for drainage)	<9.5	5	–
Washington	WSDOT (2016)	Ballast, crushed surfacing, gravel base, backfill, borrow pit, foundation materials	Container	–	20	<9.5	5	AASHTO M318 (2015)

CRT, cathode ray tube (funnel); *GC*, glass cullet.
[a]Debris can be plastic, metal pieces, papers, wood or other contaminants (definition of debris varies across states/countries); the limit of debris content varies depending on debris type, the value given here is the overall maximum.

(a) Applications

Commonly, GC aggregate has been allowed as a fill material in embankment and backfill construction applications, as well as an unbound granular material in subgrade, base and subbase construction applications. In a bound form, GC aggregate can be used only in bituminous bound applications. There appears to be no provision for the use of GC aggregate in hydraulically bound applications, except in the United Kingdom, but with approval of overseeing organisations.

(b) Glass types

Container glass is the most acceptable material for use as an aggregate in practice. Although a number of specifications have not been very clear regarding this, it is believed that container glass is implied. Flat glass is also allowed to be used in some countries/states, but this depends on the source of the glass. In Australia, New Zealand and Alaska, only window or building glass is permitted, but vehicle glass (windscreens) is prohibited. Interestingly, ceramics, which are often listed as a prohibited glass type for use as aggregate, have been allowed in New Zealand, Canada and Alaska. Other types of glass, such as CRTs, lighting and laboratory glassware, are prohibited and their presence in container glass stockpiles is normally considered as a contaminant.

(c) Maximum aggregate size

The allowable maximum GC aggregate size varies over a wide range, from less than 4.75 mm (in Victoria) to 25 mm (in Connecticut). On average, the material is required to be prepared for aggregate in sizes less than 9.5 mm, especially in the United States.

(d) Maximum aggregate content

Depending on the application and country/state, GC aggregate has been allowed to be used at as high as 90% or 100% of the total aggregate content in nonstructural fill, drainage, embankment and backfill in Oregon and Virginia (Table 10.13), or as low as 2.5% or 5% content in unbound and bituminous bound applications for road pavement construction in Australia, New Zealand, and Connecticut (Tables 10.11 and 10.13). In general, the use of GC aggregate at a content of 10% or 15% has been commonly accepted in most of the specifications.

(e) Maximum debris content

Although the definition of debris varies across the countries/states, commonly non-glass materials such as plastics, papers, metal and wood pieces are listed as examples of debris in the specifications. Depending on the specifications, the maximum limit of debris content has been specified individually for each specific type or as a whole. Overall, the limit is generally specified to be less than 5%.

10.5 Conclusions

Glass cullet derived from soda lime glass is a safe material, with the leaching of heavy metals being below the regulatory limits specified in the US Code of Federal Regulations (2017). Thus, its use as a replacement for Portland cement and natural aggregate should not present any threat to the environment. On the other hand, glass types bearing significant amounts of toxic elements, such as CRTs and LCDs, are considered to be hazardous. Such materials, therefore, are not likely to be used as construction materials.

Most of the environmental assessments on the use GC in concrete applications have focused on CRT funnel glass, as the material is known to contain a high lead content. At low inclusion levels of CRT, as raw feed in clinker production or in the ground form as a component of cement or in granular form as a component of aggregate in concrete, the amounts of lead released from the resulting products have been shown not to be of concern. Preliminary studies have also reported that the leachability of lead from CRT aggregate in concrete can be controlled by treating the material with acid or adding a biopolymer solution to the mix. Overall, the leachability of CRT glass when used as a construction material still requires further examination.

For use in geotechnical applications, it has been shown that the GC aggregate derived from container glass is alkaline and noncorrosive. Unless contaminated, contaminant constituents of container GC and the concentrations of leached elements are normally below the regulatory limits. Properly processed GC aggregate is likely to have a very low content of organic material. The environmental assessment of GC aggregate in road pavement applications has been very limited, though it has been shown that the leaching of the material is not of concern.

The case studies using GC as part of the raw feed for the manufacture of Portland cement and GGC as a cement component have been positive, showing that satisfactory performance of concrete can be achieved with up to 50% content of GGC. The use of GGC in high-performance concrete in terms of consistence, strength and chloride ingress has been explored with satisfactory performance. Field studies with up to 100% GC as fine aggregate in many concrete applications have also been encouraging, with no major performance issues reported, though in some cases the consistence of the mix was compromised. Examination of concrete containing GC fine aggregate has also indicated that no sign of ASR was found over five years, though this may not be regarded as a sufficiently adequate period to expect to see ASR effects.

The use of GC aggregate in geotechnical applications such as backfill, utility, drainage and landfill constructions has been common in the United States since as early as 1990. The material has been used for up to 100%, with maximum aggregate size sometimes up to 19 mm. In road applications, the use of GC aggregate in unbound and hydraulically bound forms has been shown to result in satisfactory performance. However, with its use in bituminous bound form, the performance of the mixes seems to be affected by the size of the material, with coarser material tending to give some technical problems.

Both the American and the European standards, as of this writing, have no provision for the use of GGC as a cement component in concrete. However, based on the evaluation programme provided in ASTM C1709 (2011), it has been shown that GGC is a safe pozzolanic material, with its effect on hardened concrete being generally similar to that of fly ash. However, further field trials are needed to complete the ASTM evaluation programme. Similarly, the material has not been accepted for use as a fine aggregate in concrete in the standards. However, some specifications in this respect have been proposed in the 2000s in Australia and the United Kingdom. In the field of geotechnics and road pavement, AASHTO M318 (2015) provides a specification for GC aggregate use as granular base course material. In addition, a number of countries have permitted the use of the material in many geotechnical and road pavement applications.

References

AASHTO M318, 2015. Standard Specification for Glass Cullet Use for Soil-Aggregate Base Course. American Association of State Highway and Transportation Officials, Washington DC, USA.

Alaska DOT & PF, 2015. Standard Specifications for Highway Construction. Alaska Department of Transportation and Public Facilities, Juneau, AK, USA. 475 pp.

Arulrajah A, Ali M M Y, Disfani M M, Piratheepan J and Bo M W, 2013. Geotechnical performance of recycled glass-waste rock blends in footpath bases. Journal of Materials in Civil Engineering 25 (5), 653–661.

Arulrajah A, Imteaz M, Horpibulsuk S, Du Y and Shen J, 2016. Recycled concrete aggregate/municipal glass blends as a low-carbon resource material for footpaths. Road Materials and Pavement Design 1–14.

AS 4439.3, 1997. Wastes, Sediments and Contaminated Soils, Part 3: Preparation of Leachates-Bottle Leaching Procedure. Standards Association of Australia, Sydney, Australia.

ASTM C1240, 2015. Standard Specification for Silica Fume Used in Cementitious Mixtures. ASTM International, West Conshohocken, PA, USA.

ASTM C150, 2012. Standard Specification for Portland Cement. ASTM International, West Conshohocken, PA, USA.

ASTM C1709, 2011. Standard Guide for Evaluation of Alternative Supplementary Cementitious Materials (ASCM) for Use in Concrete. ASTM International, West Conshohocken, PA, USA.

ASTM C33, 2016. Standard Specification for Concrete Aggregates. ASTM International, West Conshohocken, PA, USA.

ASTM C595, 2013. Standard Specification for Blended Hydraulic Cements. ASTM International, West Conshohocken, PA, USA.

ASTM C618, 2015. Standard Specification for Coal Fly Ash and Raw or Calcined Natural Pozzolan for Use in Concrete. ASTM International, West Conshohocken, PA, USA.

ASTM C989, 2016. Standard Specification for Slag Cement for Use in Concrete and Mortars. ASTM International, West Conshohocken, PA, USA.

ASTM D4793, 1993. Standard Test Method for Sequential Batch Extraction of Waste with Water. ASTM International, West Conshohocken, PA, USA.

Bell G, 2006. Full Scale Operational Trials Involving the Use of Recycled Glass in the Manufacture of Concrete Blocks. The Waste & Resources Action Programme (WRAP), Oxon, UK. Project code GLA44-055, 13 pp.

Boniface T, 2006. Full Scale Operational Trials Using Recycled Glass in Selected End Markets: Using Recycled Glass in Cementitious Products. The Waste & Resources Action Programme (WRAP), Oxon, UK. Project code GLA44-014, 24 pp.

BS 3892-1, 1997. Pulverized-Fuel Ash. Specification for Pulverized-Fuel Ash for Use with Portland Cement. BSI, London, UK.

BS 6699, 1992. Specification for Ground Granulated Blastfurnace Slag for Use with Portland Cement. BSI, London, UK.

BS 812-123, 1999. Testing Aggregates. Method for Determination of Alkali-Silica Reactivity. Concrete Prism Method. BSI, London, UK.

BS EN 12457-1, 2002. Characterisation of Waste. Leaching. Compliance Test for Leaching of Granular Waste Materials and Sludges. One Stage Batch Test at a Liquid to Solid Ratio of 2°L/kg for Materials with High Solid Content and with Particle Size below 4°mm (without or with Size Reduction). BSI, London, UK.

BS EN 12457-2, 2002. Characterisation of Waste. Leaching. Compliance Test for Leaching of Granular Waste Materials and Sludges. One Stage Batch Test at a Liquid to Solid Ratio of 10°L/kg for Materials with Particle Size below 4°mm (without or with Size Reduction). BSI, London, UK.

BS EN 12457-3, 2002. Characterisation of Waste. Leaching. Compliance Test for Leaching of Granular Waste Materials and Sludges. Two Stage Batch Test at a Liquid to Solid Ratio of 2 L/kg and 8 L/kg for Materials with a High Solid Content and with a Particle Size below 4°mm (without or with Size Reduction). BSI, London, UK.

BS EN 12457-4, 2002. Characterisation of Waste. Leaching. Compliance Test for Leaching of Granular Waste Materials and Sludges. One Stage Batch Test at a Liquid to Solid Ratio of 10°L/kg for Materials with Particle Size below 10°mm (without or with Size Reduction). BSI, London, UK.

BS EN 13043, 2002. Aggregates for Bituminous Mixtures and Surface Treatments for Roads, Airfields and Other Trafficked Areas. BSI, London, UK.

BS EN 196-1, 1995. Methods of Testing Cement. Determination of Strength. BSI, London, UK.

BS EN 196-2, 1995. Methods of Testing Cement. Chemical Analysis of Cement. BSI, London, UK.

BS EN 196-21, 1992. Methods of Testing Cement. Determination of the Chloride, Carbon Dioxide and Alkali Content of Cement. BSI, London, UK.

BS EN 196-3, 1995. Methods of Testing Cement. Determination of Setting Time and Soundness. BSI, London, UK.

BS EN 196-6, 1992. Methods of Testing Cement. Determination of Fineness. BSI, London, UK.

BS EN 450-1, 2012. Fly Ash for Concrete. Definition, Specifications and Conformity Criteria. BSI, London, UK.

Byars E A, Zhu H and Meyer C, 2003. Use of waste glass for construction products: legislative and technical issues. In: Dhir R K, Newlands M D and Halliday J E (Eds.), Proceedings of the International Symposium on Recycling and Reuse of Waste Materials, Dundee, UK. Thomas Telford, London, UK, pp. 827–838.

Byars E A, Zhu H and Morales B, 2004a. Conglasscrete I. The Waste & Resources Action Programme (WRAP), Oxon, UK. Project code: GLA2-006, 54 pp.

Byars E A, Morales B and Zhu H, 2004b. Conglasscrete II. The Waste & Resources Action Programme (WRAP), Oxon, UK. Project code: GLA0015-007, 30 pp.

Byars E A, Morales B and Zhu H, January 2004c. Waste glass as concrete aggregate and pozzolan. Concrete 41–44.

Caltrans, 2015a. Standard Special Provisions. California State Transportation Agency, Sacramento, CA, USA, p. 2015. Available from: http://www.dot.ca.gov.

Caltrans, 2015b. Standard Specifications. California State Transportation Agency, Sacramento, CA, USA, p. 2015. 1149 pp.

Chen G, Lee H, Young K L, Yue P L, Wong A, Tao T and Choi K K, 2002. Glass recycling in cement production – an innovative approach. Waste Management 22 (7), 747–753.

City of Armadale, 2010. Information Bulletin. City of Armadale, Armadale, Western Australia, Issue No. 05/2010. Available from: https://www.armadale.wa.gov.au.

Code of Federal Regulations, 2017. Title 40: Protection of Environment, Part 261-Identification and Listing of Hazardous Waste, Subpart C – Characteristics of Hazardous Waste. Available from: https://www.ecfr.gov/.

Collins R J, Dunster A and Kennedy J, 2004. Improving Specifications for Use of Recycled and Secondary Aggregates in Construction. The Waste & Resources Action Programme (WRAP), Oxon, UK. 58 pp.

ConnDOT, 2016. The Standard Specifications for Roads, Bridges, Facilities and Incidental Construction Form 816 Merged with Supplemental Specifications. Connecticut Department of Transportation, Newington, CT, USA. 605 pp.

Cosentino P J, Kalajian E H, Heck III H H H and Sheikh C S, 1995. Developing Specification for Waste Glass and Waste-to-Energy Bottom Ash as Highway Fill Materials, Volume 2 of 2 (Waste Glass). Report No. FL/DOT/RMC/06650-7754 Florida Department of Transportation, Tallahassee, FL, USA. 152 pp.

Dames and Moore Inc, 1993. Glass Feedstock Evaluation Project Engineering Suitability Evaluation. Report No. GL-93-95. Clean Washington Center, Seattle, WA, USA. 110 pp.

DBMD, 1995. Bouwstoffenbesluit, Nr. 12689/165. The Building Materials Decree (DBMD), the Netherlands. Ministry of Housing. Spatial Planning and the Environment (VROM), The Netherland.

DECC, 2007. Trial of Recycled Glass Pipe Embedment Material. Department of Environment and Climate Change NSW, Southern Sydney, NSW, Australia. 11 pp.

Department of Transport, 2004. Design Manual for Roads and Bridges. Pavement Design and Maintenance, Section 1 Preamble, Part 2 HD 35/04: Conservation and the Use of Secondary and Recycled Materials, vol. 7. HMSO, London, UK. 16 pp.

Department of Transport, 2008. Manual of Contract Documents for Highway Works. Specification for Highway Works, Series 900 Road Pavements – Bituminous Bound Materials, vol. 1. HMSO, London, UK. 68 pp.

Department of Transport, 2009. Manual of Contract Documents for Highway Works. Specification for Highway Works, Series 500 Drainage and Service Ducts, vol. 1. HMSO, London, UK. 29 pp.

Department of Transport, 2016a. Manual of Contract Documents for Highway Works. Specification for Highway Works, Series 600 Earthworks, vol. 1. HMSO, London, UK. 92 pp.

Department of Transport, 2016b. Manual of Contract Documents for Highway Works. Specification for Highway Works, Series 700 Road Pavements – General, vol. 1. HMSO, London, UK. 42 pp.

Department of Transport, 2016c. Manual of Contract Documents for Highway Works. Specification for Highway Works, Series 800 Road Pavements – Unbound, Cement and Other Hydraulically Bound Mixtures, vol. 1. HMSO, London, UK. 31 pp.

Department of Transport, 2016d. Manual of Contract Documents for Highway Works. Specification for Highway Works, Series 1000 Road Pavements – Concrete Materials, vol. 1. HMSO, London, UK. 45 pp.

Dhir R K, Dyer T D, Tang M C and Csetenyi L J, 2003a. Realising a High-Value Sustainable Recycling Solution to the Glass Cullet Surplus. Specification for Powdered Glass as a Component in Cement Combinations. Concrete Technology Unit (CTU), University of Dundee, Dundee, Scotland. 6 pp.

Dhir R K, Dyer T D, Tang M C and Csetenyi L J, 2003b. Realising a High-Value Sustainable Recycling Solution to the Glass Cullet Surplus. Specification for Crushed Glass as a Fine Aggregate or Filler Aggregate for Use in Concrete. Concrete Technology Unit (CTU), University of Dundee, Dundee, Scotland. 5 pp.

Disfani M M, Arulrajah A, Ali M M and Bo M W, 2011a. Fine recycled glass: a sustainable alternative to natural aggregates. International Journal of Geotechnical Engineering 5, 255–266.

Disfani M M, Arulrajah A, Bo M W and Hankour R R, 2011b. Recycled crushed glass in road work applications. Waste Management 31 (11), 2341–2351.

Disfani M M, Arulrajah A, Bo M W and Sivakugan N, 2012. Environmental risks of using recycled crushed glass in road applications. Journal of Cleaner Production 20, 170–179.

Dumitru I, Song T, Bornstein B, Brooks P and Moss J, 2013. Field trials using recycled glass as natural sand replacement and powdered glass as cementitious materials replacement in concrete pavement. In: Third International Conference on Sustainable Construction Materials and Technologies, Kyoto, Japan 9 pp.

EN 12620, 2008. Aggregates for Concrete. European Committee for Standardization (CEN), Brussels, Belgium.

EN 197-1, 2011. Cement. Composition Specifications and Conformity Criteria for Common Cement. European Committee for Standardization (CEN), Brussels, Belgium.

Environmental Agency, 2017. Classify Different Types of Waste: 4. Electronic and Electrical Equipment. Available from: http://www.gov.uk.

EPA Victoria, 2007. Classification of Wastes. Environmental Protection Agency of Victoria, Melbourne, Australia. Information Bulletin, Publication No. 448.

FDOT, 2017. Standard Specifications for Road and Bridge Construction. Florida Department of Transportation, Tallahassee, FL, USA. 1218 pp.

Grading & Base Unit, 2016. Grading and Base Manual. Grading & Base Unit, Minnesota Department of Transportation, St. Paul, MN, USA. 159 pp.

Grubb D G, Davis A F, Sands S C, Carnivale III M, Wartman J and Gallagher P, 2006a. Field evaluation of crushed glass-dredged material blends. Journal of Geotechnical and Geoenvironmental Engineering 132 (5), 577–590.

Grubb D G, Davis A F, Sands S C, Carnivale III M, Wartman J and Gallagher P, 2006b. Errata "Field evaluation of crushed glass-dredged material blends". Journal of Geotechnical and Geoenvironmental Engineering 133 (1), 127–128.

Grubb D G, Gallagher P M, Wartman J, Liu Y and Carnivale III M, 2006c. Laboratory evaluation of crushed glass-dredged material blends. Journal of Geotechnical and Geoenvironmental Engineering 562–576.

Grubb D G, Wartman J and Malasavage N E, 2008a. Aging of crushed glass-dredged material blend embankments. Journal of Geotechnical and Geoenvironmental Engineering 134 (11), 1676–1684.

Grubb D G, Cadden A W and Miller D D, 2008b. Crushed glass-dredged material (CG-DM) blends: role of organic matter content and DM variability on field compaction. Journal of Geotechnical and Geoenvironmental Engineering 134 (11), 1665–1675.

HDR Engineering Inc, 1997. Glass Cullet Utilization in Civil Engineering Applications. Technical Report. Nebraska State Recycling Association, Omaha, NE, USA. 30 pp.

IDT, 2012. Standard Specifications for Highway Construction – 2012. Idaho Transportation Department, Boise, ID, USA. 555 pp.

Imteaz M A, Ali M M Y and Arulrajah A, 2012. Possible environmental impacts of recycled glass used as a pavement base material. Waste Management and Research 30 (9), 917–921.

Jang H S, Jeon S H, So H S and So S Y, 2014. A study of the possibility of using TFT-LCD waste glass as an admixture fir steam-cured PHC piles. Magazine of Concrete Research 66 (4), 196–208.

Kim D, Quinlan M and Yen T F, 2009. Encapsulation of lead from hazardous CRT glass waste using biopolymer cross-linked concrete systems. Waste Management 29, 321–328.

Krivit D, 1999. Ramsey County Demonstrations of Recycled Glass as an Aggregate Supplement. Final report. Dan Krivit and Associates, St. Paul, MN, USA. 23 pp.

Lairaksa N, Moon A R and Makul N, 2013. Utilization of cathode ray tube waste: encapsulation of PbO-containing funnel glass in Portland cement clinker. Journal of Environmental Management 117, 180–186.

Larsen D A, 1989. Feasibility of Utilising Waste Glass in Pavements. Report No. 343-21-89-6. Connecticut Department of Transportation, Wethersfield, CT, USA. 27 pp.

Lee Jr. L T, 2007. Recycled Glass and Dredged Materials. Geotechnical and Structures Laboratory US Army Engineer Research and Development Center, MS, USA. ERDC TN-DOER-T8, 12 pp.

Lin K L, Wang N F, Shie J L, Lee T C and Lee C, 2008. Elucidating the hydration properties of paste containing thin film transistor liquid crystal display waste glass. Journal of Hazardous Materials 159, 471–475.

Lin K L, Huang W J, Shie J L, Lee T C, Wang K S and Lee C H, 2009. The utilization of thin film transistor liquid crystal display waste glass as a pozzolanic material. Journal of Hazardous Materials 163, 916–921.

Lin D F, Lin K L, Luo H L and Xu J Q, 2016. Acid-alkali resistance of new reclaimed tiles containing sewage sludge ash and waste glass. Materials 9 (7), 546–557.

Lin D F, Lin H L, Lin K L and Liu Z K, 2017. Effects of waste glass and foundry sand additions on reclaimed tiles containing sewage sludge ash. Environmental Technology 1–10.

Ling T C and Poon C S, 2011. Utilization of recycled glass derived from cathode ray tube glass as fine aggregate in cement mortar. Journal of Hazardous Materials 192 (2), 451–456.

Ling T C and Poon C S, 2012a. A comparative study on the feasible use of recycled beverage and CRT funnel glass as fine aggregate in cement mortar. Journal of Cleaner Production 29–30, 46–52.

Ling T C and Poon C S, 2012b. Feasible use of recycled CRT funnel glass as heavyweight fine aggregate in barite concrete. Journal of Cleaner Production 33, 42–49.

Ling T C and Poon C S, 2013. Effects of particle size of treated CRT funnel glass on properties of cement mortar. Materials and Structures 46 (1), 25–34.

Makowski C, Thomson G, Foye P and Higgins S, 2007. Broward county beach demonstration project: from beers to beaches. In: Kraus N C and Rosati J D (Eds.), Proceedings of Sixth International Symposium in Coastal Engineering and Science of Coastal Sediment Process, New Orleans, LA, USA, pp. 1–13.

Makowski C, Finkl C W and Rusenko K, 2011. Using recycled glass cullet for coastal protection: a review of geotechnical, biological and abiotic analyses. Journal of Coastal Research 64 (special issue), 1362–1366.

Makowski C, Finkl C W and Rusenko K, 2013. Suitability of recycled glass cullet as artificial dune fill along coastal environments. Journal of Coastal Research 29 (4), 772–782.

Malish W R, Day D E and Wisxon B G, 1975. Use of Domestic Waste Glass for Urban Paving. National Environmental Research Center. Report No. EPA-670/2-75-053. US EPA, Cincinnati, OH, USA. 60 pp.

Maschio S, Tonello G and Furlani E, 2013. Recycling glass cullet from waste CRTs for the production of high strength mortars. Journal of Waste Management 2013 8 pp.

MassDOT, 2015. Supplemental Specifications to the Standard Specifications for Highways and Bridges. Massachusetts Department of Transportation, Boston, MA, USA. 275 pp.

Maupin Jr. G W, 1997. Final Report: Glasphalt Test Section in Virginia. Report No. FHWA/ VTRC 98-R6. Virginia Transportation Research Council, Charlottesville, VA, USA. 10 pp.

Mirzahosseini M and Riding K A, 2014. Effect of curing temperature and glass type on the pozzolanic reactivity of glass powder. Cement and Concrete Research 58, 103–111.

MnDOT, 2016. Standard Specifications for Construction. Minnesota Department of Transportation, St. Paul, MN, USA. 695 pp.

Moncea A M, Badanoiu A, Georgescu M and Stoleriu S, 2013. Cementitious composites with glass waste from recycling of cathode ray tubes. Materials and Structures 46 (12), 2135–2144.

Moqsud M A and Hayashi S, 2007. Effectiveness of recycled waste glass as a contaminated tidal mud improvement material. Daffodil International University Journal of Science and Technology 2 (2), 53–55.

Morrison C, 2004. Reuse of CRT glass as aggregate in concrete. In: Limbachiya M C and Robert J J (Eds.), Proceedings of the International Conference on Sustainable Waste Management and Recycling: Glass Cullet, London, UK. Thomas Telford, London, UK, pp. 91–98.

MTO, 2013a. OPSS.MUNI 1010: Ontario Provincial Standard Specification. Material Specification for Aggregates – Base, Subbase, Select Subgrade and Backfill Material. Ministry of Transportation of Ontario (MTO), Ontario, Canada. 18 pp.

MTO, 2013b. OPSS.PROV 1010: Ontario Provincial Standard Specification. Material Specification for Aggregates – Base, Subbase, Select Subgrade and Backfill Material. Ministry of Transportation of Ontario (MTO), Ontario, Canada. 13 pp.

Nassar R and Soroushian P, 2011. Field investigation of concrete incorporating milled waste glass. Journal of Solid Waste Technology and Management 37 (4), 307–319.

NEN 7345, 1995. Leaching Characteristics of Solid Earthy and Stony Building and Waste Materials. Leaching Test. Determination of the Leaching of Inorganic Components from Building and Monolithic Waste Materials with the Diffusion Test. The Nederlands Normalisatie-Instituut, Delft, The Netherlands.

NHDOT, 2016. Standard Specifications for Road and Bridge Construction. New Hampshire Department of Transportation, Concord, NH, USA. 510 pp.

Nicholls J C and Lay J, 2002. Glassphalt. The Use of Recycled Glass in Asphalt by RMC Aggregates (UK) Ltd. Aggregate Information Service. Digest No. 72, 4 pp.

NJDOT, 2007. Standard Specifications for Road and Bridge Construction. New Jersey Department of Transportation, Trenton, NJ, USA. 590 pp.

Northwest EcoBuilding Guild, 2014a. Crushed Glass Structural Fill at West Bay Business Park. Available from: http://www.ecobuilding.org/.

Northwest EcoBuilding Guild, 2014b. Crushed Glass Fill at Little Rock Housing Project. Available from: http://www.ecobuilding.org/.

NSW EPA, 2014. The Recovered Glass Sand Exemption 2014. NSW Environment Protection Authority, Sydney, Australia. 4 pp.

NYCDOT, 2017. Standard Specifications (USC). New York City Department of Transportation, Albany, NY, USA. 1311 pp.

NZ Transport Agency, 2014. Specification for Dense Graded and Stone Mastic Asphalts. New Zealand Transport Agency, Wellington, New Zealand. SP/SM/010, 30 pp.

ODOT, 2015. Standard Specifications for Construction. Oregon Department of Transportation, Salem, OR, USA. 1158 pp.

Omran A and Tagnit-Hamou A, 2016. Performance of glass-powder concrete in field applications. Construction and Building Materials 109, 84–95.

Onitsuka K, Shen J, Hara Y and Sato M, 2001. Utilization of foaming waste glass as construction materials. In: Dhir R K, Limbachiya M C and Dyer T D (Eds.), Proceedings of the International Symposium on Recycling and Reuse of Glass Cullet, Dundee, Scotland, UK. Thomas Telford, London, UK, pp. 195–207.

PennDOT, 2001. Laboratory Evaluation of Select Engineering-Related Properties of Glass Cullet. Department of Transportation, Harrisburg, PA, USA. 21 pp.

PennDOT, 2004. Specifications. Pennsylvania Department of Transportation, Harrisburg, PA, USA. 1180 pp.

PennDOT, 2005. PennDOT District 5-0 SR 222 Highway Project. Pennsylvania Department of Transportation, Harrisburg, PA, USA. Strategic Recycling Program Fact Sheet, 1 pp.

PennDOT, 2006. PennDOT District 6-0, I-95 Tinicum Township Project. Pennsylvania Department of Transportation, Harrisburg, PA, USA. Strategic Recycling Program Fact Sheet, 1 pp.

PennDOT, 2016. Specifications. Pennsylvania Department of Transportation, Harrisburg, PA, USA. PUB 408/2016, 1204 pp.

Peyvandi A, Soroushian P and Nassar R, January 2013. Recycled glass concrete. Concrete International 29–32.

Polley C, Cramer S M and de la Cruz R V, 1998. Potential for using waste glass in Portland cement concrete. Journal of Materials in Civil Engineering 10 (4), 210–219.

Reindl J, 2003. Reuse/Recycling of Glass Cullet for Non-container Uses. Dane County Department of Public Works, Madison, WI, USA. 136 pp.

Remarque W, Heinz D and Schleusser C, 2001. Glass powder as a reactive addition for blast furnace cements. In: Dhir R K, Limbachiya M C and Dyer T D (Eds.), Proceedings of the International Symposium on Recycling and Reuse of Glass Cullet, Dundee, UK. Thomas Telford, London, UK, pp. 229–238.

RMS, 2011. QA Specification 3154: Granulated Glass Aggregate, first ed./Revision 0. Roads and Maritime Services, Sydney, New South Wales, Australia. 8 pp.

RMS, 2013a. QA Specification R116: Heavy Duty Dense Graded Asphalt, eighth ed./Revision 4. Roads and Maritime Services, Sydney, New South Wales, Australia. 43 pp.

RMS, 2013b. QA Specification R117: Light Duty Dense Graded Asphalt, first ed./Revision 1. Roads and Maritime Services, Sydney, New South Wales, Australia. 39 pp.

RMS, 2013c. QA Specification R118: Crumb Rubber Asphalt, first ed./Revision 0. Roads and Maritime Services, Sydney, New South Wales, Australia. 36 pp.

RMS, 2013d. QA Specification R121: Stone Mastic Asphalt, third ed./Revision 0. Roads and Maritime Services, Sydney, New South Wales, Australia. 34 pp.

RMS, 2014. QA Specification 3051: Granular Base and Subbase Materials for Surfaced Road Pavements, sixth ed./Revision 5. Roads and Maritime Services, New South Wales, Sydney, Australia. 26 pp.

Romero D, 2013. Concrete durability and Environmental Performance of Mixtures Containing Recycled Hazardous Waste Aggregates (Ph.D. thesis). University of Miami, Coral Gables, FL, USA. 207 pp.

Romero D, James J, Mora R and Hays C D, 2013. Study on the mechanical and environmental properties of concrete containing cathode ray tube glass aggregate. Waste Management 33, 1659–1666.

Sagoe-Crentsil K, Brown T and Taylor A, 2001. Guide Specification: Recycled Glass as Sand Replacement Premix Concrete. Commonwealth Scientific and Industrial Research Organization, Victoria, Australia. Confidential DBCE Doc. 01/232, 44 pp.

SCDOT, 2007. Standard Specifications for Highway. South Carolina Department of Transportation, Columbia, SC, USA. 1080 pp.

Senadheera S, Rana A and Nash P T, 2005. Characterization of the Behaviour of Granular Material Containing Glass Cullet. Available from: http://www.vegvesen.no.

SHA, 2008. Standard Specifications for Construction and Materials. Maryland State Highway Administration, Baltimore, MD, USA. 857 pp.

Shayan A and Xu A, 2006. Performance of glass powder as a pozzolanic material in concrete: a field trial on concrete slabs. Cement and Concrete Research 36, 457–468.

Siddiki N Z, Kim D and Salgado R, 2004. Use of recycled and waste materials in Indiana. Transportation Research Record: Journal of the Transportation Research Board 1874, 78–85.

Soil & Environmental Engineers Inc and Re-sourcing Associates Inc, 1998. A Tool Kit for the Use of Post-consumer Glass as a Construction Aggregate. Report No. GL-97-5. Clean Washington Centre, Seattle, Washington, USA. 92 pp.

Song T, Dumitru I and Bornstein B, 2016. Long term durability properties of concretes with glass sand and glass powder. In: Fourth International Conference on Sustainable Construction Materials and Technologies, Las Vegas, NV, USA 9 pp.

Soroushian P, 2012. Towards Broad Use of Recycled Glass Concrete on MSU Campus. Michigan State University, East Lansing, MI, USA. 49 pp.

Stevens M, 2007. Materials Recycling and Reuse – Ffinding Opportunities in Colorado Highways. Report No. CDOT-2007-12. Colorado Department of Transportation Research Branch, Denver, CO, USA. 42 pp.

Stroup-Gardiner M and Wattenberg-Komas T, 2013. Recycled materials and byproducts in highway. Manufacturing and Construction Byproducts, vol. 8. Transportation Research Board, Washington DC, USA, pp. 66–83. National Cooperative Highway Research Program Synthesis 435.

Su N and Chen J S, 2002. Engineering properties of asphalt concrete made with recycled glass. Resources, Conservation and Recycling 35, 259–274.

Sua-iam G and Makul N, 2013. Use of limestone powder during incorporation of Pb-containing cathode ray tube waste in self-compacting concrete. Journal of Environmental Management 128, 931–940.

Tagnit-Hamou A and Bengougam A, March 2012. The use of glass powder as supplementary cementitious material. Concrete International 56–61.

Tagnit-Hamou A, Soliman N, Omran A F, Mousa M T, Gauvreau N and Provencher M, March 2015. Novel ultra-high-performance glass concrete. Concrete International 41–47.

TMR, 2016a. Transport and Main Roads Specifications: MRTS35 Recycled Materials for Pavements. Department of Transport and Main Roads, Brisbane, Queensland, Australia. 33 pp.

TMR, 2016b. Transport and Main Roads Specifications: MRTS05 Unbound Pavements. Department of Transport and Main Roads, Brisbane, Queensland, Australia. 20 pp.

Transit N Z, 2006a. TNZ M/04: Specification for Basecourse Aggregate. New Zealand Transport Agency, Wellington, New Zealand. 15 pp.

Transit N Z, 2006b. TNZ M/04 Notes: Notes to the Specification for Basecourse Aggregate. New Zealand Transport Agency, Wellington, New Zealand. 22 pp.

TxDOT, 1999. Year of the Recycled Roadway Materials: February – Glass Cullet. Texas Department of Transportation, Austin, TX, USA. 29 pp.

US EPA, 1992. Method 1311: Toxicity Characteristic Leaching Procedure (TCLP). US Environmental Protection Agency, Washington DC, USA. 35 pp.

US EPA, 1994. Method 1312: Synthetic Precipitation Leaching Procedure (SPLP). US Environmental Protection Agency, Washington DC, USA. 30 pp.

US EPA, 2016. Cathode Ray Tubes (CRTs). Available from: http://www.epa.gov.

VDOT, 2016. Road and Bridge Specifications. Virginia Department of Transportation. 1091 pp.

VicRoads, 2011. Use of Recycled Materials for Road Construction. VicRoads, Melbourne, Victoria, Australia. Technical Note 107, 4 pp.

VicRoads, 2017. 400 series – Asphalt and Surface Treatments Section 407 – Hot Mix Asphalt. VicRoads, Melbourne, Victoria, Australia. 21 pp.

Wang H Y and Hou T C, 2011. A study of elevated temperatures on the strength properties of LCD glass powder cement mortars. Integrated Waste Management 1, 391–404.

Wang H Y, Zeng H H and Wu J Y, 2014. A study on the marco and micro properties of concrete with LCD glass. Construction and Building Materials 50, 664–670.

Wartman J, Grubb D G and Nasim A S M, 2004. Select engineering characteristics of crushed glass. Journal of Materials in Civil Engineering 16 (6), 526–539.

Wisconsin Transportation Information Center, 1999. Using Recovered Materials in Highway Construction. Wisconsin Transportation Bulletin No. 20, 8 pp.

Wood M, Jeffreys C and Richardon P, 2011. Waverley Council: Recycled Glass in Roads. Department of Sustainability, Environment, Water, Population and Communities, National Waste Policy, Australia. 2 pp.

WSDOT, 2016. 2016 Standard Specifications for Road, Bridge, and Municipal Construction (Amended). Washington Department of Transportation, Olympia, WS, USA. Publication No. M41-10, 1000 pp.

Zhao H, Poon C S and Ling T C, 2013. Properties of mortar prepared with recycled cathode ray tube funnel glass sand at different mineral admixture. Construction and Building Materials 40, 951–960.

Epilogue

11

The work presented in this book is different from the norm, and its uniqueness lies, as mentioned in the three previously published books (Dhir et al., 2016a,b, 2017), in the adoption of the newly developed methodology of the authors, namely, the *analytical systemisation method*. This method was used for examining effectively the large volume of globally sourced material on the subject of glass cullet (GC) and its production, characteristics and use, in the form of both cement and fine aggregate, including filler aggregate, in various aspects of construction. Subsequently this information was packaged in a manner that is easy to understand, disseminate and put into practice, so that GC can be used as a sustainable construction material.

The main purpose of this book is to encourage the added-value applications of GC in construction and to minimise repetitive research, which often confuses the issues, retards progress towards sustainability and wastes valuable resources. The work is based on a data matrix built from 751 globally sourced publications, originating from 51 countries, with major contributions from the United States, followed by the United Kingdom. The use of GC in cement, concrete, geotechnics, road pavements, ceramics and other applications (e.g., as a filtration medium and blasting abrasive) is discussed. The associated environmental issues, relevant case studies, as well as standards and specifications are also covered.

Overall, the volume of literature published and the data sourced for different applications of GC showed that the highest research activity has been focused in the area of concrete, on its use in both a finely ground form as a component of cement and a granular form as fine aggregate and filler aggregate, to which the chemical characteristics (being pozzolanic) and physical characteristics (being hard and with exceptionally low water absorption) of the material are the best suited.

Amongst the issues that caused concern when dealing with sourced data, and one that happens to be critical, is the reliability of the information provided in some of the studies and the extent of the variability and inconsistencies in the obtained results, though this is to be expected when working with a large database populated by globally sourced literature. This applied to the materials used in different studies and their characteristics, as well as the mix designs, test conditions, test methods and laboratory control. All these made the analysis and evaluation of results difficult, very time consuming and at times problematic. Lack of information on the methodologies adopted in some cases did not help. Reference to foreign standards was not possible at times and this led to some uncertainty. Although in most cases, such difficulties were resolved based on the authors' experience, there were occasions when specific

Sustainable Construction Materials: Glass Cullet. https://doi.org/10.1016/B978-0-08-100984-0.00011-4

sets of results could not be comprehended to the extent that the data had to be ignored. One of the main points to emerge from this work is that for the research in the area of construction to be more effective, it needs to go through a process of validation and coordination.

Of all the types of GC used in research, soda-lime GC, which is processed from container glass and flat glass, is much more suited to be explored for its potential use in various construction applications, because it is available in large quantities and the material is chemically safe. Although other types of glass, such as cathode ray tubes, may also offer viable options for use, their leaching behaviour and environmental assessment need further vigorous investigations.

Owing to its amorphous and silica-rich nature, GC offers an attractive option for use as a raw feed for the manufacture of Portland clinker or, in a ground form, as a constituent of cement. The cementitious characteristics of ground glass cullet (GGC) have been shown to meet the requirements specified for fly ash in the standards, such as BS EN 450-1 (2012). The use of GGC as a replacement for Portland cement (PC), at a content less than 20%, such as in CEM II/A cements in BS EN 197-1 (2011), can perform satisfactorily in most concrete applications. It would also be preferable to grind GGC finer than PC where most of the improvements in the hardened properties of concrete, particularly compressive strength and chemical durability, are specified as a requirement.

For GC use as sand in concrete, although the material can be crushed and sieved to specified particle size and distribution requirements, conforming to relevant standards such as BS EN 12620:2002+A1 (2008), the grading of GC-sand used in the various studies has been found to be coarser than that of natural sand. The grading of GC-sand needs to be considered when using it as a natural sand replacement, as it can give rise to different particle packing and void size distribution, which in turn affect the properties of the concrete. As GC-sand is harder, and has lower water absorption, than natural sand, its use as a natural sand replacement leads to improvement in the deformation and permeation properties of concrete. There has always been concern over the detrimental expansion of concrete due to alkali–silica reaction (ASR) when GC-sand is used. However, it has been shown that this ASR expansion can be nullified with the use of pozzolanic cements such as fly ash, ground granulated blast furnace slag and metakaolin, as well as finely ground GC itself. As the construction industry is rapidly moving towards the use of CEM II and CEM III cements, this lends itself to using GC-sand without the ASR concern.

The material is also suitable for use as a filler aggregate of size less than 65 μm, as per BS EN 12620:2002+A1 (2008), in concrete with various consistencies covering normal concrete mixes to self-compacting concrete. Although the relevant research work data are limited, it has been shown that the effect of GC filler in most cases has been positive.

In geotechnical applications, GC has been tested for its potential for use as a medium gravel (with a maximum size of 20 mm), fine gravel (6.3 mm) or coarse sand (2 mm). It should be noted here that using GC at coarser size fractions can be problematic, as the material tends to be flaky and elongated, which can give rise to aggregate packing problems, affecting the geotechnical engineering properties. Notwithstanding this, GC-sand and GC-gravel appear to show similar or slightly better performance compared with their natural counterparts in terms of compaction characteristics, stress–strain behaviour, shear strength, permeability and thermal conductivity.

In the area of road pavements, the performance of GC as a partial replacement for natural sand in unbound and hydraulically bound applications is generally considered satisfactory. However, in the bituminous bound form, the addition of antistripping agent is needed to enhance the resistance to rutting and moisture damage of mixtures containing GC.

The use of GC as part of the raw materials or a fluxing agent in ceramics applications, such as bricks and tiles, could also be beneficial as it allows one to achieve desirable engineering properties at a lower firing temperature. Its use in other alternative markets, including aesthetic finishing material, filtration medium, blast abrasive, epoxy composite and paint filler, has been explored, with performance generally noted as appealing.

GC derived from soda-lime glass is a safe material and not likely to present any threat to the environment provided that it is not contaminated. Despite many promising laboratory findings, the progress on the use of GC in the real construction environment has been slow. However, some successful case studies have been reported for its use in concrete applications in the United Kingdom, and its use in both geotechnical and road applications in the United States.

At present, both the American and the European standards have no provision for the use of GGC as a cement component and GC as an aggregate component in concrete. On the other hand, many national standards and specifications from Australia, Canada, New Zealand, the United Kingdom and the United States permit the use of GC as a granular material in geotechnical and road pavement applications.

References

BS EN 197-1, 2011. Cement. Part 1: Composition, Specifications and Conformity Criteria for Common Cements. BSI, London, UK.

BS EN 450-1, 2012. Fly Ash for Concrete. Part 1: Definition, Specifications and Conformity Criteria.

BS EN 12620:2002+A1, 2008. Aggregates for Concrete. BSI, London, UK.

Dhir R K, de Brito J, Mangabhai R and Lye C Q, 2016a. Sustainable Construction Materials: Copper Slag. Woodhead Publishing, Elsevier, Cambridge. 322 pp.

Dhir R K, Ghataora G S and Lynn C J, 2016b. Sustainable Construction Materials: Sewage Sludge Ash. Woodhead Publishing, Elsevier, Cambridge. 288 pp.

Dhir R K, de Brito J, Lynn C J and Silva R V, 2017. Sustainable Construction Materials: Municipal Incinerated Bottom Ash. Woodhead Publishing, Elsevier, Cambridge. 458 pp.

Index

'*Note*: Page numbers followed by "f" indicate figures, "t" indicate tables.'

A

AAS. *See* Atomic absorption spectroscopy
AASHTO. *See* American Association of State Highway and Transportation Officials
Abrasion. *See also* Los Angeles abrasion test
  in GC filler, 246, 251
  in GC-sand, 169, 209–211, 210f, 210t
  in GGC, 146–148, 152
  in PC, 148
  in RCA, 148
  of tiles, 355, 361, 362f
  water/cement ratio and, 148, 209
Absorption. *See* Water absorption
Accelerated carbonation test, 247
ACI 318-11, 186
Acid attack
  in concrete, 138, 139t
  in GC filler, 248
  in GC-sand, 169, 199–200, 200f
  in GGC, 138, 139t, 153–154
  in SCC, 248
  water/cement ratio and, 200
Adhesion, of bituminous mixtures, 300, 314, 319
AERCO, 374
Aesthetic finishing materials
  of GC colour, 371
  GC filler as, 371
  GC for, 371
Air content
  of concrete, 108–109
  of FA, 109
  of GC filler, 235
  of GC-sand, 176, 177f
  of GGC, 108–109
  of mortar, 177f
  of PC, 108
Air voids
  in bituminous mixtures, 307f, 310
  in cement, 129
Air-entrained concrete, 146
  freeze-thaw attack on, 146, 209
  GC filler in, 235
  GC-sand in, 176
  GGC in, 146
Alam, M. S., 131
El-Alami, I., 100
Ali, M. M., 23t, 273
Alkali
  in concrete, 60
  in soda lime glass, 60
  standards for, 413
Alkali-silica reaction (ASR), 60
  colour and, 203t, 205
  in concrete, 62, 141–145, 142t
  FA and, 62, 215, 299–300
  of GC, 401–402
  GC colour and, 63–65
  of GC filler, 246, 248–250, 249f, 249t
  of GC fine aggregate, 77, 404
  of GC-sand, 169–170, 202–205, 202t–203t, 206f–207f, 215
  of GGBS, 62, 142, 169, 299–300
  of GGC, 98–99, 141–145, 142t, 152–154, 169
  of hydraulically bound mixtures, 299–300, 319
  of mortar, 205
  of PC, 77
  pozzolanic reactions and, 169, 204–205, 401–402
  silica fume and, 169
  of soda lime glass, 142, 202–205
Aluminoborosilicate glass, 44, 56
Aluminosilicate glass, 56
  calcium oxide in, 58
  chemical composition of, 77
  compressive strength of, 114
  in GGC, 101f, 102
  initial and final setting times for, 101, 101f
  modulus of elasticity and hardness of, 75, 76t
  RCPT for, 198f

Aluminosilicate glass *(Continued)*
specific gravity of, 72, 74f, 175
water absorption of, 75t
Aluminum, in glaze, 369
Aluminum nitride, 335
Aluminum oxide
in GC filler, 236–238
in glass, 58, 59f, 77
for glass-ceramics, 331, 334
in soda lime glass, 53
Amber colour, 63–65, 63t–64t, 77, 111, 201–202
American Association of State Highway and Transportation Officials (AASHTO), 416–417
M145-91, 260
T294, 278
American Society for Testing and Materials (ASTM), 415
1567, 205
C33, 415
C78, 122
C150, 61, 409
C187, 99
C191, 101
C227, 205
C267-01, 200
C293, 122
C311, 103, 104f
C324, 342, 355
C373, 346–348
C618, 56, 60–61, 71–72, 77–78, 103, 410
C642, 129–130
C666, 146, 147t, 210t
C674, 343
C944, 148, 209
C989, 410
C1012, 200
C1202, 135, 197
C1260, 62, 142, 145, 169, 202, 202t–203t, 205, 206f
C1293, 203t
C1293-8, 205
C1567, 62, 142
C1709, 410, 410t–413t
D1073, 68, 68f
D2166, 274
D2434-68, 279
D2487, 67, 67f
D2850, 274
D3080, 274
D4972, 267
D5918, 282
D6927, 304
on GC fine aggregate, 65
Analytical systemisation, 2–3, 16, 435
Anderson, M., 364
Andreola, F., 366–369
Angle of friction, 276
direct shear test for, 274
dry density and, 274–275, 275f
of GC, 274, 274f–275f, 276, 277f, 290
of soils, 274
triaxial test for, 274
Angle of internal friction, 259
shear strength and, 273
Angularity
of GC, 63
of GC-sand, 214
of GGC, 100
of hydraulically bound mixtures, 299–300
Antistripping agents, for bituminous mixtures, 313t, 314, 315t–316t, 316–317
Army, US, 272
Arsenic
from GC, 317
in TFT-LCD, 148, 149t
Arulrajah, A., 23t, 24, 273, 317
ASLP. *See* Australia Standard Leaching Protocol
Asphalt. *See also specific topics*
GC in, 51
PSD for, 70
Asphalt Institute, 310
ASR. *See* Alkali-silica reaction
ASTM. *See* American Society for Testing and Materials
Atomic absorption spectroscopy (AAS), 393–394
Atterberg limits, for GC, 267–268
Australia
barriers to GC recycling in, 41–44
EPA Victoria in, 317, 320, 397–398
GC publications in, 19
geotechnical applications from, 258, 397, 418t
GGC in, 149–152
standards of, 415
Australia Standard Leaching Protocol (ASLP), 397–400

Autogenous shrinkage
of concrete, 126–127, 127t
of GC-sand, 188
of GGC, 126–127, 127t

B

Bailey and Associates, 375
Barbieri, L., 23t, 24
Barium leaching, 211, 215, 284, 317, 394
Barriers, to GC recycling, 38–40
Basalt, 241–242, 338
Bentonite, 262
Benzene, 284
Benzo(a)pyrene, 284
Bergmann, C. P., 368
Bernardo, E., 23t, 24, 329, 334
Bigot curve, 343
Biological oxygen demand (BOD), 399
Bituminous mixtures
adhesion of, 300, 314, 319
air voids in, 307f, 310
antistripping agents for, 313t, 314, 315t–316t, 316–317
case studies for, 408–409
dynamic modulus of, 314
environmental impact of, 317
with GC, 298, 299t, 300, 304–317, 307f–309f
Marshall flow of, 306f, 310
Marshall mix design for, 304–311
Marshall stability of, 305f, 310, 314, 320
moisture damage resistance of, 314–317, 315t–316t
PSD of, 310
rutting resistance of, 312–314, 313t
stiffness modulus for, 311–312, 312t
stripping in, 300, 314–316
tensile strength of, 314
VFB in, 304, 309f, 311
VMA in, 304, 308f, 310
Blaine test, for GGC, 72, 73f, 77–78, 107, 394
Blast abrasive
copper and, 375
GC as, 374–375
Bleeding
of GC fillers, 232, 234
in GC-sand, 173
in GGC, 107, 129
permeation and, 129
plastic shrinkage and, 235–236
Blue colour, 56, 63t–64t, 71–72
BOD. *See* Biological oxygen demand
Book
data matrix for, 17
layout of, 9–10
literature search and appraisal for, 17–27
methodology for, 15–34, 16f
Boric oxide, 56
Borosilicate glass, 56
calcium oxide in, 58
chemical composition of, 77
compressive strength of, 114
foam glass-ceramics from, 336t
in GC-glass, 202
glass-ceramics of, 334
for glaze, 370
modulus of elasticity and hardness of, 75, 76t
Bove, J., 372
Bragança, S. R., 368
BRE. *See* Building Research Establishment
Bricks
compaction of, 346
compressive strength of, 352–353, 353f–354f, 377
from CRTs, 342–344, 346–348, 351–352, 354
deformation of, 344–346
density of, 350–352, 351f
drying sensitivity index for, 343
drying shrinkage of, 343–344, 344f, 377
as fired clay bodies, 346–352
firing shrinkage of, 344–346, 345f
flexural strength of, 343, 354, 355t, 377
fluxing agent for, 345–346
GC in, 342–354
hygroscopicity of, 343
from LCDs, 351–352
mechanical properties of, 352–354
physical properties of, 342–343, 342t
PI for, 343
porosity of, 346–349, 347t
shrinkage of, 343–344
sintering process for, 377
temperature rise in, 353, 354f
as unfired clay bodies, 342–343, 342t
water absorption of, 346–349, 348f–351f
water for, 342
Bring-sites, for waste glass collection, 45t
Brundtland report, of United Nations, 3

Brusatin, G., 338
BS
812-123, 62, 142, 145, 202, 202t, 205, 248, 404
812-124, 282
882, 171–173
1377-3, 267
1377-7, 274
1377-8, 274
1881-102, 173
1881-208, 193
3921, 352
5930, 67–68, 67f
6717, 209
BS EN
196-1, 103, 104f
196-2, 60
196-3:2005+A1, 99, 101
197-1, 61, 120–121, 205, 436
206, 134
206-1, 153
450-1, 56, 60, 71–72, 77–78, 102–103, 411–413, 436
771-1, 349, 352
1304, 364
1990:2002+A1, 134
12350-11, 234
12390-5, 122
12457 Part 1-4, 211, 391
12620:2002+A1, 69–70, 70f, 168, 171–173, 214, 232, 436
12697-34, 304
13108-1, 68, 68f
13242:2002+A1, 266
ISO 14688-1, 264
Building Research Establishment (BRE)
Digest 330, 248
Digest 363, 240
Digest 405, 197
Bulk density. *See* Density
Byars, E., 23t, 100, 103, 404, 411–413, 415–416

C

Cadmium, 43, 56, 148, 149t, 317
Caki, M., 370
Calcium carbonate
carbonation and, 135
for foam glass-ceramics, 335
Calcium fluoride, 334
Calcium hydroxide
acid attack and, 199
carbonation and, 135, 197, 199
GC filler and, 247
in GGC, 98–99
PC and, 56
pozzolanic reactions with, 120–121
sulphate and, 140
Calcium oxide
in GC filler, 236–238
in glass, 58, 59f
in soda lime glass, 53–56
California bearing ratio (CBR)
for GC, 259, 272, 272f, 320
MDD and, 272, 272f
modified compaction test for, 272
for resilient modulus, 278, 279f
standard compaction test for, 272
for unbound mixtures, 300–301, 301f, 301t
Canada
GC publications in, 19
GGC in, 149–152
standards of, 415
Carbon dioxide
from bricks, 351
carbonation and, 135, 197
from cement production, 98
closed-loop recycling and, 40
CRTs and, 37–38
GC and, 4, 36
in GC-sand, 169
in GGC, 152
LOI and, 61–62
from PC clinker, 98
sustainable construction materials and, 2–4
Carbon monoxide, 152
Carbonation
calcium hydroxide and, 135, 197, 199
carbon dioxide and, 135, 197
in concrete, 135–138, 138t
in GC filler, 246–247, 247t
in GC-sand, 188, 197–199, 199t, 215
in GGC, 135–138, 138t
pH and, 135, 197
pozzolanic reactions and, 247
water/cement ratio and, 199
Carboniferous shale (CS), 345–346, 353

Case studies
for bituminous mixtures, 408–409
for GC, 284–286, 284t–285t, 400–409, 402t–403t
for GC filler, 251
for GC-sand, 213–214
for geotechnical applications, 404–406, 405t
for GGC, 149–152, 150t–151t
for hydraulically bound mixtures, 408
for road pavement, 318–319, 318t–319t, 406–409, 406t–407t
for unbound mixtures, 407–408
Cathode ray tubes (CRTs)
bricks from, 342–344, 346–348, 351–352, 354
carbon dioxide and, 37–38
collection of, 45t
FM from, 211
for foam glass-ceramics, 339
foam glass-ceramics from, 335–339, 336t, 342
GC-sand from, 176, 211
GGC from, 394
for glass fibre, 373
for glaze, 369–370
heavy metal leaching from, 391
leaching from, 366
lead in, 391
lead leaching from, 394, 395t
mercury in, 391
mortar from, 395t
porcelain from, 366–368
thermal conductivity of, 342
tiles from, 355, 358f, 364, 366
waste glass from, 43, 43t, 76
CBR. *See* California bearing ratio
CCD. *See* Charge-coupled device
CEM I 42.5, 238
Cement. *See also* Concrete; Portland cement; *specific topics*
air voids in, 129
carbon dioxide from, 98
GGC fineness and, 71–72
GGC in, 394, 410
LOI for, 61–62
setting times for, 101–102, 101f
Cement and Concrete Composites, 21
Cement clinker
GC in, 393, 400–401
GGC in, 396f
lead leaching from, 149
CEN. *See* Comité Européen de Normalisation
Ceramics. *See also* Bricks; Foam glass-ceramics; Glass-ceramics; Glaze; Porcelain; Tiles
GC in, 327–388, 437
Charge-coupled device (CCD), for waste glass separation, 49–50
Chemical attack
on GC filler, 246
on GC-sand, 196–205
Chemical properties, of GC, 53–62
Chen, C. H., 197
Chen, D., 50
Chesner, W. H., 283
China
cement production in, 4
GC publications in, 19
Chinese National Standard (CNS)
382-R2002, 344–345, 352
1127-R3042, 348
3763, 195
Chloride
in GC filler, 246
in GC-sand, 197
pozzolanic reactions and, 197
standards for, 413
Chloride migration coefficient
in concrete, 134–135, 137t
durability and, 134–135, 137t
for FA, 135, 137t
in GC filler, 246, 246t
in GC-sand, 169
in GGC, 134–135, 137t, 152
in SCC, 246
Chromium, 148, 149t, 284, 397–398
Clay bodies. *See* Bricks
Clean Washington Center (CWC), 18, 50, 77
on blast abrasive, 374
Clear colour. *See* Flint colour
Closed-loop recycling, 6–7, 40, 41f
CNS. *See* Chinese National Standard
Cobalt, 56
Coconut oil ethanolamine, 314–316
Code of Federal Regulations, US, 391, 394, 397, 424

Coefficient of curvature
compaction and, 264
for GC, 260, 262, 264f
PSD and, 260, 262, 264f
Coefficient of uniformity, 258
compaction and, 264
for GC, 67, 264
PSD and, 67, 260, 263f
Cohesion
of bituminous mixtures, 314
of concrete, 173, 175, 175t
GC and, 276, 276f
PI and, 276
shear strength and, 273
Collection schemes, for waste glass, 44, 45t
Collins, R. J., 415–416
Colour. *See also* Glass cullet colour; *specific colours*
ASR and, 203t, 205
contamination of, 8
of GC-sand, 201–204, 203t, 215
of GGC, 145
of glass, 56
of soda lime glass, 145
Comité Européen de Normalisation (CEN)
CEN/TC 51-draft, 210t
CEN/TS 51-draft, 146, 147t
on GC fine aggregate, 65
Compaction, 259. *See also* Modified compaction test; Standard compaction test
of bricks, 346
from dynamic loading, 282
of GC, 269–270
particle breakage from, 264
PSD and, 262–264
of soil, 269–270
Compressive strength
with age, 120–121, 120f
of bricks, 352–353, 353f–354f, 377
of concrete, 114–121, 115f–116f, 120f
elastic modulus and, 125, 126f, 186, 241–242
flexural strength and, 184
of foam glass-ceramics, 335, 340–341, 340t
of GC filler, 236–242, 237f, 245f
of GC-sand, 169, 177–181, 178f–180f, 186, 211, 214
of GGC, 114–121, 115f–116f, 120f, 140–141, 152–153
of glass-ceramics, 333–334, 333t
of PC, 114, 152
of SCC, 238
sulphate and, 140–141
sulphate attack and, 201
temperature rise and, 211
of tiles, 363–364, 364f, 377
water/cement ratio and, 117, 119f, 177, 236–238, 239f
Concrete, 168. *See also specific topics*
acid attack in, 138, 139t
air content in, 108–109
alkali in, 60
annual global production of, 168
ASR of, 62, 141–145, 142t
autogenous shrinkage of, 126–127, 127t
carbonation in, 135–138, 138t
chloride migration coefficient in, 134–135, 137t
cohesiveness of, 173, 175, 175t
compressive strength of, 114–121, 115f–116f, 120f
consistence of, 105–107
creep of, 126
cumulative heat in, 111
density of, 107–108, 109f, 234–235
diffusion of, 133–134, 133t
drying shrinkage of, 127–128, 128f
durability of, 134–148
elastic modulus of, 124–125, 126f
environmental impact of, 392–396
FA in, 103–105
flexural strength of, 122–124, 124f
freeze-thaw attack on, 146
GC colour and, 65
GC fillers in, 232–236
GC fine aggregate in, 65–66, 394–396, 401–404, 415–416
GC in, 51, 392–396, 402t–403t, 436
GC-sand in, 29, 169–176, 214
GGC in, 97–166, 401, 410t–411t, 414t
heat of hydration for, 110–111, 112t–113t
hydration reactions in, 72, 110–111
initial and final setting times for, 109–110, 110t

mechanical properties of, 111–128
moisture diffusion coefficient for, 133–134, 133t
permeability of, 131–133, 132t
permeation in, 128–134, 130t
plastic shrinkage of, 235–236, 235t
porosity of, 129
properties of, 103–111
PSD in, 65–66
rate of heat evolution for, 111
recycling of, 2
setting times for, 416
splitting tensile strength of, 121–122
stability of, 107, 108t, 173, 234
sulphate attack in, 139–141, 140t
sulphate in, 60
tensile strength of, 121–124
water absorption of, 129–131, 131f
water/cement ratio for, 117, 119f, 394
Concrete Technology Unit, of University of Dundee, 171–173
Conferences, 2
Confining pressure, 277, 278f, 290
Conradt, R., 50
Consistence (workability)
of concrete, 105–107
FA and, 100
flakiness index and, 265
of GC filler, 232–234, 233t, 238
of GC-sand, 170–173, 181, 209
of GGC, 99–100, 100f
of glass fibres, 373
of hydraulically bound mixtures, 299–300
of SCC, 238
water for, 99–100, 100f
Construction and Building Materials, 21
Container glass
contaminants in, 397, 398t
for foam glass-ceramics, 336t
GC from, 62
for road pavement, 423
for soda lime glass, 76, 436
for tiles, 365f
waste glass from, 42, 43t
Contaminants
of colour, 8
in container glass, 397, 398t
in CRTs and LCDs, 43
in GC, 8, 39–41, 76–77, 284, 397–399, 398t
in GC fine aggregate, 416
LOI and, 61–62
in open-loop recycling, 41
removal of, 48–50, 76–77
in waste glass, 49–50
water absorption and, 266
Copper
from GC, 317, 398
in TFT-LCD, 148, 149t
Copper slag (CS)
blast abrasive and, 375
GC-sand and, 182, 186
Cosentino, P. J., 399
Coventry, S., 373
Creep
of concrete, 126
elastic modulus and, 188
of GC filler, 242
of GC-sand, 186–188, 214
of GGC, 126
CRT funnel glass, lead leaching from, 149, 394
CRTs. *See* Cathode ray tubes
Crushers, for waste glass, 48, 49t
CS. *See* Carboniferous shale; Copper slag
Cube strength
elastic modulus and, 186
flexural strength and, 184, 185f
of GC filler, 242f
of GC-sand, 182, 183f, 184, 185f, 186
tensile strength and, 182, 183f
Cui, Y., 100
Cumulative heat
in concrete, 111
in GGC, 111
CWC. *See* Clean Washington Center

D

Dal Bo, M., 370
Dames and Moore, Inc., 273, 399
Data matrix
for book, 17
data mining and parking for, 27, 30f–31f
for GC publications, 27
initial sorting of literature for, 27, 28f–29f
Data mining, 27, 30f–31f
De Brito, J., 197, 199

De Castro, S., 197, 199
Deformation
of bricks, 344–346
of GC filler, 241–243
of GC-sand, 169, 185–190, 214
Marshall flow and, 306f, 310
rutting resistance and, 313
Density. *See also* Maximum dry density
of bricks, 350–352, 351f
of concrete, 107–108, 109f, 234–235
dry, 274–275, 275f
of foam glass-ceramics, 335–338, 337t
of GC filler, 234–235
of GC-sand, 175–176
of GGC, 107–108, 109f
of glass-ceramics, 329, 331t
of porcelain, 368–369
specific gravity and, 175
of tiles, 359, 360f
Depassivation, of steel, 197, 247
Design codes, 2
Design Manual for Roads and Bridges, 301
Dhir, R. K., 23t, 24, 98–99, 145, 148, 173, 184, 186, 190, 193, 197, 209, 238, 251, 411–413, 416
Diffusion
of concrete, 133–134, 133t
of GC-sand, 190, 195
of GGC, 133–134, 133t
Dima, V., 371
DIN 52108, 209
Direct shear test, 274
Disfani, M. M., 23t, 24, 397–398
Dolomite, 238
Dondi, M., 343, 346–348, 354, 356, 359, 364
Dry density. *See also* Maximum dry density
angle of friction and, 274–275, 275f
Drying sensitivity index, for bricks, 343
Drying shrinkage
of bricks, 343–344, 344f, 377
of concrete, 127–128, 128f
of GC filler, 242–243
of GC-sand, 188
of GGC, 127–128, 128f
of tiles, 356
Du, H. J., 111, 132, 200–201
Dumitru, I., 60, 127–128, 197, 209, 214
Durability
chloride migration coefficient and, 134–135, 137t
of concrete, 134–148
of GC fillers, 232, 246–251
of GC-sand, 169–170, 196–211
of GGC, 134–148
Los Angeles abrasion test for, 266
permeation and, 190
of soils, 281–282
Dyer, T. D., 23t, 24, 98–99, 145
Dynamic loading
compaction from, 282
of GC, 282
of soils, 282
Dynamic modulus, of bituminous mixtures, 314

E

E-fibreglass, 371
Elastic modulus
compressive strength and, 125, 126f, 186, 241–242
of concrete, 124–125, 126f
creep and, 188
cube strength and, 186
of GC filler, 241–242, 242f
of GC-sand, 186, 187f, 211, 214
of GGC, 124–125, 126f, 153
of glass fibre, 374
Marshall flow and, 306f, 310
temperature rise and, 211
water/cement ratio and, 186, 241
Elastomeric roof coating, GC in, 375–376
EM. *See* Etruria marl
EN. *See also* BS EN
197-1, 409
772-7, 349
1015-21, 195
12620, 415–416
Environmental impact
of bituminous mixtures, 317
of concrete, 392–396
of CRT funnel glass, 424
of GC, 283–284, 283t, 317, 390–400
of GC filler, 251
of GC-sand, 211–213
of GGC, 148–149, 149t
of glass-ceramics, 328

of hydraulically bound mixtures, 317
of road pavement, 317, 320, 399–400
of soda-lime glass, 317, 437
Environmental Protection Agency (EPA), US, 148, 149t, 391
on GC environmental impact, 284
on heavy metal leaching, 211
Environmental Protection Agency Victoria (EPA Victoria), in Australia, 317, 320, 397–398
EPA. *See also* Environmental Protection Agency
EPA Victoria. *See* Environmental Protection Agency Victoria
Epoxy, 373
Etruria marl (EM), 345–346, 353
EU Packaging and Packaging Waste Directive, 7–8
Eurocode 2, 2, 122, 125, 182
on compressive strength-elastic modulus relationship, 241–242
on elastic modulus, 186
on shrinkage, 190
on tensile strength, 182–183
Eurocode 7, 260
European Standard, 415. *See also* EN
European Union, 36, 37f
Eurostat, 36

F

FA. *See* Fly ash
Feldspar, 368–370
Ferrous metal. *See also* Iron
screening and removal of, 49
Ferrous oxide. *See* Iron oxide
Fibreglass, 371
waste glass from, 43t, 44
Fibre-reinforced concrete, GC-sand for, 211
Fillers. *See also* Glass cullet filler
for paint, 375
Filtration
GC for, 371–373
for micro-irrigation, 372
of water, 41, 372
Fineness
of GC-sand, 193
of GGC, 71–72, 414
Fineness modulus (FM)
from CRTs, 211
of GC fine aggregate, 69–70, 69f
of GC-sand, 77–78, 173–175, 175t
porosity and, 190
PSD and, 69–70, 69f
Firing shrinkage
of bricks, 344–346, 345f
with load, 345–346
of tiles, 361–363
Fitzsimons, L., 313
Flakiness index, 265
Flat glass
collection of, 45t
foam glass-ceramics from, 336t
GC from, 62
for road pavement, 423
soda lime glass from, 76, 436
waste glass from, 42, 43t
Flexural strength
of bricks, 343, 354, 355t, 377
compressive strength and, 184
of concrete, 122–124, 124f
cube strength and, 184, 185f
of foam glass-ceramics, 340–341, 340t
of GC filler, 240, 241t
of GC-sand, 169, 184, 185f, 201, 214
of GGC, 122–124, 124f, 153
of glass-ceramics, 333–334, 333t
of road pavement, 184
sulphate attack and, 201
of tiles, 356, 364–365, 365f, 377
Flint colour, 63–65, 63t–64t
of GC, 77
of GC-sand, 201–202
of GGC, 111
Float glass, tiles from, 365f
Fluxing agent
for bricks, 345–346
for porcelain, 368
Fly ash (FA), 58
acid attack in, 138
air content of, 109
ASR and, 62, 215, 299–300
autogenous shrinkage of, 127, 127t
carbonation in, 135–137
chemical composition requirements for, 58
chloride migration coefficient for, 135, 137t
in concrete, 103–105

Fly ash (FA) *(Continued)*
consistence and, 100
in CRT funnel glass, 394–396
diffusion of, 133–134
drying shrinkage of, 127–128
for foam glass-ceramics, 340
in GC-sand, 170, 213, 215
in GGC, 71–72, 102, 145, 152
initial and final setting times for, 101, 110, 110t
LOI for, 61
in PC, 98
for porcelain, 369
pozzolanic reactions of, 133–134, 205
in SCC, 107
soundness and, 102
standards for, 410
sulphate attack in, 139
water absorption of, 131
water and, 103–105
water/cement ratio for, 133
FM. *See* Fineness modulus
Foam glass-ceramics, 335–342
compressive strength of, 335, 340–341, 340t
from CRTs, 335–339, 336t, 342
density of, 335–338, 337t
flexural strength of, 340–341, 340t
foaming agent for, 335, 340–341
GC for, 335, 336t, 338–339
GGBS for, 335, 338–339
porosity of, 338–339, 339t
sintering process for, 338
from soda lime glass, 336t, 338
thermal conductivity of, 341–342, 341t
waste glass for, 377
water absorption of, 338–339, 339t
Foaming agent, for foam glass-ceramics, 335, 340–341
Francis, A. A., 338–341
Freeze-thaw attack
on air-entrained concrete, 146, 209
on concrete, 146
on GC, 282
on GC filler, 235, 246, 250, 251f
on GC-sand, 209, 215
on GGC, 146, 147t, 152–154
on tiles, 355
Friction. *See also* Angle of friction; Angle of internal friction
shear strength and, 273
Frost susceptibility
of GC, 282
of soils, 282
Furlani, E., 332–334

G

GC. *See* Glass cullet
GC filler. *See* Glass cullet filler
GC fine aggregate. *See* Glass cullet fine aggregate
GC-glass. *See* Glass cullet-glass
GC-sand. *See* Glass cullet-sand
Georgiadis, A., 238, 241
Geotechnical applications. *See also specific applications and parameters*
case studies for, 404–406, 405t
of GC, 257–296, 286t–287t, 289t, 396–399, 404–406, 405t, 416–423, 418t, 436–437
of GC fine aggregate, 67–68
PSD and, 67–68
standards for, 416–423, 418t–422t
GGBS. *See* Ground granulated blast furnace slag
GGC. *See* Ground glass cullet
Ghavidel, R., 125
Gibney, L., 313
Glass. *See also specific types*
calcium hydroxide in, 58
chemical composition of, 59f
colour of, 56
consumption of, 7–8, 8f
modulus of elasticity and hardness of, 75, 76t
production of, 6, 7f
recycling of, 36
silica in, 53, 58
silica sand in, 6–7
thermal conductivity of, 280–281
water absorption of, 75
Glass cullet (GC), 2, 6–9. *See also specific types*
added value from, 37, 435
for aesthetic finishing materials, 371
alkali in, 60
alternative markets for, 9
angle of friction of, 274, 274f–275f, 276, 277f, 290

angle of internal friction for, 259
angularity of, 63
ASR of, 62, 401–402
Atterberg limits for, 267–268
bituminous mixtures with, 298, 299t, 300, 304–317, 307f–309f
in bricks, 342–354
carbon dioxide and, 4, 36
case studies for, 284–286, 284t–285t, 400–409, 402t–403t
CBR for, 259, 272, 272f, 320
in cement clinker, 393, 400–401
cementitious properties of, 56–58
in ceramics, 327–388, 437
chemical and environmental testing of, 397, 397t
chemical composition of, 53–56, 54t–55t, 77
chemical properties of, 53–62
closed-loop recycling of, 6–7, 40, 41f
coefficient of curvature for, 260, 262, 264f
coefficient of uniformity for, 67, 264
cohesion and, 276, 276f
compaction of, 269–270
in concrete, 392–396, 402t–403t, 436
confining pressure of, 277, 278f, 290
contaminants in, 8, 39–41, 76–77, 284, 397–399, 398t
dynamic loading of, 282
in elastomeric roof coating, 375–376
environmental impact of, 283–284, 283t, 317, 390–400
as epoxy, 373
in European Union, 36, 37f
for filtration, 371–373
financial and environmental benefits of, 37–38
flakiness index for, 265
for foam glass-ceramics, 335, 336t, 338–339
freeze-thaw attack on, 282
frost susceptibility of, 282
for GC filler, 70–71, 70f, 77–78
geotechnical applications of, 257–296, 286t–287t, 289t, 396–399, 404–406, 405t, 416–423, 418t, 436–437
geotechnical characteristics of, 259–268
as glass fibres, 373–374
in glass manufacturing, 36
in glass-ceramics, 328–334, 330t
in glaze, 369–371
guidance for use of, 286–287, 288f
handling and classification of, 50–53, 51t–53t
heavy metal leaching from, 391
hydraulically bound mixtures with, 298–300, 299t
interfacial transition zones of, 63
interparticle friction of, 63
landfill charges and, 39
from LCD, 36
leaching from, 283, 283t, 317, 392f, 394, 397–399
lead in, 51
LL for, 267, 268f
LOI for, 77
Los Angeles abrasion test for, 266, 266f
market opportunities for, 40–41, 42t
material of, 391–392
MDD of, 259, 270, 270f–272f, 290
mechanical properties of, 269–281
modulus of elasticity and hardness for, 75, 76t
in mortar, 402t–403t
OMC of, 259, 270, 270f–271f, 290
open-loop recycling of, 40, 41f
organic content of, 267, 399
as paint filler, 375
particle interlocking effect of, 63
particle shape of, 264
for pedestrian pavements, 299
permeability of, 259, 279, 280f, 290
pH of, 267, 397
physical appearance of, 62–65, 63t–64t
physical properties of, 62–75
PI for, 267, 269f
PL for, 267, 268f
in porcelain, 366–369, 377
porosity of, 73–75, 78
processing into, 47–53, 47f
production and properties of, 35–96
PSD of, 260–262, 261f
recycling, barriers to, 38–40
recycling of, 36–37
re-melting of, 6–8, 40
resilient modulus of, 278, 279f
for road pavement, 297–326, 406–409, 406t–407t, 416–423, 437

Glass cullet (GC) *(Continued)*
secant modulus of, 277, 278f, 290
shear strength of, 274–275
sorting and processing of, 8
sorting by colour, 38
soundness of, 267
specific gravity of, 72–73, 74f, 265, 265f, 290
strain to failure for, 273, 273t
stress-strain behaviour of, 273
sulphate in, 77
thermal conductivity of, 280–281, 281t
for tiles, 355–366, 377
unbound mixtures with, 298, 299t, 300–301, 301f
uniformity coefficient for, 258
for walkways, 299
water absorption of, 73–75, 75t, 190–195, 266, 346–348, 349f–351f
water/cement ratio for, 133
zeolites and, 41
Glass cullet (GC) colour, 63–65, 63t–64t, 77. *See also specific colours*
aesthetic finishing materials of, 371
in bricks, 349
concrete and, 65
contamination of, 8
Glass cullet (GC) filler, 231–256, 436
abrasion in, 246, 251
acid attack in, 248
as aesthetic finishing material, 371
air content of, 235
in air-entrained concrete, 235
ASR of, 246, 248–250, 249f, 249t
bleeding of, 232, 234
calcium hydroxide and, 247
carbonation in, 246–247, 247t
case studies for, 251
chemical attack on, 246
chloride in, 246
chloride migration coefficient for, 246, 246t
compressive strength of, 236–242, 237f, 245f
in concrete, 232–236
consistence of, 232–234, 233t, 238
creep of, 242
cube strength of, 242f
deformation of, 241–243
density of, 234–235
drying shrinkage of, 242–243
durability of, 232, 246–251
elastic modulus of, 241–242, 242f
environmental impact of, 251
flexural strength of, 240, 241t
freeze-thaw attack for, 235, 246, 250, 251f
GC for, 70–71, 70f, 77–78
grinding time for, 236, 245f
limestone and, 247
for mortar, 232, 233t
oxygen permeability test for, 245
permeability of, 245–246
permeation in, 243–246
physical attack on, 246
plastic shrinkage of, 235–236
pozzolanic reactions of, 236–238, 247
PSD for, 70–71, 70f
PSD of, 232
RH and, 242–243, 245
for SCC, 232, 233t, 238–239, 240t, 243–244, 244f, 244t, 248
in SCC, 232
shrinkage of, 242–243, 243f
silica fume in, 245
slump classes of, 232, 234
sorptivity of, 244t
splitting tensile strength of, 239–240
stability of, 234
sulphate attack in, 246–248
tensile strength of, 239–240
water absorption of, 244–245
water in, 234
water/cement ratio for, 235–238, 239f, 240–241
Glass cullet (GC) fine aggregate
aggregate size of, 415–416
ASR of, 77, 404
compatibility of, 320
in concrete, 65–66, 394–396, 401–404, 415–416
contaminants in, 416
FM of, 69–70, 69f
geotechnical applications of, 67–68
PSD of, 65–68, 66f–67f
for road pavement, 68, 68f, 399–400
standards for, 415–416

Glass cullet (GC) publications, 435
analysis, evaluation and modelling of data for, 28–29
data matrix for, 27
dissemination of, 29–32
global status of, 19, 20f–21f
institutions and organizations involved in, 24, 25t
in journals, 21
researchers involved in, 23–24, 23t
subject area distribution in, 24–27, 26f
timeline for, 17–18, 19f
types of, 21, 22f
Glass cullet-glass (GC-glass)
borosilicate glass in, 202
lead glass in, 202
Pyrex in, 202
Glass cullet-sand (GC-sand), 167–230
abrasion in, 169, 209–211, 210f, 210t
acid attack in, 169, 199–200, 200f
aggregate size of, 204, 204f
air content of, 176, 177f
in air-entrained concrete, 176
angularity of, 214
ASR of, 62, 169–170, 202–205, 202t–203t, 206f–207f, 215
autogenous shrinkage of, 188
bleeding in, 173
carbonation in, 188, 197–199, 199t, 215
case studies for, 213–214
chemical attack on, 196–205
chloride in, 197
colour of, 201–204, 203t, 215
compressive strength of, 177–181, 178f–180f, 186, 211, 214
in concrete, 29, 169–176, 214
consistence of, 170–173, 181, 209
content of, 205
creep of, 186–188, 214
from CRTs, 176, 211
CS and, 182, 186
cube strength of, 182, 183f, 184, 185f, 186
deformation of, 185–190, 214
density of, 175–176
diffusion of, 190, 195
drying shrinkage of, 188
durability of, 169–170, 196–211
elastic modulus of, 186, 187f, 211, 214
environmental impact of, 211–213
FA in, 170, 215
for fibre-reinforced concrete, 211
fineness of, 193
flexural strength of, 169, 184, 185f, 201, 214
FM for, 77–78, 173–175, 175t
freeze-thaw attack on, 209, 215
GGBS and, 170
hydration reactions in, 211
ISAT for, 192–193, 193f
leaching from, 211–213, 212t
for lightweight concrete, 211
lithium in, 170
in mortar, 173, 176, 213
in non-air-entrained concrete, 176
oxygen permeability test for, 195, 196t
permeability of, 190, 195, 196t
permeation in, 169–170, 190–195, 215
pessimum size of, 204, 204f
physical attack on, 205–211
plastic shrinkage of, 188
porosity of, 190, 191t, 215
pozzolanic reactions of, 204–205
PSD for, 67
PSD of, 193, 214
RCPT for, 197, 198f
for road pavement, 214
shrinkage of, 188–190, 189f, 201
silica in, 169
slump classes of, 171–173, 172f, 174t
soda lime glass in, 202–205, 213
sorptivity of, 192, 192t
specific gravity of, 169, 173
splitting tensile strength of, 169, 181–183
stability of, 173–175
strength and deformation of, 169
sulphate attack in, 200–201, 201f
temperature rise in, 211
tensile strength of, 182, 183f, 214
water absorption of, 168–169, 182, 190, 194–195, 194f, 215
water/cement ratio for, 177, 199–200, 209, 211
Glass fibre
consistence of, 373
CRTs for, 373
GC as, 373–374
spinning temperature of, 373
viscosity of, 373

Glass Packaging Institute, 36
Glass-ceramics
 compressive strength of, 333–334, 333t
 density of, 329, 331t
 environmental impact of, 328
 experimental campaigns for, 357t
 flexural strength of, 333–334, 333t
 GC in, 328–334, 330t
 leaching from, 334
 modifying agents for, 331
 nucleation agents for, 329
 petrurgic method for, 329
 porosity of, 328, 330–332, 332t
 powder method for, 329
 shrinkage of, 332–333
 single-stage method for, 328
 sintering process for, 329
 two-stage method for, 328
 Vickers hardness of, 334
 waste glass for, 376–377
 water absorption of, 330–332, 332t
Glaze
 aluminum in, 369
 from CRTs, 369–370
 GC in, 369–371
 thermal expansion coefficient for, 371
Glüsing, A. K., 50
Gray & Osborne, Inc., 372
Green colour, 63–65, 63t–64t
 of GC, 77
 of GC-sand, 201–202, 215
 of GGC, 111
Grinding time, for GC filler, 236, 245f
Ground glass cullet (GGC)
 abrasion in, 146–148, 152
 acid attack in, 138, 139t, 153–154
 air content in, 108–109
 air-entraining admixture for, 146
 aluminosilicate glass in, 101f, 102
 angularity of, 100
 ASR of, 98–99, 141–145, 142t, 152–154, 169
 autogenous shrinkage of, 126–127, 127t
 Blaine test for, 72, 73f, 77–78, 107, 394
 bleeding in, 107, 129
 calcium hydroxide in, 98–99
 carbonation in, 135–138, 138t
 case studies for, 149–152, 150t–151t
 in cement, 394, 410
 in cement clinker, 396f
 cementitious characteristics of, 99–103
 chloride migration coefficient in, 134–135, 137t, 152
 colour of, 145
 compressive strength of, 114–121, 115f–116f, 120f, 140–141, 152–153
 in concrete, 97–166, 401, 410t–411t, 414t
 consistence of, 99–100, 100f
 creep of, 126
 from CRTs, 394
 cumulative heat in, 111
 density of, 107–108, 109f
 diffusion of, 133–134, 133t
 drying shrinkage of, 127–128, 128f
 durability of, 134–148
 elastic modulus of, 124–125, 126f, 153
 environmental impact of, 148–149, 149t
 expansion of, 143f, 145
 FA in, 71–72, 102, 145
 fineness of, 71–72, 414
 flexural strength of, 122–124, 124f, 153
 freeze-thaw attack on, 146, 147t, 152–154
 heat of hydration for, 110–111, 112t–113t
 hydration reactions in, 72, 110–111
 initial and final setting times for, 101–102, 101f, 109–110
 initial and final setting times of, 110t
 LOI for, 61–62, 61f
 mechanical properties of, 111–128
 moisture diffusion coefficient for, 133–134, 133t
 in mortar, 98–99, 105, 414t
 PC and, 98–100
 permeability of, 131–133, 132t
 permeation in, 128–134, 130t
 porosity of, 129, 153–154
 pozzolanic reactions of, 98–99, 102–103, 120–121, 132–134, 149–152, 205, 410
 PSD for, 71–72, 71f
 rate of heat evolution for, 111
 of RCA, 129, 132–133
 RCPT for, 135, 136f
 retarding admixture in, 110
 in SCC, 107
 setting times for, 415–416
 slump classes of, 105, 106f, 107t

soda lime glass in, 101, 101f, 103, 145
soundness of, 102
splitting tensile strength of, 121–122
stability of, 107, 108t
standards for, 411–413, 414t
strength activity index for, 102–103, 104f, 415
sulphate attack in, 139–141, 140t
temperature rise in, 110–111, 127t
tensile strength of, 121–124, 153
water absorption of, 100, 129–131, 131f
water penetration test for, 132–133
water/cement ratio for, 105, 106f, 117, 119f
Ground granulated blast furnace slag (GGBS), 58
acid attack in, 138
ASR of, 62, 142, 169, 299–300
carbonation in, 135–137
in CRT funnel glass, 394–396
drying shrinkage of, 127–128
for foam glass-ceramics, 335, 338–339
GC-sand and, 170, 213
initial and final setting times for, 101
pozzolanic reactions of, 205
RCPT for, 198f
standard for, 410
Grubb, D., 23t, 276
Gyratory Testing Machine, 304

H

Hammer mill, for waste glass crushing, 49t
Hanpongpun, W., 329–331
Harrison, T. W., 346, 353
Heat of hydration
for concrete, 110–111, 112t–113t
for GGC, 110–111, 112t–113t
Heavy metal leaching. *See also specific metals*
from CRTs, 391
EPA on, 211
from GC, 391
from GC-sand, 211
from LCDs, 391
from tiles, 366
Helically fluted rollers, for waste glass crushing, 49t
Highway Agency, 301
Holcroft, C., 372
Hong Kong
GC publications in, 19
GGC in, 149–152
Hreglich, S., 370, 373
Hupa, L., 370
Hveem method, 304
Hwang, J. -Y., 358, 363
Hydrated lime, as antistripping agent, 316–317
Hydration reactions, 56
acid attack and, 199
in concrete, 72, 110–111
drying shrinkage and, 127
in GC-sand, 211
in GGC, 72, 110–111
in PC, 120–121
sulphate and, 140
Hydraulically bound mixtures
ASR of, 319
case studies for, 408
environmental impact of, 317
with GC, 298–300, 299t
Hydroponics, 41
Hygroscopicity
of bricks, 343
of tiles, 356

I

Imteaz, M. A., 317, 399
Incinerated ashes, 3
India, GC publications in, 19
Initial surface absorption test (ISAT), for GC-sand, 192–193, 193f
Interfacial transition zones, of GC, 63
Iron
with GGC, 152
glass colour from, 56
Iron leaching, from GC, 398
Iron oxide
in foam glass-ceramics, 338
in GC filler, 236–238
in glass, 58, 77
ISAT. *See* Initial surface absorption test

J

JIS A 1146, 203t
Journals, 21

K

Kaolin, 276. *See also* Metakaolin
  glass-ceramics of, 334
  for porcelain, 369
  PSD of, 262
Kara, P., 23t
Kerbside collection, of waste glass, 44, 45t
Keywords, for literature search and appraisal, 17, 18t
Kim, K., 359, 368
Klevbo, G., 236–238, 250
Konig, J., 342
Korjakins, A., 236, 245–246
Kou, C. S., 23t
KS F2546, 145

L

Lairaksa, N., 149
Lamb, T. W., 277
Lancellotti, I., 374
Lanchance-Tremblay, E., 313
Landfill charges, GC and, 39
LCDs. *See* Liquid crystal displays
Le Chatelier test, for soundness, 102
Leaching. *See also* Heavy metal leaching; *specific metals*
  from CRTs, 366
  from GC, 283, 283t, 317, 392f, 394, 397–399
  from GC-sand, 211–213, 212t
  from glass-ceramics, 334
  from soda lime glass, 213, 267
  TCLP and, 391
Lead
  in CRTs and LCDs, 43, 391
  in GC, 51, 317
  in glass, 56
  specific gravity of, 74f
  standards for, 413
Lead glass
  calcium oxide in, 58
  in GC-glass, 202
  modulus of elasticity and hardness of, 75, 76t
  RCPT for, 198f
  specific gravity of, 73, 175
  waste glass from, 43t, 44
  water absorption of, 75t
Lead leaching, 149
  from CRTs, 394, 395t
  from GC, 284, 398
  from GC-sand, 211, 215
  from soda lime glass, 213
Lee, J., 109
Lee, S. T., 109
Lichtenberger, B., 282
Lightweight concrete, 211
Limbachiya, M. C., 192
Limestone
  as antistripping agent, 314–316
  carbonation in, 135–137
  compressive strength of, 125
  compressive strength-elastic modulus relationship of, 241–242
  elastic modulus of, 125
  GC filler and, 238, 247
  with GGC, 152
  PSD of, 262
  in SCC, 234, 241
  in soda lime glass, 53–56
Lin, K. -L., 23t, 344–345, 348, 351–352, 359–361, 364
Ling, T., 197
Ling, T. C., 24, 200, 209, 211
Liquid crystal displays (LCDs)
  bricks from, 351–352
  foam glass-ceramics from, 336t
  GC from, 36
  glass-ceramics from, 334
  heavy metal leaching from, 391
  lead in, 391
  mercury in, 391
  for porcelain, 366, 368
  TFT-LCD, 148, 149t
  tiles from, 359, 359f–360f, 361, 363
  waste glass from, 43, 43t, 76
Liquid limit (LL), 267, 268f
Literature search and appraisal
  for book, 17–27
  identifying and sourcing literature for, 17, 18t
  keywords for, 17, 18t
  search engines for, 17, 18t
  websites for, 17, 18t
Lithium, 170

Liu, H., 334
LL. *See* Liquid limit
LNEC E391, 197
LOI. *See* Loss on ignition
Loryuenyong, V., 346, 348, 351–352, 354
Los Angeles abrasion test, 75, 266, 266f
Loss on ignition (LOI)
for cement, 61–62
contaminants and, 61–62
for GC, 77
for GGC, 61–62, 61f
standards for, 413
Lu, J., 329
Luz, A. P., 368
Lye, C. Q., 2, 189–190, 241–242

M

Madandoust, R., 125
Magnesium oxide, 53–56
for glass-ceramics, 331, 334
Magnesium sulphate, 267, 371
Makul, N., 211
Malisch, W. R., 17
Manganese dioxide, 335, 338–339
Marangoni, M., 338
Mardani-Aghabaglou, A., 192
Marshall, Bruce, 304
Marshall flow, of bituminous mixtures, 306f, 310
Marshall mix design, for bituminous mixtures, 304–311
Marshall stability, of bituminous mixtures, 305f, 310, 314, 320
Maschio, S., 211
Matos, A. M., 247
Matteucci, F., 356, 359
Mavilia, L., 50
Maximum dry density (MDD)
CBR and, 272, 272f
of GC, 259, 270, 270f–272f, 290
Mercury, 148, 149t, 284, 317, 391
Metakaolin, 58
ASR and, 62
pozzolanic reactions of, 205
Meyer, C., 23t, 24
Micro-Deval test, for modulus of elasticity and hardness, 75
Micro-irrigation, filtration for, 372
Mirzahosseini, M., 111
Modified compaction test, 259, 262
for CBR, 272
Modifying agents, for glass-ceramics, 331
Modulus of elasticity and hardness. *See also* Elastic modulus
for GC, 75, 76t
Moisture damage resistance, of bituminous mixtures, 314–317, 315t–316t
Moisture diffusion coefficient
for concrete, 133–134, 133t
for GGC, 133–134, 133t
Molybdenum, 317
Moncea, A. M., 149
Monocyclic aromatic hydrocarbons, 284
Mortar
acid attack in, 138, 139t
air content of, 177f
ASR of, 205
from CRTs, 395t
FM for, 70
GC filler for, 232, 233t
GC in, 402t–403t
GC-sand in, 173, 176, 213
GGC in, 98–99, 105, 414t
lead leaching from, 149
permeability of, 195
strength activity index for, 103
sulphate attack in, 139, 140t
water for, 100
water/cement ratio for, 394

N

Nandi, V. S., 371
Nassar, R., 108
Neithalath, N., 60, 111, 133
NEN 7345, 394
Nepheline syenite, for porcelain, 368
New Zealand
barriers to GC recycling in, 41–44
geotechnical applications from, 418t
standards of, 415
Nickel, 317
Nitrogen dioxide, in GGC, 152
Nitrogen monoxide, in GGC, 152
Non-air-entrained concrete, 146, 176, 209
Nounu, G., 125
NT Build 443, 197
Nucleation agents, 329

O

OMC. *See* Optimum moisture content
Open-loop recycling, 40–41, 41f
Optimum moisture content (OMC), of GC, 259, 270, 270f–271f, 290
Organic content, of GC, 267, 399
Oxygen permeability test, 131–132
  for GC filler, 245
  for GC-sand, 195, 196t

P

PAHs. *See* Polycyclic aromatic hydrocarbons
Paint filler, GC as, 375
Paper, screening and removal of, 49
Parghi, A., 131
Particle interlocking effect, of GC, 63
Particle size distribution (PSD), 260–262
  of bentonite, 262
  of bituminous mixtures, 310
  coefficient of curvature and, 260, 262, 264f
  coefficient of uniformity and, 67, 260, 263f
  compaction and, 262–264
  in concrete, 65–66
  FM and, 69–70, 69f
  of GC, 260–262, 261f, 263t
  of GC filler, 70–71, 70f, 232
  of GC fine aggregate, 65–68, 66f–67f
  of GC-sand, 67, 193, 214
  geotechnical applications and, 67–68
  of GGC, 71–72, 71f
  of kaolin, 262
  of limestone, 262
  for road pavement, 68, 68f
  of soil, 260, 260t, 261f
PAS. *See* Publicly Available Standard
PC. *See* Portland cement
PCBs. *See* Printed circuit boards
Pedestrian pavements, 299
PennDOT, 405–406
Permeability
  of concrete, 131–133, 132t
  frost susceptibility and, 282
  of GC, 259, 279, 280f, 290
  of GC filler, 245–246
  of GC-sand, 190, 195, 196t
  of GGC, 131–133, 132t
  of mortar, 195
  of soils, 279, 280f
  water/cement ratio and, 195
Permeation
  in concrete, 128–134, 130t
  durability and, 190
  in GC filler, 243–246
  in GC-sand, 169–170, 190–195, 215
  in GGC, 128–134, 130t
Pessimum size, of GC-sand, 204, 204f
Petersen, R. R., 339
Petrurgic method, for glass-ceramics, 329
Pfefferkorn plasticity index, 343, 355
pH
  carbonation and, 135, 197
  of GC, 267, 397
  of soil, 267
Physical attack
  on GC filler, 246
  on GC-sand, 205–211
PI. *See* Plasticity index
Plastic limit (PL), 267, 268f
Plastic shrinkage
  of concrete, 235–236, 235t
  of GC filler, 235–236
  of GC-sand, 188
  of SCC, 235–236, 235t
Plasticity index (PI)
  for bricks, 343
  cohesion and, 276
  for GC, 267, 269f
  Marshall flow and, 306f, 310
  for tiles, 355
Plastics, screening and removal of, 49
Plot Digitizer, 27
Polley, C., 146, 214
Polycyclic aromatic hydrocarbons (PAHs), 284
Polyvinyl butyral, 49
Poon, C. S., 23t, 24, 197, 200, 209, 211
Porcelain
  from CRTs, 366–368
  density of, 368–369
  fluxing agent for, 368
  from GC, 366–369, 377
  from LCDs, 366, 368
  powder method for, 368
  sintering process for, 368
  vitrification process for, 366

water absorption of, 368–369
Porcine bone ash, for foam glass-ceramics, 340
Porosity
of bricks, 346–349, 347t
of concrete, 129
FM and, 190
of foam glass-ceramics, 338–339, 339t
of GC, 73–75, 78
of GC-sand, 190, 191t, 215
of GGC, 129, 153–154
of glass-ceramics, 328, 330–332, 332t
of tiles, 357–358, 361
water/cement ratio and, 190
Portland cement (PC)
abrasion in, 148
air content of, 108
ASR of, 77
calcium hydroxide and, 56
carbonation in, 135–137
chloride migration coefficient in, 135, 137t
compressive strength of, 114, 152
drying shrinkage of, 127
elastic modulus of, 125, 126f
FA in, 98
freeze-thaw attack on, 146
GC filler and, 236–238
GGC and, 98–100
global production of, 4, 4f
hydration reactions of, 120–121
initial and final setting times for, 110
specific gravity of, 74f
tensile strength of, 124, 125f
in unbound mixtures, 300–301
water/cement ratio for, 117, 121
Portland clinker
carbon dioxide from, 98
specific gravity of, 73
Potash, 60
Potassium
for glaze, 370
for porcelain, 368
Potassium nitrate, 338
Potassium oxide, 53–56
Powder method
for glass-ceramics, 329
for porcelain, 368
Powell, B. D., 278
Pozzolanic reactions, 56
ASR and, 169, 204–205, 401–402
with calcium hydroxide, 120–121
carbonation and, 247
chloride and, 197
of FA, 133–134
FM and, 70
of GC filler, 236–238, 247
of GC-sand, 204–205
of GGC, 71, 98–99, 102–103, 120–121, 132–134, 149–152, 410
of soda lime glass, 98
strength activity index and, 102–103
sulphate and, 140
water penetration test and, 132
Printed circuit boards (PCBs)
fibreglass for, 44
waste glass from, 76
Processing
into GC, 47–53, 47f
of waste glass, 47–53, 47f
PSD. *See* Particle size distribution
Publications. *See* Glass cullet publications
Publicly Available Standard (PAS), 44, 45f
Pudner, M., 372
Pyrex, 43t, 44, 202

Q

QGLC. *See* Quaternary glacial lacustrine clay
Quartzite, 241–242
Quaternary glacial lacustrine clay (QGLC), 345–346, 349, 350f, 353

R

Raimondo, M., 359, 366
Rajabipour, F., 192, 209
Rambaldi, E., 368
Rapid chloride penetration test (RCPT)
for GC-sand, 197, 198f
for GGC, 135, 136f
Rate of heat evolution
for concrete, 111
for GGC, 111
RCA. *See* Recycled concrete aggregate
RCPT. *See* Rapid chloride penetration test
Recycled and secondary aggregates (RSAs)
production of, 5–6, 5f
for road pavement, 298
as sustainable construction materials, 6

Recycled concrete aggregate (RCA)
abrasion in, 148
GGC of, 129, 132–133
water penetration test for, 132–133
Recycling
of concrete aggregates, 2
of GC, 36
of glass, 36
sustainable construction materials from, 4
of waste glass, 37
of waste materials, 2
Relative humidity (RH), 138t, 199t
GC filler and, 242–243, 245
Resilient modulus
CBR for, 278, 279f
of GC, 278, 279f
of soils, 278
Retarding admixture, in GGC, 110
RH. *See* Relative humidity
Ribeiro, S., 368
Riding, K. A., 111
RILEM Technical Committee
AAR-2, 203t
106-AAR, 202, 205
116, 244
162, 123–124, 184
Road pavement. *See also* Bituminous mixtures; Hydraulically bound mixtures; Unbound mixtures
case studies for, 318–319, 318t–319t, 406–409, 406t–407t
CBR for, 272
environmental impact of, 317, 320, 399–400
flexural strength of, 184
GC fine aggregate for, 68, 68f, 399–400
GC for, 297–326, 406–409, 406t–407t, 416–423, 437
GC-sand for, 214
Los Angeles abrasion test for, 266
PSD for, 68, 68f
RSAs for, 298
standards for, 416–423
Roofs. *See* Elastomeric roof coating; Tiles
Rotating breaker bar, 49t
Rotating disk and breaker bar, 49t
Rotating drum and breaker plate, 49t
Rozenstrauha, I., 332–333
RSAs. *See* Recycled and secondary aggregates
Rutting resistance, of bituminous mixtures, 312–314, 313t

S

Sagoe-Crentsil, K., 416
Samarin, A., 125
Sandstone, 241–242
SCC. *See* Self-compacting concrete
Schwarz, N., 60, 133
Search engines, 17, 18t
Secant modulus, 277, 278f, 290
Segregation, in SCC, 234
Selenium, 148, 149t, 317
Self-compacting concrete (SCC), 107, 436
acid attack in, 248
autogenous shrinkage of, 127, 127t
chloride migration coefficient for, 246
compressive strength of, 238
consistence of, 238
GC filler for, 232, 233t, 238–239, 240t, 243–244, 244f, 244t, 248
initial and final setting times of, 110t
limestone in, 234, 241
plastic shrinkage of, 235–236, 235t
segregation in, 234
shrinkage of, 243, 244f
sorptivity of, 244t
water absorption of, 244
water in, 241
water/cement ratio for, 238
Seminars, 2
Serpa, D., 209
Setting times
for cement, 101–102, 101f
for concrete, 109–110, 416
for GGC, 415–416
Sharifi, Y., 109
Shear strength, 273–276
Shi, C., 107, 109–110
Shrinkage. *See also* Autogenous shrinkage; Drying shrinkage; Firing shrinkage; Plastic shrinkage
of bricks, 343–344
of GC filler, 242–243, 243f
of GC-sand, 188–190, 189f, 201
of glass-ceramics, 332–333
of SCC, 243, 244f

sulphate attack and, 201
of tiles, 361–363, 362f–363f
water/cement ratio and, 243
Siikamäki, R., 370
Silica (silicon dioxide)
for foam glass-ceramics, 338–339
in GC, 77
in GC filler, 236–238
in GC-sand, 169
in glass, 53, 58, 59f
Silica fume, 58
ASR and, 169
drying shrinkage of, 127–128
in GC filler, 245
sulphate attack in, 139
water absorption of, 131
Silica sand
with GGC, 152
in glass, 6–7
Silicon carbide, for foam glass-ceramics, 335, 340–341
Silicon dioxide. *See* Silica
Silicon nitride, 338
Silva, R. V., 2
Silver, 317
Single-stage method, for glass-ceramics, 328
Sintering process
for bricks, 377
for foam glass-ceramics, 338
for glass-ceramics, 329
for porcelain, 368
for tiles, 358–359, 363–364
Slump classes
of GC filler, 232, 234
of GC-sand, 171–173, 172f, 174t
of GGC, 105, 106f, 107t
Smith, A. S., 349
Soda ash
to lower melting point of glass formers, 60
in soda lime glass, 53
strength activity index of, 103
Soda lime glass/GC, 53–56
alkali in, 60
ASR of, 142, 202–205
autogenous shrinkage of, 127t
for bricks, 349
chemical composition of, 77
colour of, 145
compressive strength of, 114
from container glass, 436
environmental impact of, 317, 437
from flat glass, 436
foam glass-ceramics from, 336t, 338
in GC-sand, 202–205, 213
in GGC, 101, 101f, 103, 145
glass-ceramics of, 334
heavy metal leaching from, 366
initial and final setting times for, 101, 101f
leaching from, 213, 267
modulus of elasticity and hardness of, 75, 76t
in porcelain, 368
pozzolanic reactions of, 98
RCPT for, 198f
specific gravity of, 72, 74f, 175, 234–235
for tiles, 363, 365
as waste glass, 76
water absorption of, 75t
Sodium sulphate, 267, 282
Soil
angle of friction of, 274
compaction of, 269–270
density of, 274–275
durability of, 281–282
dynamic loading of, 282
frost susceptibility of, 282
permeability of, 279, 280f
pH of, 267
PSD of, 260, 260t, 261f
resilient modulus of, 278
shear strength of, 273–276
stress-strain behaviour of, 273
Soil modulus, 277
Song, T., 404
Soroushian, P., 108
Sorptivity
of GC filler, 244t
of GC-sand, 192, 192t
of SCC, 244t
Soundness
of GC, 267
of GGC, 102
South Korea, 149–152, 401
Specific gravity
density and, 175
of GC, 72–73, 74f, 265, 265f, 290
of GC-sand, 169, 173
of soda lime glass, 234–235

Spinning temperature, 373
Splitting tensile strength
of concrete, 121–122
of GC filler, 239–240
of GC-sand, 169, 181–183
of GGC, 121–122
water/cement ratio and, 181
Stability
of concrete, 107, 108t, 173, 234
of GC filler, 234
of GC-sand, 173–175
of GGC, 107, 108t
Marshall, 305f, 310, 314, 320
Standard compaction test, 262, 272
Standards, 409–423. *See also specific standards*
for alkali, 413
for chloride, 413
for FA, 410
for GC fine aggregate, 415–416
for geotechnical applications, 416–423, 418t–422t
for GGBS, 410
for GGC, 411–413, 414t
for lead, 413
for LOI, 413
for road pavement, 416–423
for sulphate, 413
Steel
corrosion of, 3
depassivation of, 197, 247
Stiffness modulus, for bituminous mixtures, 311–312, 312t
Strain to failure, for GC, 273, 273t
Strength. *See* Compressive strength; Cube strength; Flexural strength; Shear strength; Splitting tensile strength; Temperature rise; Tensile strength
Strength activity index
for GGC, 102–103, 104f, 415
pozzolanic reactions and, 102–103
Stress-strain behaviour
of GC, 273
of soils, 273
Stress-strain curve, elastic modulus and, 241
Stripping, in bituminous mixtures, 300, 314–316
Sua-iam, G., 211
Sulphate
compressive strength and, 140–141
in concrete, 60
in GC, 77
magnesium sulphate, 267, 371
pozzolanic reactions and, 140
sodium sulphate, 267, 282
standards for, 413
Sulphate attack
in concrete, 139–141, 140t
in GC filler, 246–248
in GC-sand, 169, 200–201, 201f
in GGC, 139–141, 140t
water/cement ratio and, 240
Sulphur dioxide, 152
Sustainable construction materials, 3–6
carbon dioxide and, 2–4
design codes and, 2
from recycling, 4
RSAs as, 6

T

Tagnit-Hamou, A., 23t, 24
Taha, B., 125
Taiwan, 19
Tan, K. H., 111, 132, 200–201
Tang, A., 23t, 209, 240
Tang, M. C., 140
Tarvornpanich, T., 368
TCLP. *See* Toxicity characteristics leaching procedure
Temperature rise
in bricks, 353, 354f
in GC-sand, 211
in GGC, 110–111, 127t
water/cement ratio and, 211
Tensile strength. *See also* Splitting tensile strength
of bituminous mixtures, 314
of concrete, 121–124
cube strength and, 182, 183f
of GC filler, 239–240
of GC-sand, 182, 183f, 214
of GGC, 121–124, 153
of glass fibre, 374
TFT-LCD. *See* Thin-film-transistor liquid-crystal display
Thermal conductivity, 280–281
of CRTs, 342
of foam glass-ceramics, 341–342, 341t

of GC, 280–281, 281t
of glass, 280–281
Thermal expansion coefficient, for glaze, 371
Thermal resistance, of borosilicate glass, 56
Thin-film-transistor liquid-crystal display (TFT-LCD), 148, 149t
Tiles
abrasion of, 355, 361, 362f
compressive strength of, 363–364, 364f, 377
from CRTs, 355, 358f, 364, 366
density of, 359, 360f
drying shrinkage of, 356
fired, 356–365
firing shrinkage of, 361–363
flexural strength of, 356, 364–365, 365f, 377
freeze-thaw attack for, 355
GC for, 355–366, 377
heavy metal leaching from, 366
hygroscopicity of, 356
from LCDs, 359, 359f–360f, 361, 363
mechanical performance of, 363–365
PI for, 355
porosity of, 357–358, 361
shrinkage of, 361–363, 362f–363f
sintering process for, 358–359, 363–364
unfired, 355–356, 356t
Vickers hardness of, 359–361, 361f
vitrification process for, 358–359
water absorption of, 357–358, 358f–359f, 361
Titanium dioxide, 329, 375
Toplan, H. Ö., 369
Toxicity characteristics leaching procedure (TCLP), 391, 394
Triaxial test, for angle of friction, 274
Tricalcium aluminate, 102
sulphate and, 140
Tricalcium silicate, 102
Tsai, C. L., 50
Tucci, A., 369
Tuncan, M., 131
Turkey, GC publications in, 19
Two-stage method, for glass-ceramics, 328

U

UK-India Education Research Initiative (UKIERI), 2
Unbound mixtures, 319
case studies for, 407–408
CBR for, 301, 301t
with GC, 298, 299t, 300–301, 301f
Unfired clay bodies. *See* Bricks
Uniformity. *See* Coefficient of uniformity
United Kingdom
barriers to GC recycling in, 41–44
Design Manual for Roads and Bridges in, 301
GC for road pavement in, 298
GC publications in, 19
GC-sand in, 213
geotechnical applications from, 419t
GGC in, 149–152
PAS 101 in, 44, 45f
pH in, 267
standards of, 415
United Nations, Brundtland report of, 3
United States (US). *See also* Environmental Protection Agency
Army of, 272
barriers to GC recycling in, 41–44
Code of Federal Regulations of, 391, 394, 397, 424
GC case studies from, 284–285
GC for road pavement in, 298
GC publications in, 19
geotechnical applications from, 258, 397, 419t–422t
GGC in, 149–152
University of Dundee, 171–173
US. *See* United States

V

Vebe test, 105
Vertical shaft impactor, 49t
VFB. *See* Voids filled with bitumen
Vicat test, 99
Vickers hardness
of glass-ceramics, 334
of tiles, 359–361, 361f
Viscosity, 373
Vitrification process
for porcelain, 366
for tiles, 358–359
VMA. *See* Voids in the mineral aggregate
Voids filled with bitumen (VFB), 304, 309f, 311
Voids in the mineral aggregate (VMA), 304, 308f, 310

W

Walkways, 299
Wang, H., 23t
Wang, H. Y., 100, 138
Wannagon, A., 369
Wartman, J., 23t
Waste glass. *See also* Glass cullet
  classification and handling of, 44–47, 46t
  collection schemes for, 44, 45t
  colour of, 49–50
  contaminants in, 49–50
  from CRTs, 43, 43t, 76
  crushers for, 48, 49t
  for foam glass-ceramics, 377
  for glass-ceramics, 376–377
  manual separation of, 48
  processing of, 47–53, 47f
  screening of, 49–50
  soda lime glass as, 76
  types of, 41–44, 43t
  visual inspection of, 48
Water
  for bricks, 342
  for consistence, 99–100, 100f
  FA and, 103–105
  filtration of, 41, 372
  in GC filler, 234
  for mortar, 100
  in SCC, 241
Water absorption
  of bricks, 346–349, 348f–351f
  of concrete, 129–131, 131f
  of foam glass-ceramics, 338–339, 339t
  of GC, 73–75, 75t, 190–195, 266, 346–348, 349f–351f
  of GC filler, 244–245
  of GC-sand, 168–169, 182, 190, 194–195, 194f, 215
  of GGC, 100, 129–131, 131f
  of glass, 75
  of glass-ceramics, 330–332, 332t
  of porcelain, 368–369
  of SCC, 244
  of tiles, 357–358, 358f–359f, 361
  water/cement ratio and, 129–130, 131f
Water penetration test, 132–133
Water/cement ratio
  abrasion and, 148, 209
  acid attack and, 200
  carbonation and, 199
  compressive strength and, 117, 119f, 177, 236–238, 239f
  for concrete, 117, 119f, 394
  drying shrinkage and, 127
  elastic modulus and, 186, 241
  for FA, 133
  for GC, 133
  for GC filler, 235–238, 239f, 240–241
  for GC-sand, 177, 199–200, 209, 211
  for GGC, 105, 106f, 117, 119f
  for mortar, 394
  permeability and, 195
  porosity and, 190
  RCPT and, 135
  for SCC, 238
  shrinkage and, 243
  splitting tensile strength and, 181
  sulphate attack and, 240
  temperature rise and, 211
  water absorption and, 129–130, 131f
Websites, for literature search and appraisal, 17, 18t
Wheel tracking test, 313
Whitman, R. V., 277
Workability. *See* Consistence
Workshops, 2
Wright, L., 375
Wu, Y., 107, 109–110

Y

Yang, J., 329
Youssef, N. F., 358–359
Yürüyen, S., 369

Z

Zeolites, 41
Zhang, W., 334
Zhao, H., 211
Zhu, H. Y., 23t
Zinc, 148, 149t, 317, 398
Zycosoil, 314–316

Kolibri 360 GmbH
Ferdinand-Jühlke-Straße 7
99095 Erfurt, Deutschland
produktsicherheit@kolibri360.de